Lecture Notes in Computer Science 8668

Commenced Publication in 1973
Founding and Former Series Editors:
Gerhard Goos, Juris Hartmanis, and Jan van Leeuwen

T0212691

Elena Barcucci   Andrea Frosini
Simone Rinaldi (Eds.)

Discrete Geometry for Computer Imagery

18th IAPR International Conference, DGCI 2014
Siena, Italy, September 10-12, 2014
Proceedings

 Springer

Volume Editors

Elena Barcucci
Andrea Frosini
Università degli Studi di Firenze
Dipartimento di Matematica e Informatica
Viale Morgagni 65
50134 Firenze, Italy
E-mail:{elena.barcucci, andrea.frosini}@unifi.it

Simone Rinaldi
Università degli Studi di Siena
Dipartimento di Ingegneria dell'Informazione
e Scienze Matematiche
Viao Roma, 56
53100 Siena, Italy
E-mail: rinaldi@unisi.it

ISSN 0302-9743   e-ISSN 1611-3349
ISBN 978-3-319-09954-5   e-ISBN 978-3-319-09955-2
DOI 10.1007/978-3-319-09955-2
Springer Cham Heidelberg New York Dordrecht London

Library of Congress Control Number: 2014947226

LNCS Sublibrary: SL 6 – Image Processing, Computer Vision, Pattern Recognition,
and Graphics

Typesetting: Camera-ready by author, data conversion by Scientific Publishing Services, Chennai, India

Printed on acid-free paper

Springer is part of Springer Science+Business Media (www.springer.com)

Preface

This volume collects the full papers presented at the 18th IAPR International Conference on Discrete Geometry for Computer Imagery, DGCI 2014, that was held in Siena, Italy in September 2014, and jointly organized by the researchers in Discrete Mathematics of the Universities of Siena and Firenze.

As in the previous editions, the conference attracted researchers from different countries attesting the international relevance of the event. After an accurate reviewing process that, from this edition, supported the decisions with a final rebuttal phase, 34 papers were accepted, out of about 60 submissions.

The 22 papers scheduled in a single-track of oral presentations were organized in topical sections on Models for Discrete Geometry; Discrete and Combinatorial Topology; Geometric Transforms; Discrete Shape Representation, Recognition and Analysis; Discrete Tomography; Morphological Analysis; Discrete Modeling and Visualization; Discrete and Combinatorial Tools for Image Segmentation and Analysis. The remaining 12 papers on these same topics were grouped into a single poster session.

As in the previous two editions, the conference included a demonstration session, in the intent of providing the opportunity to present and share effective applications, new tools, and libraries related to the mainstream of the image processing.

Three internationally well-known researchers provided invited lectures: Professor Peter Gritzmann, from Technische Universität of München, Germany, Professor Lorenzo Robbiano from the University of Genova, Italy, and Professor Marco Gori, from the Univerity of Siena, Italy.

DGCI 2014 was supported by the International Association of Pattern Recognition (IAPR), and constituted the main event associated with the Technical Committee on Discrete Geometry IAPR-TC18.

We would like to thank the main sponsoring institutions: the Dipartimento di Matematica ed Informatica of the University of Firenze, and the Dipartimento di Ingegneria dell'Informazione e Scienze Matematiche of the University of Siena who also hosted the conference and provided all the necessary facilities.

We are grateful to the members of the Steering Committee for their valuable support and the inspiring discussions, with a special mention to David Coeurjolly who helped us through all the steps of the reviewing process.

Special thanks to the authors of the submitted contributions whose researches confirmed the high level standard of the conference, to the Program Committee and all the reviewers for their accurate and proficient work, to the local Organizing Committee for its tireless support, and to all the participants attending the conference, who contributed to make this event a success.

September 2014   Elena Barcucci
                                                                   Andrea Frosini
                                                                  Simone Rinaldi

Organization

Organizing Committee

Elena Barcucci	University of Firenze, Italy (Chair)
Andrea Frosini	University of Firenze, Italy (Chair)
Simone Rinaldi	University of Siena, Italy (Chair)

Local organizers, University of Siena, Italy

Daniela Battaglino	Stefano Brocchi
Veronica Guerrini	Andrea Machetti
Rita Nugari	Simonetta Palmas
Chiara Pappadopulo	Alessandro Rossi
Elena Sbrocca	Samanta Socci

Steering Committee

David Coeurjolly (President)	CNRS, University of Lyon, France
Eric Andres	University of Poitiers, France
Gunilla Borgefors	University of Uppsala, Sweden
Srečko Brlek	University of Quebec, Canada
Jean-Marc Chassery	CNRS, University of Grenoble, France
Isabelle Debled-Rennesson	LORIA, University of Nancy, France
Ullrich Köthe	University of Heidelberg, Germany
Annick Montanvert	University of Grenoble, France
Kálmán Palágyi	University of Szeged, Hungary
Gabriella Sanniti di Baja	Institute of Cybernetics Eduardo Caianiello, CNR, Italy
Isabelle Sivignon	CNRS, University of Grenoble, France
Maria Jose Jimenez	University of Seville, Spain

Program Committee

Joost Batenburg	Centrum Wiskunde & Informatica, The Netherlands
Gilles Bertrand	ESIEE Cité Descartes, France
Isabelle Bloch	CNRS, France
Sara Brunetti	Department of Ingegneria dell'Informazione e Scienze Matematiche, Italy
Michel Couprie	ESIEE Cité Descartes, France
Guillaume Damiand	LIRIS - CNRS, France
Laurent Fuchs	Département XLIM-SIC UMR - CNRS, France

Referees

Table of Contents

Models for Discrete Geometry

Discrete and Combinatorial Topology

Geometric Transforms

Discrete Shape Representation, Recognition and Analysis

Discrete Tomography

Morphological Analysis

Discrete Modelling and Visualization

Discrete and Combinatorial Tools for Image Segmentation and Analysis

Facet Connectedness of Discrete Hyperplanes with Zero Intercept: The General Case

Eric Domenjoud[1], Xavier Provençal[2], and Laurent Vuillon[2]

[1] CNRS, Loria, UMR CNRS 7503, Nancy, France
Eric.Domenjoud@loria.fr
[2] Université de Savoie, LAMA, UMR CNRS 5127, Chambéry, France
{Xavier.Provencal,Laurent.Vuillon}@univ-savoie.fr

Abstract. A digital discrete hyperplane in $\mathbb{Z}^d$ is defined by a normal vector $\mathbf{v}$, a shift μ, and a thickness θ. The set of thicknesses θ for which the hyperplane is connected is a right unbounded interval of $\mathbb{R}^+$. Its lower bound, called the *connecting thickness* of $\mathbf{v}$ with shift μ, may be computed by means of the fully subtractive algorithm. A careful study of the behaviour of this algorithm allows us to give exhaustive results about the connectedness of the hyperplane at the connecting thickness in the case $\mu = 0$. We show that it is connected if and only if the sequence of vectors computed by the algorithm reaches in finite time a specific set of vectors which has been shown to be Lebesgue negligible by Kraaikamp & Meester.

Keywords: discrete hyperplane, connectedness, connecting thickness, fully subtractive algorithm.

1 Preliminaries

In order to prevent any ambiguity, we denote by $\mathbb{N}_0$ the set of nonnegative integers ($\mathbb{N}_0 = \{0, 1, 2, \dots\}$), and by $\mathbb{N}_1$ the set of positive integers ($\mathbb{N}_1 = \{1, 2, \dots\}$). We denote by $\mathbb{R}^+$ the set of non-negative real numbers. Given $d \in \mathbb{N}_1$, $(\mathbf{e}_1, \dots, \mathbf{e}_d)$ denotes the canonical basis of $\mathbb{R}^d$. The usual scalar product on $\mathbb{R}^d$ is denoted by $\langle \, . \, , \, . \, \rangle$. For any vector $\mathbf{v} \in \mathbb{R}^d$, $\|\mathbf{v}\|_1$ and $\|\mathbf{v}\|_\infty$ denote respectively the usual 1-norm and ∞-norm of $\mathbf{v}$, which means $\|\mathbf{v}\|_1 = \sum_{i=1}^d |v_i|$, and $\|\mathbf{v}\|_\infty = \max_i |v_i|$. Given $\mathbf{v} \in \bigcup_{d \geq 1} \mathbb{R}^d$, we denote by $\#\mathbf{v}$ the dimension of the space $\mathbf{v}$ belongs to, that is to say, $\mathbf{v} \in \mathbb{R}^{\#\mathbf{v}}$. Given $\mathbf{v} = (v_1, \dots, v_d) \in \mathbb{R}^d$ we denote by $\dim_{\mathbb{Q}}(v_1, \dots, v_d)$, or simply by $\dim_{\mathbb{Q}}(\mathbf{v})$ the dimension of $v_1 \mathbb{Q} + \cdots + v_d \mathbb{Q}$ as a vector space over $\mathbb{Q}$. If $\dim_{\mathbb{Q}}(v_1, \dots, v_d) = 1$ then we denote by $\gcd(v_1, \dots, v_d)$, or simply by $\gcd(\mathbf{v})$, the greatest real number γ such that v_i/γ is an integer for all i. Two distinct points $\mathbf{x}$ and $\mathbf{y}$ in $\mathbb{Z}^d$ are *facet-neighbours* (neighbours for short) if $\|\mathbf{x} - \mathbf{y}\|_1 = 1$ or equivalently if $\mathbf{x} - \mathbf{y} = \pm\mathbf{e}_i$ for some $i \in \{1, \dots, d\}$. This notion of neighbouring refers to the representation of a point in $\mathbb{Z}^d$ as a *voxel*, i.e as a unit cube centred at the point. Two points are facet-neighbours if the voxels representing them share a facet. A *path* in $\mathbb{Z}^d$ is a sequence $(\mathbf{x}^1, \dots, \mathbf{x}^n)$

E. Barcucci et al. (Eds.): DGCI 2014, LNCS 8668, pp. 1–12, 2014.
© Springer International Publishing Switzerland 2014

such that $\mathbf{x}^{i-1}$ and $\mathbf{x}^i$ are neighbours for all $i \in \{2, \ldots, n\}$. A subset S of $\mathbb{Z}^d$ is *connected* if it is not empty, and for all pairs of points $\mathbf{x}$ and $\mathbf{y}$ in S, there exists a path $(\mathbf{x}^1, \ldots, \mathbf{x}^n)$ in S such that $\mathbf{x}^1 = \mathbf{x}$ and $\mathbf{x}^n = \mathbf{y}$.

Given a vector $\mathbf{v} \in \mathbb{R}^d \setminus \{\mathbf{0}\}$, and two real numbers μ and θ, the *arithmetic discrete hyperplane* with *normal vector* $\mathbf{v}$, *shift* μ, and *thickness* θ [1,11], denoted by $\mathbb{P}(\mathbf{v}, \mu, \theta)$, is the subset of $\mathbb{Z}^d$ defined by

$$\mathbb{P}(\mathbf{v}, \mu, \theta) = \{\mathbf{x} \in \mathbb{Z}^d \mid 0 \leqslant \langle \mathbf{v}, \mathbf{x} \rangle + \mu < \theta\}. \tag{1}$$

Given a vector $\mathbf{v} \in \mathbb{R}^d \setminus \{\mathbf{0}\}$ and a shift $\mu \in \mathbb{R}$, we are interested in the set of values θ for which $\mathbb{P}(\mathbf{v}, \mu, \theta)$ is connected. This set is known to be a right-unbounded interval of $\mathbb{R}^+$ [5]. Its lower bound is called the *connecting thickness of* $\mathbf{v}$ *with shift* μ, and is denoted by $\Omega(\mathbf{v}, \mu)$. By definition, $\mathbb{P}(\mathbf{v}, \mu, \theta)$ is connected if $\theta > \Omega(\mathbf{v}, \mu)$, and disconnected if $\theta < \Omega(\mathbf{v}, \mu)$. The question that we address in this work is the connectedness of $\mathbb{P}(\mathbf{v}, \mu, \theta)$ at the critical thickness $\Omega(\mathbf{v}, \mu)$, i.e. whether $\mathbb{P}(\mathbf{v}, \mu, \Omega(\mathbf{v}, \mu))$ is connected or not. In most cases, it is easily shown that the answer is negative. Only for a specific class of vectors, the answer was unknown up to now, although some partial results have been established [2,3]. We present here the general case when $\mu = 0$.

The problem of computing the connecting thickness has already been addressed several times [4,5,6,8,9,10]. This computation may be performed by means of the *fully subtractive algorithm* [12]. We recall it briefly in the next section, and by the way, we give some useful properties.

2   Computation of the Connecting Thickness

Let us start by giving some bounds on $\Omega(\mathbf{v}, \mu)$ which will be useful later.

Theorem 1 ([5]). *Let $d \geqslant 1$, $\mathbf{v} \in \mathbb{R}^d \setminus \{\mathbf{0}\}$, and $\mu, \theta \in \mathbb{R}$.*

- *If $\mathbf{v}$ has exactly one non zero coordinate v_i, then $\mathbb{P}(\mathbf{v}, \mu, \theta)$ is connected if and only if $\theta > \mu \bmod |v_i| = \mu \bmod \gcd(\mathbf{v})$. Therefore, for all μ, we have $\Omega(\mathbf{v}, \mu) = \mu \bmod \gcd(\mathbf{v})$.*
- *If $\mathbf{v}$ has at least two non zero coordinates then let $\xi(\mathbf{v}) = \min\{|v_i| \mid v_i \neq 0\}$.*
  - *If $\theta \leqslant \|\mathbf{v}\|_\infty$ then $\mathbb{P}(\mathbf{v}, \mu, \theta)$ is disconnected for all μ.*
  - *If $\theta \geqslant \|\mathbf{v}\|_\infty + \xi(\mathbf{v})$ then $\mathbb{P}(\mathbf{v}, \mu, \theta)$ is connected for all μ.*
  Therefore, for all μ, we have $\|\mathbf{v}\|_\infty \leqslant \Omega(\mathbf{v}, \mu) \leqslant \|\mathbf{v}\|_\infty + \xi(\mathbf{v})$.

The problem of computing the connecting thickness may first be simplified thanks to the following relation [5] which allows us to get rid of the shift μ in the computations:

$$\Omega(\mathbf{v}, \mu) = \begin{cases} \Omega(\mathbf{v}, 0) + (\mu \bmod \gcd(\mathbf{v})) & \text{if } \dim_\mathbb{Q}(\mathbf{v}) = 1; \\ \Omega(\mathbf{v}, 0) & \text{if } \dim_\mathbb{Q}(\mathbf{v}) \geqslant 2. \end{cases}$$

We are then left to compute $\Omega(\mathbf{v}, 0)$ that we simply denote by $\Omega(\mathbf{v})$. For convenience, we shall usually write $\Omega(v_1, \ldots, v_d)$ instead of $\Omega((v_1, \ldots, v_d))$. The problem may be further simplified thanks to the observation that for any permutation

σ of $\{1, \ldots, d\}$, $\mathbb{P}(\mathbf{v}, \mu, \theta)$ is connected if and only if $\mathbb{P}((|v_{\sigma(1)}|, \ldots, |v_{\sigma(d)}|), \mu, \theta)$ is connected [5]. We may therefore assume that $\mathbf{v}$ belongs to $(\mathbb{R}^+)^d \setminus \{\mathbf{0}\}$, and that its coordinates are suitably ordered. In the sequel, we denote by $\mathcal{O}_d^+$ the set of vectors $\mathbf{v} \in \mathbb{R}^d \setminus \{\mathbf{0}\}$ such that $0 \leqslant v_1 \leqslant \cdots \leqslant v_d$. Finally, if $\mathbf{v} \in \mathcal{O}_d^+$ and $v_1 = 0$, then $\mathbb{P}((0, v_2, \ldots, v_d), \mu, \theta)$ is connected if and only if $\mathbb{P}((v_2, \ldots, v_d), \mu, \theta)$ is connected.

The following theorem is the key to the computation of $\Omega(\mathbf{v})$. It appears under various forms and in various contexts in the literature.

Theorem 2 ([5,6,8,9,10]). *Let $d \geqslant 2$, $\mathbf{v} \in \mathcal{O}_d^+$, and $\mathbf{v}' = (v_1, v_2 - v_1, \ldots, v_d - v_1)$. For all $\mu, \theta \in \mathbb{R}$, $\mathbb{P}(\mathbf{v}, \mu, \theta)$ is connected if and only if $\mathbb{P}(\mathbf{v}', \mu, \theta - v_1)$ is connected. Therefore, $\Omega(\mathbf{v}, \mu) = v_1 + \Omega(\mathbf{v}', \mu)$.*

We define π and σ as:

$$\pi : \bigcup_{d \geqslant 2} \mathbb{R}^d \quad \to \quad \bigcup_{d \geqslant 2} \mathbb{R}^{d-1}$$
$$(v_1, v_2, \ldots, v_d) \quad \mapsto \quad (v_2, \ldots, v_d);$$

$$\sigma : \bigcup_{d \geqslant 2} \mathbb{R}^d \quad \to \quad \bigcup_{d \geqslant 2} \mathbb{R}^d$$
$$(v_1, v_2, \ldots, v_d) \quad \mapsto \quad (v_1, v_2 - v_1, \ldots, v_d - v_1).$$

Thanks to Th. 2 and to the preceding remarks, given $\mathbf{v} \in \mathcal{O}_d^+$, we may compute $\Omega(\mathbf{v})$ recursively as follows:

$$\Omega(\mathbf{v}) = \begin{cases} 0 & \text{if } \#\mathbf{v} = 1, \\ \Omega(\pi(\mathbf{v})) & \text{if } \#\mathbf{v} \geqslant 2 \text{ and } v_1 = 0, \\ v_1 + \Omega(\text{sort}(\sigma(\mathbf{v}))) & \text{if } \#\mathbf{v} \geqslant 2 \text{ and } v_1 > 0, \end{cases}$$

where 'sort' orders the coordinates of its argument in non decreasing order.

The algorithm deduced from these equations is known as the *Ordered Fully Subtractive algorithm (OFS)* [12]. In effect, OFS computes a possibly infinite sequence of pairs $(\mathbf{v}^n, \Omega^n)_{n \geqslant 1}$ defined by:

$$\begin{array}{l} (\mathbf{v}^1, \Omega^1) = (\text{sort}(\mathbf{v}), 0) \text{ and for all } n \geqslant 1 \text{ such that } \#\mathbf{v}^n \geqslant 2, \\[4pt] (\mathbf{v}^{n+1}, \Omega^{n+1}) = \begin{cases} (\pi(\mathbf{v}^n), \Omega^n) & \text{if } v_1^n = 0; \\ (\text{sort}(\sigma(\mathbf{v}^n)), \Omega^n + v_1^n) & \text{if } v_1^n > 0. \end{cases} \end{array}$$

If $\#\mathbf{v}^n = 1$ for some n, then OFS actually terminates, and the sequence is finite. This sequence has the following properties. For all $\theta, \mu \in \mathbb{R}$ and all $n \geqslant 1$ such that $(\mathbf{v}^n, \Omega^n)$ is defined:

- $\Omega^n = \sum_{i=1}^{n-1} v_1^i$;
- $\Omega(\mathbf{v}, \mu) = \Omega^n + \Omega(\mathbf{v}^n, \mu)$;
- $\mathbb{P}(\mathbf{v}, \mu, \theta)$ is connected if and only if $\mathbb{P}(\mathbf{v}^n, \mu, \theta - \Omega^n)$ is connected;
- $\mathbb{P}(\mathbf{v}, \mu, \Omega(\mathbf{v}, \mu))$ is connected if and only if $\mathbb{P}(\mathbf{v}^n, \mu, \Omega(\mathbf{v}^n, \mu))$ is connected.

At each step of the algorithm, either $\mathbf{v}$ decreases componentwise, or the number of coordinates of $\mathbf{v}$ decreases, which may happen only finitely many times. Also, Ω^n is increasing and bounded by $\Omega(\mathbf{v})$. OFS is therefore convergent in the sense that $\mathbf{v}^n$ and Ω^n tend towards limits $\mathbf{v}^\infty$ and Ω^∞. However, it terminates if and only if $\dim_{\mathbb{Q}}(v_1, \dots, v_d) = 1$.

If OFS terminates, which means that at some step n_0, we have $\#\mathbf{v}^{n_0} = 1$, then $\Omega(\mathbf{v}) = \Omega^{n_0}$, and for all $\mu \in \mathbb{R}$,

$$\Omega(\mathbf{v}, \mu) = \Omega^{n_0} + (\mu \bmod \gcd(\mathbf{v})). \tag{2}$$

If OFS does not terminate, then for all $\mu \in \mathbb{R}$, we have

$$\Omega(\mathbf{v}, \mu) = \Omega(\mathbf{v}) = \Omega^\infty + \|\mathbf{v}^\infty\|_\infty. \tag{3}$$

Indeed, for all $n \geq 1$, we have $\Omega(\mathbf{v}, \mu) = \Omega^n + \Omega(\mathbf{v}^n, \mu)$, and, by Th. 1, we have $\|\mathbf{v}^n\|_\infty \leq \Omega(\mathbf{v}^n, \mu) \leq \|\mathbf{v}^n\|_\infty + \xi(\mathbf{v}^n)$. Since $\lim_{n\to\infty} \xi(\mathbf{v}^n) = 0$, we get $\Omega(\mathbf{v}, \mu) = \lim_{n\to\infty}(\Omega^n + \Omega(\mathbf{v}^n, \mu)) = \Omega^\infty + \|\mathbf{v}^\infty\|_\infty$. In this case, there must exist n_0 such that $\#\mathbf{v}^n \geq 2$, and $v_1^n > 0$ for all $n \geq n_0$. Let $d_\infty = \#\mathbf{v}^{n_0}$. Note that $d_\infty \geq 2$. Then for all $n \geq n_0$, we have $\|\mathbf{v}^{n+1}\|_1 = \|\mathbf{v}^n\|_1 - (d_\infty - 1) v_1^n$, so that

$$\Omega^\infty = \Omega^{n_0} + \sum_{n=n_0}^\infty v_1^n = \Omega^{n_0} + \sum_{n=n_0}^\infty \frac{\|\mathbf{v}^n\|_1 - \|\mathbf{v}^{n+1}\|_1}{d_\infty - 1} = \Omega^{n_0} + \frac{\|\mathbf{v}^{n_0}\|_1 - \|\mathbf{v}^\infty\|_1}{d_\infty - 1}.$$

For all $\mu \in \mathbb{R}$, we get

$$\Omega(\mathbf{v}, \mu) = \Omega(\mathbf{v}) = \Omega^{n_0} + \frac{\|\mathbf{v}^{n_0}\|_1 - \|\mathbf{v}^\infty\|_1}{d_\infty - 1} + \|\mathbf{v}^\infty\|_\infty. \tag{4}$$

In particular, if $v_1^n > 0$ for all n, then $n_0 = 1$ and $d_\infty = d$, so that

$$\Omega(\mathbf{v}) = \frac{\|\mathbf{v}\|_1 - \|\mathbf{v}^\infty\|_1}{d - 1} + \|\mathbf{v}^\infty\|_\infty. \tag{5}$$

Theorem 3 ([9,10]). *Let $d \geq 2$ and $\mathbf{v} \in \mathcal{O}_d^+$. If $v_1^n > 0$ for all n, then*

$$\lim_{n\to\infty} \mathbf{v}^n = \mathbf{0} \text{ if and only if } \|\mathbf{v}^n\|_\infty \leq \|\mathbf{v}^n\|_1/(d - 1) \text{ for all } n.$$

According to this theorem, if $\|\mathbf{v}^n\|_\infty \leq \|\mathbf{v}^n\|_1/(d_\infty - 1)$ for all $n \geq n_0$, then Eq. (4) becomes $\Omega(\mathbf{v}) = \Omega^{n_0} + \|\mathbf{v}^{n_0}\|_1/(d_\infty - 1)$. In particular, if $n_0 = 1$, we get

$$\Omega(\mathbf{v}) = \frac{\|\mathbf{v}\|_1}{d - 1}. \tag{6}$$

3   Connectedness at the Connecting Thickness

By definition of $\Omega(\mathbf{v}, \mu)$, $\mathbb{P}(\mathbf{v}, \mu, \theta)$ is disconnected for all $\theta < \Omega(\mathbf{v}, \mu)$, and connected for all $\theta > \Omega(\mathbf{v}, \mu)$. The question we want to answer now is whether $\mathbb{P}(\mathbf{v}, \mu, \Omega(\mathbf{v}, \mu))$ is connected. This question has an easy answer when OFS terminates, which means when $\dim_{\mathbb{Q}}(\mathbf{v}) = 1$.

Theorem 4. *If* $\dim_{\mathbb{Q}}(\mathbf{v}) = 1$ *then* $\mathbb{P}(\mathbf{v}, \mu, \Omega(\mathbf{v}, \mu))$ *is disconnected for all* $\mu \in \mathbb{R}$.

Proof. When $\dim_{\mathbb{Q}}(\mathbf{v}) = 1$, $\gcd(\mathbf{v}^n)$ is obviously an invariant of OFS. Therefore, when the halting condition is reached, we have $\mathbf{v}^n = \gamma \, \mathbf{e}_1$ where $\gamma = \gcd(v_1, \ldots, v_d)$. Then $\mathbb{P}(\mathbf{v}, \mu, \Omega(\mathbf{v}, \mu))$ is connected if and only if $\mathbb{P}(\mathbf{v}^n, \mu, \Omega(\mathbf{v}^n, \mu))$ is connected. But $\Omega(\mathbf{v}^n, \mu) = \mu \bmod \gamma$, and $\mathbb{P}(\mathbf{v}^n, \mu, \Omega(\mathbf{v}^n, \mu)) = \{x \in \mathbb{Z} \mid 0 \leqslant \gamma \, x + \mu < \mu \bmod \gamma\}$. This set is empty, hence disconnected. $\qquad \square$

If OFS does not terminate, then $\Omega(\mathbf{v}, \mu) = \Omega(\mathbf{v})$, and after some step n_0, no coordinate of the vector vanishes anymore, meaning that $v_1^n > 0$ for all $n \geqslant n_0$. Then $\mathbb{P}(\mathbf{v}, \mu, \Omega(\mathbf{v}))$ is connected if and only if $\mathbb{P}(\mathbf{v}^{n_0}, \mu, \Omega(\mathbf{v}^{n_0}))$ is connected. Therefore, we shall now study the case of vectors such that $v_1^n > 0$ for all n.

Theorem 5. *If* $v_1^n > 0$ *for all* n, *and* $\|\mathbf{v}^n\|_{\infty} \geqslant \|\mathbf{v}^n\|_1/(d-1)$ *for some* n, *then* $\mathbb{P}(\mathbf{v}, \mu, \Omega(\mathbf{v}))$ *is disconnected for all* $\mu \in \mathbb{R}$.

The following lemma will be useful for the proof of this theorem.

Lemma 6. *Let* $d \geqslant 3$ *and* $\mathbf{v} \in \mathcal{O}_d^+$. *If there exists* $r \in \{2, \ldots, d-1\}$ *such that* $\dim_{\mathbb{Q}}(v_1, \ldots, v_r) \geqslant 2$, *and* $v_{r+1} \geqslant \Omega(v_1, \ldots, v_r)$ *then* $\Omega(\mathbf{v}) = v_d$.

Proof. If $v_1 = 0$ then we have $r \geqslant 3$ because $\dim_{\mathbb{Q}}(v_1, \ldots, v_r) \geqslant 2$, and the conditions of the theorem still hold for $(v_2, \ldots, v_d)$, taking $d' = d - 1$ and $r' = r - 1$. We may therefore assume, without loss of generality, that $v_1 > 0$. Then

$$\Omega(\mathbf{v}) = v_1 + \Omega(v_1, v_2 - v_1, \ldots, v_r - v_1, v_{r+1} - v_1, \ldots, v_d - v_1),$$

and

$$\Omega(v_1, \ldots, v_r) = v_1 + \Omega(v_1, v_2 - v_1, \ldots, v_r - v_1).$$

Let $\mathbf{v}' = (v_1, v_2 - v_1, \ldots, v_d - v_1)$ and $\mathbf{v}'' = \text{sort}(\mathbf{v}')$. We have $v_r' > 0$ since otherwise we would have $v_1 = \cdots = v_r$, hence $\dim_{\mathbb{Q}}(v_1, \ldots, v_r) = 1$. We also have $\dim_{\mathbb{Q}}(v_1', \ldots, v_r') = \dim_{\mathbb{Q}}(v_1, \ldots, v_r) \geqslant 2$. Hence, $(v_1', \ldots, v_r')$ has at least two non zero coordinates. By hypothesis, we have $v_{r+1}' = v_{r+1} - v_1 \geqslant \Omega(v_1, v_2, \ldots, v_r) - v_1 = \Omega(v_1, v_2 - v_1, \ldots, v_r - v_1) = \Omega(v_1', v_2', \ldots, v_r')$, which, by Th. 1, implies $v_{r+1}' \geqslant \|(v_1', \ldots, v_r')\|_{\infty}$. Hence, $(v_1'', \ldots, v_r'') = \text{sort}(v_1', \ldots, v_r')$, and $(v_{r+1}'', \ldots, v_d'') = (v_{r+1}', \ldots, v_d') = (v_{r+1} - v_1, \ldots, v_d - v_1)$. Thus, $\mathbf{v}''$ still satisfies the conditions of the theorem. Furthermore, when we apply OFS to $\mathbf{v}$, for all n, we have $\|\mathbf{v}^n\|_{\infty} = v_{\#\mathbf{v}^n}^n = v_d - \Omega^n$. Therefore, $\Omega(\mathbf{v}) = \Omega^{\infty} + \|\mathbf{v}^{\infty}\|_{\infty} = \Omega^{\infty} + (v_d - \Omega^{\infty}) = v_d$. $\qquad \square$

Proof of Th. 5 (sketch). For all $n \geqslant 1$ and all $r \in \{2, \ldots, d\}$, we consider $K_r^n = (r-1) v_r^n - (v_1^n + \cdots + v_r^n)$. We check that, as long as $v_1^n > 0$, K_r^n is always increasing, both in r and in n. We define r_0 as the smallest index r such that $K_r^n \geqslant 0$ for some n, and n_0 as the smallest index n such that $K_{r_0}^n \geqslant 0$. Since $K_2^n = -v_1^n < 0$, we have $r_0 \geqslant 3$.

We prove that $\dim_{\mathbb{Q}}(v_1^{n_0}, \ldots, v_{r_0-1}^{n_0}) \geqslant 2$, and $\Omega(v_1^{n_0}, \ldots, v_{r_0-1}^{n_0}) = (v_1^{n_0} + \cdots + v_{r_0-1}^{n_0})/(r_0 - 2) \leqslant v_{r_0}^{n_0}$. By Lemma 6, we get $\Omega(\mathbf{v}^{n_0}) = v_d^{n_0} = \|\mathbf{v}^{n_0}\|_{\infty}$. Then, by Th. 1, $\mathbb{P}(\mathbf{v}^{n_0}, \mu, \Omega(\mathbf{v}^{n_0}))$, hence $\mathbb{P}(\mathbf{v}, \mu, \Omega(\mathbf{v}))$, is disconnected for all $\mu \in \mathbb{R}$. $\qquad \square$

We are now left to consider the case where $v_1^n > 0$ and $\|\mathbf{v}^n\|_1 > (d-1)\|\mathbf{v}^n\|_\infty$ for all n. That is to say, where $v_1^n > 0$ for all n but $\mathbf{v}$ does not satisfy the second condition of Th. 5. For $d \geqslant 2$, We note $\mathcal{K}_d$ the set of such vectors. Kraaikamp & Meester [9] have shown that $\mathcal{K}_d$ is Lebesgue negligible for all $d \geqslant 3$. If $\mathbf{v} \in \mathcal{K}_d$ then, by Th. 3, $\mathbf{v}^\infty = 0$ so that $\Omega(\mathbf{v}) = \|\mathbf{v}\|_1/(d-1)$. From what precedes, it is the only case where $v_1^n > 0$ for all n, and $\mathbb{P}(\mathbf{v}, \mu, \Omega(\mathbf{v}))$ could possibly be connected. Some partial results have already been published in the literature.

A first result was obtained in [2] for the case where $\mathbf{v} = (\alpha, \alpha + \alpha^2, 1)$, and α is the inverse of the Tribonacci number, which means $\alpha^3 + \alpha^2 + \alpha - 1 = 0$. We have $\mathbf{v} \in \mathcal{K}_3$, and it has been shown that $\mathbb{P}(\mathbf{v}, 0, \Omega(\mathbf{v}))$ is connected, and $\mathbb{P}(\mathbf{v}, \Omega(\mathbf{v}), \Omega(\mathbf{v}))$ is disconnected.

For the case $d = 3$, it has been shown in [3] that $\mathbb{P}(\mathbf{v}, 0, \Omega(\mathbf{v}))$ is connected for all vectors in $\mathcal{K}_3$. The proof relies on techniques from the field of substitutions on planes, and seems difficult to extend to higher dimensions.

In the sequel, we shall address the general case, and prove that $\mathbb{P}(\mathbf{v}, 0, \Omega(\mathbf{v}))$ is connected for all $\mathbf{v} \in \cup_{d \geqslant 2}\mathcal{K}_d$. Note that in the case $d = 2$, the condition of Kraaikamp & Meester becomes $v_2^n < v_1^n + v_2^n$ which always holds if $v_1^n > 0$ for all n, i.e. if $\dim_\mathbb{Q}(v_1, v_2) = 2$. In this case, we have $\Omega(v_1, v_2) = v_1 + v_2 = \|\mathbf{v}\|_1$. By Th. 1, $\mathbb{P}(\mathbf{v}, \mu, \Omega(\mathbf{v}))$ is connected for all $\mu \in \mathbb{R}$.

4   Main Connectedness Result

In order to establish the main result of this work, we need to study carefully the behaviour of the fully subtractive algorithm. To do so, we consider an unordered version of this algorithm, which means that the coordinates of the vector are not ordered anymore. This works as follows: as long as $\#\mathbf{v} \geqslant 2$, if some coordinate is zero, it is erased, otherwise, a minimal coordinate is subtracted from all other ones. We call this algorithm UFS for *Unordered Fully Subtractive*.

For all $k \geqslant 1$, we define σ_k and π_k as:

$$\sigma_k : \bigcup_{d \geqslant k} \mathbb{R}^d \qquad \to \bigcup_{d \geqslant k} \mathbb{R}^d$$
$$(v_1, \ldots, v_k, \ldots, v_d) \mapsto (v_1 - v_k, \ldots, v_{k-1} - v_k, v_k, v_{k+1} - v_k, \ldots, v_d - v_k);$$

$$\pi_k : \bigcup_{d \geqslant k} \mathbb{R}^d \qquad \to \bigcup_{d \geqslant k} \mathbb{R}^{d-1}$$
$$(v_1, \ldots, v_k, \ldots, v_d) \mapsto (v_1, \ldots, v_{k-1}, v_{k+1}, \ldots, v_d).$$

UFS computes a possibly infinite sequence of pairs $(\mathbf{v}^n, \Omega^n)_{n \geqslant 1}$ defined by:

$$\left(\mathbf{v}^1, \Omega^1\right) = (\mathbf{v}, 0), \text{ and for all } n \geqslant 1 \text{ such that } \#\mathbf{v}^n \geqslant 2,$$
$$\left(\mathbf{v}^{n+1}, \Omega^{n+1}\right) = \begin{cases} \left(\pi_{i_0}(\mathbf{v}^n), \Omega^n\right) & \text{if } v_{i_0}^n = 0; \\ \left(\sigma_{i_0}(\mathbf{v}^n), \Omega^n + v_{i_0}^n\right) & \text{if } v_{i_0}^n = \min_i v_i^n > 0. \end{cases}$$

Note that UFS is not deterministic since several coordinates of $\mathbf{v}^n$ could be minimal. However, OFS and UFS generate the same sequence $(\Omega^n)_{n \geqslant 1}$ and if

$(\mathbf{v}^n)_{n \geqslant 1}$ and $(\mathbf{v}'^n)_{n \geqslant 1}$ are the sequences of vectors generated respectively by OFS and UFS, we have $\mathbf{v}^n = \text{sort}(\mathbf{v}'^n)$ for all n.

From now on, we consider a vector $\mathbf{v}$ in $(\mathbb{R}^+)^d$ such that UFS never erases a coordinate, meaning that $\min_i v_i^n > 0$ for all n. We consider the infinite sequence $\Delta(\mathbf{v}) = (\delta_n)_{n \geqslant 1} \in \{1, ..., d\}^\omega$ where δ_n is the index of the minimal coordinate of $\mathbf{v}^n$ which is subtracted from all other ones. For all $n \geqslant 1$, we have $\Delta(\mathbf{v}) = \delta_1 \cdots \delta_{n_0 - 1} \Delta(\mathbf{v}^n)$, $\mathbf{v}^n = \sigma_{\delta_{n-1}} \cdots \sigma_{\delta_1}(\mathbf{v})$, and $\Omega^n = v_{\delta_1}^1 + \cdots + v_{\delta_{n-1}}^{n-1}$. We set $\theta_n = v_{\delta_n}^n$ so that $\Omega^n = \sum_{i=1}^{n-1} \theta_i$. Note that, by hypothesis, $\theta_n > 0$ for all n. We have

$$\theta_n = v_{\delta_n}^n = \langle \mathbf{v}^n, \mathbf{e}_{\delta_n} \rangle = \langle \sigma_{\delta_{n-1}} \cdots \sigma_{\delta_1}(\mathbf{v}), \mathbf{e}_{\delta_n} \rangle = \langle \mathbf{v}, {}^t\sigma_{\delta_1} \cdots {}^t\sigma_{\delta_{n-1}}(\mathbf{e}_{\delta_n}) \rangle$$

where ${}^t\sigma$ denotes the transpose of σ. We set $\mathbf{T}_n = {}^t\sigma_{\delta_1} \cdots {}^t\sigma_{\delta_{n-1}}(\mathbf{e}_{\delta_n})$ so that, for all n, we have $\langle \mathbf{v}, \mathbf{T}_n \rangle = \theta_n$. The sequence $(\mathbf{T}_n)_{n \geqslant 1}$ has some nice properties.

Lemma 7 ([7]). *If the first occurrence in $\Delta(\mathbf{v})$ of some $k \in \{1, ..., d\}$ is at position n, meaning that $\delta_n = k$, and $\delta_i \neq k$ for all $i < n$, then $\mathbf{T}_1 + \cdots + \mathbf{T}_n = \mathbf{e}_k$.*

Lemma 8 ([7]). *If m and n are the positions of two consecutive occurrences in $\Delta(\mathbf{v})$ of some $k \in \{1, ..., d\}$, meaning that $m < n$, $\delta_m = \delta_n = k$ and $\delta_i \neq k$ for all $i \in \{m+1, ..., n-1\}$, then $\mathbf{T}_{m+1} + \cdots + \mathbf{T}_n = \mathbf{T}_m$.*

Lemma 9 ([7]). *If $i \leqslant j < k$ and $\delta_j \neq \delta_k$ then $\langle \mathbf{v}, (\mathbf{T}_1 + \cdots + \mathbf{T}_k) + \mathbf{T}_i \rangle \geqslant \Omega^\infty$. Therefore, $(\mathbf{T}_1 + \cdots + \mathbf{T}_k) + \mathbf{T}_i \notin \mathbb{P}(\mathbf{v}, 0, \Omega(\mathbf{v}))$.*

We recall now the construction of the *geometric palindromic closure* of $\Delta(\mathbf{v})$ [7]. This construction builds incrementally a connected subset of $\mathbb{Z}^d$, which is easily shown to be included in $\mathbb{P}(\mathbf{v}, 0, \Omega(\mathbf{v}))$. We shall show that it is in fact exactly $\mathbb{P}(\mathbf{v}, 0, \Omega(\mathbf{v}))$ when $\mathbf{v} \in \mathcal{K}_d$.

We define a sequence $(\mathbb{P}_n)_{n \geqslant 0}$ of subsets of $\mathbb{Z}^d$ by:

$$\mathbb{P}_0 = \{0\}, \quad \text{and} \quad \mathbb{P}_n = \mathbb{P}_{n-1} \cup (\mathbb{P}_{n-1} + \mathbf{T}_n) \quad \text{for } n \geqslant 1.$$

Theorem 10 ([7]). $\mathbb{P}_\infty = \lim_{n \to \infty} \mathbb{P}_n$ *is connected.*

The set $\mathbb{P}_\infty$ is the *geometric palindromic closure* of $\Delta(\mathbf{v})$ in $\mathbb{Z}^d$ [7]. From the definition of $\mathbb{P}_n$, we get the following characterisation:

$$\mathbb{P}_n = \left\{ \sum_{i \in I} \mathbf{T}_i \mid I \subseteq \{1, ..., n\} \right\} \qquad \text{for all } n \geqslant 0;$$

$$\mathbb{P}_\infty = \left\{ \sum_{i \in I} \mathbf{T}_i \mid I \subset \mathbb{N}_1, \, |I| < \infty \right\}.$$

The inclusion $\mathbb{P}_\infty \subseteq \mathbb{P}(\mathbf{v}, 0, \Omega(\mathbf{v}))$ is straightforward. From what precedes, each $\mathbf{x}$ in $\mathbb{P}_\infty$ may be written as $\mathbf{x} = \sum_{i \in I} \mathbf{T}_i$, for some finite subset I of $\mathbb{N}_1$. Then $\langle \mathbf{v}, \mathbf{x} \rangle = \sum_{i \in I} \langle \mathbf{v}, \mathbf{T}_i \rangle = \sum_{i \in I} \theta_i \in [0; \Omega(\mathbf{v})[$. Hence, $\mathbf{x}$ belongs to $\mathbb{P}(\mathbf{v}, 0, \Omega(\mathbf{v}))$. In the sequel, we prove that we have also $\mathbb{P}(\mathbf{v}, 0, \Omega(\mathbf{v})) \subseteq \mathbb{P}_\infty$, provided that each $k \in \{1, ..., d\}$ occurs infinitely many times in $\Delta(\mathbf{v})$. The lemma below states that it is the case if $\mathbf{v} \in \mathcal{K}_d$.

Lemma 11. *If* $\mathbf{v} \in \mathcal{K}_d$ *then each* $k \in \{1, \ldots, d\}$ *occurs infinitely many times in* $\Delta(\mathbf{v})$.

Proof. Assume, by contradiction, that $\mathbf{v} \in \mathcal{K}_d$, and some $k \in \{1, \ldots, d\}$ does not occur anymore in $\Delta(\mathbf{v})$ for $n \geqslant n_0$. Since $\mathbf{v} \in \mathcal{K}_d \iff \mathbf{v}^{n_0} \in \mathcal{K}_d$, and $\Delta(\mathbf{v}) = \delta_1 \cdots \delta_{n_0-1} \Delta(\mathbf{v}^{n_0})$, we may assume, without loss of generality, that $n_0 = 1$, i.e. that k never occurs in $\Delta(\mathbf{v})$. Then for all $n > 1$, we have $v_k^n = v_k^{n-1} - \theta_{n-1} = v_k - \sum_{i=1}^{n-1} \theta_i = v_k - \Omega^n$. Since $\mathbf{v} \in \mathcal{K}_d$, we have $\lim_{n \to \infty} \mathbf{v}^n = \mathbf{0}$, so that $v_k = \lim_{n \to \infty} \Omega^n = \Omega^{\infty} = \|\mathbf{v}\|_1/(d-1)$. But $v_k \leqslant \|\mathbf{v}\|_{\infty}$ and, by assumption, $\|\mathbf{v}\|_{\infty} < \|\mathbf{v}\|_1/(d-1)$. Hence a contradiction. $\square$

From now on, we assume that $\mathbf{v}$ belongs to $\mathcal{K}_d$, so that each $k \in \{1, \ldots, d\}$ occurs infinitely many times in $\Delta(\mathbf{v})$. To prove our main theorem, we still need some additional technical results. The first lemma below is an immediate consequence of the proof of Th. 13 in [7].

Lemma 12 ([7]). *For all* $\mathbf{x}, \mathbf{y}$ *in* $\mathbb{P}_{\infty}$, *we have* $\mathbf{x} - \mathbf{y} \in \mathbb{P}_{\infty}$ *or* $\mathbf{y} - \mathbf{x} \in \mathbb{P}_{\infty}$.

Theorem 13. *Each* $\mathbf{x}$ *in* $\mathbb{Z}^d$ *may be written as* $\pm(\alpha_1 \mathbf{T}_1 + \cdots + \alpha_m \mathbf{T}_m)$ *for some* $m \geqslant 0$ *and* $\alpha_1, \ldots, \alpha_m \in \mathbb{N}_0$.

Proof. Let $\mathbf{x} \in \mathbb{Z}^d$. We have $\mathbf{x} = x_1 \mathbf{e}_1 + \cdots + x_d \mathbf{e}_d$. From lemma 7, each $\mathbf{e}_k$ may be written as $\sum_{j=1}^{n_k} \mathbf{T}_j$ where n_k is the index of the first occurrence of k in $\Delta(\mathbf{v})$. Then $\mathbf{x} = \sum_{j=1}^{r} y_j \mathbf{T}_j$ for some $r \geqslant 0$ and $y_1, \ldots, y_r \in \mathbb{Z}$. Since $\mathbf{T}_i$'s belong to $\mathbb{P}_{\infty}$, this last sum may always be decomposed as $(\mathbf{U}_1 + \cdots + \mathbf{U}_p) - (\mathbf{V}_1 + \cdots + \mathbf{V}_q)$ for some $p, q \geqslant 0$ where $\mathbf{U}_i$'s and $\mathbf{V}_i$'s belong to $\mathbb{P}_{\infty}$. If $p > 0$ and $q > 0$ then either $\mathbf{U}_1 - \mathbf{V}_1 = 0$, in which case we may simply *remove* $\mathbf{U}_1$ and $\mathbf{V}_1$ from the sum, or by lemma 12, we have either $\mathbf{U}_1 - \mathbf{V}_1 = \mathbf{U}'_1$ where $\mathbf{U}'_1 \in \mathbb{P}_{\infty}$ or $\mathbf{U}_1 - \mathbf{V}_1 = -\mathbf{V}'_1$ where $\mathbf{V}'_1 \in \mathbb{P}_{\infty}$. Replacing $\mathbf{U}_1 - \mathbf{V}_1$ with either $\mathbf{U}'_1$ or $-\mathbf{V}'_1$, the number of terms in the sum decreases. Repeating this process as long as $p > 0$ and $q > 0$, yields an expression of $\mathbf{x}$ as $\pm(\mathbf{W}_1 + \cdots + \mathbf{W}_s)$ for some $s \geqslant 0$ and $\mathbf{W}_1, \ldots, \mathbf{W}_s \in \mathbb{P}_{\infty}$. Now, each $\mathbf{W}_i$ may be written as a finite sum of $\mathbf{T}_j$'s. Doing so, and collecting the $\mathbf{T}_j$'s yields the result. $\square$

Corollary 14. *Each* $\mathbf{x}$ *in* $\mathbb{P}(\mathbf{v}, 0, \Omega(\mathbf{v}))$ *may be written as* $\alpha_1 \mathbf{T}_1 + \cdots + \alpha_m \mathbf{T}_m$ *for some* $m \geqslant 0$ *and* $\alpha_1, \ldots, \alpha_m \in \mathbb{N}_0$.

Proof. We have $\mathbf{x} = \varepsilon \times (\alpha_1 \mathbf{T}_1 + \cdots + \alpha_m \mathbf{T}_m)$ for some $\varepsilon = \pm 1$, $m \geqslant 0$, and $\alpha_1, \ldots, \alpha_m \in \mathbb{N}_0$. Then $\langle \mathbf{v}, \mathbf{x} \rangle = \varepsilon \times (\alpha_1 \langle \mathbf{v}, \mathbf{T}_1 \rangle + \cdots + \alpha_m \langle \mathbf{v}, \mathbf{T}_m \rangle) = \varepsilon \times (\alpha_1 \theta_1 + \cdots + \alpha_m \theta_m)$. Since $\mathbf{x} \in \mathbb{P}(\mathbf{v}, 0, \Omega(\mathbf{v}))$ we have $\langle \mathbf{v}, \mathbf{x} \rangle \geqslant 0$, so that $\varepsilon = +1$ because $\theta_i > 0$ for all i. $\square$

Given a nonempty subset X of $\mathbb{Z}$, we denote by $X^{\star} 0^{\omega}$ the set of infinite sequences of which the terms belong to X, and containing only finitely many non-zero terms. We define a linear mapping ψ from the $\mathbb{Z}$-module $\mathbb{Z}^{\star} 0^{\omega}$ to $\mathbb{Z}^d$ by $\psi(W) = \sum_{i \geqslant 1} w_i \mathbf{T}_i$. Then $\mathbb{P}_{\infty} = \psi(\{0, 1\}^{\star} 0^{\omega}) = \{\psi(W) \mid W \in \{0, 1\}^{\star} 0^{\omega}\}$. Next lemma is the immediate reformulation of Lemmas 7 and 8 in terms of ψ.

Lemma 15

- If n is the index of the first occurrence of k in $\Delta(\mathbf{v})$, then $\psi(1^n\,0^\omega) = e_k$.
- If m and n are the indexes of two consecutive occurrences of k in $\Delta(\mathbf{v})$, then
  $\psi(0^{m-1}\,0\,1^{n-m}\,0^\omega) = \psi(0^{m-1}\,1\,0^{n-m}\,0^\omega)$.

To prove that $\mathbb{P}(\mathbf{v}, 0, \Omega(\mathbf{v})) = \mathbb{P}_\infty$, it is sufficient to prove that each $\mathbf{x}$ in $\mathbb{P}(\mathbf{v}, 0, \Omega(\mathbf{v}))$ may be written as $\psi(W)$ for some W in $\{0,1\}^\star 0^\omega$. Thanks to Cor. 14, we may find $W \in \mathbb{N}_0^\star 0^\omega$ such that $\mathbf{x} = \psi(W)$. Using Lemma 15, we shall transform W into $W' \in \{0,1\}^\star 0^\omega$ such that $\psi(W') = \psi(W)$. We define two transformations on $\mathbb{N}_0^\star 0^\omega$ as follows.

$$
\begin{array}{ccccc}
i & = & m & & n \\
\Delta & = & \cdots \quad k & \cdots\cdots\cdots\cdots & k & \cdots
\end{array}
$$

Reduction

$$
\begin{array}{ccccc}
W & = & \cdots \quad 0 & w'_{m+1} + 1 & \cdots & w'_n + 1 & \cdots \\
\to W' & = & \cdots \quad 1 & w'_{m+1} & & \cdots \quad w'_n & \cdots
\end{array}
$$

Expansion

$$
\begin{array}{ccccc}
W & = & \cdots \quad u + 2 & w_{m+1} & \cdots & w_n & \cdots \\
\to W' & = & \cdots \quad u + 1 & w_{m+1} + 1 & \cdots & w_n + 1 & \cdots
\end{array}
$$

where $m < n$, and $u, w_{m+1}, \ldots, w_n, w'_{m+1}, \ldots, w'_n \geqslant 0$, and $\delta_i \neq k$ for all $i \in \{m+1, \ldots, n-1\}$.

Since ψ is linear, according to Lemma 15, both these transformations preserve $\psi(W)$. Given $W \in \mathbb{N}_0^\star 0^\omega$, we apply these two transformations with the following strategy.

First apply **Reduction** as much as possible.

Then, as long as $w_i \geqslant 2$ for some i:
1. apply **Expansion** once at the last position m such that $w_m \geqslant 2$;
2. apply **Reduction** as much as possible.

This strategy, if it terminates, obviously yields a sequence in $\{0,1\}^\star 0^\omega$ since otherwise **Expansion** would still apply.

Theorem 16. *If $W \in \mathbb{N}_0^\star 0^\omega$ and $\psi(W) \in \mathbb{P}(\mathbf{v}, 0, \Omega(\mathbf{v}))$, then applying **Reduction** and **Expansion** to W with the strategy above, terminates and yields $W' \in \{0,1\}^\star 0^\omega$ such that $\psi(W') = \psi(W)$.*

Before proving this theorem, let us introduce some more notation. Given a sequence W in $\mathbb{N}_0^\star$, $|W|$ is the length of W. If $W \in \mathbb{N}_0^\star 0^\omega$ then $|W|$ is the index of the last non-zero term in W, or 0 if $W = 0^\omega$.

Proof of Th. 16 (sketch). We consider the multiset M of all terms in W which are greater than 2. Our strategy ensures that M never increases. It decreases each time an expansion is performed on $w_m \cdots w_n$ if $w_m \geqslant 3$. It also decreases when a reduction is performed on $w_m \cdots w_n$ if a position i exists between $m+1$ and n such that $w_i \geqslant 3$. Hence, after finitely many transformation steps, for each transformation, we have $w_i \leqslant 2$ for all $i \in [m; n]$, and M does not evolve anymore. It is therefore sufficient to prove termination when $W \in \{0, 1, 2\}^*0^\omega$.

We observe that each transformation decreases W in the lexicographic ordering induced by $1 < 2 < 0$, so that the transformation process never loops. Although this ordering is not a well-order on infinite sequences, it is on sequences with bounded length. Each transformation which does not increase $|W|$, may be seen as operating on $\{0, 1, 2\}^{|W|}$. Therefore, there may be only finitely many reductions between two expansions, and the transformation process may not terminate only if it performs infinitely many expansions which increase $|W|$.

Finally, let $\rho_2(W)$ be the number of maximal sub-sequences in W containing no 0, and containing at least one 2. A careful examination of the transformation rules shows that $\rho_2(W)$ never increases. A case analysis shows that each expansion step which increases $|W|$ is eventually followed by a reduction step which reduces $\rho_2(W)$. The transformation process therefore terminates. $\qquad\square$

From this theorem and Cor. 14, we deduce that each $\mathbf{x}$ in $\mathbb{P}(\mathbf{v}, 0, \Omega(\mathbf{v}))$ may be written as $\psi(W)$ with $W \in \{0, 1\}^*0^\omega$, and therefore $\mathbb{P}(\mathbf{v}, 0, \Omega(\mathbf{v})) \subseteq \mathbb{P}_\infty$.

Theorem 17. *Let $d \geqslant 2$. For all $\mathbf{v}$ in $\mathcal{K}_d$, $\mathbb{P}(\mathbf{v}, 0, \Omega(\mathbf{v}))$ is connected.*

As a corollary we get the following result.

Corollary 18. *For all $\mathbf{v} \in \mathcal{K}_d$, we have $\dim_\mathbb{Q}(\mathbf{v}) = d$.*

Proof. Assume that $\dim_\mathbb{Q}(\mathbf{v}) < d$. Then there exists $\mathbf{p} \neq \mathbf{0}$ in $\mathbb{Z}^d$ such that $\langle \mathbf{v}, \mathbf{p} \rangle = 0$. Thus, $\mathbf{p} \in \mathbb{P}(\mathbf{v}, 0, \Omega(\mathbf{v}))$, hence $\mathbf{p} = \psi(U)$ for some $U \in \{0, 1\}^*0^\omega$. Since $\mathbf{p} \neq \mathbf{0}$, we have $U \neq 0^\omega$. Then $\langle \mathbf{v}, \mathbf{p} \rangle = \sum_{i \geqslant 1} u_i \langle \mathbf{v}, \mathbf{T}_i \rangle = \sum_{i \geqslant 1} u_i \theta_i > 0$. Hence a contradiction. $\qquad\square$

It should be noted that Kraaikamp & Meester actually proved Th. 3 with the assumption that $\dim_\mathbb{Q}(\mathbf{v}) = d$. However, they used this hypothesis only to ensure that $v_1^n > 0$ for all n. As a matter of fact, an earlier version of this theorem exists [10] with only the assumption that $v_1^n > 0$ for all n. With this weaker assumption, we get $\dim_\mathbb{Q}(\mathbf{v}) = d$ as a corollary.

5   Connectedness of Hyperplanes with Non-zero Shift

We have established that for each $d \geqslant 2$, $\mathbb{P}(\mathbf{v}, 0, \Omega(\mathbf{v}))$ is connected for all $\mathbf{v} \in \mathcal{K}_d$. The question which arises naturally is whether $\mathbb{P}(\mathbf{v}, \mu, \Omega(\mathbf{v}))$ is still connected when $\mu \neq 0$. We already know that $\mathbb{P}(\mathbf{v}, \mu, \Omega(\mathbf{v}))$ is connected for all $\mu \in \mathbb{R}$ if $\mathbf{v} \in \mathcal{K}_2$. For $d \geqslant 3$, we don't have a general result. However, it has been established in [2], for a specific vector in $\mathcal{K}_3$, that $\mathbb{P}(\mathbf{v}, \mu, \Omega(\mathbf{v}))$ is disconnected if $\mu = \Omega(\mathbf{v})$. Theorem 20 below shows that this holds for all $\mathbf{v} \in \mathcal{K}_d$, for all $d \geqslant 3$.

Lemma 19 ([7]). *Let $\Delta \in \{1, \ldots, d\}^\omega$ and $\mathbb{P}_\infty$ be the geometric palindromic closure of Δ. Then $\mathbb{P}_\infty \setminus \{\mathbf{0}\}$ has exactly as many connected components as the cardinal of $\{\delta_i \mid i \in \mathbb{N}_1\}$.*

If $\mathbf{v} \in \mathcal{K}_d$ then each $k \in \{1, \ldots, d\}$ occurs in $\Delta(\mathbf{v})$. Thus, $\mathbb{P}_\infty \setminus \{\mathbf{0}\}$ has exactly d connected components, and is therefore disconnected since $d \geqslant 2$.

Theorem 20. *If $d \geqslant 3$ and $\mathbf{v} \in \mathcal{K}_d$ then $\mathbb{P}(\mathbf{v}, \Omega(\mathbf{v}), \Omega(\mathbf{v}))$ is disconnected.*

Proof. We have

$$
\begin{aligned}
&\mathbb{P}(\mathbf{v}, \Omega(\mathbf{v}), \Omega(\mathbf{v})) \\
&= \{\mathbf{x} \in \mathbb{Z}^d \mid 0 \leqslant \langle \mathbf{v}, \mathbf{x} \rangle + \Omega(\mathbf{v}) < \Omega(\mathbf{v})\} \\
&= \{\mathbf{x} \in \mathbb{Z}^d \mid -\Omega(\mathbf{v}) \leqslant \langle \mathbf{v}, \mathbf{x} \rangle < 0\} \\
&= -\{\mathbf{x} \in \mathbb{Z}^d \mid 0 < \langle \mathbf{v}, \mathbf{x} \rangle \leqslant \Omega(\mathbf{v})\} \\
&= -\big((\mathbb{P}(\mathbf{v}, \Omega(\mathbf{v}), \Omega(\mathbf{v})) \setminus \{\mathbf{x} \in \mathbb{Z}^d \mid \langle \mathbf{v}, \mathbf{x} \rangle = 0\}) \cup \{\mathbf{x} \in \mathbb{Z}^d \mid \langle \mathbf{v}, \mathbf{x} \rangle = \Omega(\mathbf{v})\}\big).
\end{aligned}
$$

Since $\mathbf{v} \in \mathcal{K}_d$, we have $\Omega(\mathbf{v}) = (v_1 + \cdots + v_d)/(d-1)$, and by Cor. 18, $\dim_\mathbb{Q}(\mathbf{v}) = d$. Hence the only solution in $\mathbb{Q}^d$ of $\langle \mathbf{v}, \mathbf{x} \rangle = 0$ is $\mathbf{x} = \mathbf{0}$, and the only solution of $\langle \mathbf{v}, \mathbf{x} \rangle = \Omega(\mathbf{v})$ is $x_1 = \cdots = x_d = \frac{1}{d-1}$. Therefore, if $d \geqslant 3$, the equation $\langle \mathbf{v}, \mathbf{x} \rangle = \Omega(\mathbf{v})$ has no solution in $\mathbb{Z}^d$. Hence, $\mathbb{P}(\mathbf{v}, \Omega(\mathbf{v}), \Omega(\mathbf{v})) = -\mathbb{P}(\mathbf{v}, 0, \Omega(\mathbf{v})) \setminus \{\mathbf{0}\}$, which, by Lemma 19, is disconnected. $\square$

6   Summary of Results and Perspectives

Let us now summarise the results of the previous sections about the connectedness of $\mathbb{P}(\mathbf{v}, \mu, \Omega(\mathbf{v}, \mu))$.

Theorem 21. *Given $d \geqslant 2$ and $\mathbf{v} \in (\mathbb{R}^+)^d \setminus \{\mathbf{0}\}$, we have the following results.*

- *If $\dim_\mathbb{Q}(\mathbf{v}) = 1$ then $\mathbb{P}(\mathbf{v}, \mu, \Omega(\mathbf{v}, \mu))$ is disconnected for all $\mu \in \mathbb{R}$.*
- *If $\dim_\mathbb{Q}(\mathbf{v}) \geqslant 2$ then let $(\mathbf{v}^n)_{n \geqslant 1}$ be the sequence of vectors computed by the fully subtractive algorithm applied to $\mathbf{v}$. We have $\Omega(\mathbf{v}, \mu) = \Omega(\mathbf{v})$ for all $\mu \in \mathbb{R}$, and $\mathbb{P}(\mathbf{v}, 0, \Omega(\mathbf{v}))$ is connected if and only if $\mathbf{v}^n \in \mathcal{K}_{d'}$ for some n, and some $d' \leqslant d$. In this case:*
  - *If $d' = 2$ then $\mathbb{P}(\mathbf{v}, \mu, \Omega(\mathbf{v}))$ is connected for all $\mu \in \mathbb{R}$.*
  - *If $d' \geqslant 3$ then $\mathbb{P}(\mathbf{v}, \mu, \Omega(\mathbf{v}))$ is disconnected for some values of μ.*
  In particular, $\mathbb{P}(\mathbf{v}, \Omega(\mathbf{v}), \Omega(\mathbf{v}))$ is disconnected.

This theorem provides a complete characterisation of the connectedness of $\mathbb{P}(\mathbf{v}, 0, \theta)$ at the critical thickness $\theta = \Omega(\mathbf{v}, 0)$. For the case $d = 3$, Th. 5.1 in [3] already established the connectedness of $\mathbb{P}(\mathbf{v}, 0, \Omega(\mathbf{v}, 0))$ for all $\mathbf{v} \in \mathcal{K}_3$. However, the *only if* part of that theorem is wrong. The authors claim falsely that the vectors in $\mathcal{K}_3$ are the only ones in $(\mathbb{R}^+)^3$ for which $\mathbb{P}(\mathbf{v}, 0, \Omega(\mathbf{v}))$ is connected. This is obviously false since $\mathbb{P}(\mathbf{v}, 0, \Omega(\mathbf{v}))$ is connected for all vectors of the form $(0, v_2, v_3)$ such that $\dim_\mathbb{Q}(v_2, v_3) = 2$.

For $\mathbf{v} \in \mathcal{O}_3^+ \setminus \mathcal{K}_3$, if $\dim_\mathbb{Q}(\mathbf{v}) > 1$, we have eventually $v_3^n \geqslant v_1^n + v_2^n$ for some n. The mistake in the proof lies in the argument that this implies $\Omega(\mathbf{v}^n) = v_3^n = \|\mathbf{v}^n\|_\infty$. Then, by Th. 1, $\mathbb{P}(\mathbf{v}^n, 0, \Omega(\mathbf{v}^n))$ would be disconnected. Actually, we have $\Omega(\mathbf{v}^n) = v_3^n$ only if $\dim_\mathbb{Q}(v_1^n, v_2^n) = 2$. When $\dim_\mathbb{Q}(v_1^n, v_2^n) = 1$, we have eventually $v_1^m = 0$ for some m, and then $\mathbf{v}^{m+1} = (v_2^m, v_3^m)$. Since $\dim_\mathbb{Q}(\mathbf{v}^{m+1}) =$

$\dim_{\mathbb{Q}}(\mathbf{v}) > 1$, we have $\mathbf{v}^{m+1} \in \mathcal{K}_2$. Then $\mathbb{P}(\mathbf{v}^{m+1}, 0, \Omega(\mathbf{v}^{m+1}))$ is connected, and $\mathbb{P}(\mathbf{v}, 0, \Omega(\mathbf{v}))$ as well. In this case, $\mathbb{P}(\mathbf{v}, \mu, \Omega(\mathbf{v}))$ is actually connected for all $\mu \in \mathbb{R}$. As a matter of fact, we have $\Omega(\mathbf{v}^n) = v_3^n + \gcd(v_1^n, v_2^n)$. Take for instance $\mathbf{v} = (1, 1, \sqrt{2}+1)$. This vector does not belong to $\mathcal{K}_3$, but after one application of OFS, we get $\mathbf{v}^2 = (0, 1, \sqrt{2})$ which reduces to $\mathbf{v}^3 = (1, \sqrt{2})$. Now $\mathbb{P}(\mathbf{v}^3, 0, \Omega(\mathbf{v}^3))$ is connected so that $\mathbb{P}(\mathbf{v}, 0, \Omega(\mathbf{v}))$ is connected.

In order to effectively determine whether $\mathbb{P}(\mathbf{v}, 0, \Omega(\mathbf{v}))$ is connected, in addition to Th. 21, we still need a way to decide whether the fully subtractive algorithm will eventually reach some $\mathcal{K}_d$. At present, we are not even able to decide whether a given vector $\mathbf{v} \in \mathbb{R}^d$ belongs to $\mathcal{K}_d$. This lets some open questions and some research directions. Given $\mathbf{v} \in \mathbb{R}^d \setminus \{0\}$ we are interested in the following questions:

- decide whether OFS will erase some coordinate, i.e. whether $v_1^n = 0$ for some n;
- if not, decide whether $\mathbf{v}$ belongs to $\mathcal{K}_d$;
- if $\mathbf{v} \in \mathcal{K}_d$, characterise the values of μ for which $\mathbb{P}(\mathbf{v}, \mu, \Omega(\mathbf{v}))$ is connected.

References

1. Andrès, E., Acharya, R., Sibata, C.: Discrete analytical hyperplanes. CVGIP: Graphical Model and Image Processing 59(5), 302–309 (1997)
2. Berthé, V., Domenjoud, E., Jamet, D., Provençal, X.: Fully subtractive algorithm, tribonacci numeration and connectedness of discrete planes. RIMS Lecture notes 'Kokyuroku Bessatu' (to appear, 2014)
3. Berthé, V., Jamet, D., Jolivet, T., Provençal, X.: Critical connectedness of thin arithmetical discrete planes. In: Gonzalez-Diaz, R., Jimenez, M.-J., Medrano, B. (eds.) DGCI 2013. LNCS, vol. 7749, pp. 107–118. Springer, Heidelberg (2013)
4. Brimkov, V.E., Barneva, R.P.: Connectivity of discrete planes. Theor. Comput. Sci. 319(1-3), 203–227 (2004)
5. Domenjoud, E., Jamet, D., Toutant, J.-L.: On the connecting thickness of arithmetical discrete planes – extended version (in preparation, 2014)
6. Domenjoud, E., Jamet, D., Toutant, J.-L.: On the connecting thickness of arithmetical discrete planes. In: Brlek, S., Reutenauer, Ch., Provençal, X. (eds.) DGCI 2009. LNCS, vol. 5810, pp. 362–372. Springer, Heidelberg (2009)
7. Domenjoud, E., Vuillon, L.: Geometric Palindromic Closures. Uniform Distribution Theory 7(2), 109–140 (2012)
8. Jamet, D., Toutant, J.-L.: Minimal arithmetic thickness connecting discrete planes. Discrete Applied Mathematics 157(3), 500–509 (2009)
9. Kraaikamp, C., Meester, R.W.J.: Ergodic properties of a dynamical system arising from percolation theory. Ergodic Theory and Dynamical Systems 15(04), 653–661 (1995)
10. Meester, R.W.J.: An algorithm for calculating critical probabilities and percolation functions in percolation models defined by rotations. Ergodic Theory and Dynamical Systems 9, 495–509 (1989)
11. Réveillès, J.-P.: Géométrie discrète, calcul en nombres entiers et algorithmique. Thèse d'état, Université Louis Pasteur (Strasbourg, France) (1991)
12. Schweiger, F.: Multidimensional continued fractions. Oxford Science Publications. Oxford University Press, Oxford (2000)

About the Maximum Cardinality of the Digital Cover of a Curve with a Given Length

Yan Gérard and Antoine Vacavant

ISIT, UMR 6284 CNRS / Université d'Auvergne
Clermont-Ferrand, France
{yan.gerard,antoine.vacavant}@udamail.fr

Abstract. We prove that the number of pixels -with pixels as unit lattice squares- of the digitization of a curve Γ of Euclidean length l is less than $3\lfloor \frac{l}{\sqrt{2}} \rfloor + 4$ which improves by a ratio of $\frac{4\sqrt{2}}{3}$ the previous known bound in $4\lfloor l \rfloor$ [3]. This new bound is the exact maximum that can be reached. Moreover, we prove that for a given number of squares n, the Minimal Length Covering Curves of n squares are polygonal curves with integer vertices, an appropriate number of diagonal steps and 0, 1 or 2 vertical or horizontal steps. It allows to express the functions $N(l)$, the maximum number of squares that can be crossed by a curve of length l, and $L(n)$, the minimal length necessary to cross n squares. Extensions of these results are discussed with other distances, in higher dimensions and with other digitization schemes.

Keywords: length, cover, digitization, curve, complexity, cardinality.

1 Introduction

1.1 Multiresolution and Fixed Resolution Bounds for the Size of the Digital Representation of a Shape

The questions that we investigate in this paper came to our attention as we were trying to bound the size of the multiresolution representation, with quadtrees or octrees, of a real 2D or 3D shape. Such a bound is known in the case of a 2D shape [3]. This bound on the number of quads of the representation of a shape is related with the maximal number of squares of size $\frac{1}{2^n}$ that a curve can cross. It means that the bound in a multiresolution framework comes from a bound, which is computed at a fixed resolution. The fixed resolution bound used in [3] is $NumberOfPixels(\Gamma) \leq 4\lceil l_2(\Gamma) \rceil$ where Γ is a rectifiable curve of Euclidean length $l_2(\Gamma)$ and where pixels are a set of unit squares tiling the plane. This bound comes simply from the fact that a piece of curve of length 1 can not cross 4 new unit squares. Is this bound tight? Not exactly. And -even if it is not the purpose of this paper- the reader should keep in mind that a better bound for this number of pixels (crossed by a curve of given length) implies better bounds for the size of the representation of a shape in a multiresolution framework (the construction of the multiresolution bound is described in [3]). As far as we know

E. Barcucci et al. (Eds.): DGCI 2014, LNCS 8668, pp. 13–24, 2014.
© Springer International Publishing Switzerland 2014

and except in previous references, this task to bound the number of squares crossed by a curve of given length is original. Our goal is to close the problem by computing the tightest possible bound, namely the exact maximum $N(l)$ of the number of pixels of the digitization of a curve Γ in function of its euclidean length l (Fig. 1). The result that we provide is better from a factor $2\sqrt{2}$ with respect to the previous known result in $6\lceil l_2(\Gamma)\rceil$.

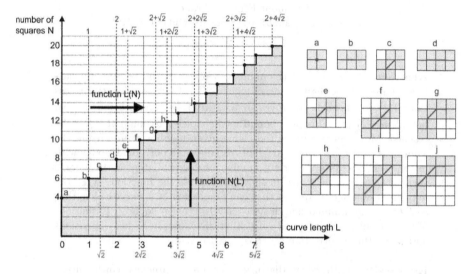

Fig. 1. The functions $N(l)$ and $L(n)$ giving respectively the maximum number of squares in the cover of a curve of length l and the minimal length necessary to cover n squares. Expressions are given in Theorem 1. Some examples of optimal curves (a, b,..., j) corresponding to the saillant points of the staircase function are drawn on the right.

1.2   Notations and Problem Statement

We start to work with closed squares $[i, i+1] \times [j, j+1]$ as pixels where i and j are integers. Given any real subset $P \subset \mathbb{R}^2$, the set S of squares which contains at least one point of P is usually called the cover of P (see Fig. 2). We denote it by $cover(P)$. We are interested in the cardinality $|cover(\Gamma)|$ of the cover of a rectifiable curve $\Gamma : [0, 1] \to \mathbb{R}^2$ with respect to its Euclidean length l (we recall that a rectifiable curve has a finite length which is defined as the upper bound of the lengths of the polygonal lines having ordered vertices on the curve). As an alternative terminology, we consider the number of squares crossed by the curve Γ. Other digitization schemes will be considered in the following but the results that we will provide for them are just consequences of the ones that we have with closed squares.

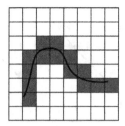

Fig. 2. A curve Γ and its cover $cover(\Gamma)$ namely the set of squares crossed by Γ either by one of their edges, or by one of their vertices

It follows from definition that the cover of a single point can have 1, 2 or 4 squares, 4 in the case of an integer point (i, j). We can notice that an horizontal path of integer length $l = k$ with integer vertices crosses $4 + 2\lfloor l \rfloor$ squares while a diagonal path of length $l = k\sqrt{2}$ crosses exactly $4 + 3\frac{l}{\sqrt{2}}$ squares (see for instance Fig. 3) which is more for $k \geq 2$. Is it the maximum cardinality of the cover of a curve with a given length? This question is solved in Theorem 1 but let us introduce before some notations: We denote by $N(l)$ the maximum number of squares n that a curve Γ of length l can cross and by $L(n)$ the minimal length l necessary to cross n squares (existence of the minimum is proved in Lemma 2). As the function $N(l)$ takes discrete values as input, it is a staircase while $L(n)$ is the discrete function obtained by inverting the axes (see Fig. 1). Our main objective is to provide the expressions of $N(l)$ and $L(n)$. This result is obtained in two steps, a first lemma stating the existence of Minimum Length Covering Curves of a given *set* of squares (Sec. 2) and a second lemma about Minimum Length Covering Curves of a given *number* of squares (Sec. 3). Then, the main theorem provides the expressions of $N(l)$ and $L(n)$ as well as the shape of the optimal curves. We end with some extensions of these results (Sec. 4).

2   Minimum Length Covering Curves of a Set of Squares

Let us consider a finite set of squares S (not necessarily connected). We are interested in the lengths l of the curves Γ which crosses S namely with inclusion $S \subset cover(\Gamma)$. The set of the possible lengths l is an interval since if there exists a curve of length l crossing S, then for any $\epsilon \geq 0$, we can build a new curve Γ' crossing S with a length $L + \epsilon$ just by adding a segment of length ϵ. But is this interval of the form $[L, +\infty[$ or $]L, +\infty[$? In other words, do these curves Γ have a minimal length or is it a lower bound? Lemma 1 states the existence of curves of minimal length. We call them *Minimal Length Covering Curves -MLCC* for short- of S (see Fig. 4).

Before going further, notice here that we have just the inclusion $S \subset cover(\Gamma)$ and not equality. It makes a difference with polygonal curves known in the specific framework of closed digital curves as Minimum Length Polygon [1] [2] but the principle of length minimality is the same.

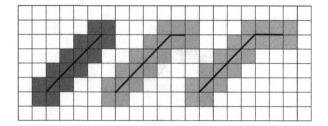

Fig. 3. Theorem 1 proves that these curves are optimal in the sense that no shorter curve has a cover with so much squares (or crosses so much squares)

Lemma 1. *Given a finite set of squares S, among all continuous curves Γ that cross all the squares of S (S ⊂ cover(Γ)), there exists at least one with a minimal length (called Minimal Length Covering Curves of S). MLCC are polygonal curves with vertices on the edges of the squares of S.*

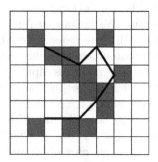

Fig. 4. A finite set of squares S and its Minimal Length Covering Curve Γ. Notice here that we have just the inclusion $S \subset cover(\Gamma)$ and not equality. We can also notice that MLCC can have non integral vertices.

Proof. The lemma states two things: first the existence of the Minimal Length Covering Curves, and secondly a structural property of these curves. Let us start with the second point: the structure of MLCC.

Assume that a MLCC is not a polygonal curve, then there exists in its cover a square in which it is not polygonal. The curve goes in this square by a point x on its boundary and goes outside of it by another point y. The shortest path to go from x to y is -of course- the segment $[xy]$. It means that by replacing the arc of Γ from x to y by this segment, we reduce its length, which contradicts the hypothesis of length minimality (see Fig. 5).

It follows that the curve is polygonal. It follows also from previous remark that it cannot have vertices in the interior of the squares. The vertices are necessarily on the edges of the squares.

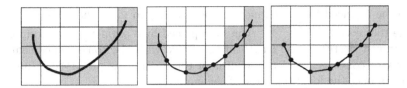

Fig. 5. A curve of minimal length crossing a set of pixel is necessarily polygonal. Otherwise, shorter curves would exist.

It remains to prove the first point of the lemma: the existence of a Minimum Length Covering Curve of S, in other words the fact that the lower bound of the lengths of the curves crossing S is a minimum. We consider a sequence of curves Γ_i verifying $S \subset cover(\Gamma_i)$ with a length L_i that converges to the lower bound of the lengths l of the curves crossing the n squares of S. Our goal is now to build a MLCC of S, namely a curve Γ of length l covering S.

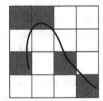
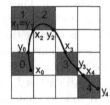

Fig. 6. A set of (red) squares S, a curve Γ and the corresponding ordering of the squares (the ambiguity between the indices 1 and 2 is removed with the rule *east* before *west*). The curve is decomposed into 5 pieces $\Gamma_i^{s_k}$ in the red squares (in the square of index 1, this piece is reduced to a point) and 4 pieces $\Gamma_i^{link(s_k \to s_{k+1})}$ from the square s_k to s_{k+1} (the second piece is also reduced to a point).

Let us introduce properly the order of the squares of S crossed by Γ_i (see Fig. 6). We define it as the sequence of the squares in the order where the curve arrives in them, but some precisions are required: the curve can reach an integer point or follow an edge with the consequence that in some cases, we may have some ambiguities for the choice of the next square. We can avoid these small difficulties just by using in this case a rule such that *north* before *south* and in a second time *east* before *west*. It provides a decomposition of the curve $\Gamma_i = \bigcup_k (\Gamma_i^{s_k} \bigcup \Gamma_i^{link(s_k \to s_{k+1})})$ where $\Gamma_i^{s_k}$ is a continuous arc of Γ_i in the square s_k of S and where $\Gamma_i^{link(s_k \to s_{k+1})}$ is a continuous arc going from the square s_k to the square s_{k+1} (in the case of neighboring squares in S, these arcs may be reduced to points). We can also define the point x_k of entry of the curve Γ_i in its kth square s_k and y_k its kth point of exit (in the case where the initial point starts in a square of S as in Fig. 6, we choose it as x_0 and the same for the end point). We can already notice that without loss of generality, the piece of curve

Γ_i^k can be replaced by the segment going from x_k to y_k in the square of index k while the piece $\Gamma_i^{link(s_k \to s_{k+1})}$ can be replaced by a segment going from y_k to x_{k+1}. It just decreases the length of the curve with the consequence that the property of convergence of the lengths of these curves is preserved. Moreover, we still respect the local topology of the curve.

There is nevertheless a possible unwanted property: the sequence of the squares s_k may be infinite. It is easy to see that the curve cannot cross an infinite number of different squares, since its length is bounded but the sequence may be infinite due to repetitions (if there are some kinds of accumulations). In this case, we are going to replace the curve Γ_i by a shorter curve Γ_i' with, this time, a finite sequence of squares. As it is shorter, the convergence of the length $L_2(\Gamma_i')$ is again preserved. How do we build Γ_i'? Just by cutting useless loops:

If Γ_i crosses a square s, it may cross it several other times for indices k, k', k''... to have a path going to another square of S which has not been crossed before. It means that if the square s is crossed more than n times (we recall that n is the cardinality of the set S), it makes some loops and some of these loops are useless because they don't cross "new" squares of S (new in the sense that they were never crossed before). These useless loops going outside from s are replaced in a new curve Γ_i' by short segments in s without loosing any square of S. Then after this cut, Γ_i' is shorter than Γ_i and it remains at most n loops for each square and thus n repetitions. Hence, the sequence of squares of S crossed by the curve Γ_i' is finite and bounded by n^2.

We know now that for all curves Γ_i, our sequences of the squares contain at most n^2 squares of S. It follows that the set of the orders of the squares s_i^k of S crossed by the curves Γ_i is finite and one of these orders appears an infinite number of times in the sequences of the orders of the curves Γ_i. We consider now the subsequence of curves Γ_{i_j} with this order: they have all exactly the same sequence of squares s_k for $1 \le k \le n'$ (with $n' \le n^2$). We can now focus on the points of entry $x_{j,k}$ and exit $y_{j,k}$ for the curve Γ_i in the square s_k. Then we have $2n'$ sequences of points of index j. As they move in compact sets (they belong to the boundary of a square) and the product of a finite number (here $2n' \le 2n^2$) of compacts is still compact, we can extract from these $2n'$ sequences a unique sequence such that $\lim_{j \to +\infty} x_{j,k} = x_k$ and $\lim_{j \to +\infty} y_{j,k} = y_k$. It just remains to lie the points x_k to y_k and y_k to x_{k+1} by a segment in order to obtain a new curve Γ. What can we say about the length of Γ? The length of Γ or Γ_{i_j} is the sum of the lengths of all their pieces in the squares s_k and between the squares:

$$L(\Gamma_{i_j}) = \sum_{k=0}^{n'} d(x_{j,k}, y_{i,k}) + \sum_{k=0}^{n'-1} d(y_{j,k}, x_{i+1,k})$$

$$L(\Gamma) = \sum_{k=0}^{n'} d(x_k, y_k) + \sum_{k=0}^{n'-1} d(y_k, x_k).$$

It follows by continuity of the distance $\lim_{j\to+\infty} L(\Gamma_{i_j}) = L(\Gamma)$. It proves that the length of the curve Γ is the lower bound and thus: we have built a curve Γ covering S of minimal length.

In many cases, the vertices of the MLCC are not only on the edges but on the integer points of the grid. Nevertheless, in some particular cases, it can occur that a vertex of the minimal curve is not an integer point but on one of the edges (see an example Fig. 6).

Lemma 1 deals with a finite set of squares S. If S has now an infinite cardinality, then all covering curves are of infinite length since the diameter of S is itself infinite. MLCC are not defined in this case.

At last, we can notice that Lemma 1 holds in higher dimensions, with hypercubes instead of squares, and curves of dimension 1.

3 Minimum Length Covering Curves of n Squares

3.1 About Their Existence

Previous lemma proves the existence of a Minimum Length Covering Curve for a given finite set of squares S. Let us now relax the condition on S by assuming that we know only its cardinality. Hence, we provide a number of squares n and ask about the existence of a minimal length for the curves crossing n squares or more.

Lemma 2. *Given a natural integer n, there exists curves Γ of minimal length crossing n squares (for any curve P crossing n squares or more, $L(P) \geq L(\Gamma)$). We call them Minimal Length Covering Curve of n squares.*

Proof. Given a number of squares n (just the number), we can consider all the possible configurations of sets of n squares. We can reduce ourselves to 4-connected configurations, with the consequence that, up to a translation, there are only a finite number of such configurations. Thus, due to Lemma 1, the lower bound of the lengths of curves crossing n squares is in fact a minimum that is obtained for some polygonal curves with vertices on the edges of the squares.

3.2 MLCC of n Squares and Expressions of $L(n)$ and $N(l)$

We can at last express the functions $N(l)$ and $L(n)$ and provide the exact structure of Minimal Length Covering Curves of n squares.

Theorem 1. *Minimal Length Covering Curves of n squares are made of 0, 1 or 2 horizontal or vertical steps and an arbitrary number of diagonal steps (Fig. 1).*

- *If $l \bmod \sqrt{2} < 2 \bmod \sqrt{2}$, then $N(l) = 3\lfloor \frac{l}{\sqrt{2}} \rfloor + 4$ (optimal curves have $\lfloor \frac{l}{\sqrt{2}} \rfloor$ diagonal steps).*

- If $2 mod \sqrt{2} \leq l mod \sqrt{2} < 1$ and $l > 1$, then $N(l) = 3\lfloor \frac{l}{\sqrt{2}} \rfloor + 5$ (optimal curves have $\lfloor \frac{l}{\sqrt{2}} \rfloor - 1$ diagonal steps and 2 horizontal or vertical steps). If $2 - \sqrt{2} \leq l < 1$, we have still $N(l) = 4$ (the first step of the staircase is broken).
- If $1 \leq l mod \sqrt{2}$, then $N(l) = 3\lfloor \frac{l}{\sqrt{2}} \rfloor + 6$ (optimal curves have $\lfloor \frac{l}{\sqrt{2}} \rfloor$ diagonal steps and 1 horizontal or vertical step).

Conversely, function $L(n)$ is:

- $L(1) = L(2) = L(3) = L(4) = 0$, $L(5) = 1$.
- If $n mod 3 = 0$ (and $n \geq 6$), then $L(n) = 1 + (\frac{n}{3} - 2)\sqrt{2}$.
- If $n mod 3 = 1$ (and $n \geq 6$), then $L(n) = (\lfloor \frac{n}{3} \rfloor - 1)\sqrt{2}$.
- If $n mod 3 = 2$ (and $n \geq 6$), then $L(n) = 2 + (\lfloor \frac{n}{3} \rfloor - 2)\sqrt{2}$.

Proof. The first step is to prove that a MLCC of n squares is necessarily a polygonal curve with integer vertices. Let Γ be a MLCC of n squares. It is of course a MLCC of its cover $cover(\Gamma)$. We assume that Γ has a non integral vertex, and we are going to build another curve Γ' with the same length crossing at least the same number n of squares. Then Γ' optimality will have consequences on the structure of Γ.

If we consider that the first or the last vertices are not integral, it is easy to see that it contradicts length minimality (there is only the particular case of a single horizontal or vertical segment but even in this case, it is clear that such segments with non integral vertices are not MLCC of n squares). Thus the extremities of a MLCC are integral points. Now, let us consider a non integral intermediary vertex v of Γ on the edge e between two squares s_i and s_{i+1} and obtain a contradiction. If we consider the quadrants of its edges, only one configuration among four is compatible with length minimality (Fig. 7).

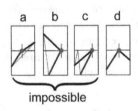

Fig. 7. If we consider a non integral vertex of Minimum Length Covering Curve, we can imagine four cases a), b), c) or d) according to the quadrant of the direction of the next edge towards the initial one. Among these four cases represented here, only one is possible since the three others contradict the minimality of the length of the curve.

This only possible configuration allows to unfold the curve by a sequence of symmetries so that the cardinality of the cover of the images of each segment is preserved at each step (Fig. 8). At the end, as the curve is completely unfolded,

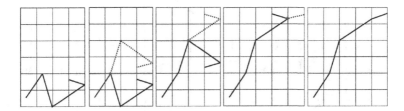

Fig. 8. The fact that only case from Fig. 7-d) can occur in a MLCC allows to unfold any MLCC with a sequence of symetries that preserve the cardinality of the covers of each edge. Hence the unfolded curve Γ' has the same length than Γ (MLCC of n squares), crosses at least the same number of squares -it is still a MLCC of n squares- and due to monotonicity, case from Fig. 7-d) cannot occur anymore.

we obtain a monotonic curve Γ' such that its cover has at least the same number of squares than Γ. As Γ' has the same length and crosses at least the same number of squares, it is still a MLCC of n squares. Its length is minimal and due to monicity, there can exist no more intermediary vertices such as in Fig. 7. It follows that all intermediary vertices of Γ' are integer points.

Let us now consider a segment of Γ' going for instance from $(0,0)$ to (a,b) where (a,b) are coprime positive integers. The initial point $(0,0)$ is in 4 squares. The last one too. And the segment crosses $a-1$ vertical edges and $b-1$ horizontal edges. As there is no integer point in the inner part of the segment, it makes $5+a+b$ squares. Now let us assume that $0 < b < a$ and compare with a horizontal segment $(a,0)$. It crosses $4 + 2a$ squares which is at least equal to the $5 + a + b$ squares of the segment (a,b). And $(a,0)$ is shorter. It follows that Γ' cannot have a segment (a,b) different from $(1,0)$, $(1,1)$, or $(0,1)$ because otherwise, by replacing the segment (a,b) by $(a,0)$, we obtain a shorter path crossing at least the same number of squares. If we come back to our initial MLCC Γ, it proves that a MLCC of n squares can contain only segments in horizontal, vertical and diagonal directions, and whatever the order of these steps, as soon as the cover has no loop, it does not change the number of squares. It follows that any MLCC of n squares can be unfolded in a polygonal curve with a horizontal segment $(a,0)$ and a diagonal one (b,b). The length is $a + b\sqrt{2}$ and the number of squares is $4 + 2a + 3b$. Knowing these structural properties, the MLCC of n squares are characterized by the minimization problem: minimize $a + b\sqrt{2}$ under constraints $n \leq 4 + 2a + 3b$ with positive integer values a and b. This minimum provides the value of the function $L(n)$. The other way to consider the question is the converse one: the maximum of $4+2a+3b$ subject to constraint $a+b\sqrt{2} \leq l$, with a and b positive integers provides the value of the function $N(l)$ (see Fig. 9).

The critical values are given by integer points (a,b) with left triangle of slopes $-\frac{2}{3}$ and $-\frac{\sqrt{2}}{2}$ which does not contain any other integer point. It follows that only three values of a are possible: 0, 1 or 2 (if $a \geq 3$, then $(a',b') = (a-3, b+2)$ verifies also $N \leq 4 + 2a' + 3b'$ with a better score i.e smaller length $a' + b'\sqrt{2} =$

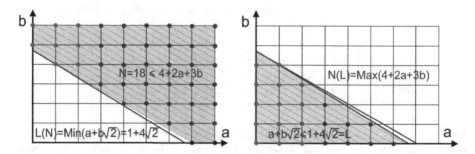

Fig. 9. On the left, $L(n)$ is the minimum of $a+b\sqrt{2} \leq l$ under constraint $n \leq 4+2a+3b$ with $(a,b) \in \mathbb{N}^2$. On the right, $N(l)$ is the maximum of $4+2a+3b$ under constraint $a+b\sqrt{2} \leq l$. Due to the respective slopes of the extremal lines $-\frac{\sqrt{2}}{2}$ and $-\frac{2}{3}$, the optimal points have coordinates a equal to 0, 1 or 2 (the red points).

$a+b-3+2\sqrt{2}$). Then the optimal curves are mainly diagonal with 0, 1 or 2 horizontal or vertical steps. Without more details here, it leads to the expressions of functions $L(n)$ and $N(l)$ expressed in Theorem 1.

4   Extensions of the Results

Let us give an overview of the consequences and extensions of Theorem 1.

4.1   With Other Kinds of Covers

In the case of some other kinds of covers, for instance by considering squares $[i, i+1[\times[j, j+1[$ whose union is a partition of the plane $\mathbb{R}^2$, Minimal Length Covering Curves no more exist since the lower bound is not a minimum anymore. Nevertheless, the difference for the functions $N(l)$ remains small : if the curve Γ is a MLCC of a set of closed squares S (as previously), then we just have to add a little loop around each integer point or each edge that it crosses to build a curve of length $L(\Gamma) + \epsilon$ which has exactly the same cover (see Fig. 10). It follows in this case

Theorem 2. *With the squares $[i, i+1[\times[j, j+1[$, we have*
If $l\,mod\sqrt{2} \leq 2mod\sqrt{2}$, then $N(l) = 3\lfloor\frac{l}{\sqrt{2}}\rfloor + 4$
If $2mod\sqrt{2} < l\,mod\sqrt{2} \leq 1$, then $N(l) = 3\lfloor\frac{l}{\sqrt{2}}\rfloor + 5$
If $1 < l\,mod\sqrt{2}$, then $N(l) = 3\lfloor\frac{l}{\sqrt{2}}\rfloor + 6$.

4.2   With Closed Curves

If we restrict ourselves to closed curves, we have clearly $N_{closed}(l) \leq N(l)$ while $L_{closed}(n) \geq L(n)$. We can also notice that a square in the diagonal direction

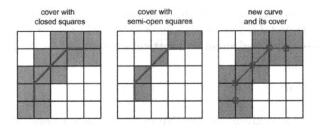

cover with closed squares   cover with semi-open squares   new curve and its cover

Fig. 10. Given a curve Γ with its cover S in the sens of closed squares $[i, i+1] \times [j, j+1]$, for any $\epsilon > 0$, we can build new curves Γ' of length $L(\Gamma') \leq L(\Gamma) + \epsilon$ with the same cover S but now in the sens of semi-open squares $[i, i+1[\times [j, j+1[$

with sides of length $k\sqrt{2}$ and integer vertices is a closed curve of length $l = 4k\sqrt{2}$ which crosses exactly $12k = 3\lfloor \frac{l}{\sqrt{2}} \rfloor \leq 12\lfloor \frac{l}{4\sqrt{2}} \rfloor$ squares. It follows

$$12\lfloor \frac{l}{4\sqrt{2}} \rfloor \leq N_{closed}(l) \leq N(l) \leq 6 + 12\lfloor \frac{l}{4\sqrt{2}} \rfloor.$$

It means in particular that the difference between the values of the functions for closed and general curves remains bounded by a small constant.

4.3   In Dimension d

The results presented here could be generalized in dimension d: the first intuition is to consider that the MLCC of n hypercubes in $\mathbb{R}^d$ will follow paths in the "hyper"-diagonal direction of the hypercubes. A path of k hyperdiagonal steps has a length $k\sqrt{d}$ for a number of hypercubes equal to $2^d + (2^d - 1)k$. The ratio is $\frac{2^d-1}{\sqrt{d}}$. This should be compared with a path which remains for instance in the hyperplane $x_d = 0$ and an hyperdiagonal direction in this plane. Such a path crosses $2(2^{d-1} + (2^{d-1} - 1)k)$ hypercubes (the coefficient 2 comes from the hypercubes centered with $x_d = +-\frac{1}{2}$) for k steps of length $\sqrt{d-1}$. The ratio is $2\frac{2^{d-1}-1}{\sqrt{d-1}}$. The comparison between these two ratios shows that the second one is greater than the first one except for $d = 2$. It follows that some MLCC in dimension d are in fact MLCC drawn in dimension 2 with all other coordinates equal to integers. If we notice $N^d(l)$ the maximal number of hypercubes crossed by a curve of length l in dimension d, then $N^d(l) \geq 2^{d-2}N^2(l)$ which makes more hypercubes than the $2^d + \frac{2^d-1}{\sqrt{d}}l$ hypercubes crossed by an hyperdiagonal path.

4.4   With Other Distances

With L_1 norm instead of Euclidean norm for the length of the curve, the expressions of $N(l)$ and $L(n)$ are quite straightforward. We just have to notice that following the curve, new squares appear each time that a line $x = i$ or $y = j$ is crossed. The L_1 distance necessary to cross two consecutive lines is 1. It follows $N(l) = 4 + 2\lfloor l \rfloor$ and $L(n) = \lfloor \frac{n-3}{2} \rfloor$ (for $n > 2$). MLCC of n squares are just polygonal curves with horizontal and vertical segments.

5   Conclusion

In this article, we have given a precise expression of two functions, $N(l)$ and $L(n)$, which respectively calculates the maximum number of squares in the cover of a curve of length l, and the minimal length to cover n squares. These results were obtained by constructing MLCC, curves of minimal length for a given number of squares n. We have also extracted some possible extensions of this work, with other distances, higher dimensions or particular classes of curves.

In our study, we aim at using these results in a multiresolution scheme, based on quadtrees or octrees. Like in this paper, we have to determine the values of the functions $N(l)$ and $L(n)$, guided this time by appropriate tree traversal strategies.

The next future work is to consider *surfaces* instead of curves in any arbitrary dimension $d > 2$. However, it is no more possible to bound the number of hypercubes covered by a surface in function of its area since there exists surfaces of area as small as necessary crossing all the hypercubes of compact domain. It makes the MACS (Minimum Area Covering Surfaces) problem of n hypercubes completely different. Several ideas could be followed by considering specific classes of surfaces or another characteristic of the surface that may be related again with curves length.

References

1. Montanari, U.: A note on minimal length polygonal approximation to a digitized contour. Communications of the ACM 13(1), 41–47 (1970)
2. Sklansky, J., Chazin, R.L., Hansen, B.J.: Minimum perimeter polygons of digitized silhouettes. IEEE Trans. Computers 21(3), 260–268 (1972)
3. Hunter, G., Steiglitz, K.: Operations on images using quad trees. IEEE Trans. PAMI 1(2), 145–153 (1979)

Binary Pictures with Excluded Patterns

Daniela Battaglino[1], Andrea Frosini[2], Veronica Guerrini[1],
Simone Rinaldi[1], and Samanta Socci[1]

[1] Università di Siena, Dipartimento di Matematica e Informatica,
Pian dei Mantellini 44, 53100 Siena
[2] Università di Firenze, Dipartimento di Sistemi e Informatica,
viale Morgagni 65, 50134 Firenze

Abstract. The notion of a pattern within a binary picture (polyomino) has been introduced and studied in [3], and resembles the notion of pattern containment within permutations. The main goal of this paper is to extend the studies of [3] by adopting a more geometrical approach: we use the notion of pattern avoidance in order to recognize or describe families of polyominoes defined by means of geometrical constraints or combinatorial properties. Moreover, we extend the notion of pattern in a polyomino, by introducing generalized polyomino patterns, so that to be able to describe more families of polyominoes known in the literature.

1 Patterns in Binary Pictures and Polyomino Classes

In recent years a considerable interest in the study of the notion of *pattern* within a *combinatorial structure* has grown. This kind of research started with patterns in permutations [12], while in the last few years it is being carried on in several directions. One of them is to define and study analogues of the concept of pattern in permutations in other combinatorial objects such as set partitions [11,14], words, trees [13]. The works [3,4] fit into this research line, in particular [4] introduces and studies the notion of *pattern* in *finite binary pictures* (specifically, in *polyominoes*).

A finite binary picture is an $m \times n$ matrix of 0's and 1's. Intuitively speaking, 1's correspond to black pixels (which constitute the *image*) and the 0's correspond to white pixels (which form the *background*). Often, the studied images should fulfill several additional properties like symmetry, connectivity, or convexity. In particular, an image is *connected* if the set of black pixels is connected with respect to the edge-adjacency relation. A connected image is usually called a *polyomino* (see Figure 1).

The work [3], from which we borrow most of the basic definitions and notations, uses an algebraic setting to provide a unified framework to describe and handle some families of binary pictures (in particular polyominoes), by the avoidance of patterns. Therefore, in order to fruitfully present our paper, we need to recall some definitions and the main results from [3].

Let $\mathfrak{M}$ be the class of binary pictures (or *matrices*). We denote by $\preccurlyeq$ the usual subpicture (or submatrix) order on $\mathfrak{M}$, *i.e.* $M' \preccurlyeq M$ if M' may be obtained from M by deleting any collection of rows and/or columns.

E. Barcucci et al. (Eds.): DGCI 2014, LNCS 8668, pp. 25–38, 2014.

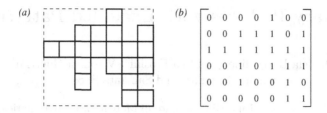

Fig. 1. A polyomino and its representation as a binary picture (or matrix)

Notice that, in a binary picture representing a polyomino the first (resp. the last) row (resp. column) should contain at least a 1. We can consider the restriction of the submatrix order $\preccurlyeq$ on the set of polyominoes $\mathfrak{P}$. This defines the poset $(\mathfrak{P}, \preccurlyeq_P)$ and the pattern order between polyominoes: a polyomino P is a *pattern* of a polyomino Q (which we denote $P \preccurlyeq_P Q$) when the binary picture representing P is a submatrix of that representing Q. We point out that the order $\preccurlyeq_P$ has already been studied in [8] under the name of *subpicture order*, where the authors – among other things – proved that $(\mathfrak{P}, \preccurlyeq_P)$ contains infinite antichains, and it is a graded poset (the rank function being the semi-perimeter of the bounding box of the polyominoes).

This allows to introduce a natural analogue of permutation classes for polyominoes: a *polyomino class* is a set of polyominoes $\mathcal{C}$ that is downward closed for $\preccurlyeq_P$: for all polyominoes P and Q, if $P \in \mathcal{C}$ and $Q \preccurlyeq_P P$, then $Q \in \mathcal{C}$. Basing on the results obtained in [1], in [3] the authors proved that some of the most famous families of polyominoes, including: the *bargraphs*, the *convex*, the *column-convex*, the *L-convex*, the *directed-convex* polyominoes, are indeed polyomino classes. On the other side, there are also well-known families of polyominoes which are not polyomino classes, like: the family of polyominoes having a square shape, the family of polyominoes having exactly $k > 1$ columns, or the directed polyominoes (see Section 4).

Similarly to the case of permutations, for any set $\mathcal{B}$ of polyominoes, let us denote by $Av_P(\mathcal{B})$ the set of all polyominoes that do not contain any element of $\mathcal{B}$ as a pattern. Every such set $Av_P(\mathcal{B})$ of polyominoes defined by pattern avoidance is a polyomino class. Conversely, like for permutation classes, every polyomino class may be characterized in this way [3].

Proposition 1. *For every polyomino class $\mathcal{C}$, there is a unique antichain $\mathcal{B}$ of polyominoes such that $\mathcal{C} = Av_P(\mathcal{B})$. The set $\mathcal{B}$ consists of all minimal polyominoes (in the sense of $\preccurlyeq_P$) that do not belong to $\mathcal{C}$.*

We call $\mathcal{B}$ the *polyomino-basis* (or *p-basis* for short), to distinguish from other kinds of bases. We observe that, denoting $Av_M(\mathcal{M})$ the set of binary matrices that do not have any submatrix in $\mathcal{M}$, we have $Av_P(\mathcal{M}) = Av_M(\mathcal{M}) \cap \mathfrak{P}$.

On the other side, it is quite natural to describe classes of polyominoes by the avoidance of submatrices, then we introduce the notion of *matrix-basis* (or *m-basis*) of a polyomino class $\mathcal{C}$, which is every antichain $\mathcal{M}$ of matrices such that $\mathcal{C} = Av_P(\mathcal{M})$. Differently from the *p*-basis, the *m*-basis needs not be unique.

Example 1 (Injections). Let $\mathcal{I}$ be the class of *injections*, i.e. polyominoes having at most a zero entry for each row and column such as, for instance

$$
\begin{array}{cc}
1 & 0 \\
1 & 1 \\
1 & 1
\end{array}
\qquad
\begin{array}{c}
1 \\
1 \\
1
\end{array}
\qquad
\begin{array}{cc}
1 & 1 \\
1 & 1 \\
1 & 0 \\
1 & 1
\end{array}
\qquad
\begin{array}{cccc}
1 & 1 & 0 & 1 \\
0 & 1 & 1 & 1 \\
1 & 1 & 1 & 0 \\
1 & 0 & 1 & 1
\end{array}
$$

The set $\mathcal{I}$ is clearly a polyomino class, and its *p*-basis is given by the minimal polyominoes which are not injections, i.e. the twelve polyominoes on the top of Fig. 2. An *m*-basis of $\mathcal{I}$ is clearly given by set

$$
\mathcal{M} = \left\{ [0\,0], \begin{bmatrix} 0 \\ 0 \end{bmatrix} \right\}.
$$

Moreover, consider the sets:

$$
\mathcal{M}_1 = \left\{ [0\,1\,0], [1\,0\,0], [0\,0\,1], \begin{bmatrix} 0 \\ 1 \\ 0 \end{bmatrix}, \begin{bmatrix} 0 \\ 0 \\ 1 \end{bmatrix}, \begin{bmatrix} 1 \\ 0 \\ 0 \end{bmatrix} \right\}, \quad \mathcal{M}_2 = \mathcal{M}_1 \cup \left\{ \begin{bmatrix} 0 \\ 0 \\ 0 \end{bmatrix} \right\}.
$$

We may easily check that $\mathcal{M}_1$ and $\mathcal{M}_2$ are antichains (see Fig. 2), and that their avoidance characterizes injections: $\mathcal{I} = Av_P(\mathcal{M}_1) = Av_P(\mathcal{M}_2)$. So, also $\mathcal{M}_1$ and $\mathcal{M}_2$ are *m*-bases, although $\mathcal{M}_1 \subset \mathcal{M}_2$.

We recall [3] that the *p*-basis and an *m*-basis of a polyomino class are related by the following.

Proposition 2. *Let $\mathcal{C}$ be a polyomino class, and let $\mathcal{M}$ be an m-basis of $\mathcal{C}$. Then the p-basis of $\mathcal{C}$ consists of all polyominoes that contain a submatrix in $\mathcal{M}$, and that are minimal (w.r.t. $\preccurlyeq_P$) for this property.*

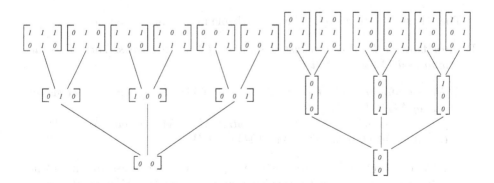

Fig. 2. The *p*-basis and some *m*-bases of $\mathcal{I}$

The reader can check the previous property in Fig. 2 for the case of the class $\mathcal{I}$ of injections.

The main goal of this paper is to extend the studies of [3,4] by adopting a more geometrical approach: we use the notion of pattern avoidance in order to recognize or describe families of polyominoes defined by means of geometrical constraints or combinatorial properties. In particular, we will develop the following research topics:

i) *robust polyomino classes*, i.e. polyomino classes $\mathcal{C}$ where there is an m-basis containing the p-basis. We will show that in this case, the p-basis is the minimal antichain $\mathcal{M}$ for set inclusion, and for $\preccurlyeq$, such that $Av_M(\mathcal{M}) = \mathcal{C}$.
ii) given a set of patterns $\mathcal{M}$, study the class of polyominoes avoiding the patterns of $\mathcal{M}$ as submatrices, and give a characterization of this class in terms of the geometrical/combinatorial properties of its elements.
iii) extend the notion of pattern in a polyomino, by introducing *generalized polyomino patterns*, so that to be able to describe more families of polyominoes known in the literature. Such a generalization resembles what has been done for pattern avoiding permutations with the introduction of *vincular, bivincular* patterns [5].

For brevity sake, some of the proofs will be omitted. The interested reader can find all the proofs of the paper in Chapter 3 of [2], and the proof of Proposition 10 in the Appendix.

2   Robust Polyomino Classes

Every polyomino class is equipped with (at least) two basis, the p-basis and the m-basis. A natural question is to investigate the relation between the p-basis and the m-basis, and in particular to understand the conditions that render the p-basis the most compact way to describe a polyomino class.

Definition 1. *A class is* robust *when all m-bases contain the p-basis.*

The p-basis of a robust class has remarkable minimality properties.

Proposition 3. *Let $\mathcal{C}$ be a robust class, and let $\mathcal{P}$ be its p-basis. Then, $\mathcal{P}$ is the unique m-basis $\mathcal{M}$ which satisfies:*

(1.) *$\mathcal{M}$ is a minimal subset subject to $\mathcal{C} = Av_P(\mathcal{M})$, i.e. for every strict subset $\mathcal{M}'$ of $\mathcal{M}$, $\mathcal{C} \neq (\mathcal{M}')$;*
(2.) *for every submatrix M' of some matrix $M \in \mathcal{M}$, we have $M' = M$ or $\mathcal{C} \neq Av_P(\mathcal{M}'))$, with $\mathcal{M}' = \mathcal{M} \setminus \{M\} \cup \{M'\}$.*

Proof. Condition (1.) follows directly by Proposition 2. Let us assume that Condition (2.) does not hold, i.e. there exists a proper submatrix M' of some matrix $M \in \mathcal{P}$ such that $\mathcal{C} = Av_P(\mathcal{P}')$, with $\mathcal{P}' = \mathcal{P} \setminus \{M\} \cup \{M'\}$. So we have that $\mathcal{P}' \preccurlyeq_P \mathcal{P}$ and $\mathcal{P}'$ is an m-basis of $\mathcal{C}$. Since $\mathcal{C}$ is a robust class we have that

$\mathcal{P} \preccurlyeq_P \mathcal{P}'$ and then $\mathcal{P} = \mathcal{P}'$, in particular $M = M'$. Suppose that there exists another m-basis $\mathcal{M} \neq \mathcal{P}$ satisfying (1.) and (2.). By Proposition 2, every pattern of $\mathcal{M}$ is contained in some pattern of $\mathcal{P}$, thus $\mathcal{P}$ contains $\mathcal{M}$. Since $\mathcal{C}$ is a robust class, then $\mathcal{P} \subseteq \mathcal{M}$, so $\mathcal{P} = \mathcal{M}$.   □

We point out that Condition (1.) ensures minimality in the sense of inclusion, while Condition (2.) ensures minimality for $\preccurlyeq$.

Example 2. Let be $\mathcal{C} = Av_P(P, P')$, where P, P' are depicted in Figure 3. The class $\mathcal{C}$ is not robust, in fact there is an m-basis M disjoint from the p-basis:

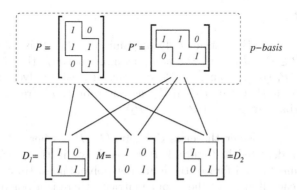

Fig. 3. A non robust class

In practice, P and P' are precisely the minimal polyominoes which contain M as a pattern, then by Proposition 2, $Av_P(P, P') = Av_P(M)$.

In this section, we try to establish some criteria to test the robustness of a class of polyominoes. First, we prove that it is easy to test robustness of a class whose basis is made of just one element:

Proposition 4. *Let M be a pattern. Then, $Av_P(M)$ is robust if and only if M is a polyomino.*

Proof. If M is not a polyomino, then its p-basis is clearly different from P, so $Av_P(M)$ is not robust. On the other side, let us assume that M is a polyomino and that $Av_P(M)$ is not robust. Let us assume that an m-basis of $Av(M)$ is made of a (non polyomino) matrix M' such that $M' \preccurlyeq_P M$. Since M' is not a polyomino then it contains at least two disconnected elements B and C, and there are at least two possible ways to connect B and C (by rows or by columns). So, there exists at least another polyomino $P \neq M$ such that $M' \preccurlyeq_P P$, and P belongs to the p-basis of $Av_P(M')$. Thus, $Av_P(M) \subseteq Av_P(M')$. The same technique can be used to prove that an m-basis of $Av(M)$ canont be made of more than one matrix.   □

Now, we aim at extending the previous result to a generic set of polyominoes, i.e. find sufficient and necessary conditions such that, given set of polyominoes $\mathcal{P}$, the class $Av_P(\mathcal{P})$ is robust.

Proposition 5. *Let be P_1, P_2 two polyominoes and let be $\mathcal{C} = Av_P(P_1, P_2)$. If for every element $\overline{P}$ in $P_1 \wedge P_2$ we have that:*

(1) *$\overline{P}$ is a polyomino, or*
(2) *every chain from $\overline{P}$ to P_1 (resp. from $\overline{P}$ to P_2) contains at least a polyomino P' (resp. P''), different from P_1 (resp. P_2), such that $\overline{P} \preccurlyeq_P P' \preccurlyeq_P P_1$ (resp. $\overline{P} \preccurlyeq_P P'' \preccurlyeq_P P_2$),*

then $\mathcal{C}$ is robust.

Proof. Clearly, if $P_1 \wedge P_2$ contains only polyominoes, then $\mathcal{C}$ is robust. On the other side, let $\overline{P} \in P_1 \wedge P_2$, with $\overline{P}$ a non polyomino pattern; then by (2.) every chain from $\overline{P}$ to P_1 (resp. from $\overline{P}$ to P_2) contains at least a polyomino P' (resp. P''), different from P_1 (resp. P_2). If $\mathcal{C}$ was not robust, P' (resp. P'') should belong to the p-basis in place of P_1 (resp. P_2).   □

Example 3. Let us consider the class $\mathcal{C} = Av_P(P_1, P_2)$, where P_1 and P_2 are the polyominoes depicted in Figure 4. Here, as shown in the picture, $P_1 \wedge P_2$ contains six elements, and four of them are not polyominoes. However, one can check that, for each item $\overline{P}$ of these four matrices, there is a polyomino in the chain from $\overline{P}$ to P_1 (resp. from $\overline{P}$ to P_2). Thus, by Proposition 5, the class $\mathcal{C}$ is robust.

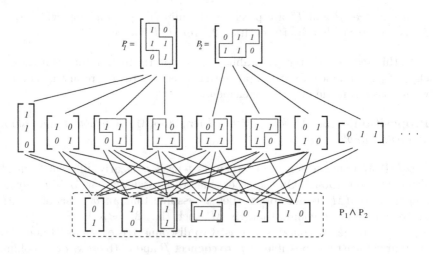

Fig. 4. A robust class

However, the statement of Proposition 5 cannot be inverted, as we can see in the following example.

Example 4 (Parallelogram polyominoes). We recall that a *parallelogram polyomino* is a polyomino whose boundary can be decomposed in two paths, the upper and the lower paths, which are made of north and east unit steps and meet only at their starting and final points (see Fig. 5 (c)). We can easily prove that parallelogram polyominoes can be represented by the avoidance of the submatrices:

$$M_1 = \begin{bmatrix} 1 & 0 \\ 1 & 1 \end{bmatrix}, \ M_2 = \begin{bmatrix} 1 & 1 \\ 0 & 1 \end{bmatrix}.$$

These two patterns form a *p*-basis for the class $\mathcal{P}$ of parallelogram polyominoes. Clearly

$$M_1 \wedge M_2 = \left\{ \begin{bmatrix} 1 & 1 \end{bmatrix}, \ \begin{bmatrix} 1 \\ 1 \end{bmatrix}, \ \begin{bmatrix} 0 \end{bmatrix} \right\}.$$

If $\mathcal{P}$ was not robust, then $M = \begin{bmatrix} 0 \end{bmatrix}$ should belong to the an *m*-basis of $\mathcal{P}$; precisely, we should have $Av_P(M) = \mathcal{P}$. But this is not true, since clearly $Av_P(M)$ is the class of rectangles. Thus, $\mathcal{P}$ is robust. Observe that the set $\{M_1, M_2, [1\,0\,1]\}$ forms an *m*-basis of the class, but it is not minimal w.r.t. set inclusion.

3   Classes of Polyominoes Defined by Submatrix Avoidance

As we have mentioned, several families of polyominoes covered in the literature can be characterized in terms of submatrix avoidance. In particular, if the family of polyominoes is defined by imposing geometric constraints on its elements, then these constraints can be naturally represented by the avoidance of some matrix patterns. For instance, in [4] it was proved that the convexity constraint can be represented by the avoidance of the two submatrices:

$$H = \begin{bmatrix} 1 & 0 & 1 \end{bmatrix} \quad \text{and} \quad V = \begin{bmatrix} 1 \\ 0 \\ 1 \end{bmatrix}.$$

Similarly, in [3] it was proved that the families of directed-convex, column-convex, stack polyominoes are polyomino classes. In this section, we consider some polyomino classes which can be represented by the avoidance of submatrices, and deal with the problem of giving a combinatorial/geometrical characterization to these classes. Most of these classes have not been considered yet in the literature, and they show quite simple characterizations and interesting combinatorial properties.

Polyominoes avoiding rectangles. Let $O_{m,n}$ be set of rectangles – binary pictures with all the entries equal to 1 – of dimension $m \times n$ (see Figure 6 (a)). With $n = m = 2$ these objects (also called *snake-like polyominoes*) have a simple geometrical characterization.

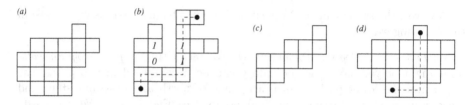

Fig. 5. (a) a convex polyomino; (b) a directed polyomino; (c) a parallelogram polyomino; (d) an L-convex polyomino

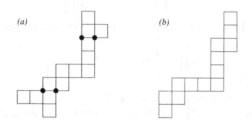

Fig. 6. (a) a snake-like polyomino; (b) a snake

Proposition 6. *Every snake-like polyomino can be uniquely decomposed into three parts: a unimodal staircase polyomino oriented with respect to two axis-parallel directions d_1 and d_2 and two (possibly empty) L-shaped polyominoes placed at the extremities of the staircase. These two L-shaped polyominoes have to be oriented with respect to d_1, d_2.*

We have studied the classes $Av_P(O_{m,n})$, for other values of m, n, obtaining similar characterizations which here are omitted for brevity.

Snakes. Let us consider the family of *snake-shaped* polyominoes (briefly, *snakes*) – as that shown in Fig. 6 (b):

Proposition 7. *The family of snakes is a polyomino class, which can be described by the avoidance of the following polyomino patterns:*

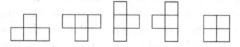

Hollow stacks. Let us recall that a stack polyomino is a convex polyomino containing two adjacent corners of its minimal bounding rectangle (see Fig. 7 (a)). Stack polyominoes clearly form a polyomino class, described by the avoidance of the patterns:

$$\begin{bmatrix} 1 & 1 \\ 1 & 0 \end{bmatrix} \quad \begin{bmatrix} 1 & 1 \\ 0 & 1 \end{bmatrix} \quad \begin{bmatrix} 1 & 0 & 1 \end{bmatrix}$$

A *hollow stack* (polyomino) is a polyomino obtained from a stack polyomino P by removing from P a stack polyomino P' which is geometrically contained in P and whose basis lie on the basis of the minimal bounding rectangle of P. Figure 7 (b), (c) depict two hollow stacks.

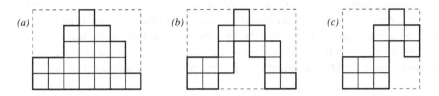

Fig. 7. (a) a stack polyomino; (b), (c): hollow stacks

Proposition 8. *The family $\mathcal{H}$ of hollow stack polyominoes forms a polyomino class with p-basis given by:*

Rectangles with rectangular holes. Let $\mathcal{R}$ be the class of polyominoes obtained from a rectangle by removing sets of cells which have themselves a rectangular shape, and such that there is no more than one connected set of 0's for each row and column. The family $\mathcal{R}$ can easily be proved to be a polyomino class, and moreover:

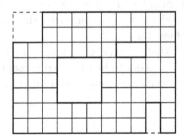

Fig. 8. A rectangle with rectangular holes

Proposition 9. *The class $\mathcal{R}$ can be described by the avoidance of the patterns:*

$$[0\,1\,0], \quad \begin{bmatrix} 0 \\ 1 \\ 0 \end{bmatrix} \begin{bmatrix} 1\,0 \\ 0\,0 \end{bmatrix} \begin{bmatrix} 0\,1 \\ 0\,0 \end{bmatrix} \begin{bmatrix} 0\,0 \\ 1\,0 \end{bmatrix} \begin{bmatrix} 0\,0 \\ 0\,1 \end{bmatrix}.$$

4   Generalized Matrix Patterns

As already pointed out, there are several families of polyominoes that are not polyomino classes. Amongst them, we have mentioned directed polyominoes and polyominoes without holes. By Proposition 1, these families of polyominoes cannot be expressed in terms of submatrix avoidance. In order to overcome this problem, we extend the notion of pattern in a polyomino, by introducing *generalized polyomino patterns*, so that to be able to describe more families of polyominoes. Our generalization resembles what has been done for pattern avoiding permutations with the introduction of *vincular, bivincular* patterns [5].

L-convex polyominoes. A convex polyomino is *k-convex* if every pair of its cells can be connected by a monotone path with at most k changes of direction, and k is called the *convexity degree* of the polyomino [8] . For $k = 1$ we have the *L*-convex polyominoes, where any two cells can be connected by a path with at most one change of direction (see Fig. 5 (*d*)). Recently, *L*-convex polyominoes have been considered from several points of view: in [9,10] the authors solve the main enumeration problems for *L*-convex polyominoes, while in [7] they approach them from a language-theoretical perspective. In [3] it was shown that *L*-convex polyominoes form a polyomino class, and they can be represented by the avoidance of the submatrices:

$$H = \begin{bmatrix} 1\,0\,1 \end{bmatrix}, \ V = \begin{bmatrix} 1 \\ 0 \\ 1 \end{bmatrix}, \ S_1 = \begin{bmatrix} 1\,0 \\ 0\,1 \end{bmatrix}, \ S_2 = \begin{bmatrix} 0\,1 \\ 1\,0 \end{bmatrix} .$$

2-convex polyominoes. Differently from *L*-convex polyominoes, 2-convex polyominoes do not form a polyomino class. As a matter of fact, the 2-convex polyomino in Figure 9 (*a*) contains the 3-convex polyomino (*b*) as a pattern, so the class is not downward closed w.r.t. $\preccurlyeq_P$. Similarly, the set of k-convex polyominoes is not a polyomino class, for $k \geq 2$.

In practice, this means that 2-convex polyominoes cannot be described in terms of pattern avoidance. In order to be able to represent 2-convex polyominoes we extend the notion of pattern avoidance, introducing the *generalized pattern*

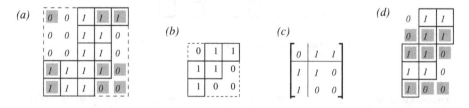

Fig. 9. (*a*) a 2-convex polyomino P; (*b*) a pattern of P that is not a 2-convex polyomino; (*c*) a generalized pattern, which is not contained in (*a*), but is contained in the 3-convex polyomino (non 2-convex) (*d*)

avoidance. Our extension consists in imposing the adjacency of two columns or rows by introducing special symbols, i.e. vertical/horizontal lines: with A being a pattern, a vertical line between two columns of A, c_i and c_{i+1} (a horizontal line between two rows r_i and r_{i+1}), will read that c_i and c_{i+1} (respectively r_i and r_{i+1}) must be adjacent. When the vertical (resp. horizontal) line is external, it means that the adjacent column (resp. row) of the pattern must touch the minimal bounding rectangle of the polyomino. Moreover, we will use the $*$ symbol to denote 0 or 1 indifferently.

Proposition 10. *The class of 2-convex polyominoes can be can be described by the avoidance of the set $\mathcal{M}$ of generalized patterns:*

$$\begin{bmatrix} 1 & 0 & 1 \end{bmatrix} \quad \begin{bmatrix} 1 \\ 0 \\ 1 \end{bmatrix} \quad \begin{bmatrix} 0 & * & 1 \\ * & 1 & 0 \\ 1 & 0 & 0 \end{bmatrix} \quad \begin{bmatrix} 0 & 0 & 1 \\ 0 & 1 & * \\ 1 & * & 0 \end{bmatrix} \quad \begin{bmatrix} 1 & * & 0 \\ 0 & 1 & * \\ 0 & 0 & 1 \end{bmatrix} \quad \begin{bmatrix} 1 & 0 & 0 \\ * & 1 & 0 \\ 0 & * & 1 \end{bmatrix}$$

$$\begin{bmatrix} 1 & * & 0 & 0 \\ * & 1 & * & 0 \\ 0 & * & 1 & * \\ 0 & 0 & * & 1 \end{bmatrix} \quad \begin{bmatrix} 0 & 0 & * & 1 \\ 0 & * & 1 & * \\ * & 1 & * & 0 \\ 1 & * & 0 & 0 \end{bmatrix}$$

The proof of Proposition 10 is reported in the Appendix. Let us just observe, referring to Fig. 9, that the pattern (c) is not contained in the 2-convex polyomino (a), but it is contained in the 3-convex polyomino (d). It is possible generalize the previous result and give a characterization of the class of k-convex polyominoes, with $k > 2$, using generalized patterns.

Directed polyominoes. A polyomino P is *directed* when every cell of P can be reached from a distinguished cell (called the source) by a path – internal to the polyomino – that uses only north and east steps. Figure 5 (b) depicts a directed polyomino. The reader can simply check that the class of the directed polyominoes is not a polyomino class by observing that – in the picture – the four marked cells represent a polyomino which is not directed.

Proposition 11. *The class of directed polyominoes can be represented as the class of polyominoes avoiding the following patterns*

$$\begin{bmatrix} 1 \\ 0 \end{bmatrix} \qquad \begin{bmatrix} 0 & 1 \\ * & 0 \end{bmatrix}$$

This proof is analogous to that of Proposition 10, and also relies on the set of patterns determined in [3], whose avoidance describes the class of directed-convex polyominoes. We would like to point out that there are families of polyominoes which cannot be described, even using generalized pattern avoidance. For instance, the reader can easily check that one of these families is that of polyominoes having a square shape.

References

1. Barcucci, E., Frosini, A., Rinaldi, S.: On directed-convex polyominoes in a rectangle. Discrete Math. 298(1-3), 62–78 (2005)
2. Battaglino, D.: Enumeration of polyominoes defined in terms of pattern avoidance or convexity constraints, PhD thesis (University of Sienna), arXiv:1405.3146v1 (2014)
3. Battaglino, D., Bouvel, M., Frosini, A., Rinaldi, S.: Permutation classes and polyomino classes with excluded submatrices. ArXiv 1402.2260 (2014)
4. Battaglino, D., Bouvel, M., Frosini, A., Rinaldi, S., Socci, S., Vuillon, L.: Pattern avoiding polyominoes. In: Proceedings di Italian Conference Theoretical Computer Science, Palermo, Settembre 9-11 (2013)
5. Bona, M.: Combinatorics of permutations. Chapman-Hall and CRC Press (2004)
6. Brändén, P., Claesson, A.: Mesh patterns and the expansion of permutation statistics as sums of permutation patterns. Electr. J. of Combin. 18(2) (2011) (The Zeilberger Festschrift volume)
7. Brocchi, S., Frosini, A., Pinzani, R., Rinaldi, S.: A tiling system for the class of L-convex polyominoes. Theor. Comput. Sci. 457, 73–81 (2013)
8. Castiglione, G., Restivo, A.: Ordering and Convex Polyominoes. In: Margenstern, M. (ed.) MCU 2004. LNCS, vol. 3354, pp. 128–139. Springer, Heidelberg (2005)
9. Castiglione, G., Frosini, A., Munarini, E., Restivo, A., Rinaldi, S.: Combinatorial aspects of L-convex polyominoes. European J. Combin. 28, 1724–1741 (2007)
10. Castiglione, G., Frosini, A., Restivo, A., Rinaldi, S.: Enumeration of L-convex polyominoes by rows and columns. Theor. Comput. Sci. 347, 336–352 (2005)
11. Klazar, M.: On abab-free and abba-free sets partitions. European J. Combin. 17, 53–68 (1996)
12. Knuth, D.E.: The Art of Computer Programming, 2nd edn. Volume 1: Fundamental algorithms; Addison-Wesley Series in Computer Science and Information Processing. Addison-Wesley Publishing Co., Reading (1975)
13. Rowland, E.: Pattern avoidance in binary trees. J. Combin. Theory Ser. A 117, 741–758 (2010)
14. Sagan, B.E.: Pattern avoidance in set partitions. Ars Combin. 94, 79–96 (2010)

Appendix

Proof of Proposition 10. We recall that in a 2-convex polyomino, for each two cells, there is a path connecting them, which uses only two types of steps among n, s, e, w (north, south, east and west unit steps, respectively) and has at most two changes of direction. Moreover, for any two cells c_1 and c_2 of a polyomino, the minimal number of changes of direction from c_1 to c_2 can be computed from just two paths, starting with a vertical and a horizontal step, respectively, in which every side has maximal length. We will refer to these as the *extremal paths* connecting c_1 and c_2.

($\Rightarrow$) If P is a 2-convex polyomino then P avoids $\mathcal{M}$.

Let us assume by contradiction that P is a 2-convex polyomino containing one of the patterns of $\mathcal{M}$, clearly not H and V, by convexity. For simplicity sake, we will consider only the two patterns of $\mathcal{M}$,

$$Z_1 = \begin{bmatrix} 0 & * & 1 \\ * & 1 & 0 \\ 1 & 0 & 0 \end{bmatrix} \quad \text{and} \quad Z_2 = \begin{bmatrix} 0 & 0 & * & 1 \\ 0 & * & 1 & * \\ * & 1 & * & 0 \\ 1 & * & 0 & 0 \end{bmatrix},$$

since the proof for the other patterns can be obtained by symmetry. If P contains Z_1 (resp. Z_2) then it has to contain a submatrix P' (resp. P'') of the form:

$$P' = \begin{array}{|c|ccccc} 0 & * & \ldots & 1 \\ \hline * & 1 & \ldots & 0 \\ \cdot & & & \\ \cdot & & & \\ \hline 1 & 0 & \ldots & 0 \end{array} \qquad P'' = \begin{array}{ccc|ccc} 0 & \ldots & 0 & * & \ldots & 1 \\ \hline 0 & \ldots & * & 1 & \ldots & * \\ * & \ldots & 1 & * & \ldots & 0 \\ \hline 1 & \ldots & * & 0 & \ldots & 0 \end{array},$$

where the $0, 1, *$ are the elements of Z_1 (resp. Z_2) and the dots can be replaced by $0, 1$ indifferently, clearly in agreement with the convexity and polyomino constraints.

Among all the polyominoes which can be obtained from P' (resp. P''), the one having the minimal convexity degree is that, called $\overline{P}'$ (resp. $\overline{P}''$), having the maximal number of 1 entries. It is easy to verify that the minimal number of changes of direction requested to connect the 1 entries in boldface of $\overline{P}'$ (resp. $\overline{P}''$) is three, so $\overline{P}'$ (resp. $\overline{P}''$) is a 3-convex polyomino, which contradicts our assumption.

($\Leftarrow$) If P avoids $\mathcal{M}$ then P is a 2-convex polyomino.

Again by contradiction let us assume that P avoids $\mathcal{M}$ and it is a 3-convex polyomino, *i.e.* there exist two cells of P, c_1 and c_2, such that any path from c_1 to c_2 requires at least three changes of direction.

Let us take into consideration the two extremal paths from c_1 to c_2. The only possible cases are the following (up to rotation):

- the two extremal paths are distinct, Fig. 10 (a);

- one of the extremal paths does not exist, see Fig. 10 (b);
- the two extremal paths coincide after the first change of direction, see Fig. 10 (c);
- the two extremal paths coincide after the second change of direction, see Fig. 10 (d).

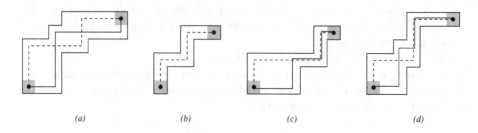

(a)   (b)   (c)   (d)

Fig. 10. The possible cases of extremal paths connecting the cells c_1 and c_2

Here, we will consider only the first case, since the others are abalogous: as sketched in the picture below, the polyomino P of Fig. 10 (a) has to contain a submatrix P' – given by the boldface entries – of the form:

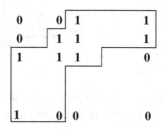

It is easy to see that such a submatrix is one of those that we can obtain replacing appropriately the symbol $*$ in the pattern Z_2. So, P contains Z_2 against the hypothesis. We point out that the pattern Z_1, and its rotations, can be obtained from the pattern Z_2 (or its rotation) replacing appropriately the $*$ entries, but we need to consider them in order to include the 3-convex polyominoes with three rows or columns. □

2D Topological Map Isomorphism for Multi-Label Simple Transformation Definition

Guillaume Damiand, Tristan Roussillon, and Christine Solnon

Université de Lyon, CNRS, LIRIS, UMR5205, INSA-Lyon, F-69622 France

Abstract. A 2D topological map allows one to fully describe the topology of a labeled image. In this paper we introduce new tools for comparing the topology of two labeled images. First we define 2D topological map isomorphism. We show that isomorphic topological maps correspond to homeomorphic embeddings in the plane and we give a polynomial-time algorithm for deciding of topological map isomorphism. Then we use this notion to give a generic definition of multi-label simple transformation as a set of transformations of labels of pixels which does not modify the topology of the labeled image. We illustrate the interest of multi-label simple transformation by generating look-up tables of small transformations preserving the topology.

Keywords: Combinatorial maps, 2D topological maps isomorphism, labeled image, simple points, simple sets.

1 Introduction

Image processing often needs to group pixels into clusters having some common properties (which can be colorimetric, semantic, geometric. . .). One way to describe these clusters is to use labeled images where a label is associated with each pixel. Given two labeled images, it is interesting to be able to decide if they have the same topology. This can be useful for example for object tracking or image analysis. A related question is to decide whether a labeled image can be transformed into another one while preserving the topology of the partition. This second question is interesting to propose a deformable model with a topological control for example. In this paper, these two problems are addressed.

Many data-structures were proposed to describe the topology of labeled images. A well-known one is the Region Adjacency Graph (RAG) [12]: the image is partitioned into regions corresponding to sets of connected pixels; the RAG associates a node with every region and an edge with every pair of adjacent regions. RAGs are used in different image processings like image segmentation [14] or object recognition [9]. However a RAG does not fully describe the topology of a partition in regions: it does not represent multi-adjacency relations nor the order of adjacent regions when turning around a given region. Thus two partitions having different topologies may be described by isomorphic RAGs.

2D topological maps [3] are more powerful data structures for describing the topology of subdivided objects as they fully describe the topology of labeled

E. Barcucci et al. (Eds.): DGCI 2014, LNCS 8668, pp. 39–50, 2014.

images. They combine 2D combinatorial maps (2-maps [10]), describing the topology of subdivided objects (multi-adjacency relations as well as the order of adjacent regions when turning around a given region) with enclosure trees, describing region enclosure relations.

In this paper, we define isomorphism between topological maps and we show that two labeled images are homeomorphic if their associated topological maps are isomorphic. We describe a polynomial-time algorithm for checking whether two topological maps are isomorphic, thus providing an efficient way of deciding whether two labeled images are homeomorphic. Then we use the isomorphism definition of topological maps in order to give a simple definition of multi-label simple transformation (called *ML-simple*), i.e. a set of modifications of labels of pixels that preserve the topology of the whole partition. Lastly we use all these tools in order to generate different look-up tables of ML-simple transformations allowing to test in amortized constant time if two local configurations are equivalent, and allowing to retrieve in linear time in the size of the output all the configurations that are equivalent.

In Sect. 2 we introduce all the preliminary notions on labeled images, combinatorial maps, isomorphism and homotopic deformation. In Sect. 3 we define topological map isomorphism and present the algorithm allowing to test if two topological maps are isomorphic. Section 4 introduces the definition of ML-simple transformation and shows that it is possible to restrict the test of topological map isomorphism on bounding boxes around the modified pixels. In Sect. 5 we present the construction of look-up tables of ML-simple transformations. We conclude and give some perspectives in Sect. 6.

2   Preliminary Notions

2.1   Labeled Images and Partitions into Regions

A 2D *labeled image* is a triple (I_d, L, l), where $I_d \subseteq \mathbb{Z}^2$ is a set of pixels (the image domain), L is a finite set of labels, and $l : \mathbb{Z}^2 \to L$ is a labeling function which associates a label with every pixel. Two pixels $p_1 = (x, y)$ and $p_2 = (x', y')$ are 4-*adjacent* (resp. 8-*adjacent*) if $|x-x'|+|y-y'| = 1$ (resp. $\max(|x-x'|, |y-y'|) = 1$). A k-*path* (with $k = 4$ or 8) is a sequence of pixels such that two consecutive pixels of the sequence are k-adjacent. A set of pixels S is k-*connected* if for each pair of pixels $(p_1, p_2) \in S^2$ there is a k-path from p_1 to p_2 having all its pixels in S.

A *region* in a labeled image $i = (I_d, L, l)$ is a maximal set of 4-connected pixels having the same label. An additional region is defined, denoted *infinite(i)*, which is the complement of I_d, i.e., $infinite(i) = \mathbb{Z}^2 \setminus I_d$. The set of regions of i, including *infinite(i)*, is denoted *regions(i)* and is a partition of $\mathbb{Z}^2$.

A region R is *enclosed* in another region R' if all 8-paths from one pixel of R to a pixel of *infinite(i)* contains at least one pixel of R'. Region R is *directly enclosed* in R' if there is no region $R'' \neq R'$ such that R is enclosed in R'' and R'' is enclosed in R'. Every region except the infinite one has exactly one direct enclosing region whereas it may have 0, 1 or more directly enclosed regions. Direct enclosure relations may be described by an enclosure tree rooted in *infinite*(i) (see Fig. 1 for an example of a partition).

Digital contours of regions are made explicit by using the *interpixel* topology [8]. In interpixel topology, the cellular decomposition of the euclidean space $\mathbb{R}^2$ into regular elements is considered. *Pixels* are 2-dimensional elements (unit squares), *linels* are 1-dimensional elements (unit segments) and *pointels* are 0-dimensional elements (points). Two linels are connected if they share a pointel in their boundary. A *frontier* between two regions R and R' is a maximal set of connected linels separating pixels belonging to R and R'.

The *boundary* of a region R is the set of linels which separate pixels of R from 4-adjacent pixels not in R. This boundary is partitioned in two sets: the *external boundary* contains the linels separating R from non enclosed regions; *internal boundaries* contain the linels separating R from its enclosed regions. Note that every region except the infinite one has a non empty external boundary whereas it may have an empty internal boundary. The infinite region has an empty external boundary but its internal boundary is not empty (unless $I_d = \emptyset$).

2.2   Combinatorial Maps

Combinatorial maps [10] were defined to describe the subdivision of objects in cells (vertices, edges, faces...) plus the incidence and adjacency relations between these cells. In 2D, a combinatorial map (2-map) can be seen as a graph where each edge is cut in two darts (also known as half-edges). Darts are oriented and two relations are defined on the set of darts: $\beta_1(d)$ is the dart following d when turning around the face which contains d and $\beta_2(d)$ is the dart opposite to d in the face adjacent to the face which contains d. More formally, a 2-map is defined by a triple $M = (D, \beta_1, \beta_2)$ such that D is a finite set of darts, β_1 is a permutation on D, and β_2 is an involution on D. A 2-map is *connected* if for every pair of darts $(d, d') \in D^2$, there exists a sequence of darts $(d_1, \ldots, d_n)$ such that $d_1 = d$, $d_n = d'$, and $\forall 1 \le i < n, d_{i+1} = \beta_1(d_i)$ or $d_{i+1} = \beta_2(d_i)$.

An example of 2-map is given in Fig. 1(b). This combinatorial map contains 16 darts (drawn by oriented curves). Two darts linked by β_1 are drawn consecutively (e.g. $\beta_1(10) = 11$). Note that a dart may be linked with itself in case of loops (e.g. $\beta_1(1) = 1$). Two darts linked by β_2 are drawn in parallel and have reverse orientations (e.g. $\beta_2(1) = 2$ or $\beta_2(10) = 3$). This 2-map is not connected and is composed of 3 different connected components.

2.3   Topological Maps

A topological map is a combinatorial data-structure which fully describes the topology of a partition into regions of a labeled image. It is composed of three parts: a combinatorial map describing the adjacency relations between regions in an ordered way, an enclosure tree describing the direct enclosure relations between regions and an interpixel matrix describing the geometry of the different contours. In this paper, we focus on topology and do not use geometry so that we do not consider the interpixel matrix.

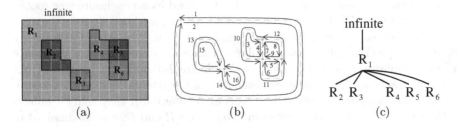

Fig. 1. Example of topological map. (a) A labeled image i. (b) The minimal combinatorial map describing i. (c) The region enclosure tree.

Definition 1 (2D topological map). *Given a 2D labeled image $i = (I_d, L, l)$, its 2D topological map is defined by $TM(i) = (M, T, r, d)$ where:*

- $M = (D, \beta_1, \beta_2)$ *is a 2-map such that each face of M corresponds to a boundary of a region of i, and β_2 describes adjacency relationships between the faces. Furthermore, M is minimal in its number of darts;*
- $T = (N, E)$ *is an enclosure tree of regions: each node of T corresponds to a region of i, and has a child for every region which is directly enclosed in it; the root of T is infinite(i);*
- $r : D \rightarrow N$ *associates each dart $d \in D$ with the region $r(d) \in N$ whose boundary contains d;*
- $d : N \rightarrow D$ *associates each region $R \in N$ with a dart $d(R) \in D$ which is a dart of the face corresponding to the external boundary of R, except for the infinite region which does not have an external boundary so that $d(\text{infinite}(i))$ belongs to its internal boundary.*

Given a 2D labeled image, its topological map is unique (up to isomorphism between 2-maps). Indeed, the 2-map is minimal in number of darts. Thus, each pair of darts $(d, \beta_2(d))$ describes a frontier between two regions R and R' (i.e. a maximal set of connected linels separating pixels belonging to R and R').

In a 2D topological map, each region is represented as a node in the enclosure tree and as face(s) in the 2-map (see example in Fig. 1). Two 4-adjacent regions share a common frontier represented as pair(s) of darts linked by β_2 in the 2-map. In Fig. 1, darts 8 and 12 represent the frontier between regions R_1 and R_5. The relation between two 8-adjacent but not 4-adjacent regions is implicitly represented by a third region, which is 4-adjacent to both. In Fig. 1, the two consecutive darts 13 and 14, linked by β_1, which represent one internal boundary of region R_1, are respectively linked by β_2 to darts 15 and 16, which represent the external boundary of regions R_2 and R_3.

2.4   Isomorphisms and Signatures

Two 2-maps $M = (D, \beta_1, \beta_2)$ and $M' = (D', \beta_1', \beta_2')$ are *isomorphic* if there exists a bijection $f : D \rightarrow D'$ (called *map isomorphism function*) such that $\forall d \in D$,

$\forall i \in \{1, 2\}$, $f(\beta_i(d)) = \beta_i'(f(d))$ [10]. In [7], map signatures are defined such that two connected 2-maps are isomorphic iff their signatures are equal. Given a connected map $M = (D, \beta_1, \beta_2)$ such that $|D| = k$, the signature of M is a sequence of $2k$ integer values $\sigma(M) = < d_1, d_2, \ldots, d_{2k} >$, with $d_i \in [1; k]$ for all $i \in [1; 2k]$. This signature may be used to define the *canonical form* of M: $canonical(M) = (D', \beta_1', \beta_2')$ with $D' = \{1, \ldots, k\}$ and $\forall i \in D', \beta_1'(i) = d_{2i-1}$ and $\beta_2'(i) = d_{2i}$ (see [7] for more details).

Two trees $T = (N, E)$ and $T' = (N', E')$ are *isomorphic* if there exists a bijection $f : N \to N'$ (called *tree isomorphism function*) such that $\forall u, v \in N, (u, v) \in E \Leftrightarrow (f(u), f(v)) \in E'$. If the trees are rooted in r and r', respectively, then f must map the roots of the trees, i.e., $f(r) = r'$. If the complexity of graph isomorphism is still an open question, the complexity of tree isomorphism is polynomial and [1] describes an algorithm in $\mathcal{O}(|N|)$ which associates signatures with nodes of the trees. The algorithm of [1] may be extended to integrate labels in signatures: the signature of a leaf is its label; the signature of a node which is not a leaf is the concatenation of its label with the sorted sequence of its children signatures concatenated with the corresponding edge labels.

2.5   Transformations Preserving Topology

In digital topology, the topology of a binary image is usually defined from a pair of adjacency relations (e.g. 4 for one label, 8 for the other) so that a digital version of the Jordan Curve theorem holds [13]. The drawback of this popular solution is that we have two adjacency relations for one partition. Its topology may change if we interchange the two colors, because the chosen adjacency pair is not an intrinsic feature of the partition, but depends on the object to represent. There is no good choice if the topology of the object is intricate (e.g. there are many nested connected components) or if both labels represent regions of interest. This is especially true for images of more than two labels.

In a binary image, a point is *simple* if its label can be changed without changing the connectedness properties of the support of either label [13]. This concept can be extended to labeled images if we independently consider the support of each label and its complement. However, to take into account the adjacency relations between regions, it is proposed in [2] to also consider the union of two labels. A generalization of this idea may be found in [11]. In [6], a new definition of multi-label simple point is proposed to guarantee that the topology of a partition is preserved when a pixel is flipped from a region to an adjacent region. The drawback of these methods is that only elementary transformations are taken into account. However, two partitions can be homeomorphic even if there does not exist a sequence of elementary transformations mapping the two partitions.

In this work, we represent a labeled image by a topological map, which is an intrinsic feature of its partition into regions. Isomorphism between two topological maps, which is equivalent to homeomorphism between two partitions, provides a way of deciding whether a global transformation preserves the topology of the partition or not.

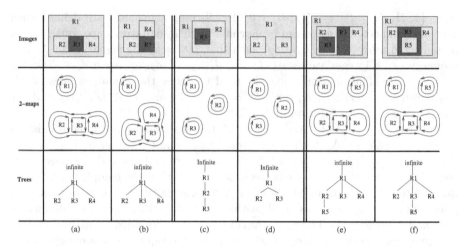

Fig. 2. Examples of non isomorphic topological maps. (a) and (b) have isomorphic trees but not isomorphic 2-maps (R_2 and R_4 are not adjacent in (a) whereas they are 8-adjacent in (b)). (c) and (d) have isomorphic 2-maps but not isomorphic trees. (e) and (f) have isomorphic trees and 2-maps, but isomorphism functions are not compatible.

3   Topological Map Isomorphism

In [5], 2-map isomorphism is extended to consider plane isomorphism, i.e. isomorphism between connected 2-maps drawn on the plane. This preliminary definition is extended here to 2D topological maps by considering additional information given by region enclosure trees.

Definition 2 (2D topological map isomorphism). *Two 2D topological maps* $TM = (M = (D, \beta_1, \beta_2), T = (N, E), r, d)$ *and* $TM' = (M' = (D', \beta_1', \beta_2'), T' = (N', E'), r', d')$ *are isomorphic iff: (1) there exists a map isomorphism function* $f_m : D \to D'$; *(2) there exists a tree isomorphism function* $f_t : N \to N'$; *(3) two faces of* M *and* M' *are matched by* f_m *iff the corresponding regions of* T *and* T' *are matched by* f_t, *i.e.,* $\forall d_i \in D$: $f_t(r(d_i)) = r'(f_m(d_i))$.

As illustrated in Fig. 2, the three conditions of Def. 2 are necessary to ensure topological map isomorphism. Moreover, Def. 2 leads to the following result:

Theorem 1. *Let* i *and* i' *be two labeled images, and* $TM(i)$ *and* $TM(i')$ *be their associated 2D topological maps.* $TM(i)$ *and* $TM(i')$ *are isomorphic iff the partitions into regions of* i *and* i' *are homeomorphic embeddings in the plane.*

Two embeddings are homeomorphic in the plane when they describe the same regions and the same adjacency relations and region enclosure relations.

In [15] the definition of maptree is given. This is a 2D combinatorial map plus a black and white adjacency tree. This tree has one white node for each region of the partition and one black node for each connected component of the combinatorial map (no two adjacent nodes have the same color). The white father (resp. sons) of a black node is the region (resp. regions) corresponding to the external face (resp. internal faces) incident to a same connected component of the combinatorial map. It is thus easy to see that a maptree is equivalent to a topological map.

Proposition 1 in [15] proves that a maptree provides a unique representation (up to homeomorphism of the sphere) of the embedding of a non-connected combinatorial map. Moreover, if the maptree is rooted, the representation is unique up to homeomorphism of the plane. Theorem 1 is a direct consequence of this proposition.

Now, let us describe an algorithm for deciding whether two 2D topological maps are isomorphic or not. Let us consider a 2D topological map $TM = (M, T = (N, E), r, d)$. Each node $R \in N$ of the region enclosure tree corresponds to an image region and is associated with a dart $d(R)$ which belongs to a face of a connected component of M. Let us note F_R the face of M which contains $d(R)$, and M_R the connected component of M which contains $d(R)$. We label the tree $T = (N, E)$ by defining the labeling function λ as follows:

- every region $R \in N$ is labeled with the signature of the connected component of M which contains $d(R)$, i.e., $\lambda(R) = \sigma(M_R)$;
- every edge $(R, R') \in E$ is labeled with the smallest dart of the face corresponding to F_R in $canonical(M_R)$.

The labeling of the region enclosure trees provides a way of deciding whether two 2D topological maps are isomorphic or not:

Theorem 2. *Two 2D topological maps are isomorphic iff their labeled region enclosure trees are isomorphic.*

Indeed, the labels associated with the nodes ensure that two regions mapped by a tree isomorphism function actually belong to isomorphic connected components of the 2-maps; the labels associated with the edges ensure that two regions mapped by a tree isomorphism function are enclosed in regions which correspond to a same face in the canonical form of the corresponding connected components of the 2-maps.

Hence, to decide whether two 2D topological maps are isomorphic we first compute the labeling functions of the region enclosure trees and then use the algorithm of [1] to decide whether the two trees are isomorphic. The time complexity of the construction of the labeling function mainly involves computing the signature and the canonical form of each connected component of the 2-map. This may be done in $\mathcal{O}(k \cdot t^2)$ where k is the number of connected components of the 2-map and t is the maximum number of darts in a connected component (see Property 10 of [7]). The time complexity for deciding whether two labeled trees are isomorphic is linear w.r.t. the number of nodes, i.e., the number of regions.

Our contribution here is the definition of 2D topological map isomorphism (which could be also given for maptrees) and the description of a polynomial time algorithm for deciding of isomorphism. The second main contribution of this paper, given in the next section, is the definition of ML-simple transformation.

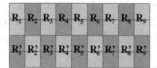
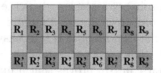

Fig. 3. Two partitions having homeomorphic embeddings in the plane. In the left image, there is no ML-simple transformation of less than 9 pixels. A possible ML-simple transformation is the one modifying simultaneously the labels of the 9 pixels of a same line and giving the right image.

4   ML-simple Transformation

Given a labeled image (I_d, L, l), the goal is to modify the label of some pixels while preserving the topology of the partition. To achieve this goal, we start to define what is a transformation in a labeled image.

Definition 3 (Transformation). *Given a 2D labeled image $i = (I_d, L, l)$, a transformation T is a set of pairs $\{(p_1, v_1), \ldots, (p_n, v_n)\}$, with $\forall k \in \{1, \ldots, n\}$, $p_k \in I_d$ a pixel and $v_k \in L$ a label. The labeled image obtained by applying transformation T on i is the labeled image $T(i) = (I_d, L, l')$ such that $\forall k \in \{1, \ldots, n\}$, $l'(p_k) = v_k$ and $\forall p \in \mathbb{Z}^2 \setminus \{p_1, \ldots, p_n\}$, $l'(p) = l(p)$.*

Now, thanks to the definition of transformation and the definition of topological map isomorphism, it is straightforward to define the notion of *ML-simple Transformation* (*ML* stands for multi-label). Intuitively a transformation is ML-simple if the two topological maps of the initial image and of the transformed image are isomorphic.

Definition 4 (ML-simple transformation). *Let $i = (I_d, L, l)$ be a 2D labeled image. A transformation T on i is ML-simple if $TM(i)$ is isomorphic to $TM(T(i))$.*

Theorem 1 says that two isomorphic topological maps represent two homeomorphic embeddings in the plane, thus an ML-simple transformation does not modify the topology of the embedding in the plane. Note that a ML-simple transformation that modifies only one pixel corresponds to the notion of ML-simple point [6]. Figure 3 shows that transformation of many pixels could be required in some configurations. In this example, there is no ML-simple transformation of sets having less than 9 pixels.

The definition of ML-simple transformation gives a straightforward way to test if a transformation is ML-simple: this may be done by computing the two topological maps and testing if they are isomorphic. This simple algorithm can be improved thanks to Theorem 3, which shows that it is enough to compare subimages surrounding the pixels of the transformation.

Given a transformation T on a labeled image (I_d, L, l), let $(I_{d|T}, L, l)$ be the subimage concerned by the transformation. More precisely, $min(I_{d|T})$

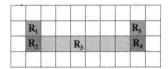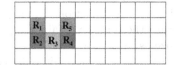

Fig. 4. Two partitions having homeomorphic embeddings in the plane. There is no sequence of flips of ML-simple points allowing to transform the image to the left in the image to the right. However there is a sequence of ML-simple transformations of four pixels which transforms the left image into the right one.

$= (min\{p_1.x, \ldots, p_k.x\}-1, min\{p_1.y, \ldots, p_k.y\}-1)$ and $max(I_{d|T}) = (max\{p_1.x, \ldots, p_k.x\}+1, max\{p_1.y, \ldots, p_k.y\}+1)$ are the two extremal pixels of the bounding box lying at distance 1 from the extremal points of T. In other words, $(I_{d|T}, L, l)$ is the subimage enclosing all the pixels of the transformation by a border of 1 pixel in thickness.

Theorem 3 shows that it is enough to compare the topological map of the subimage $(I_{d|T}, L, l)$ before and after the transformation in order to test if a transformation is ML-simple.

Theorem 3. *Let $i = (I_d, L, l)$ be a 2D labeled image and T be a transformation on i. T is ML-simple for i iff T is ML-simple for $(I_{d|T}, L, l)$.*

The main argument of the proof uses the fact that the labels of the pixels in the complement of $I_{d|T}$ are not modified by the transformation, nor the pixels in the border of $I_{d|T}$ (because the bounding box has been enlarged by a band of 1 pixel in thickness). Intuitively, this band of pixels around the bounding box guarantees that the frontiers are preserved between $I_{d|T}$ and its complement.

5   Look-up Tables of ML-simple Transformations

One interest of ML-simple transformations is to allow more deformations preserving the topology than ML-simple points. This is illustrated in the example given in Fig. 4. Both partitions are two homeomorphic embeddings in the plane, however there is no sequence of flips of ML-simple points allowing to transform the image on the left into the image on the right. This becomes possible if we use ML-simple transformations instead of flips of ML-simple points.

Deforming a labeled image while preserving the topology of the partition in regions can be done by searching for a ML-simple transformation and applying it. Given a local configuration of pixels, the question is thus to find all the possible ML-simple transformations of the current configuration. Indeed Theorem 3 ensures that the ML-simple transformation test can be restricted to the local window around the modified pixels.

To answer this question in an efficient way, we generated look-up tables of ML-simple transformations of three sizes (s_x, s_y): 1×1 (which is thus equivalent to the notion of ML-simple point), 2×1 and 1×2 (modification of two 4-adjacent pixels) and 2×2 (modification of four pixels forming a square).

To generate these look-up tables, we first generated all the possible labeled images of size $(s_x + 2, s_y + 2)$. We associate to each image i a bitset $b(i)$ having $2 \times (s_x + 2) \times (s_y + 2)$ bits. Bit number $2 \times (x + y \times (s_x + 2))$ is equal to 1 if pixel (x, y) belongs to the same region as pixel $(x - 1, y)$, and bit number $1 + 2 \times (x + y \times (s_x + 2))$ is equal to 1 if pixel (x, y) belongs to the same region as pixel $(x, y - 1)$. Each bitset describes a labeled image up to relabeling of labels, which is what we want since configurations do not consider the value of the labels but only the partition in regions.

The different bitsets are stored in two data-structures. The first one is an array of lists of bitsets, each element of the array being the list of all the bitsets having the same values in their boundaries and having isomorphic topological maps. Each list contains all the possible ML-simple transformations of a given configuration, i.e. all the local configurations belonging to the same equivalence class. The second data structure is an associative array (for example a hash-table) allowing to retrieve, given a bitset, the index of its equivalence class in the array of lists.

Thanks to these two data-structures, it is easy to test if two local configurations (of a given size) are equivalent by computing their two bitsets (by a linear scan of the pixels of the two images) and computing their two indices of their equivalence class. If the two indices are equal, the two configurations are topologically equivalent. Thanks to the hash-table, this can be done in amortized constant time.

Moreover, given a local configuration, we can directly apply successively all the possible ML-simple transformations (of a given size) while preserving the topology of the partition. We first compute the bitset of the local configuration and get the index of its equivalence class. Thanks to this index, we have a direct access to the list of all the bitsets belonging to the same equivalence class, i.e. having the same topology. For each bitset in the list, we can locally modify the labeled image according to the label of pixels not modified by the transformation (belonging to the boundary of the image) and we are sure that the modified image has the same topology as the initial one.

Table 1 shows some information on the generated look-up tables with size 1×1, 2×1 and 2×2. The number of configurations increases quickly as the size of the ML-transformation increases. The number of equivalence classes is significant (between 75% and 84% of the total number of configurations). Indeed there are many non ML-simple configurations. Lastly, we can notice the important time required to compute the look-up tables (about 2 hours for 2×2 pixels), however this generation was done only once as a preprocessing step.

An upper bound on the number of configurations can be computed. For a transformation of size (s_x, s_y), there are $p = (s_x + 2) \times (s_y + 2)$ pixels. If we denote by k the number of different labels, each pixel can be labeled with an integer between 1 and k ($k \in \{1, \ldots, p\}$). Thus the number of different configurations with k labels is $\frac{p!}{(p-k)!}$ and the total number of configurations is $\sum_{k=1}^{p} \frac{p!}{(p-k)!}$,

Table 1. Generation of look-up tables of all ML-simple transformations of size 1×1, 2×1 and 2×2. *nb configs* is the total number of configurations, *nb classes* the number of equivalence classes, *time* the time spent to compute the look-up tables and *max/classe* the maximal number of elements belonging to the same class.

size	nb configs	nb classes	time	max/classe
1×1	1 002	850	0.04s	3
2×1	16 239	13 211	5.4s	6
2×2	756 436	567 728	7042.s	30

that is 986 409 (resp. 1 302 061 344 and 56 874 039 553 216) for size 1×1 (resp. 2×1 and 2×2). We can observe a major difference between these results and the experimental ones because the previous formula gives an upper bound and not exactly the number of configurations. Indeed, different permutations of labels can give the same configuration (i.e. the same partition into regions) while they could be recounted several times in the formula. Note also that it is very difficult to estimate the number of classes, and to give a better upper bound than the total number of configurations.

6   Conclusion

In this paper, we defined the isomorphism between 2D topological maps and showed that two topological maps are isomorphic if and only if the corresponding images are homeomorphic embeddings in the plane. Thanks to this definition, we defined the ML-simple transformation as a set of modifications of labels of pixels that preserve the topology of the partition. Thanks to these notions, we were able to generate different look-up tables of ML-simple transformations for different sizes of modified pixels.

Topological map isomorphism could be helpful in work that need to verify that modifications preserve the topology of the partition. This is for example the case for rigid transformation, where it is possible to use topological map isomorphism as a control tool to check the results of the transformation.

The look-up tables will serve in order to propose more deformations in our framework of deformable partition [4]. We hope that these new deformations will improve the previous results since with ML-simple points, final results of the deformation process are sometimes blocked in local configurations that can not be modified with ML-simple points. Moreover, we want to increase the speed of the simulation thanks to the access in amortized constant time to the elements of the table. Note that it is possible to use directly the topological map isomorphism in order to test if a transformation is ML-simple. This possibility is particularly interesting in order to use transformations of a large number of pixels.

Our first perspective is the extension of this work to 3D, where topological maps and signatures are already defined. The only problem to solve is the position of fictive edges in 3D topological maps. We must propose a canonical representation for these edges in order to retrieve the property that two maps are isomorphic if the corresponding images are homeomorphic embeddings in the

3D Euclidean space. Our second perspective is the generation of look-up tables with bigger size. The problem is then the memory storage of these tables while keeping an efficient access to all the elements.

Acknowledgement. This work has been partially supported by the French National Agency (ANR), projects DIGITALSNOW ANR-11-BS02-009 and SOLSTICE ANR-13-BS02-01.

References

1. Aho, A.V., Hopcroft, J.E., Ullman, J.D.: The Design and Analysis of Computer Algorithms. Addison-Wesley (1974)
2. Bazin, P.-L., Ellingsen, L.M., Pham, D.L.: Digital homeomorphisms in deformable registration. In: Karssemeijer, N., Lelieveldt, B. (eds.) IPMI 2007. LNCS, vol. 4584, pp. 211–222. Springer, Heidelberg (2007)
3. Damiand, G., Bertrand, Y., Fiorio, C.: Topological model for two-dimensional image representation: Definition and optimal extraction algorithm. Computer Vision and Image Understanding 93(2), 111–154 (2004)
4. Damiand, G., Dupas, A., Lachaud, J.-O.: Combining topological maps, multi-label simple points, and minimum-length polygons for efficient digital partition model. In: Aggarwal, J.K., Barneva, R.P., Brimkov, V.E., Koroutchev, K.N., Korutcheva, E.R. (eds.) IWCIA 2011. LNCS, vol. 6636, pp. 56–69. Springer, Heidelberg (2011)
5. Damiand, G., Solnon, C., de la Higuera, C., Janodet, J.-C., Samuel, E.: Polynomial algorithms for subisomorphism of nd open combinatorial maps. Computer Vision and Image Understanding 115(7), 996–1010 (2011)
6. Dupas, A., Damiand, G., Lachaud, J.-O.: Multi-label simple points definition for 3D images digital deformable model. In: Brlek, S., Reutenauer, C., Provençal, X. (eds.) DGCI 2009. LNCS, vol. 5810, pp. 156–167. Springer, Heidelberg (2009)
7. Gosselin, S., Damiand, G., Solnon, C.: Efficient search of combinatorial maps using signatures. Theoretical Computer Science 412(15), 1392–1405 (2011)
8. Khalimsky, E., Kopperman, R., Meyer, P.R.: Computer graphics and connected topologies on finite ordered sets. Topology and its Applications 36, 1–17 (1990)
9. Le Bodic, P., Locteau, H., Adam, S., Héroux, P., Lecourtier, Y., Knippel, A.: Symbol detection using region adjacency graphs and integer linear programming. In: Proc. of ICDAR, Barcelona, Spain, pp. 1320–1324. IEEE Computer Society (July 2009)
10. Lienhardt, P.: N-Dimensional generalized combinatorial maps and cellular quasi-manifolds. International Journal of Computational Geometry and Applications 4(3), 275–324 (1994)
11. Mazo, L.: A framework for label images. In: Ferri, M., Frosini, P., Landi, C., Cerri, A., Di Fabio, B. (eds.) CTIC 2012. LNCS, vol. 7309, pp. 1–10. Springer, Heidelberg (2012)
12. Rosenfeld, A.: Adjacency in digital pictures. Information and Control 26(1), 24–33 (1974)
13. Rosenfeld, A.: Digital Topology. The American Mathematical Monthly 86(8), 621–630 (1979)
14. Trémeau, A., Colantoni, P.: Regions adjacency graph applied to color image segmentation. IEEE Transactions on Image Processing 9, 735–744 (2000)
15. Worboys, M.: The maptree: A fine-grained formal representation of space. In: Xiao, N., Kwan, M.-P., Goodchild, M.F., Shekhar, S. (eds.) GIScience 2012. LNCS, vol. 7478, pp. 298–310. Springer, Heidelberg (2012)

Isthmus-Based Parallel and Asymmetric 3D Thinning Algorithms*

Michel Couprie and Gilles Bertrand

Université Paris-Est, LIGM, Équipe A3SI, ESIEE Paris, France
{michel.couprie,gilles.bertrand}@esiee.fr

Abstract. Critical kernels constitute a general framework settled in the context of abstract complexes for the study of parallel thinning in any dimension. We take advantage of the properties of this framework, to propose a generic thinning scheme for obtaining "thin" skeletons from objects made of voxels. From this scheme, we derive algorithms that produce curvilinear or surface skeletons, based on the notion of 1D or 2D isthmus.

1 Introduction

When dealing with skeletons, one has to face two main problems: topology preservation, and preservation of meaningful geometrical features. Here, we are interested in the skeletonization of objects that are made of voxels (unit cubes) in a regular 3D grid, *i.e.*, in a binary 3D image. In this context, topology preservation is usually obtained through the iteration of thinning steps, provided that each step does not alter the topological characteristics. In sequential thinning algorithms, each step consists of detecting and choosing a so-called simple voxel, that may be characterized locally (see [1,2]), and removing it. Such a process usually involves many arbitrary choices, and the final result may depend, sometimes heavily, on any of these choices. This is why parallel thinning algorithms are generally preferred to sequential ones. However, removing a set of simple voxels at each thinning step, in parallel, may alter topology. The framework of critical kernels, introduced by one of the authors in [3], provides a condition under which we have the guarantee that a subset of voxels can be removed without changing topology. This condition is, to our knowledge, the most general one among the related works. Furthermore, critical kernels indeed provide a method to design new parallel thinning algorithms, in which the property of topology preservation is built-in, and in which any kind of constraint may be imposed (see [4,5]).

Among the different parallel thinning algorithms that have been proposed in the literature, we can distinguish symmetric from asymmetric algorithms. Symmetric algorithms (see *e.g.* [6,7,8]) (also known as fully parallel algorithms) produce skeletons that are invariant under 90 degrees rotations. They consist of the iteration of thinning steps that are made of 1) the identification and selection

---

* This work has been partially supported by the "ANR-2010-BLAN-0205 KIDICO" project.

E. Barcucci et al. (Eds.): DGCI 2014, LNCS 8668, pp. 51–62, 2014.

of a set of voxels that satisfy certain conditions, independently of orientation or position in space, and 2) the removal, in parallel, of all selected voxels from the object. Symmetric algorithms, on the positive side, produce a result that is uniquely defined: no arbitrary choice is needed. On the negative side, they generally produce thick skeletons, see Fig. 1.

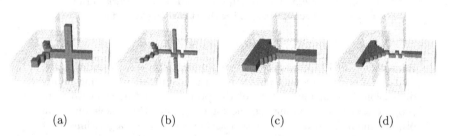

(a)   (b)   (c)   (d)

Fig. 1. Different types of skeletons. (a): Curvilinear skeleton, symmetric. (b): Curvilinear skeleton, asymmetric. (c): Surface skeleton, symmetric. (d): Surface skeleton, asymmetric.

Asymmetric skeletons, on the opposite, are preferred when thinner skeletons are required. The price to pay is a certain amount of arbitrary choices to be made. In all existing asymmetric parallel thinning algorithms, each thinning step is divided into a certain number of substeps. In the so-called directional algorithms (see *e.g.* [9,10,11]), each substep is devoted to the detection and the deletion of voxels belonging to one "side" of the object: all the voxels considered during the substep have, for example, their south neighbor inside the object and their north neighbor outside the object. The order in which these directional substeps are executed is set beforehand, arbitrarily. Subgrid (or subfield) algorithms (see *e.g.* [12,13]) form the second category of asymmetric parallel thinning algorithms. There, each substep is devoted to the detection and the deletion of voxels that belong to a certain subgrid, for example, all voxels that have even coordinates. Considered subgrids must form a partition of the grid. Again, the order in which subgrids are considered is arbitrary.

Subgrid algorithms are not often used in practice because they produce artifacts, that is, waving skeleton branches where the original object is smooth or straight. Directional algorithms are the most popular ones. Most of them are implemented through sets of masks, one per substep. A set of masks is used to characterize voxels that must be kept during a given substep, in order to 1) preserve topology, and 2) prevent curves or surfaces to disappear. Thus, topological conditions and geometrical conditions cannot be easily distinguished, and the slightest modification of any mask involves the need to make a new proof of the topological correctness.

Our approach is radically different. Instead of considering single voxels, we consider cliques. A clique is a set of mutually adjacent voxels. Then, we identify the critical kernel of the object, according to some definitions, which is a union of cliques. The main theorem of the critical kernels framework [3,5] states that we can remove in parallel any subset of the object, provided that we keep at least one voxel of every clique that constitutes the critical kernel, and this guarantees topology preservation. Here, as we try to obtain thin skeletons, our goal is to keep, whenever possible, exactly one voxel in every such clique. This leads us to propose a generic parallel asymmetric thinning scheme, that may be enriched by adding any sort of geometrical constraint. For example, we define the notions of 1D and 2D isthmuses. A 1D (resp. 2D) isthmus is a voxel that is "locally like a piece of curve" (resp. surface). From our generic scheme, we easily derive, by adding the constraint to preserve isthmuses, specific algorithms that produce curvilinear or surface skeletons.

2 Voxel Complexes

In this section, we give some basic definitions for voxel complexes, see also [14,1].

Let $\mathbb{Z}$ be the set of integers. We consider the families of sets $\mathbb{F}_0^1$, $\mathbb{F}_1^1$, such that $\mathbb{F}_0^1 = \{\{a\} \mid a \in \mathbb{Z}\}$, $\mathbb{F}_1^1 = \{\{a, a+1\} \mid a \in \mathbb{Z}\}$. A subset f of $\mathbb{Z}^n$, $n \geq 2$, that is the Cartesian product of exactly d elements of $\mathbb{F}_1^1$ and $(n-d)$ elements of $\mathbb{F}_0^1$ is called a *face* or an *d-face* of $\mathbb{Z}^n$, d is the *dimension of f*. In the illustrations of this paper except Fig. 6, a 3-face (resp. 2-face, 1-face, 0-face) is depicted by a cube (resp. square, segment, dot), see *e.g.* Fig. 4.

A 3-face of $\mathbb{Z}^3$ is also called a *voxel*. A finite set that is composed solely of voxels is called a *(voxel) complex* (see Fig. 2). We denote by $\mathbb{V}^3$ the collection of all voxel complexes.

We say that two voxels x, y are *adjacent* if $x \cap y \neq \emptyset$. We write $\mathcal{N}(x)$ for the set of all voxels that are adjacent to a voxel x, $\mathcal{N}(x)$ is the *neighborhood of x*. Note that, for each voxel x, we have $x \in \mathcal{N}(x)$. We set $\mathcal{N}^*(x) = \mathcal{N}(x) \setminus \{x\}$.

Let $d \in \{0, 1, 2\}$. We say that two voxels x, y are *d-neighbors* if $x \cap y$ is a d-face. Thus, two distinct voxels x and y are adjacent if and only if they are d-neighbors for some $d \in \{0, 1, 2\}$.

Let $X \in \mathbb{V}^3$. We say that X is *connected* if, for any $x, y \in X$, there exists a sequence $\langle x_0, ..., x_k \rangle$ of voxels in X such that $x_0 = x$, $x_k = y$, and x_i is adjacent to x_{i-1}, $i = 1, ..., k$.

3 Simple Voxels

Intuitively a voxel x of a complex X is called a simple voxel if its removal from X "does not change the topology of X". This notion may be formalized with the help of the following recursive definition introduced in [5], see also [15,16] for other recursive approaches for simplicity.

Definition 1. Let $X \in \mathbb{V}^3$.
We say that X is *reducible* if either:
i) X is composed of a single voxel; or
ii) there exists $x \in X$ such that $\mathcal{N}^*(x) \cap X$ is reducible and $X \setminus \{x\}$ is reducible.

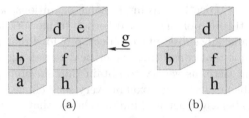

Fig. 2. (a) A complex X which is made of 8 voxels, (b) A complex $Y \subseteq X$, which is a thinning of X

Definition 2. Let $X \in \mathbb{V}^3$. A voxel $x \in X$ is *simple for* X if $\mathcal{N}^*(x) \cap X$ is reducible. If $x \in X$ is simple for X, we say that $X \setminus \{x\}$ is an *elementary thinning of* X.

Thus, a complex $X \in \mathbb{V}^3$ is reducible if and only if it is possible to reduce X to a single voxel by iteratively removing simple voxels. Observe that a reducible complex is necessarily non-empty and connected.

In Fig. 2 (a), the voxel a is simple for X ($\mathcal{N}^*(a) \cap X$ is made of a single voxel), the voxel d is not simple for X ($\mathcal{N}^*(d) \cap X$ is not connected), the voxel h is simple for X ($\mathcal{N}^*(h) \cap X$ is made of two voxels that are 2-neighbors and is reducible).

In [5], it was shown that the above definition of a simple voxel is equivalent to classical characterizations based on connectivity properties of the voxel's neighborhood [17,18,19,20,2]. An equivalence was also established with a definition based on the operation of collapse [21], this operation is a discrete analogue of a continuous deformation (a homotopy), see also [15,3,2].

The notion of a simple voxel allows one to define thinnings of a complex, see an illustration Fig. 2 (b).

Let $X, Y \in \mathbb{V}^3$. We say that Y *is a thinning of* X or that X is *reducible to* Y, if there exists a sequence $\langle X_0, ..., X_k \rangle$ such that $X_0 = X$, $X_k = Y$, and X_i is an elementary thinning of X_{i-1}, $i = 1, ..., k$.

Thus, a complex X is reducible if and only if it is reducible to a single voxel.

4   Critical Kernels

Let X be a complex in $\mathbb{V}^3$. It is well known that, if we remove simultaneously (in parallel) simple voxels from X, we may "change the topology" of the original object X. For example, the two voxels f and g are simple for the object X depicted Fig. 2 (a). Nevertheless $X \setminus \{f, g\}$ has two connected components whereas X is connected.

In this section, we recall a framework for thinning in parallel discrete objects with the warranty that we do not alter the topology of these objects [3,4,5]. This method is valid for complexes of arbitrary dimension.

Let $d \in \{0, 1, 2, 3\}$ and let $C \in \mathbb{V}^3$. We say that C is a *d-clique* or a *clique* if $\cap \{x \in C\}$ is a d-face. If C is a d-clique, d is the *rank* of C.

If C is made of solely two distinct voxels x and y, we note that C is a d-clique if and only if x and y are d-neighbors, with $d \in \{0, 1, 2\}$.

Let $X \in \mathbb{V}^3$ and let $C \subseteq X$ be a clique. We say that C is *essential for X* if we have $C = D$ whenever D is a clique such that:

i) $C \subseteq D \subseteq X$; and

ii) $\cap\{x \in C\} = \cap\{x \in D\}$.

Observe that any complex C that is made of a single voxel is a clique (a 3-clique). Furthermore any voxel of a complex X constitutes a clique that is essential for X.

In Fig. 2 (a), $\{f, g\}$ is a 2-clique that is essential for X, $\{b, d\}$ is a 0-clique that is not essential for X, $\{b, c, d\}$ is a 0-clique essential for X, $\{e, f, g\}$ is a 1-clique essential for X.

Definition 3. Let $S \in \mathbb{V}^3$. The $\mathcal{K}$-*neighborhood of S*, written $\mathcal{K}(S)$, is the set made of all voxels that are adjacent to each voxel in S. We set $\mathcal{K}^*(S) = \mathcal{K}(S) \setminus S$.

We note that we have $\mathcal{K}(S) = \mathcal{N}(x)$ whenever S is made of a single voxel x. We also observe that we have $S \subseteq \mathcal{K}(S)$ whenever S is a clique.

Definition 4. Let $X \in \mathbb{V}^3$ and let C be a clique that is essential for X. We say that the clique C is *regular for X* if $\mathcal{K}^*(C) \cap X$ is reducible. We say that C is *critical for X* if C is not regular for X.

Thus, if C is a clique that is made of a single voxel x, then C is regular for X if and only if x is simple for X.

In Fig. 2 (a), the cliques $C_1 = \{b, c, d\}$, $C_2 = \{f, g\}$, and $C_3 = \{f, h\}$ are essential for X. We have $\mathcal{K}^*(C_1) \cap X = \emptyset$, $\mathcal{K}^*(C_2) \cap X = \{e, h\}$, and $\mathcal{K}^*(C_3) \cap X = \{g\}$. Thus, C_1 and C_2 are critical for X, while C_3 is regular for X.

The following result is a consequence of a general theorem that holds for complexes of arbitrary dimensions [3,5], see an illustration Fig. 2 (a) and (b) where the complexes X and Y satisfy the condition of Th. 5.

Theorem 5. *Let $X \in \mathbb{V}^3$ and let $Y \subseteq X$.*
The complex Y is a thinning of X if any clique that is critical for X contains at least one voxel of Y.

5   A Generic 3D Parallel and Asymmetric Thinning Scheme

Our goal is to define a subset Y of a voxel complex X that is guaranteed to include at least one voxel of each clique that is critical for X. By Th. 5, this subset Y will be a thinning of X.

Let us consider the complex X depicted Fig. 3 (a). There are precisely three cliques that are critical for X:

- the 0-clique $C_1 = \{b, c\}$ (we have $\mathcal{K}^*(C_1) \cap X = \emptyset$);
- the 2-clique $C_2 = \{a, b\}$ (we have $\mathcal{K}^*(C_2) \cap X = \emptyset$);
- the 3-clique $C_3 = \{b\}$ (the voxel b is not simple).

Suppose that, in order to build a complex Y that fulfills the condition of Th. 5, we select arbitrarily one voxel of each clique that is critical for X. Following such a strategy, we could select c for C_1, a for C_2, and b for C_3. Thus, we would have $Y = X$, no voxel would be removed from X. Now, we observe that the complex $Y' = \{b\}$ satisfies the condition of Th. 5. This complex is obtained by considering first the 3-cliques before selecting a voxel in the 2-, 1-, or 0 cliques.

The complex X of Fig. 3 (b) provides another example of such a situation. There are precisely three cliques that are critical for X:

- the 1-clique $C_1 = \{e, f, g, h\}$ (we have $\mathcal{K}^*(C_1) \cap X = \emptyset$);
- the 1-clique $C_2 = \{e, d, g\}$ (we have $\mathcal{K}^*(C_2) \cap X = \emptyset$);
- the 2-clique $C_3 = \{e, g\}$ ($\mathcal{K}^*(C_3) \cap X$ is not connected).

If we select arbitrarily one voxel of each critical clique, we could obtain the complex $Y = \{f, d, g\}$. On the other hand, if we consider the 2-cliques before the 1-cliques, we obtain either $Y' = \{e\}$ or $Y'' = \{g\}$. In both cases the result is better in the sense that we remove more voxels from X.

This discussion motivates the introduction of the following 3D asymmetric and parallel thinning scheme AsymThinningScheme(see also [4,5]). The main features of this scheme are the following:

- Taking into account the observations made through the two previous examples, critical cliques are considered according to their decreasing ranks (step 4). Thus, each iteration is made of four sub-iterations (steps 4-8). Voxels that have been previously selected are stored in a set Y (step 8). At a given sub-iteration, we consider voxels only in critical cliques included in $X \setminus Y$ (step 6).
- *Select* is a function from $\mathbb{V}^3$ to V^3, the set of all voxels. More precisely, *Select* associates, to each set S of voxels, a unique voxel x of S. We refer to such a function as a *selection function*. This function allows us to select a voxel in a given critical clique (step 7). A possible choice is to take for $Select(S)$, the first pixel of S in the lexicographic order of the voxels coordinates.
- In order to compute curvilinear or surface skeletons, we have to keep other voxels than the ones that are necessary for the preservation of the topology of the object X. In the scheme, the set K corresponds to a set of features that we want to be preserved by a thinning algorithm (thus, we have $K \subseteq X$). This set K, called *constraint set*, is updated dynamically at step 10. $Skel_X$ is a function from X on $\{True, False\}$ that allows us to keep some *skeletal voxels* of X, e.g., some voxels belonging to parts of X that are surfaces or curves. For example, if we want to obtain curvilinear skeletons, a popular solution is to set $Skel_X(x) = True$ whenever x is a so-called *end voxel* of X: an end voxel is a voxel that has exactly one neighbor inside X ; see Fig. 7(a) a skeleton obtained in this way. However, this solution is limited and does not permit to obtain surface skeletons. Better propositions for such a function will be introduced in section 6.

By construction, at each iteration, the complex Y at step 9 satisfies the condition of Th. 5. Thus, the result of the scheme is a thinning of the original complex X. Observe also that, except step 4, each step of the scheme may be computed in parallel.

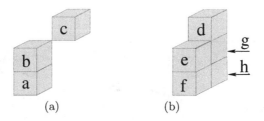

Fig. 3. Two complexes

Algorithm 1: AsymThinningScheme($X, Skel_X$)

Data: $X \in \mathbb{V}^3$, $Skel_X$ is a function from X on $\{True, False\}$
Result: X

1   $K := \emptyset$;
2   **repeat**
3      $Y := K$;
4   **for** $d \leftarrow 3$ **to** 0 **do**
5         $Z := \emptyset$;
6   **foreach** d-*clique* $C \subseteq X \setminus Y$ *that is critical for* X **do**
7            $Z := Z \cup \{Select(C)\}$;
8         $Y := Y \cup Z$;
9      $X := Y$;
10   **foreach** *voxel* $x \in X \setminus K$ *such that* $Skel_X(x) = True$ **do** $K := K \cup \{x\}$;
11 **until** *stability* ;

Fig. 4 provides an illustration of the scheme AsymThinningScheme. Let us consider the complex X depicted in (a). We suppose in this example that we do not keep any skeletal voxel, *i.e.*, for any $x \in X$, we set $Skel_X(x) = False$. The traces of the cliques that are critical for X are represented in (b), the *trace of a clique* C is the face $f = \cap\{x \in C\}$. Thus, the set of the cliques that are critical for X is precisely composed of six 0-cliques, two 1-cliques, three 2-cliques, and one 3-clique. In (c) the different sub-iterations of the scheme are illustrated (steps 4-8):
- when $d = 3$, only one clique is considered, the dark grey voxel is selected whatever the selection function;
- when $d = 2$, all the three 2-cliques are considered since none of these cliques contains the above voxel. Voxels that could be selected by a selection function are depicted in medium grey;
- when $d = 1$, only one clique is considered, a voxel that could be selected is depicted in light grey;
- when $d = 0$, no clique is considered since each of the 0-cliques contains at least one voxel that has been previously selected.
After these sub-iterations, we obtain the complex depicted in (d). The figures (e) and (f) illustrate the second iteration, at the end of this iteration the complex

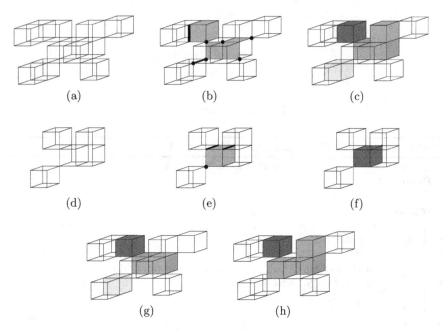

Fig. 4. (a): A complex X made of precisely 12 voxels. (b): The traces of the cliques that are critical for X. (c): Voxels that have been selected by the algorithm. (d): The result Y of the first iteration. (e): The traces of the 4 cliques that are critical for Y. (f): The result of the second iteration. (g) and (h): Two other possible selections at the first iteration.

Fig. 5. Ultimate asymmetric skeletons obtained by using `AsymThinningScheme`

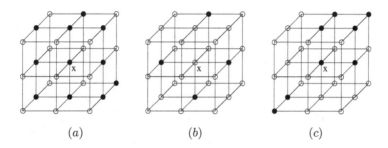

$$(a) \qquad (b) \qquad (c)$$

Fig. 6. In this figure, a voxel is represented by its central point. (a): A voxel x and the set $\mathcal{N}(x) \cap X$ (black points). (b): A set S which is a 1-surface, $\mathcal{N}^*(x) \cap X$ is reducible to S, thus x is a 2-isthmus. for X. (c): A voxel x and the set $\mathcal{N}(x) \cap X$ (black points). The voxel x is a 1-isthmus for X.

is reduced to a single voxel. In (g) and (h) two other possible selections at the first iteration are given.

Of course, the result of the scheme may depend on the choice of the selection function. This is the price to be paid if we try to obtain thin skeletons. For example, some arbitrary choices have to be made for reducing a two voxels wide ribbon to a simple curve.

In the sequel of the paper, we take for $Select(S)$, the first pixel of S in the lexicographic order of the voxels coordinates.

Fig. 5 shows another illustration, on bigger objects, of `AsymThinningScheme`. Here also, for any $x \in X$, we have $Skel_X(x) = False$ (no skeletal voxel). The result is called an ultimate asymmetric skeleton.

6   Isthmus-Based Asymmetric Thinning

In this section, we show how to use our generic scheme `AsymThinningScheme` in order to get a procedure that computes either curvilinear or surface skeletons. This thinning procedure preserves a constraint set K that is made of "isthmuses".

Intuitively, a voxel x of an object X is said to be a 1-isthmus (resp. a 2-isthmus) if the neighborhood of x corresponds - up to a thinning - to the one of a point belonging to a curve (resp. a surface) [5].

We say that $X \in \mathbb{V}^3$ is a 0-*surface* if X is precisely made of two voxels x and y such that $x \cap y = \emptyset$.

We say that $X \in \mathbb{V}^3$ is a 1-*surface* (or a *simple closed curve*) if:
i) X is connected; and ii) For each $x \in X$, $\mathcal{N}^*(x) \cap X$ is a 0-surface.

Definition 6. Let $X \in \mathbb{V}^3$, let $x \in X$.
We say that x is a 1-*isthmus for X* if $\mathcal{N}^*(x) \cap X$ is reducible to a 0-surface.
We say that x is a 2-*isthmus for X* if $\mathcal{N}^*(x) \cap X$ is reducible to a 1-surface.
We say that x is a 2$^+$-*isthmus for X* if x is a 1-isthmus or a 2-isthmus for X.

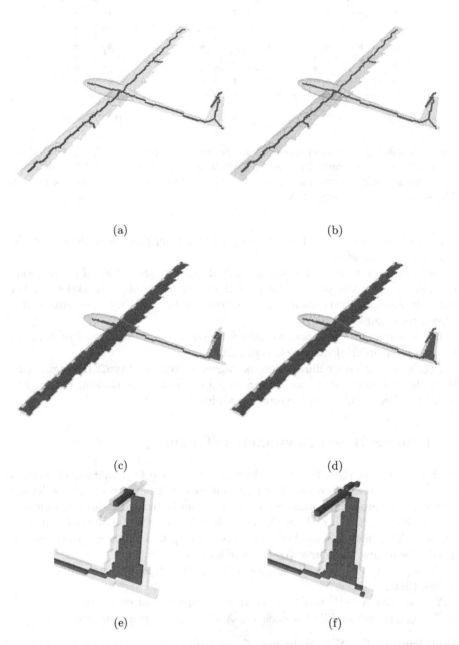

(a)   (b)

(c)   (d)

(e)   (f)

Fig. 7. Asymmetric skeletons obtained by using `AsymThinningScheme`. (a): the function $Skel_X$ is based on end voxels. (b,c,d): the function $Skel_X$ is based on k-isthmuses, with $k = 1, 2$ and 2^+ respectively. (e,f): detail of (c,d) respectively.

Our aim is to thin an object, while preserving a constraint set K that is made of voxels that are detected as k-isthmuses during the thinning process. We obtain curvilinear skeletons with $k = 1$, surface skeletons with $k = 2$, and surface/curvilinear skeletons with $k = 2^+$. These three kinds of skeletons may be obtained by using `AsymThinningScheme`, with the function $Skel_X$ defined as follows:

$$Skel_X(x) = \begin{cases} True & \text{if } x \text{ is a } k\text{-isthmus,} \\ False & \text{otherwise,} \end{cases}$$

with $k \in \{1, 2, 2^+\}$.

Observe there is the possibility that a voxel belongs to a k-isthmus at a given step of the algorithm, but not at further steps. This is why previously detected isthmuses are stored (see line 10 of `AsymThinningScheme`).

In Fig. 7(b-f), we show a curvilinear skeleton, a surface skeleton and a surface/curvilinear skeleton obtained by our method from the same object.

7   Conclusion

We introduced an original generic scheme for asymmetric parallel topology-preserving thinning of 3D objects made of voxels, in the framework of critical kernels. We saw that from this scheme, one can easily derive several thinning operators having specific behaviours, simply by changing the definition of skeletal points. In particular, we showed that ultimate, curvilinear, surface, and surface/curvilinear skeletons can be obtained, based on the notion of 1D/2D isthmuses.

A key point, in the implementation of the algorithms proposed in this paper, is the detection of critical cliques and isthmus voxels. In [5], we showed that it is possible to detect critical cliques thanks to a set of masks, in linear time. We also showed that the configurations of 1D and 2D isthmuses may be pre-computed by a linear-time algorithm and stored in lookup tables. Finally, based on a breadth-first strategy, the whole method can be implemented to run in $O(n)$ time, where n is the number of voxels of the input 3D image.

In an extended paper, in preparation, we will show how to deal with the robustness to noise issue thanks to the notion of isthmus persistence. We will also compare our method with all existing asymmetric parallel skeletonization algorithms acting in the 3D cubic grid.

References

1. Kong, T.Y., Rosenfeld, A.: Digital topology: introduction and survey. Comp. Vision, Graphics and Image Proc. 48, 357–393 (1989)
2. Couprie, M., Bertrand, G.: New characterizations of simple points in 2D, 3D and 4D discrete spaces. IEEE Transactions on Pattern Analysis and Machine Intelligence 31(4), 637–648 (2009)
3. Bertrand, G.: On critical kernels. Comptes Rendus de l'Académie des Sciences, Série Math. I(345), 363–367 (2007)

4. Bertrand, G., Couprie, M.: Two-dimensional thinning algorithms based on critical kernels. Journal of Mathematical Imaging and Vision 31(1), 35–56 (2008)
5. Bertrand, G., Couprie, M.: Powerful Parallel and Symmetric 3D Thinning Schemes Based on Critical Kernels. Journal of Mathematical Imaging and Vision 48(1), 134–148 (2014)
6. Manzanera, A., Bernard, T., Prêteux, F., Longuet, B.: n-dimensional skeletonization: a unified mathematical framework. Journal of Electronic Imaging 11(1), 25–37 (2002)
7. Lohou, C., Bertrand, G.: Two symmetrical thinning algorithms for 3D binary images. Pattern Recognition 40, 2301–2314 (2007)
8. Palágyi, K.: A 3D fully parallel surface-thinning algorithm. Theoretical Computer Science 406(1-2), 119–135 (2008)
9. Tsao, Y., Fu, K.: A parallel thinning algorithm for 3D pictures. CGIP 17(4), 315–331 (1981)
10. Palágyi, K., Kuba, A.: A parallel 3D 12-subiteration thinning algorithm. Graphical Models and Image Processing 61(4), 199–221 (1999)
11. Lohou, C., Bertrand, G.: A 3D 6-subiteration curve thinning algorithm based on p-simple points. Discrete Applied Mathematics 151, 198–228 (2005)
12. Bertrand, G., Aktouf, Z.: A three-dimensional thinning algorithm using subfields. In: Vision Geometry III, vol. 2356, pp. 113–124. SPIE (1996)
13. Németh, G., Kardos, P., Palágyi, K.: Topology preserving 3D thinning algorithms using four and eight subfields. In: Campilho, A., Kamel, M. (eds.) ICIAR 2010. LNCS, vol. 6111, pp. 316–325. Springer, Heidelberg (2010)
14. Kovalevsky, V.: Finite topology as applied to image analysis. Computer Vision, Graphics and Image Processing 46, 141–161 (1989)
15. Kong, T.Y.: Topology-preserving deletion of 1's from 2-, 3- and 4-dimensional binary images. In: Ahronovitz, E., Fiorio, C. (eds.) DGCI 1997. LNCS, vol. 1347, pp. 3–18. Springer, Heidelberg (1997)
16. Bertrand, G.: New notions for discrete topology. In: Bertrand, G., Couprie, M., Perroton, L. (eds.) DGCI 1999. LNCS, vol. 1568, pp. 218–228. Springer, Heidelberg (1999)
17. Bertrand, G., Malandain, G.: A new characterization of three-dimensional simple points. Pattern Recognition Letters 15(2), 169–175 (1994)
18. Bertrand, G.: Simple points, topological numbers and geodesic neighborhoods in cubic grids. Pattern Recognition Letters 15, 1003–1011 (1994)
19. Saha, P., Chaudhuri, B., Chanda, B., Dutta Majumder, D.: Topology preservation in 3D digital space. Pattern Recognition 27, 295–300 (1994)
20. Kong, T.Y.: On topology preservation in 2-D and 3-D thinning. International Journal on Pattern Recognition and Artificial Intelligence 9, 813–844 (1995)
21. Whitehead, J.: Simplicial spaces, nuclei and m-groups. Proceedings of the London Mathematical Society 45(2), 243–327 (1939)

Completions and Simple Homotopy*

Gilles Bertrand

Université Paris-Est, Laboratoire d'Informatique Gaspard-Monge,
Equipe A3SI, ESIEE Paris

Abstract. We propose an extension of simple homotopy by considering *homotopic pairs*. Intuitively, a homotopic pair is a couple of objects
(X, Y) such that X is included in Y and (X, Y) may be transformed to
a trivial couple by simple homotopic deformations that keep X inside Y.
Thus, these objects are linked by a "relative homotopy relation".

We formalize these notions by means of completions, which are inductive properties expressed in a declarative way. In a previous work,
through the notion of a *dyad*, we showed that completions were able to
handle couples of objects that are linked by a certain "relative homology
relation".

The main result of the paper is a theorem that makes clear the link
between homotopic pairs and dyads. Thus, we prove that, in the unified
framework of completions, it is possible to handle notions relative to both
homotopy and homology.

Keywords: Simple homotopy, combinatorial topology, simplicial complexes, completions.

1 Introduction

Simple homotopy, introduced by J. H. C. Whitehead in the early 1930's, may
be seen as a refinement of the concept of homotopy [1]. Two complexes are
simple homotopy equivalent if one of them may be obtained from the other by
a sequence of elementary collapses and anti-collapses.

Simple homotopy plays a fundamental role in combinatorial topology [1–7].
Also, many notions relative to homotopy in the context of computer imagery
rely on the collapse operation. In particular, this is the case for the notion of
a simple point, which is crucial for all image transformations that preserve the
topology of the objects [8–10], see also [11–13].

In this paper, we propose an extension of simple homotopy by considering
homotopic pairs. Intuitively, a homotopic pair is a couple of objects (X, Y) such
that X is included in Y and (X, Y) may be transformed to a trivial couple by
simple homotopic deformations that keep X inside Y. Thus, these objects are
linked by a "relative homotopy relation".

---

* This work has been partially supported by the "ANR-2010-BLAN-0205 KIDICO"
project.

E. Barcucci et al. (Eds.): DGCI 2014, LNCS 8668, pp. 63–74, 2014.

We formalize these notions by means of completions, which are inductive properties expressed in a declarative way [14]. In a previous work, we introduced the notions of a *dendrite* and a *dyad* [15], which were also formalized by means of completions. A dendrite is an acyclic object, a theorem asserts that an object is a dendrite if and only if it is acyclic in the sense of homology. Intuitively, a dyad is a couple of objects (X, Y), with $X \subseteq Y$, such that the cycles of X are "at the right place with respect to the ones of Y". A theorem provides a relation between dendrites and dyads. Thus, these results show that completions are able to handle couples of objects that are linked by a certain "relative homology relation".

The main result of the paper is a theorem that makes clear the link between homotopic pairs and dyads. In particular, this theorem indicates that a subset of the completions that describe dyads allows for a complete characterization of homotopic pairs. Thus, we prove that, in the unified framework of completions, it is possible to handle notions relative to both homotopy and homology.

The paper is organized as follows. First, we give some basic definitions for simplicial complexes (Sec. 2). Then, we recall some basic facts relative to the notion of a completion (Sec. 3). We recall the definitions of the completions that describe dendrites and dyads, and we introduce our notion of homotopic pairs (Sec. 4). In the following section, we introduce some tools that are necessary to prove our results (Sec. 5). We establish the theorem that provides a relation between homotopic pairs and dyads in Sec. 6. In Sec. 7, we give a result linking homotopic pairs and the more classical notion of simple homotopy.

The paper is self contained. In particular, almost all proofs are included.

2   Basic Definitions for Simplicial Complexes

Let X be a finite family composed of finite sets. The *simplicial closure of X* is the complex $X^- = \{y \subseteq x \mid x \in X\}$. The family X is a *(finite simplicial) complex* if $X = X^-$. We write $\mathbb{S}$ for the collection of all finite simplicial complexes. Note that $\emptyset \in \mathbb{S}$ and $\{\emptyset\} \in \mathbb{S}$, $\emptyset$ is the *void complex*, and $\{\emptyset\}$ is the *empty complex*.

Let $X \in \mathbb{S}$. An element of X is *a simplex of X* or *a face of X*. A *facet of X* is a simplex of X that is maximal for inclusion.

A *simplicial subcomplex* of $X \in \mathbb{S}$ is any subset Y of X that is a simplicial complex. If Y is a subcomplex of X, we write $Y \preceq X$.

Let $X \in \mathbb{S}$. The *dimension* of $x \in X$, written $dim(x)$, is the number of its elements minus one. The *dimension of X*, written $dim(X)$, is the largest dimension of its simplices, the *dimension of $\emptyset$* being defined to be -1.

A complex $A \in \mathbb{S}$ is *a cell* if $A = \emptyset$ or if A has precisely one non-empty facet x. We write $\mathbb{C}$ for the collection of all cells. A cell $\alpha \in \mathbb{C}$ is *a vertex* if $dim(\alpha) = 0$.

The *ground set* of $X \in \mathbb{S}$ is the set $\underline{X} = \cup\{x \in X \mid dim(x) = 0\}$. We say that $X \in \mathbb{S}$ and $Y \in \mathbb{S}$ are *disjoint*, or that *X is disjoint from Y*, if $\underline{X} \cap \underline{Y} = \emptyset$. Thus, X and Y are disjoint if and only if $X \cap Y = \emptyset$ or $X \cap Y = \{\emptyset\}$.

If $X \in \mathbb{S}$ and $Y \in \mathbb{S}$ are disjoint, the *join of X and Y* is the simplicial complex XY such that $XY = \{x \cup y \mid x \in X, y \in Y\}$. Thus, $XY = \emptyset$ if $Y = \emptyset$ and $XY = X$ if $Y = \{\emptyset\}$. The join αX of a vertex α and a complex $X \in \mathbb{S}$ is a *cone*.

Important convention. In this paper, if $X, Y \in \mathbb{S}$, we implicitly assume that X and Y have disjoint ground sets whenever we write XY.

We recall now some basic definitions related to the collapse operator introduced by J.H.C. Whitehead ([1], see also [16]).

Let $X \in \mathbb{S}$ and let x, y be two distinct faces of X. The couple (x, y) is a *free pair for X* if y is the only face of X that contains x. If (x, y) is a free pair for X, $Y = X \setminus \{x, y\}$ is an *elementary collapse of X* and X is an *elementary expansion of Y*. We say that X *collapses onto Y*, or that Y *expands onto X*, if there exists a sequence $\langle X_0, ..., X_k \rangle$ such that $X_0 = X$, $X_k = Y$, and X_i is an elementary collapse of X_{i-1}, $i \in [1, k]$. The complex X is *collapsible* if X collapses onto $\emptyset$. We say that X is *(simply) homotopic to Y*, or that X and Y are *(simply) homotopic*, if there exists a sequence $\langle X_0, ..., X_k \rangle$ such that $X_0 = X$, $X_k = Y$, and X_i is an elementary collapse or an elementary expansion of X_{i-1}, $i \in [1, k]$. The complex X is *(simply) contractible* if X is simply homotopic to $\emptyset$.

Let $X, Y \in \mathbb{S}$ and let $x, y \in Y \setminus X$. The pair (x, y) is free for $X \cup Y$ if and only if (x, y) is free for Y. Thus, by induction, we have the following proposition.

Proposition 1. *Let $X, Y \in \mathbb{S}$. The complex Y collapses onto $X \cap Y$ if and only if $X \cup Y$ collapses onto X.*

3 Completions

We give some basic definitions for completions, a completion may be seen as a rewriting rule that permits to derive collections of sets. See [14] for more details.

Let $\mathbf{S}$ be a given collection and let $\mathcal{K}$ be an arbitrary subcollection of $\mathbf{S}$. Thus, we have $\mathcal{K} \subseteq \mathbf{S}$. In the sequel of the paper, the symbol $\mathcal{K}$, with possible superscripts, will be a dedicated symbol (a kind of variable).

Let κ be a binary relation on $2^{\mathbf{S}}$, thus $\kappa \subseteq 2^{\mathbf{S}} \times 2^{\mathbf{S}}$. We say that κ is *finitary*, if $\mathbf{F}$ is finite whenever $(\mathbf{F}, \mathbf{G}) \in \kappa$.

Let $\langle K \rangle$ be a property that depends on $\mathcal{K}$. We say that $\langle K \rangle$ is a *completion (on $\mathbf{S}$)* if $\langle K \rangle$ may be expressed as the following property:
$\rightarrow$ If $\mathbf{F} \subseteq \mathcal{K}$, then $\mathbf{G} \subseteq \mathcal{K}$ whenever $(\mathbf{F}, \mathbf{G}) \in \kappa$. $\langle \kappa \rangle$
where κ is an arbitrary finitary binary relation on $2^{\mathbf{S}}$.

If $\langle K \rangle$ is a property that depends on $\mathcal{K}$, we say that a given collection $\mathbf{X} \subseteq \mathbf{S}$ *satisfies $\langle K \rangle$* if the property $\langle K \rangle$ is true for $\mathcal{K} = \mathbf{X}$.

Theorem 1. [14] *Let $\langle K \rangle$ be a completion on $\mathbf{S}$ and let $\mathbf{X} \subseteq \mathbf{S}$. There exists, under the subset ordering, a unique minimal collection that contains $\mathbf{X}$ and that satisfies $\langle K \rangle$.*

If $\langle K \rangle$ is a completion on $\mathbf{S}$ and if $\mathbf{X} \subseteq \mathbf{S}$, we write $\langle \mathbf{X}; K \rangle$ for the unique minimal collection that contains $\mathbf{X}$ and that satisfies $\langle K \rangle$.

Let $\langle K \rangle$ be a completion expressed as the above property $\langle \kappa \rangle$. By a fixed point property, the collection $\langle \mathbf{X}; K \rangle$ may be obtained by starting from $\mathcal{K} = \mathbf{X}$, and by iteratively adding to $\mathcal{K}$ all the sets $\mathbf{G}$ such that $(\mathbf{F}, \mathbf{G}) \in \kappa$ and $\mathbf{F} \subseteq \mathcal{K}$ (see [14]). Thus, if $\mathbf{C} = \langle \mathbf{X}; K \rangle$, then $\langle \mathbf{X}; K \rangle$ may be seen as a dynamic structure that describes $\mathbf{C}$, the completion $\langle K \rangle$ acts as a generator, which, from $\mathbf{X}$, makes it possible to enumerate all elements in $\mathbf{C}$. We will see now that $\langle K \rangle$ may in fact be composed of several completions.

Let $\langle K_1 \rangle, \langle K_2 \rangle, ..., \langle K_k \rangle$ be completions on $\mathbf{S}$. We write $\wedge$ for the logical "and". It may be seen that $\langle K \rangle = \langle K_1 \rangle \wedge \langle K_2 \rangle ... \wedge \langle K_k \rangle$ is a completion. In the sequel, we write $\langle K_1, K_2, ..., K_k \rangle$ for $\langle K \rangle$. Thus, if $\mathbf{X} \subseteq \mathbf{S}$, the notation $\langle \mathbf{X}; K_1, K_2, ..., K_k \rangle$ stands for the smallest collection that contains $\mathbf{X}$ and that satisfies each of the properties $\langle K_1 \rangle, \langle K_2 \rangle, ..., \langle K_k \rangle$.

Remark 1. If $\langle K \rangle$ and $\langle Q \rangle$ are two completions on $\mathbf{S}$, then we have $\langle \mathbf{X}; K \rangle \subseteq \langle \mathbf{X}; K, Q \rangle$ whenever $\mathbf{X} \subseteq \mathbf{S}$. Furthermore, we have $\langle \mathbf{X}; K \rangle = \langle \mathbf{X}; K, Q \rangle$ if and only if the collection $\langle \mathbf{X}; K \rangle$ satisfies the property $\langle Q \rangle$.

4   Completions on Simplicial Complexes

The notion of a dendrite was introduced in [14] as a way for defining a collection made of acyclic complexes. Let us consider the collection $\mathbf{S} = \mathbb{S}$, and let $\mathcal{K}$ denotes an arbitrary collection of simplicial complexes.

We define the two completions $\langle D1 \rangle$ and $\langle D2 \rangle$ on $\mathbb{S}$: For any $S, T \in \mathbb{S}$,

$\rightarrow$ If $S, T \in \mathcal{K}$, then $S \cup T \in \mathcal{K}$ whenever $S \cap T \in \mathcal{K}$. $\langle D1 \rangle$
$\rightarrow$ If $S, T \in \mathcal{K}$, then $S \cap T \in \mathcal{K}$ whenever $S \cup T \in \mathcal{K}$. $\langle D2 \rangle$
Let $\mathbb{D} = \langle \mathbb{C}; D1, D2 \rangle$. Each element of $\mathbb{D}$ is a *dendrite*.

Remark 2. Let κ be the binary relation on $2^{\mathbf{S}}$ such that $(\mathbf{F}, \mathbf{G}) \in \kappa$ iff there exist $S, T \in \mathbb{S}$, with $\mathbf{F} = \{S, T, S \cap T\}$ and $\mathbf{G} = \{S \cup T\}$. We see that κ is finitary and that $\langle D1 \rangle$ may be expressed as the property $\langle \kappa \rangle$ given in the preceding section. Thus $\langle D1 \rangle$ is indeed a completion, and so is $\langle D2 \rangle$.

The collection $\mathbb{T}$ of all trees (*i.e.*, all connected acyclic graphs) provides an example of a collection of dendrites. It may be checked that $\mathbb{T}$ satisfies both $\langle D1 \rangle$ and $\langle D2 \rangle$, and that we have $\mathbb{T} \subseteq \mathbb{D}$. In fact, we have the general result [14]:

A complex is a dendrite if and only if it is acyclic in the sense of homology.
As a consequence, any contractible complex is a dendrite but there exist some dendrites that are not contractible. The punctured Poincaré homology sphere provides an example of this last fact. Note also that each complex in $\langle \mathbb{C}; D1 \rangle$ is contractible [6], but there exist some contractible complexes that are not in $\langle \mathbb{C}; D1 \rangle$. The dunce hat [18] provides an example of this last fact. It follows that it is not possible, using only the two completions $\langle D1 \rangle$ and $\langle D2 \rangle$, to characterize precisely the collection composed of all contractible complexes.

The aim of this paper is to make clear the link between (simple) homotopy and completions. By the previous remarks, we will have to consider other completions

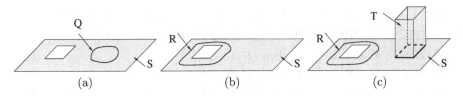

Fig. 1. (a) (Q,S) is not a dyad. (b) (R,S) is a dyad. (c) Suppose (R,S) and $(S\cap T,T)$ are dyads. Then, by $\langle \ddot{Y}1\rangle$, $(S,S\cup T)$ must be a dyad. Furthermore, by $\langle \ddot{T}\rangle$, $(R,S\cup T)$ must also be a dyad.

than the two above ones. To achieve our goal, we will proceed by using some completions that describe *dyads* [15].

Intuitively, a dyad is a couple of complexes (X,Y), with $X \preceq Y$, such that the cycles of X are "at the right place with respect to the ones of Y". For example, the couple (Q,S) of Fig. 1 (a) is not a dyad, while the couple (R,S) of Fig. 1 (b) is a dyad.

We set $\ddot{\mathbb{S}} = \{(X,Y) \mid X,Y \in \mathbb{S}, X \preceq Y\}$ and $\ddot{\mathbb{C}} = \{(A,B) \in \ddot{\mathbb{S}} \mid A,B \in \mathbb{C}\}$.

In the sequel of the paper, $\ddot{\mathcal{K}}$ will denote an arbitrary subcollection of $\ddot{\mathbb{S}}$.

We define five completions on $\ddot{\mathbb{S}}$ (the symbols $\ddot{T}$, $\ddot{U}$, $\ddot{L}$ stand respectively for "transitivity", "upper confluence", and "lower confluence"):

For any $S, T \in \mathbb{S}$,

-> If $(S\cap T,T) \in \ddot{\mathcal{K}}$, then $(S,S\cup T) \in \ddot{\mathcal{K}}$. $\qquad\qquad \langle \ddot{Y}1\rangle$

-> If $(S,S\cup T) \in \ddot{\mathcal{K}}$, then $(S\cap T,T) \in \ddot{\mathcal{K}}$. $\qquad\qquad \langle \ddot{Y}2\rangle$

For any $(R,S),(S,T),(R,T) \in \ddot{\mathbb{S}}$,

-> If $(R,S) \in \ddot{\mathcal{K}}$ and $(S,T) \in \ddot{\mathcal{K}}$, then $(R,T) \in \ddot{\mathcal{K}}$. $\qquad\qquad \langle \ddot{T}\rangle$

-> If $(R,S) \in \ddot{\mathcal{K}}$ and $(R,T) \in \ddot{\mathcal{K}}$, then $(S,T) \in \ddot{\mathcal{K}}$. $\qquad\qquad \langle \ddot{U}\rangle$

-> If $(R,T) \in \ddot{\mathcal{K}}$ and $(S,T) \in \ddot{\mathcal{K}}$, then $(R,S) \in \ddot{\mathcal{K}}$. $\qquad\qquad \langle \ddot{L}\rangle$

We set $\ddot{\mathbb{X}} = \langle \ddot{\mathbb{C}}; \ddot{Y}1, \ddot{Y}2, \ddot{T}, \ddot{U}, \ddot{L}\rangle$. Each element of $\ddot{\mathbb{X}}$ is a *dyad* [1].

See Fig. 1 (c) for an illustration of the completions $\langle \ddot{Y}1\rangle$ and $\langle \ddot{T}\rangle$. In [15], the following relation between dyads and dendrites was given.

Theorem 2. *Let $(X,Y) \in \ddot{\mathbb{S}}$ and let α be a vertex such that $\alpha X \cap Y = X$. The couple (X,Y) is a dyad if and only if $\alpha X \cup Y$ is a dendrite. In particular, a complex $Y \in \mathbb{S}$ is a dendrite if and only if $(\emptyset, Y)$ is a dyad.*

Intuitively, this result indicates that (X,Y) is a dyad if we cancel out all cycles of Y (*i.e.*, we obtain an acyclic complex), whenever we cancel out those of X (by the way of a cone).

In Fig. 1 (b), we see that it is possible to continuously deform R onto S, this deformation keeping R inside S. It follows the idea to introduce the following notions in order to make a link between dyads and simple homotopy.

---

[1] In [15], a different but equivalent definition of a dyad was given. See Th. 2 of [15].

If $X, Y \in \mathbb{S}$, we write $X \xmapsto{E} Y$, whenever Y is an elementary expansion of X. We define four completions on $\mathbb{\ddot{S}}$: For any (R, S), (R, T), (S, T) in $\mathbb{\ddot{S}}$,

-> If $(R, S) \in \ddot{\mathcal{K}}$ and $S \xmapsto{E} T$, then $(R, T) \in \ddot{\mathcal{K}}$. $\qquad \langle \ddot{\mathrm{H}}1 \rangle$

-> If $(R, T) \in \ddot{\mathcal{K}}$ and $S \xmapsto{E} T$, then $(R, S) \in \ddot{\mathcal{K}}$. $\qquad \langle \ddot{\mathrm{H}}2 \rangle$

-> If $(R, T) \in \ddot{\mathcal{K}}$ and $R \xmapsto{E} S$, then $(S, T) \in \ddot{\mathcal{K}}$. $\qquad \langle \ddot{\mathrm{H}}3 \rangle$

-> If $(S, T) \in \ddot{\mathcal{K}}$ and $R \xmapsto{E} S$, then $(R, T) \in \ddot{\mathcal{K}}$. $\qquad \langle \ddot{\mathrm{H}}4 \rangle$

We set $\ddot{\mathbb{I}} = \{(X, X) \mid X \in \mathbb{S}\}$ and $\ddot{\mathbb{H}} = \langle \ddot{\mathbb{I}}; \ddot{\mathrm{H}}1, \ddot{\mathrm{H}}2, \ddot{\mathrm{H}}3, \ddot{\mathrm{H}}4 \rangle$. Each element of $\ddot{\mathbb{H}}$ is a *homotopic pair*.

Note that we have $\langle \ddot{\mathbb{I}}; \ddot{\mathrm{H}}1 \rangle = \langle \ddot{\mathbb{I}}; \ddot{\mathrm{H}}1, \ddot{\mathrm{H}}4 \rangle$. Furthermore, $(X, Y) \in \langle \ddot{\mathbb{I}}; \ddot{\mathrm{H}}1 \rangle$ if and only if Y collapses onto X.

If (X', Y') is obtained from (X, Y) by applying one of the completions $\langle \ddot{\mathrm{H}}1 \rangle$, $\langle \ddot{\mathrm{H}}2 \rangle$, $\langle \ddot{\mathrm{H}}3 \rangle$, $\langle \ddot{\mathrm{H}}4 \rangle$, then X' is homotopic to X, and Y' is homotopic to Y. Since, for generating the collection $\ddot{\mathbb{H}}$, we start from $\ddot{\mathbb{I}}$, we have the following.

Proposition 2. *If $(X, Y) \in \ddot{\mathbb{H}}$, then X is homotopic to Y.*

Observe that, if X is homotopic to Y and if $X \preceq Y$, then we have not necessarily $(X, Y) \in \ddot{\mathbb{H}}$. See the couple (Q, S) of Fig. 1 (a).

5   Product

In this section we give some notions that are essential for the proofs of the main results of this paper. In particular, we introduce the notion of a product of a simplicial complex by a copy of this complex. Intuitively, this product has the structure of a Cartesian product of an object by the unit interval.

Let $Z, Z' \in \mathbb{S}$. We say that Z and Z' are *isomorphic* if there exists a bijection $\lambda : Z \rightarrow Z'$ such that, for all $x, y \in Z$, we have $\lambda(x) \subseteq \lambda(y)$ if and only if $x \subseteq y$. In this case, we also say that Z' is a *copy of* Z, we write $\lambda x = \lambda(x)$, and we set $\lambda X = \{\lambda x \mid x \in X\}$ whenever $X \preceq Z$. Thus, λZ stands for Z'. If $T \preceq Z$, we say that λZ is a *copy of Z with T fixed* if $\lambda T = T$.

In the sequel, we denote by $Cop(Z)$ the collection of all copies of a complex Z, and we denote by $Cop(Z; T)$ the collection of all copies of Z with T fixed.

Let $Z \in \mathbb{S}$ and let $\lambda Z \in Cop(Z)$. If $X \preceq Z$ and $Y \preceq Z$, we note that $\lambda(X \cup Y) = \lambda X \cup \lambda Y$, $\lambda(X \cap Y) = \lambda X \cap \lambda Y$. If $XY \preceq Z$, we also have $\lambda(XY) = \lambda X \lambda Y$.

Remark 3. Let $T, Z \in \mathbb{S}$, with $T \preceq Z$, and let $\lambda Z \in Cop(Z)$. Suppose that Z collapses onto T. Then, the complex λZ collapses onto λT. Nevertheless, if $T \preceq \lambda Z$, then λZ does not necessarily collapses onto T (T may be not "at the right place" w.r.t λZ). Of course if $\lambda Z \in Cop(Z; T)$, then λZ collapses onto T.

Let $X \in \mathbb{S}$ and let $\lambda X \in Cop(X)$ disjoint from X. The *product of X by λX* is the simplicial complex $X \otimes \lambda X$ such that $X \otimes \lambda X = \{x \cup \lambda y \mid x \cup y \in X\}$. Observe that z is a facet of $X \otimes \lambda X$ if and only if there exists a facet x of X such that $z = x \cup \lambda x$. Note also that we have $dim(X \otimes \lambda X) = 2dim(X) + 1$.

Let $Z \in \mathbb{S}$ and $\lambda Z \in Cop(Z)$ disjoint from Z. If $X \preceq Z$, $Y \preceq Z$, then:
- $(X \cup Y) \otimes \lambda(X \cup Y) = (X \otimes \lambda X) \cup (Y \otimes \lambda Y)$; and

- $(X \cap Y) \otimes \lambda(X \cap Y) = (X \otimes \lambda X) \cap (Y \otimes \lambda Y)$.
If $XY \preceq Z$, we have $(XY) \otimes \lambda(XY) = (X \otimes \lambda X)(Y \otimes \lambda Y)$.
If $A \in \mathbb{C}$ and if $A \preceq Z$, then $A \otimes \lambda A = A\lambda A$.

The proofs of the two following propositions will be given in an extended version of this paper.

Proposition 3. *Let* $(X, Y) \in \ddot{\mathbb{S}}$, *let* $\lambda X \in Cop(X)$ *disjoint from* Y, *and let* $Z \preceq X$. *The complex* $(X \otimes \lambda X) \cup Y$ *collapses onto* $(Z \otimes \lambda Z) \cup Y$. *In particular,* $(X \otimes \lambda X) \cup Y$ *collapses onto* Y *and* $(X \otimes \lambda X)$ *collapses onto* X.

Proposition 4. *Let* $(X, Y) \in \ddot{\mathbb{S}}$ *and let* $\lambda X \in Cop(X)$ *disjoint from* Y.
Let $(\lambda X, Z) \in \ddot{\mathbb{S}}$ *such that* Z *is disjoint from* Y. *If* X *collapses onto* X', *then* $(X \otimes \lambda X) \cup Y \cup Z$ *collapses onto* $(X' \otimes \lambda X') \cup Y \cup Z$.

Let $X, Y \in \mathbb{S}$. We say that Y is *independent from* X if a simplex $x \in Y$ is necessarily in X whenever $x \subseteq \underline{X}$. In other words, Y is independent from X if any cell that is included in Y but not in X, contains a vertex that is included in Y but not in X.

Observe that Y is independent from X if and only if $X \cup Y$ is independent from X. Also, a product such that $X \otimes \lambda X$ is independent from X and from λX.

The proof of the following proposition is easy.

Proposition 5. *Let* $X, Y \in \mathbb{S}$ *and* $(X, Z) \in \ddot{\mathbb{S}}$. *If* Z *is independent from* X, *then there exists* $\lambda Z \in Cop(Z; X)$ *such that* $\lambda Z \cap Y = X \cap Y$.

Remark 4. Let $X, Y \in \mathbb{S}$ such that X and Y are disjoint, and let $(X, Z) \in \ddot{\mathbb{S}}$. Then, there exists $\lambda Z \in Cop(Z; X)$ such that λZ and Y are disjoint. In other words, in this particular case, Prop. 5 is satisfied even if the complex Z is not independent from X.

6   Completions and Homotopic Pairs

In this section, we establish a link between dyads and homotopic pairs (Th. 3). For that purpose, we give first the following characterization of $\ddot{\mathbb{H}}$.

If $X, Y \in \mathbb{S}$, we write $X \overset{*E}{\longmapsto} Y$, whenever X expands onto Y.

Proposition 6. *Let* $(X, Y) \in \ddot{\mathbb{S}}$. *We have* $(X, Y) \in \ddot{\mathbb{H}}$ *if and only if there exists a complex* Z *independent from* Y *such that* $X \overset{*E}{\longmapsto} Z$ *and* $Y \overset{*E}{\longmapsto} Z$.

Proof. Let $(X, Y) \in \ddot{\mathbb{S}}$.
i) Suppose $X \overset{*E}{\longmapsto} Z$ and $Y \overset{*E}{\longmapsto} Z$. Then, we may derive (X, Z) from (Z, Z) by repeated applications of $\langle \ddot{H}4 \rangle$ and we may derive (X, Y) from (X, Z) by repeated applications of $\langle \ddot{H}2 \rangle$. Thus, $(X, Y) \in \ddot{\mathbb{H}}$.
ii) We proceed by induction on the four completions that describe $\ddot{\mathbb{H}}$. If $Y = X$, then Y is independent from X, $X \overset{*E}{\longmapsto} Y$, and $Y \overset{*E}{\longmapsto} Y$. Suppose $Y \neq X$ and suppose there exists a complex Z independent from Y such that $X \overset{*E}{\longmapsto} Z$ and

$Y \overset{*E}{\longmapsto} Z$. Thus, we have $\lambda X \overset{*E}{\longmapsto} \lambda Z$ and $\lambda Y \overset{*E}{\longmapsto} \lambda Z$, whenever $\lambda Z \in Cop(Z)$.

1) Let T such that $Y \overset{E}{\longmapsto} T$. Let $\lambda Z \in Cop(Z)$ disjoint from T. We consider the complex $Z' = T \cup (Y \otimes \lambda Y) \cup \lambda Z$, Z' is independent from T. By Prop. 1 and 3, we have:
$$T \overset{*E}{\longmapsto} T \cup (Y \otimes \lambda Y) \overset{*E}{\longmapsto} T \cup (Y \otimes \lambda Y) \cup \lambda Z, \text{ and}$$
$$X \overset{*E}{\longmapsto} (X \otimes \lambda X) \overset{*E}{\longmapsto} (X \otimes \lambda X) \cup \lambda Z \overset{*E}{\longmapsto} (Y \otimes \lambda Y) \cup \lambda Z \overset{*E}{\longmapsto} T \cup (Y \otimes \lambda Y) \cup \lambda Z.$$
Thus, $T \overset{*E}{\longmapsto} Z'$ and $X \overset{*E}{\longmapsto} Z'$.

2) Let T such that $T \overset{E}{\longmapsto} Y$ and $X \preceq T$. Let $\lambda Z \in Cop(Z)$ disjoint from Y. Thus, we have $\lambda T \overset{*E}{\longmapsto} \lambda Y$. We consider the complex $Z' = (T \otimes \lambda T) \cup \lambda Z$, Z' is independent from T. By Prop. 1 and 3, we have:
$$T \overset{*E}{\longmapsto} (T \otimes \lambda T) \overset{*E}{\longmapsto} (T \otimes \lambda T) \cup \lambda Y \overset{*E}{\longmapsto} (T \otimes \lambda T) \cup \lambda Z, \text{ and}$$
$$X \overset{*E}{\longmapsto} (X \otimes \lambda X) \overset{*E}{\longmapsto} (X \otimes \lambda X) \cup \lambda Z \overset{*E}{\longmapsto} (T \otimes \lambda T) \cup \lambda Z.$$
Thus, $T \overset{*E}{\longmapsto} Z'$ and $X \overset{*E}{\longmapsto} Z'$.

3) Let T such that $X \overset{E}{\longmapsto} T$ and $T \preceq Y$. Let $\lambda Z \in Cop(Z)$ disjoint from Y. We consider the complex $Z' = (Y \otimes \lambda Y) \cup \lambda Z$, Z' is independent from Y. By Prop. 1, 3, 4, we have:
$$Y \overset{*E}{\longmapsto} (Y \otimes \lambda Y) \overset{*E}{\longmapsto} (Y \otimes \lambda Y) \cup \lambda Z, \text{ and}$$
$$T \overset{*E}{\longmapsto} T \cup (X \otimes \lambda X) \overset{*E}{\longmapsto} T \cup (X \otimes \lambda X) \cup \lambda Z \overset{*E}{\longmapsto} (T \otimes \lambda T) \cup \lambda Z \overset{*E}{\longmapsto} (Y \otimes \lambda Y) \cup \lambda Z.$$
Thus, $Y \overset{*E}{\longmapsto} Z'$ and $T \overset{*E}{\longmapsto} Z'$.

4) Let T such that $T \overset{E}{\longmapsto} X$. The complex Z is independent from Y, and we have $Y \overset{*E}{\longmapsto} Z$ and $T \overset{*E}{\longmapsto} X \overset{*E}{\longmapsto} Z$.   □

As a direct consequence of Prop. 6, we have the following result. Observe that the expression $\ddot{\mathbb{H}} = \langle \ddot{\mathbb{I}}; \ddot{H}1, \ddot{H}2 \rangle$ means that a pair (X, Y) may be detected as a homotopic pair by using only transformations that keep the complex X fixed.

Proposition 7. *We have* $\ddot{\mathbb{H}} = \langle \ddot{\mathbb{I}}; \ddot{H}1, \ddot{H}2 \rangle = \langle \ddot{\mathbb{I}}; \ddot{H}2, \ddot{H}4 \rangle$.

Proof. Let $\ddot{\mathbb{H}}' = \langle \ddot{\mathbb{I}}; \ddot{H}1, \ddot{H}2 \rangle$ and $\ddot{\mathbb{H}}'' = \langle \ddot{\mathbb{I}}; \ddot{H}2, \ddot{H}4 \rangle$. We have $\ddot{\mathbb{H}}' \subseteq \ddot{\mathbb{H}}$ and $\ddot{\mathbb{H}}'' \subseteq \ddot{\mathbb{H}}$ (see Remark 1). Let $(X, Y) \in \ddot{\mathbb{H}}$. By Prop. 6, there exists Z such that $X \overset{*E}{\longmapsto} Z$ and $Y \overset{*E}{\longmapsto} Z$. We have $(X, X) \in \ddot{\mathbb{H}}'$. Thus, by $\langle \ddot{H}1 \rangle$, $(X, Z) \in \ddot{\mathbb{H}}'$ and, by $\langle \ddot{H}2 \rangle$, $(X, Y) \in \ddot{\mathbb{H}}'$. We also have $(Z, Z) \in \ddot{\mathbb{H}}''$. Thus, by $\langle \ddot{H}4 \rangle$, $(X, Z) \in \ddot{\mathbb{H}}''$ and, by $\langle \ddot{H}2 \rangle$, $(X, Y) \in \ddot{\mathbb{H}}''$. It follows that $\ddot{\mathbb{H}} \subseteq \ddot{\mathbb{H}}'$ and $\ddot{\mathbb{H}} \subseteq \ddot{\mathbb{H}}''$.   □

Lemma 1. *If* $X \in \mathbb{S}$ *and* α *is a vertex, then* $(\emptyset, \alpha X) \in \langle \ddot{\mathbb{C}}; \ddot{Y}1, \ddot{T}, \ddot{U} \rangle$.

Proof. Let $\ddot{\mathbb{H}}' = \langle \ddot{\mathbb{C}}; \ddot{Y}1, \ddot{T}, \ddot{U} \rangle$. If $Card(X) \leq 2$, then αX is a cell. In this case the property is true since $(\emptyset, \alpha X) \in \ddot{\mathbb{C}}$. Let $k \geq 3$. Suppose the property is true whenever $Card(X) < k$ and let X such that $Card(X) = k$. If X has a single facet then, again, αX is a cell and $(\emptyset, \alpha X) \in \ddot{\mathbb{C}}$. If X has more than one facet, then there exists X' and X'' such that $\alpha X = \alpha X' \cup \alpha X''$ with $Card(X') < k$, $Card(X'') < k$, and $Card(X' \cap X'') < k$. By the induction hypothesis, we have $(\emptyset, \alpha X') \in \ddot{\mathbb{H}}'$, $(\emptyset, \alpha X'') \in \ddot{\mathbb{H}}'$, and $(\emptyset, \alpha(X' \cap X'')) \in \ddot{\mathbb{H}}'$. By $\langle \ddot{U} \rangle$, we have

$(\alpha(X' \cap X''), \alpha X'') = (\alpha X' \cap \alpha X'', \alpha X'') \in \ddot{\mathbb{H}}'$. By $\langle \ddot{Y}1 \rangle$, we obtain $(\alpha X', \alpha X) \in \ddot{\mathbb{H}}'$. Now, by $\langle \ddot{T} \rangle$, we conclude that $(\emptyset, \alpha X) \in \ddot{\mathbb{H}}'$. $\square$

Lemma 2. Let $X, Y \in \mathbb{S}$. If $X \overset{E}{\longmapsto} Y$, then $(X, Y) \in \langle \ddot{\mathbb{C}}; \ddot{Y}1, \ddot{T}, \ddot{U} \rangle$.

Proof. If A is a cell, with $A \neq \emptyset$, we set $\partial A = A \setminus \{x\}$, where x is the unique facet of A. Let $\ddot{\mathbb{H}}' = \langle \ddot{\mathbb{C}}; \ddot{Y}1, \ddot{T}, \ddot{U} \rangle$. Suppose $X \overset{E}{\longmapsto} Y$. If $X = \emptyset$, then Y is a vertex, and $(X, Y) \in \ddot{\mathbb{C}}$. Otherwise, there exists a vertex α and a cell $A \in \mathbb{C}$, with $A \neq \emptyset$, such that $Y = X \cup \alpha A$ and $X \cap \alpha A = \alpha \partial A$ (the free pair is $(\underline{A}, \underline{\alpha A})$). By Lemma 1, we have $(\emptyset, \alpha \partial A) \in \ddot{\mathbb{H}}'$ and $(\emptyset, \alpha A) \in \ddot{\mathbb{H}}'$. Thus, by $\langle \ddot{U} \rangle$, we have $(\alpha \partial A, \alpha A) \in \ddot{\mathbb{H}}'$. By $\langle \ddot{Y}1 \rangle$, we obtain $(X, Y) \in \ddot{\mathbb{H}}'$. $\square$

The following theorem shows that four of the five completions that describe dyads allow for a characterization of the collection made of all homotopic pairs.

Theorem 3. We have $\ddot{\mathbb{H}} = \langle \ddot{\mathbb{C}}; \ddot{Y}1, \ddot{T}, \ddot{U}, \ddot{L} \rangle$.

Proof. Let $\ddot{\mathbb{H}}' = \langle \ddot{\mathbb{C}}; \ddot{Y}1, \ddot{T}, \ddot{U}, \ddot{L} \rangle$.

i) Setting $T = \emptyset$ in the definition of $\ddot{Y}1$, we see that $\ddot{\mathbb{I}} \subseteq \ddot{\mathbb{H}}'$. We have $(X, Y) \in \ddot{\mathbb{H}}'$ whenever $X \overset{E}{\longmapsto} Y$ (Lemma 2 and Remark 1). Now, for any $(R, S), (R, T), (S, T)$ in $\ddot{\mathbb{S}}$:
- If $(R, T) \in \ddot{\mathbb{H}}'$ and $S \overset{E}{\longmapsto} T$, then $(S, T) \in \ddot{\mathbb{H}}'$ and, by $\ddot{L}$, we have $(R, S) \in \ddot{\mathbb{H}}'$;
- If $(S, T) \in \ddot{\mathbb{H}}'$ and $R \overset{E}{\longmapsto} S$, then $(R, S) \in \ddot{\mathbb{H}}'$ and, by $\ddot{T}$, we have $(R, T) \in \ddot{\mathbb{H}}'$.
By induction, since $\ddot{\mathbb{H}} = \langle \ddot{\mathbb{I}}; \ddot{H}2, \ddot{H}4 \rangle$ (Prop. 7), it follows that $\ddot{\mathbb{H}} \subseteq \ddot{\mathbb{H}}'$.

ii) If $(X, Y) \in \ddot{\mathbb{C}}$, then it may be checked that Y collapses onto X. Thus $\ddot{\mathbb{C}} \subseteq \ddot{\mathbb{H}}$.
    - Let $S, T \in \mathbb{S}$. Suppose $(S \cap T, T) \in \ddot{\mathbb{H}}$. Thus, there exists K independent from T such that $S \cap T \overset{*E}{\longmapsto} K$ and $T \overset{*E}{\longmapsto} K$ (Prop. 6). Then, there exists a copy $\lambda K \in Cop(K; T)$ such that $\lambda K \cap S = S \cap T$ (Prop. 5). Since $S \cap T \overset{*E}{\longmapsto} \lambda K$, we have $S \overset{*E}{\longmapsto} S \cup \lambda K$ (Prop. 1). We have also $\lambda K \cap (S \cup T) = T$. Thus, since $T \overset{*E}{\longmapsto} \lambda K$, we have $S \cup T \overset{*E}{\longmapsto} S \cup \lambda K$ (Prop. 1). Therefore $(S, S \cup T) \in \ddot{\mathbb{H}}$.
    - Let $(R, S), (S, T), (R, T) \in \ddot{\mathbb{S}}$. Suppose $(R, S) \in \ddot{\mathbb{H}}$ and $(S, T) \in \ddot{\mathbb{H}}$. There exists K such that $S \overset{*E}{\longmapsto} K$, and $T \overset{*E}{\longmapsto} K$ (Prop. 6). By $\langle \ddot{H}1 \rangle$, we have $(R, K) \in \ddot{\mathbb{H}}$. Thus, by $\langle \ddot{H}2 \rangle$ we have $(R, T) \in \ddot{\mathbb{H}}$.
    - Let $(R, S), (S, T), (R, T) \in \ddot{\mathbb{S}}$. Suppose $(R, S) \in \ddot{\mathbb{H}}$ and $(R, T) \in \ddot{\mathbb{H}}$. By Prop. 6, there exists K independent from S such that $R \overset{*E}{\longmapsto} K$, and $S \overset{*E}{\longmapsto} K$. By Prop. 5, there exists $\lambda K \in Cop(K; S)$ such that $\lambda K \cap T = S$. Since $S \overset{*E}{\longmapsto} \lambda K$, we have $T \overset{*E}{\longmapsto} T \cup \lambda K$ (Prop. 1). By $\langle \ddot{H}1 \rangle$, we have $(R, T \cup \lambda K) \in \ddot{\mathbb{H}}$. Since $R \overset{*E}{\longmapsto} \lambda K$, by $\langle \ddot{H}3 \rangle$, we have $(\lambda K, T \cup \lambda K) \in \ddot{\mathbb{H}}$ and, by $\langle \ddot{H}4 \rangle$, $(S, T \cup \lambda K) \in \ddot{\mathbb{H}}$. By $\langle \ddot{H}2 \rangle$, we get $(S, T) \in \ddot{\mathbb{H}}$.
    - Let $(R, S), (S, T), (R, T) \in \ddot{\mathbb{S}}$. Suppose $(R, T) \in \ddot{\mathbb{H}}$ and $(S, T) \in \ddot{\mathbb{H}}$. There exists K such that $S \overset{*E}{\longmapsto} K$, and $T \overset{*E}{\longmapsto} K$ (Prop. 6). By $\langle \ddot{H}1 \rangle$, we have $(R, K) \in \ddot{\mathbb{H}}$. Thus, by $\langle \ddot{H}2 \rangle$, we have $(R, S) \in \ddot{\mathbb{H}}$.

Thus, by induction, we have $\ddot{\mathbb{H}}' \subseteq \ddot{\mathbb{H}}$.   □

By Th. 3, the only difference between the collection $\ddot{\mathbb{X}}$ of dyads and the collection $\ddot{\mathbb{H}}$ of homotopic pairs is the completion $\langle \ddot{Y}2 \rangle$. This difference may be illustrated by the following classical construction. Let P be the punctured Poincaré homology sphere. The complex P is not contractible since the fundamental group of P is not trivial, thus $(\emptyset, P) \notin \ddot{\mathbb{H}}$. Let α and β be two distinct vertices and let $S = \alpha P \cup \beta P$ be a suspension of P. Now the fundamental group of S is trivial and S is contractible. So we have $(\emptyset, S) \in \ddot{\mathbb{H}}$. Since $\ddot{\mathbb{H}} \subseteq \ddot{\mathbb{X}}$, we get $(\emptyset, S) \in \ddot{\mathbb{X}}$. But $(\emptyset, \alpha P) \in \ddot{\mathbb{X}}$ (Prop. 2 of [15]). Thus, by $\langle \ddot{U} \rangle$, it follows that $(\alpha P, S) \in \ddot{\mathbb{X}}$. By $\langle \ddot{Y}2 \rangle$, we deduce that $(\alpha P \cap \beta P, \beta P) \in \ddot{\mathbb{X}}$. We obtain $(P, \beta P) \in \ddot{\mathbb{X}}$. Since $(\emptyset, \beta P) \in \ddot{\mathbb{X}}$, by $\langle \ddot{L} \rangle$, we conclude that $(\emptyset, P) \in \ddot{\mathbb{X}}$, *i.e.*, that P is a dendrite (Th.ñ2).

Remark 5. Let us consider the following completion on $\ddot{\mathbb{S}}$: For any $S, T \in \mathbb{S}$,

-> If $S \xrightarrow{E} T$, then $(S, T) \in \ddot{\mathcal{K}}$. $\langle \ddot{E} \rangle$

Using Th. 3, we can check that we have $\ddot{\mathbb{H}} = \langle \ddot{\mathbb{I}}; \ddot{E}, \ddot{T}, \ddot{U}, \ddot{L} \rangle$.

7   Completions and Simple Homotopy

In the preceding section we have established a link between dyads and homotopic pairs. Here, we will clarify the relation between homotopic pairs and the more classical notion of simple homotopy (Th. 4). For that purpose, we introduce the following relation.

We denote by $\overset{H}{\sim}$ the binary relation on $\mathbb{S}$ such that, for all $X, Y \in \mathbb{S}$, we have $X \overset{H}{\sim} Y$ if and only if:
i) The complexes X and Y are disjoint; and
ii) There exists $K \in \mathbb{S}$ such that $(X, K) \in \ddot{\mathbb{H}}$ and $(Y, K) \in \ddot{\mathbb{H}}$.

For example, if $X \in \mathbb{S}$ and if $\lambda X \in Cop(X)$ is disjoint from X, then, by Prop. 3, we have $X \xrightarrow{*E} X \otimes \lambda X$ and $\lambda X \xrightarrow{*E} X \otimes \lambda X$. Thus, we have $X \overset{H}{\sim} \lambda X$.

Proposition 8. *Let $X, Y \in \mathbb{S}$ be disjoint complexes. We have $X \overset{H}{\sim} Y$ if and only if there exists $Z \in \mathbb{S}$ such that $X \xrightarrow{*E} Z$ and $Y \xrightarrow{*E} Z$.*

Proof. The "if" part is straightforward. Suppose there exists $K \in \mathbb{S}$ such that $(X, K) \in \ddot{\mathbb{H}}$ and $(Y, K) \in \ddot{\mathbb{H}}$. By Prop. 6, there exists Z' such that $X \xrightarrow{*E} Z'$ and $K \xrightarrow{*E} Z'$. By $\langle \ddot{\mathbb{H}}1 \rangle$, we have $(Y, Z') \in \ddot{\mathbb{H}}$. Then, again by Prop. 6, there exists Z such that $Y \xrightarrow{*E} Z$ and $Z' \xrightarrow{*E} Z$. Thus, we have $X \xrightarrow{*E} Z$ and $Y \xrightarrow{*E} Z$.   □

Lemma 3. *Let $X, Y \in \mathbb{S}$. If X and Y are simply homotopic, then there exists $\lambda Y \in Cop(Y)$ disjoint from X, and there exists $K \in \mathbb{S}$ such that $X \xrightarrow{*E} K$ and $\lambda Y \xrightarrow{*E} K$.*

Proof. Let $X, Y \in \mathbb{S}$.
i) If $X = Y$, then there exists $\lambda Y \in Cop(Y)$ disjoint from X. Let $K = X \otimes \lambda Y$.

By Prop. 3, the complex K satisfies the above condition.

ii) Suppose λY and K satisfy the above condition.

- Let X' such that $X \stackrel{E}{\longmapsto} X'$. Let $\mu K \in Cop(K)$ disjoint from K and X'. We have $\mu \lambda Y \preceq \mu K$ and $\mu \lambda Y$ is a copy of Y disjoint from X'. We set $K' = X' \cup (X \otimes \mu X) \cup \mu K$. We have $X' \stackrel{*E}{\longmapsto} X' \cup (X \otimes \mu X) \stackrel{*E}{\longmapsto} X' \cup (X \otimes \mu X) \cup \mu K = K'$, and $\mu \lambda Y \stackrel{*E}{\longmapsto} \mu K \stackrel{*E}{\longmapsto} (X \otimes \mu X) \cup \mu K \stackrel{*E}{\longmapsto} X' \cup (X \otimes \mu X) \cup \mu K = K'$.

- Let X' such that $X' \stackrel{E}{\longmapsto} X$. Let $\mu K \in Cop(K)$ disjoint from K. We have $\mu \lambda Y \preceq \mu K$ and $\mu \lambda Y$ is a copy of Y disjoint from X'. We set $K' = (X' \otimes \mu X') \cup \mu K$. We have $X' \stackrel{*E}{\longmapsto} (X' \otimes \mu X') \stackrel{*E}{\longmapsto} (X' \otimes \mu X') \cup \mu K = K'$, and $\mu \lambda Y \stackrel{*E}{\longmapsto} \mu K \stackrel{*E}{\longmapsto} (X' \otimes \mu X') \cup \mu K = K'$.

The proof is complete by induction on the number of elementary collapses and expansions that allow us to transform X into Y. $\qquad \square$

Theorem 4. *Let $X, Y \in \mathbb{S}$ such that X and Y are disjoint. The complexes X and Y are simply homotopic if and only if $X \stackrel{H}{\sim} Y$.*

Proof. Let $X, Y \in \mathbb{S}$ be disjoint complexes.

i) If $X \stackrel{H}{\sim} Y$, then, by Prop. 8, X and Y are simply homotopic.

ii) Suppose X and Y are simply homotopic. Then, there exists $\lambda Y \in Cop(Y)$ disjoint from X, and there exists $K \in \mathbb{S}$ such that $X \stackrel{*E}{\longmapsto} K$ and $\lambda Y \stackrel{*E}{\longmapsto} K$ (Lemma 3). Furthermore, there exists $\mu K \in Cop(K; X)$ disjoint from Y (see Remark 4). Thus, $X \stackrel{*E}{\longmapsto} \mu K$ and $\mu \lambda Y \stackrel{*E}{\longmapsto} \mu K$. Let $K' = \mu K \cup (Y \otimes \mu \lambda Y)$. We have $X \stackrel{*E}{\longmapsto} \mu K \stackrel{*E}{\longmapsto} K'$ and $Y \stackrel{*E}{\longmapsto} Y \otimes \mu \lambda Y \stackrel{*E}{\longmapsto} K'$. It follows that $(X, K') \in \ddot{\mathbb{H}}$ and $(Y, K') \in \ddot{\mathbb{H}}$. Therefore $X \stackrel{H}{\sim} Y$. $\qquad \square$

Since simple homotopy corresponds to a transitive relation, the following is a corollary of Th. 4.

Corollary 1. *Let $X, Y, Z \in \mathbb{S}$ be three mutually disjoint complexes. If $X \stackrel{H}{\sim} Y$ and $Y \stackrel{H}{\sim} Z$, then $X \stackrel{H}{\sim} Z$.*

If $X \in \mathbb{S}$ and if $\lambda X \in Cop(X)$, then X and λX are simply homotopic. Thus, we also have the following immediate consequence of Th. 4 and Prop. 8.

Corollary 2. *Let $X, Y \in \mathbb{S}$ and let $\lambda Y \in Cop(Y)$ disjoint from X. The complexes X and Y are simply homotopic if and only if there exists $K \in \mathbb{S}$ such that $X \stackrel{*E}{\longmapsto} K$ and $\lambda Y \stackrel{*E}{\longmapsto} K$.*

Remark 6. In [1], the following result was given (see Th. 4 of [1]):
Let $X, Y \in \mathbb{S}$. If X and Y are simply homotopic, then there exists $K \in \mathbb{S}$ and there exists a stellar sub-division $\widetilde{Y}$ of Y, such that $X \stackrel{*E}{\longmapsto} K$ and $\widetilde{Y} \stackrel{*E}{\longmapsto} K$.
Cor. 2 shows that we can have the same relationship between two homotopic complexes without involving sub-divisions, which change the structure of a complex. Only the notion of a copy is necessary. Observe that this result has been made possible thanks to the notion of a product, this construction allows us to have "more room" to perform homotopic transforms.

8 Conclusion

We proposed an extension of simple homotopy by considering homotopic pairs. The notion of a homotopic pair was formalized by means of completions. One of the main results of the paper (Th. 3) shows that a subset of the five completions that describe dyads allows for a complete characterization of homotopic pairs. Since dyads are linked to homology, we have a unified framework where a link between some notions relative to homotopy and some notions relative to homology may be expressed. It should be noted that such a link is not obvious in the classical framework [19].

In the future, we will further investigate the possibility to use completions for deriving results related to combinatorial topology.

References

1. Whitehead, J.H.C.: Simplicial spaces, nuclei, and m-groups. Proc. London Math. Soc. (2) 45, 243–327 (1939)
2. Björner, A.: Topological methods. In: Graham, R., Grötschel, M., Lovász, L. (eds.) Handbook of Combinatorics, pp. 1819–1872. North-Holland, Amsterdam (1995)
3. Hachimori, M.: Nonconstructible simplicial balls and a way of testing constructibility. Discrete Comp. Geom. 22, 223–230 (1999)
4. Kahn, J., Saks, M., Sturtevant, D.: A topological approach to evasiveness. Combinatorica 4, 297–306 (1984)
5. Welker, V.: Constructions preserving evasiveness and collapsibility. Discrete Math. 207, 243–255 (1999)
6. Jonsson, J.: Simplicial Complexes of Graphs. Springer (2008)
7. Kalai, G.: Enumeration of Q-acyclic simplicial complexes. Israel Journal of Mathematics 45(4), 337–351 (1983)
8. Kong, T.Y.: Topology-preserving deletion of 1's from 2-, 3- and 4-dimensional binary images. In: Ahronovitz, E., Fiorio, C. (eds.) DGCI 1997. LNCS, vol. 1347, pp. 3–18. Springer, Heidelberg (1997)
9. Couprie, M., Bertrand, G.: New characterizations of simple points in 2D, 3D and 4D discrete spaces. IEEE Transactions on Pattern Analysis and Machine Intelligence 31(4), 637–648 (2009)
10. Bertrand, G.: On critical kernels. Comptes Rendus de l'Académie des Sciences, Série Math. (345), 363–367 (2007)
11. Rosenfeld, A.: Digital topology. Amer. Math. Monthly, 621–630 (1979)
12. Kovalevsky, V.: Finite topology as applied to image analysis. Comp. Vision Graphics, and Im. Proc. 46, 141–161 (1989)
13. Kong, T.Y., Rosenfeld, A.: Digital topology: introduction and survey. Comp. Vision, Graphics and Image Proc. 48, 357–393 (1989)
14. Bertrand, G.: Completions and simplicial complexes, HAL-00761162 (2012)
15. Bertrand, G.: New structures based on completions. In: Gonzalez-Diaz, R., Jimenez, M.-J., Medrano, B. (eds.) DGCI 2013. LNCS, vol. 7749, pp. 83–94. Springer, Heidelberg (2013)
16. Giblin, P.: Graphs, surfaces and homology. Chapman and Hall (1981)
17. Bing, R.H.: Some aspects of the topology of 3-manifolds related to the Poincaré Conjecture. Lectures on Modern Mathematics II, pp. 93–128. Wiley (1964)
18. Zeeman, E.C.: On the dunce hat. Topology 2, 341–358 (1964)
19. Hatcher, A.: Algebraic topology. Cambridge University Press (2001)

2D Subquadratic Separable Distance Transformation for Path-Based Norms[*]

David Coeurjolly

CNRS, LIRIS, UMR5205, F-69621, France

Abstract. In many applications, separable algorithms have demonstrated their efficiency to perform high performance and parallel volumetric computations, such as distance transformation or medial axis extraction. In the literature, several authors have discussed about conditions on the metric to be considered in a separable approach. In this article, we present generic separable algorithms to efficiently compute Voronoi maps and distance transformations for a large class of metrics. Focusing on path based norms (chamfer masks, neighborhood sequences, ...), we detail a subquadratic algorithm to compute such volumetric transformation in dimension 2. More precisely, we describe a $O(\log^2 m \cdot N^2)$ algorithm for shapes in a $N \times N$ domain with chamfer norm of size m.

Keywords: Digital Geometry, Distance Transformation, Path-based Norms.

1   Introduction

Since early works on digital geometry, distance transformation has been playing an important role in many applications [16,15]. Given a finite input shape $X \subset \mathbb{Z}^n$, the distance transformation labels each point in X with the distance to its closest point in $\mathbb{Z}^n \setminus X$. Labeling each point by the closest background point leads to Voronoi maps. Since such characterization is parametrized by a distance function, many authors address this distance transformation problem with trade-offs between algorithmic performances and the *accuracy* of the digital distance function with respect to the Euclidean one. Hence, authors have considered distances based on chamfer masks [15,2,6] or sequences of chamfer masks [16,11,17,13]; the vector displacement based Euclidean distance [5,14]; Voronoi diagram based Euclidean distance [3,9] or the square of the Euclidean distance [7,10]. For the Euclidean metric, separable volumetric computations have demonstrated to be very efficient: optimal $O(n \cdot N^n)$ time algorithms for shapes in N^n domains, optimal multi-thread/GPU implementation... (please refer to [4] for a discussion). For path-based metrics (chamfer mask, -weighted- neighborhood sequences,...), two main techniques exist to compute the distance transformation. The first one considers a weighted graph formulation of the problem and

---

[*] This work has been mainly funded by ANR-11-BS02-009 and ANR-11-IDEX-0007-02 PALSE/2013/21 research grants.

E. Barcucci et al. (Eds.): DGCI 2014, LNCS 8668, pp. 75–87, 2014.
© Springer International Publishing Switzerland 2014

Dijkstra-like algorithms on weighted graphs to compute distances. If m denotes the size of the chamfer mask, computational cost could be in $O(m \cdot N^n)$ using a cyclic bucket data structure [19]. Another approach consists in a raster scan of the domain: first the chamfer mask is decomposed into disjoint sub-masks. Then the domain grid points are scanned in a given order (consistent with the sub-mask construction) and a local computation is performed before being propagated [16,2]. Scanning the domain several times (one per sub-mask) leads to the distance transformation values. Again, we end up with a $O(m \cdot N^n)$ computational cost.

Beside specific applications which use the anisotropic nature of the chamfer mask, rotational dependency is usually ensured increasing the mask size m together with optimizing weights. In this context and for arbitrarily large N, both Dijsktra-like and raster scan approaches have a quadratic computational cost with respect to N and m. In practical situations, $m \ll N$ but m still needs to be in $O(N^n)$ to have accurate asymptotic DT.

Please note that Dijkstra's graph approach allows us to defined constrained distance transformation (*i.e.* geodesic metric), both separable approaches and raster-scan for path-based metrics are only dedicated to compact convex domains (usually hyper-rectangular) distance transformation.

Contributions. In this article, we first describe the generic framework for separable distance transformation and metric conditions to be consistent with this model. Then, we describe subquadratic and parallel algorithms in dimension 2 to compute error-free distance transformation and Voronoi map for chamfer norms and other path-based metrics. Overall computational costs can be summarized as follows (see 3.2 for predicate definitions):

Metric	Closest	HiddenBy	Sep. Voronoi Map	Reference
L_2	$O(1)$	$O(1)$	$\Theta(n \cdot N^n)$	[7]
L_∞	$O(1)$	$O(1)$	$\Theta(n \cdot N^n)$	[10]
L_1	$O(1)$	$O(1)$	$\Theta(n \cdot N^n)$	[10]
L_p (exact pred.)	$O(\log p)$	$O(\log p \cdot \log N)$	$O(n \cdot N^n \cdot \log p \cdot \log N)$	Lem. 1, [1]
L_p (inexact pred.)	$O(1)$	$O(\log N)$	$O(n \cdot N^n \cdot \log N)$	Lem. 1, [1]
2D Chamfer norm	$O(\log m)$	$O(\log^2 m)$	$O(\log^2 m \cdot N^2)$	Theorem 1
2D Neig. seq. norm	$O(\log f)$	$O(\log^2 f)$	$O(\log^2 f \cdot N^2)$	Th. 1 and [13]

2   Preliminaries

Definition 1 (Norm and metric induced by a norm). *Given a vector space EV, a norm is a map g from EV to a sub-group F of $\mathbb{R}$ such that $\forall \boldsymbol{x}, \boldsymbol{y} \in EV$,*

$$(non\text{-}negative) \quad g(\boldsymbol{x}) \geq 0 \tag{1}$$

$$(identity\ of\ indiscernibles) \quad g(\boldsymbol{x}) = 0 \Leftrightarrow \boldsymbol{x} = \boldsymbol{0} \tag{2}$$

$$(triangular\ inequality) \quad g(\boldsymbol{x} + \boldsymbol{y}) \leq g(\boldsymbol{x}) + g(\boldsymbol{y}) \tag{3}$$

$$(homogeneity) \quad \forall \lambda \in \mathbb{R}, \quad g(\lambda \cdot \boldsymbol{x}) = |\lambda| \cdot g(\cdot \boldsymbol{x}) \tag{4}$$

$d(a, b) := g(b - a)$ is the metric induced by the norm g. (E, F, d) is called a metric space if $d : E \rightarrow F$ (with E such that for $a, b \in E$, $(b - a) \in EV$).

Note that the above definition can be extended from vector spaces to *modules* on a commutative ring ($\mathbb{Z}^n$ being a module on $\mathbb{Z}$ but not a vector space) [18]. Path-based approaches (chamfer masks, -weighted- neighborhood sequences...) aim at defining *digital* metrics induced by norms in metric spaces $(\mathbb{Z}^n, \mathbb{Z}, d)$. Note that (weighted, with $w_i \geq 0$) L_p metrics

$$d_{L_p}(a, b) = \left(\sum_{k=1}^{n} w_k |a_k - b_k|^p \right)^{\frac{1}{p}},\qquad(5)$$

define metric spaces $(\mathbb{Z}^n, \mathbb{R}, d_{L_p})$ which are not *digital*. However, rounding up the distance function $(\mathbb{Z}^n, \mathbb{Z}, \lceil d_{L_p} \rceil)$ is a digital metric space [8].

Definition 2 (Distance Transformation and Voronoi Map). *The distance transform DT_X associated with a digital metric space $(\mathbb{Z}^n, \mathbb{Z}, d)$ is a map $X \rightarrow \mathbb{Z}$ such that, for $a \in X$ $DT_X(a) = \min_{b \in \mathbb{Z}^n \backslash X} \{d(a, b)\}$. The Voronoi map is the map $X \rightarrow \mathbb{Z}^n$: $\Pi_X(a) = \arg\min_{b \in \mathbb{Z}^n \backslash X} \{d(a, b)\}$.*

Voronoi map Π_X corresponds to the intersection between the continuous Voronoi diagram for the metric d of points $\mathbb{Z}^n \setminus X$ and the lattice $\mathbb{Z}^n$. If a digital point a belongs to a Voronoi diagram d−facet ($0 \leq d < n$), a is equidistant to 2 or more points in $\mathbb{Z}^n \setminus X$ but only one is considered in $\Pi_X(a)$ this choice has no influence on DT_X.

Definition 3 (Chamfer Mask). *A weighted vector is a pair $(\boldsymbol{v}, w)$ with $\boldsymbol{v} \in \mathbb{Z}^n$ and $w \in \mathbb{N}^*$. A chamfer mask $\mathcal{M}$ is a central-symmetric set of weighted vectors with no null vectors and containing at least a basis of $\mathbb{Z}^n$.*

Many authors have proposed algorithmic and/or analytic approaches to construct chamfer masks approximating the Euclidean metric. In the following, we focus on such *chamfer norms* which are chamfer metric induced by a norm. To evaluate distances between two digital points for a given chamfer metric, direct formulations have been proposed with simple geometrical interpretation:

Definition 4 (Rational ball, minimal H-representation [18,12]). *Given a Chamfer norm $\mathcal{M}$, the rational ball associated with $\mathcal{M}$ is the polytope*

$$\mathcal{B}_R = conv \left\{ \frac{\boldsymbol{v}_k}{w_k}; \ (\boldsymbol{v}_k, w_k) \in \mathcal{M} \right\}.\qquad(6)$$

Rational ball $\mathcal{B}_R$ can also be described as the H-representation of polytope with minimal parameter [13]: $P = \{x \in \mathbb{Z}^n; Ax \leq y\}$ such that $\forall k \in [1 \dots f]$, $\exists x \in P$ $A_k x = y_k$.[1] f is the number of rows in A and the number of facets in $\mathcal{B}_R$, and is thus related to $|\mathcal{M}|$.

---

[1] A_k being the kth row of A.

An important result for distance computation can be formalized as follows:

Proposition 1 (Direct Distance Computation [12]). *Given a chamfer mask $\mathcal{M}$ induced by a norm and (A, y) its minimal parameter H-representation, then for any $a \in \mathbb{Z}^n$, the chamfer distance of a from the origin is*

$$d_{\mathcal{M}}(O, a) = \max_{1 \leq k \leq f} \{A_k a^T\}. \tag{7}$$

Among path-based digital metric, (weighted) neighborhood sequences have been proposed to have better approximation of the Euclidean metric from sequences of elementary chamfer masks [16,11,17,13]. A key result have been demonstrated in [13] stating that for such distance functions, a minimal parameter polytope representation exists and that distances can be obtained from a expression similar to (7):

$$d(O, a) = \max_{1 \leq k \leq f} \{f_k(A_k a^T)\}, \tag{8}$$

f_k being some integer sequence characterizing the neighborhood sequence metric. In the following and for the sake of simplicity, we describe our algorithms focusing on chamfer norms but similar results can be obtained for more generic path-based metric such as neighborhood sequences.

3 Separable Distance Transformation

3.1 Voronoi Map from Separable Approach and Metric Conditions

In [7,3,10,9], several authors have described optimal in time and separable techniques to compute error-free Voronoi maps or distance transformations for L_2 and L_p metrics. Separability means that computations are performed dimension by dimension. In the following, we consider *Voronoi Map* approach as defined in [3]: Let us first define an hyper-rectangular image $I_X : [1..N_1] \times \ldots \times [1..N_n] \to \{0,1\}$ such that $I_X(a) = 1$ for $a \in [1..N_1] \times \ldots \times [1..N_n]$ iff $a \in X$ ($I_X(a) = 0$ otherwise). In dimension 2, each row of the input image are processed to create independent 1D Voronoi maps along the first dimension for the metric. Then, for each further dimension, the partial Voronoi map Π_X is updated using one dimensional independent processes on image spans along the ith dimension. Algorithm 1 describes the 1D processes to perform on each row, column and higher dimensional image span[2]. In this process, metric information are embedded in the following predicates (see Fig. 1): CLOSEST(a, b, c), given three points $a, b, c \in \mathbb{Z}^n$ this predicate returns true if $d(a, b) < d(a, c)$. HIDDENBY(a, b, c, S), given three points $a, b, c \in \mathbb{Z}^n$ such that $a_i < b_i < c_i$[3] and a 1D image span S, this predicates returns true if there is no $s \in S$ such that

$$d(b, s) < d(a, s) \text{ and } d(b, s) < d(c, s). \tag{9}$$

---

[2] An image span S along the ith direction is a vector of N_i points with same coordinates except at their i^{th} one.

[3] Subscript a_i denotes the ith coordinate of point $a \in \mathbb{Z}^n$.

Algorithm 1. Voronoi map construction on 1D image span S along the ith dimension.

Data: Input binary map I_X if $i = 1$ or partial Voronoi map Π_X obtained for dimensions lower than i, and a 1D span S with points $\{s^1, \ldots, s^{N_i}\}$ sorted by their i^{th} coordinate.

Result: Updated partial Voronoi map Π_X along S.

1   **if** $i == 1$;   // Special case for the first dimension

2   **then**

3      $k = 0$;

4   **foreach** *point s in S* **do**

5   **if** $I_X(s) == 1$ **then**

6             $L_S[k] = s$;

7             $k + +$;

8   **else**

9      $L_S[0] = \Pi_X(s^1)$;

10     $L_S[1] = \Pi_X(s^2)$;

11     $k = 2$, $l = 3$;

12   **while** $l \leq N_i$ **do**

13        $w = \Pi_X(s^l)$;

14   **while** $k \geq 2$ *and* HIDDENBY$(L_S[k - 1], L_S[k], w, S)$ **do**

15           $k - -$;

16        $k + +$; $l + +$;

17        $L_S[k] = w$;

18   **foreach** *point s in S by increasing i^{th} coordinate* **do**

19   **while** *($k < |L_S|$) and not(*CLOSEST$(s,\ L_S[k],\ L_S[k + 1])$*)* **do**

20        $k + +$;

21     $\Pi_X[s] = L_S[k]$;

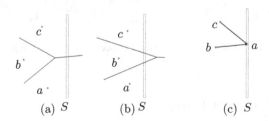

Fig. 1. Geometrical predicates for Voronoi map construction: HIDDENBY(a, b, c, S) returns true in (a) and false in (b) (straight segments correspond to Voronoi diagram edges). (c) illustrates the CLOSEST(a, b, c) predicate for $c \in S$.

In other words, HIDDENBY returns true if and only if the Voronoi cells of sites a and c *hide* the Voronoi cell of b along S. For L_1, L_2 and L_∞ metrics, CLOSEST and HIDDENBY predicates can be computed in $O(1)$ [3,7,10]. Hence, Algorithm 1 is in $O(N_i)$ for dimension i, leading to an overall computational time for the

Voronoi map and distance transformation problem in $\Theta(n \cdot N^n)$ (if we assume that $\forall i \in [1 \ldots n], N_i = N$).

In [7] or [9], authors discussed about conditions on the metric d to ensure that Algorithm 1 is correct. The key property can be informally describe as follows: given two points $a, b \in \mathbb{Z}^n$ such that $a_i < b_i$ and a straight line l along the ith direction and if we denote by $v_l(a)$ (resp. $v_l(b)$) the intersection between the Voronoi cell of a (resp. b) and l, then $v_l(a)$ and $v_l(b)$ are simply connected Euclidean segments and $v_l(a)$ appears *before* $v_l(b)$ on l (so called *monotonicity property* in [9] and is related to *quadrangle inequality* in [7]). To sum up these contributions, we have the following sufficient conditions on the metric:

Proposition 2 (Metric conditions [7]). *Let d be a metric induced by a **norm whose unit ball is symmetric with respect to grid axes** and if distance comparison predicate is exact, Algorithm 1 is correct and returns a Voronoi map Π_X.*

When implementing Algorithm 1, the distance comparison predicate is exact if we can compare two distances, *e.g.* CLOSEST predicate, without error.

For L_p norms Algorithm 1 provides exact Voronoi map computation. Indeed, distance comparison predicate can be error-free implemented from integer number comparisons considering the p power of the distance function $\left(d_{L_p}\right)^p$.

Proposition 2 also implies that most chamfer norms and neighborhood sequence based norms can also be considered in separable Algorithm 1 (see Fig. 2). We just need algorithmic tools to efficiently implement both CLOSEST and HIDDENBY predicates.

Fig. 2. Balls for different metrics satisfying Proposition 2: *(from left to right)* L_1, $L_{1.4}$, L_2, L_4, $L_{43.1}$, $\mathcal{M}_{3-4}$ and $\mathcal{M}_{5-7-11}$

3.2   A First Generic Adapter

We first detail the overall computational cost of Algorithm 1:

Lemma 1. *Let $(\mathbb{Z}^n, F, d)$ be a metric space induced by a norm with axis symmetric unit ball. If C denotes the computational cost of CLOSEST predicate and H is the computational cost of the HIDDENBY predicate, then Algorithm 1 is in $O(n \cdot N^n \cdot (C + H))$.*

From [3,10], $C = H = O(1)$ for L_1, L_2 and L_∞ norms. For given norm d, we first define generic Algorithms 2, 3 and 4: From some evaluations of d, HIDDENBY predicate is obtained by a dichotomic search on the 1D image span S to localize the abscissa of Voronoi edges of sites $\{a, b\}$ and $\{b, c\}$ (see Fig. 3).

Algorithm 2. Generic CLOSEST($a, b, c \in \mathbb{Z}^n$).

---

1 **return** $d(a, b) < d(a, c)$;

Algorithm 3. Generic VORONOIEDGE($a, b, s^i, s^j \in \mathbb{Z}^n$), $a_i < b_i$.

---

1 **if** $(j - i = 1)$ **then**
2   **if** $i = 1$ *and* CLOSEST(s^i, b, a) **then**
3         $\lfloor$ **return** $-\infty$;
4   **if** $i = N_i$ *and* CLOSEST(s^i, a, b) **then**
5         $\lfloor$ **return** ∞;
6   **return** i;
7 $mid = i + (j - i)/2$;
8 **if** CLOSEST(s^{mid}, a, b) **then**
     // s_{mid} closer to a
9   **return** VORONOIEDGE(a, b, s^{mid}, s^j)
10 **else**
     // s_{mid} closer to b
11   **return** VORONOIEDGE(a, b, s^i, s^{mid})

Algorithm 4. Generic HIDDENBY($a, b, c \in \mathbb{Z}^n$; S in the ith direction) $a_i < b_i < c_i$.

---

1 $v_{ab} =$ VORONOIEDGE(a, b, s^1, s^{N_i});
2 $v_{bc} =$ VORONOIEDGE(b, c, s^1, s^{N_i});
3 **return** $(v_{ab} > v_{bc})$;

Lemma 2. *Let $(\mathbb{Z}^n, F, d)$ be a metric space induced by a norm with axis symmetric unit ball, from Algorithms 2 and 4, we have $H = O(C \cdot \log N)$.*

Proof. First, the generic VORONOIEDGE is dichotomic with $O(\log N)$ steps and each step is in $O(C)$. VORONOIEDGE is in $O(C \cdot \log N)$. To prove the correctness of VORONOIEDGE (and thus Alg. 1), we use the convexity of the metric and the quadrangle property: since $a_i < b_i$, all grid points closer to a than b in S (if exist) are *lower* than all pixels on S closer to b than to a (if exist too). Thanks to the test in line 8, recursive call maintain this invariant. Note that tests on lines $2 - 5$ handle the fact that the edge may not belong to S. □

If we consider a chamfer norm with a rational ball of f facets, Eq. (7) suggests that $C = O(f)$. Hence, we have the following corollary:

Corollary 1. *Let $\mathcal{M}$ be a chamfer norm whose rational ball has f facets, separable exact Voronoi map Π_X can be obtained in $O(n \cdot N^n \cdot f \cdot \log N)$.*

Please remember that naive implementation of chamfer mask distance transformation using raster scan approach would lead to a $O(f \cdot N^n)$ computational cost. In the following sections, we use the convex structure of $\mathcal{B}_R$ to design a parallel subquadratic algorithm for chamfer norms.

3.3   Subquadratic Algorithm in Dimension 2

Let us consider a 2D chamfer norm $\mathcal{M}$ with m weighted vectors (note that $f := |\mathcal{B}_R| = m$ in 2D). We suppose that vectors $\{v^k\}_{k=1\ldots m}$ are sorted counter-clockwise. We define a wedge as a pair (v^k, v^{k+1}) of vectors. To each wedge is associated a row A_k in the minimal H-representation of A (A_k can also be seen as a −non-unitary− normal vector to $\mathcal{B}_R$ facets [12]). Using similar notations, [18,17] demonstrate that the distance evaluation of point a can be obtained in two steps: First, we compute the wedge (v^k, v^{k+1}) a belongs to. Then,

$$d_\mathcal{M}(O, a) = A_k \cdot a^T \tag{10}$$

Lemma 3. *Given a chamfer norm $\mathcal{M}$ in dimension 2 with m vectors, the distance computation and thus the* CLOSEST *predicate are in $O(\log m)$.*

Proof. Since vectors are sorted counter-clockwise, (v^k, v^{k+1}) wedge can be obtained by a dichotomic search with $O(\log m)$ steps. At each step, we compute the local orientation of point a w.r.t. a direction which is in $O(1)$. Once the wedge has been obtained, Eq. (10) returns the distance value in $O(1)$.   □

Please note that in practical implementations, we can use symmetries in $\mathcal{M}$ to only work on restrictions of chamfer mask directions, so called *generator* in the literature. To optimize the HIDDENBY predicate, we focus on the VORONOIEDGE function. Given two points a and b ($a_i < b_i$) and a 1D image span S along the ith dimension, we have to find the lowest abscissa e_i of the point e on S such that $d(a, e) < d(b, e)$ and $d(a, e') \geq d(b, e')$ for any e' with $e'_i > e_i$. Let us first suppose that we do not know e but we know the wedge (v^k, v^{k+1}) (resp. (v^j, v^{j+1})) to which the vector $(e - a)^T$ (resp. $(e - b)^T$) belongs to (see Fig. 3−(c)). In this situation, we know that e is the solution of

$$A_k \cdot (e - a)^T = A_j \cdot (e - b)^T . \tag{11}$$

(since $e \in S$, we have one linear equation with only one unknown e_i). As a consequence, if we know the two wedges the Voronoi edge belongs to, we have the abscissa in $O(1)$ (see Algorithm 5 and Fig. 3−(c)). To obtain both wedges, we use a dichotomic search similar to Algorithm 3: Algorithm 6 returns the wedge associated with a containing the Voronoi edge with respect to b. Applying this algorithm to obtain the wedge associated with b with respect to a defines Algorithm 5. The dichotomic search shrinks the set of vectors

---

Algorithm 5. 2D chamfer norm VORONOIEDGE$(a, b, s^i, s^j \in \mathbb{Z}^2)$.

---

1 (v^k, v^{k+1}) = VORONOIEDGEWEDGE(a, b, v^1, v^m, S);
2 (v^j, v^{j+1}) = VORONOIEDGEWEDGE(b, a, v^1, v^m, S);
3 Compute the abscissa e_i of the point e such that $A_k \cdot (e - a)^T = A_j \cdot (e - b)^T$;
4 **return** e_i;

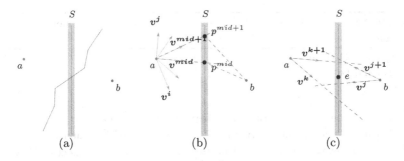

Fig. 3. VORONOIEDGEWEDGE and VORONOIEDGE: (a) initial problem, we want to compute the intersection between S and the Voronoi edge of a and b (in red). (b) an internal step of VORONOIEDGEWEDGE to reduce the set of directions of $\mathcal{M}$ at a (here the next recursive call will be on (v^i, v^{mid})).(c) final step of VORONOIEDGE where both wedges have been obtained and thus e can be computed.

$\{v^i, \ldots, v^j\}$ to end up with a wedge (v^k, v^{k+1}) such that the intersection point between the straight line $(a + v^k)$ and S is in the Voronoi cell of a and such that the intersection between $(a + v^{k+1})$ and S is in the Voronoi cell of b (see Fig. 3−(c)). Algorithm 6 thus first computes the intersection points associated with a wedge $(v^{i+(j-i)/2}, v^{i+(j-i)/2+1})$ (lines 5 − 6); evaluates the distances at these points (lines 7 − 8) and then decides which set $\{v^i, \ldots, v^{i+(j-i)/2}\}$ or $\{v^{i+(j-i)/2}, \ldots, v^j\}$ has to be considered for the recursive call (lines 14 − 20 and Fig. 3−(b)).

Theorem 1. *Let $\mathcal{M}$ be a 2D chamfer norm with axis symmetric unit ball and m weighted vectors, then we have: (i) Algorithm 5 is in $O(\log^2 m)$; (ii) Algorithm 1 (with predicates from Algorithm 5 and Lemma 3), computes a Voronoi map Π_X and thus the distance transformation of X for metric $d_{\mathcal{M}}$ in $O(\log^2 m \cdot N^2)$.*

Proof. Let us first consider (i). As described above, Algorithm 6 performs $\log m$ recursive calls and each step is in $O(\log m)$. Indeed, p^{mid} and p^{mid+1} are given by the intersections between two rational slope straight lines plus a rounding operations on rational fractions, which are assumed to be $O(1)$. Then, line 8 requires two $O(\log m)$ computations by Lemma 3. Hence, Eq. (11) leads to $O(1)$ computations, $O(\log^2 m)$ is required for Algorithm 5. (ii) is a direct consequence of (i) and Lemma 2 with $n = 2$. $\qquad\square$

4 Implementation Details and Experimental Analysis

In this section, we give some implementation details and experimental results for chamfer norm Voronoi map in dimension 2. First of all, most algorithms presented here are available in the DGTAL library [1]. For L_p metrics, we have implemented several CLOSEST and HIDDENBY predicates: If $p = \{1, 2\}$, exact computations are proposed and all predicates are in $O(1)$ with only integer number computations [7,9,10]; If $p \in \mathbb{R}$, $p \geq 1$, we have approximated computations

---

Algorithm 6. VORONOIEDGEWEDGE($a, b \in \mathbb{Z}^2; \boldsymbol{v}^i, \boldsymbol{v}^j$ in $\mathcal{M}$; S along the ith direction).

---

1 **if** $(j - i = 1)$ **then**
2 $\quad$ **return** $(\boldsymbol{v}^i, \boldsymbol{v}^{i+1})$;
3 **else**
4 $\quad$ $mid = i + (j - i)/2$;
5 $\quad$ Let p^{mid} be the intersection point between $(a + \boldsymbol{v}^{mid})$ and S;
6 $\quad$ Let p^{mid+1} be the intersection point between $(a + \boldsymbol{v}^{mid+1})$ and S;
$\quad$ // O(1) evaluation of distances w.r.t. a $d^a_{p^{mid}} = A_{mid} \cdot (p^{mid} - e)^T$;
7 $\quad$ $d^a_{p^{mid+1}} = A_{mid+1} \cdot (p^{mid+1} - e)^T$;
$\quad$ // O(log m) evaluation of distances w.r.t. b $d^b_{p^{mid}} = d_{\mathcal{M}}(b, p^{mid})$;
8 $\quad$ $d^b_{p^{mid+1}} = d_{\mathcal{M}}(b, p^{mid+1})$;
9 $\quad$ Let b_{mid} be true if $d^a_{p^{mid}} < d^b_{p^{mid}}$; false otherwise;
10 $\quad$ Let b_{mid+1} be true if $d^a_{p^{mid+1}} < d^b_{p^{mid+1}}$; false otherwise;
11 $\quad$ **if** $b_{mid} \neq b_{mid+1}$; $\qquad$ // we have the Voronoi edge wedge
12 $\quad$ **then**
13 $\quad\quad$ **return** $(\boldsymbol{v}^{mid}, \boldsymbol{v}^{mid+1})$;
14 $\quad$ **if** $b_{mid} = b_{mid+1} = true$; $\qquad$ // Both points are in a's cell
15 $\quad$ **then**
16 $\quad\quad$ **if** $a_i < b_i$ **then**
17 $\quad\quad\quad$ **return** VORONOIEDGEWEDGE($a, b, \boldsymbol{v}^{mid}, \boldsymbol{v}^j, S$);
18 $\quad\quad$ **else**
19 $\quad\quad\quad$ **return** VORONOIEDGEWEDGE($a, b, \boldsymbol{v}^i, \boldsymbol{v}^{mid}, S$);
20 $\quad$ **if** $b_{mid} = b_{mid+1} = false$; $\qquad$ // Both points are in b's cell
21 $\quad$ **then**
22 $\quad\quad$ **if** $a_i < b_i$ **then**
23 $\quad\quad\quad$ **return** VORONOIEDGEWEDGE($a, b, \boldsymbol{v}^i, \boldsymbol{v}^{mid}, S$);
24 $\quad\quad$ **else**
25 $\quad\quad\quad$ **return** VORONOIEDGEWEDGE($a, b, \boldsymbol{v}^{mid}, \boldsymbol{v}^j, S$);

---

on real numbers (*double*) and we consider the Generic HIDDENBY predicate in $O(\log N)$ (Alg. 4). Since predicates are based on floating point computations, numerical issues may occur. If $p \in \mathbb{Z}$, $p \geq 3$, we use exact integer number based computations of distances storing sum of power p quantities (which can be computed in $O(\log p)$ thanks to exponentiation by squaring). The HIDDENBY predicate is also based on Algorithm 4. Beside these predicates for L_p metrics, DGTAL also contains a generic metric adapter: if the user specifies a distance function (taking two points and returning a value) corresponding to a norm with axis symmetric unit ball, generic CLOSEST and HIDDENBY predicates can be automatically constructed. Please note that since all algorithms are separable, the generic framework provided in DGTAL allow us to have a free multi-thread implementation [4].

To implement efficient predicates leading to subquadratic algorithm in dimension 2 (Alg. 5 and 6), we store the chamfer norm weighted vectors $\mathcal{M}$ in a random access container sorted counterclockwise to be able to get the mid-vector $\boldsymbol{v}^{mid}$ in $O(1)$. When implementing Algorithms 5 and 6, few special cases have to be taken into account. For instance, we have to handle situations where a, b or c belong to S in Alg. 5 and 6. Furthermore, Eq. (11) has a solution iff $A_k \neq A_j$. Thanks to the geometrical representation of the dichotomic process (Fig. 3), such special cases are easy to handle. Fig. 4-(a) illustrates some results on a small domain.

To evaluate experimentally the computational cost given in Theorem 1, we use the following setting: given a mask size m, we generate m distinct random

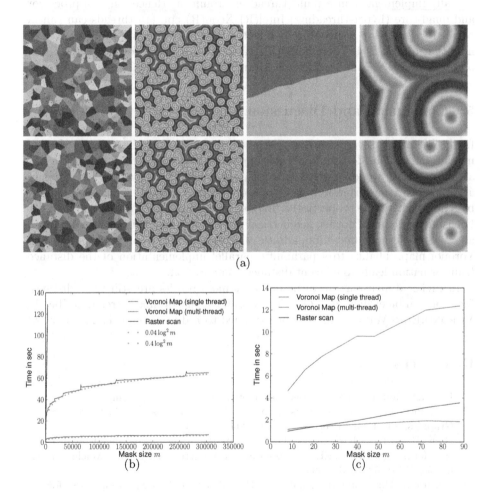

(a)

(b)   (c)

Fig. 4. (a) Separable Voronoi map and distance transformation for $\mathcal{M}_{5-7-11}$ and L_2: on a 256^2 domain with 256 and 2 random seeds, the first row corresponds to $\mathcal{M}_{5-7-11}$ and the second one to L_2. (b) and zoom in (c): Experimental evaluation of subquadratic chamfer norm Voronoi map computation.

vectors $(x, y)^T$ with $gcd(x, y) = 1$ (extracted from Farey series for instance). For a general mask of size m, we do not optimize the weights to approximate as best as possible the Euclidean metric. Indeed, weights do not play any role in the computational analysis, we just use trivial ones that ensure that $\mathcal{M}$ is a norm with axis symmetric unit ball. In Fig.4-$(b - c)$, we have considered a 2D domain 2048^2 with 2048 random sites. First, we observe that fixing N, the $\log^2 m$ term is clearly visible in the computational cost of the Voronoi map (single thread curve). Bumps in the single thread curve may be due to memory cache issues. Please note that if we consider classical chamfer norm DT from raster scan (and sub-masks), the computational cost is in $O(m \cdot N^2)$ and thus has a linear behavior in Fig. 4-(c). Since we have a separable algorithm, we can trivially implement it in a multi-thread environment. Hence, on a bi-processor and quad-core (hyper-threading) Intel(R) Xeon(R) cpu (16 threads can run in parallel), we observe a speed-up by a factor 10 (blue curve in Fig. 4-(b)). Please note that on this 2048^2 domain with 2048 sites, Euclidean Voronoi Map (L_2) is obtained in 954.837 milliseconds on a single core and 723.196 msec on 16 cores.

5   Conclusion and Discussion

In the literature, several authors discussed about the fact that a large class of metrics can be considered in separable approaches for volumetric analysis. In this article, we have proposed several algorithms to efficiently solve the Voronoi map and distance transformation: given a user-specified distance function (induced by a norm with some properties) a first generic separable algorithm can be used. Focusing on chamfer norms, geometrical interpretation of this generic approach allows us to design a first subquadratic algorithm in dimension 2 to compute the Voronoi map. Thanks to separability, parallel implementation of the distance transformation leads to efficient distance computation.

In higher dimensions, it turns out that most results are still true: distance function can be evaluated in $O(n \cdot \log m)$ and the dichotomic search described in VORONOIEDGEWEDGE can also be extended to n-dimensional chamfer norms.

References

1. DGTAL: Digital geometry tools and algorithms library, http://dgtal.org
2. Borgefors, G.: Distance transformations in digital images. Computer Vision, Graphics, and Image Processing 34(3), 344–371 (1986)
3. Breu, H., Gil, J., Kirkpatrick, D., Werman, M.: Linear time euclidean distance transform algorithms. IEEE Transactions on Pattern Analysis and Machine Intelligence 17(5), 529–533 (1995)
4. Coeurjolly, D.: Volumetric Analysis of Digital Objects Using Distance Transformation: Performance Issues and Extensions. In: Köthe, U., Montanvert, A., Soille, P. (eds.) WADGMM 2010. LNCS, vol. 7346, pp. 82–92. Springer, Heidelberg (2012)
5. Danielsson, P.E.: Euclidean distance mapping. Computer Graphics and Image Processing 14, 227–248 (1980)

6. Fouard, C., Malandain, G.: 3-D chamfer distances and norms in anisotropic grids. Image and Vision Computing 23, 143–158 (2005)
7. Hirata, T.: A unified linear-time algorithm for computing distance maps. Information Processing Letters 58(3), 129–133 (1996)
8. Klette, R., Rosenfeld, A.: Digital Geometry: Geometric Methods for Digital Picture Analysis. Series in Computer Graphics and Geometric Modeling. Morgan Kaufmann (2004)
9. Maurer, C., Qi, R., Raghavan, V.: A Linear Time Algorithm for Computing Exact Euclidean Distance Transforms of Binary Images in Arbitrary Dimensions. IEEE Trans. Pattern Analysis and Machine Intelligence 25, 265–270 (2003)
10. Meijster, A., Roerdink, J.B.T.M., Hesselink, W.H.: A general algorithm for computing distance transforms in linear time. In: Mathematical Morphology and its Applications to Image and Signal Processing, pp. 331–340. Kluwer (2000)
11. Mukherjee, J., Das, P.P., Kumar, M.A., Chatterji, B.N.: On approximating euclidean metrics by digital distances in 2D and 3D. Pattern Recognition Letters 21(6-7), 573–582 (2000)
12. Normand, N., Évenou, P.: Medial axis lookup table and test neighborhood computation for 3D chamfer norms. Pattern Recognition 42(10), 2288–2296 (2009)
13. Normand, N., Strand, R., Evenou, P.: Digital distances and integer sequences. In: Gonzalez-Diaz, R., Jimenez, M.-J., Medrano, B. (eds.) DGCI 2013. LNCS, vol. 7749, pp. 169–179. Springer, Heidelberg (2013)
14. Ragnemalm, I.: Contour processing distance transforms, pp. 204–211. World Scientific (1990)
15. Rosenfeld, A., Pfaltz, J.: Distance functions on digital pictures. Pattern Recognition 1, 33–61 (1968)
16. Rosenfeld, A., Pfaltz, J.: Sequential operations in digital picture processing. Journal of the ACM (JACM) 13, 471–494 (1966),
http://portal.acm.org/citation.cfm?id=321357
17. Strand, R.: Distance Functions and Image Processing on Point-Lattices With Focus on the 3D Face- and Body-centered Cubic Grids. Phd thesis, Uppsala Universitet (2008)
18. Thiel, E.: Géométrie des distances de chanfrein. Ph.D. thesis, Aix-Marseille 2 (2001)
19. Verwer, B.J.H., Verbeek, P.W., Dekker, S.T.: An efficient uniform cost algorithm applied to distance transforms. IEEE Transactions on Pattern Analysis and Machine Intelligence 11(4), 425–429 (1989)

Anti-Aliased Euclidean Distance Transform on 3D Sampling Lattices

Elisabeth Linnér and Robin Strand

Center for Image Analysis, Uppsala University, Sweden
{elisabeth,robin}@cb.uu.se

Abstract. The Euclidean distance transform (EDT) is used in many essential operations in image processing, such as basic morphology, level sets, registration and path finding. The anti-aliased Euclidean distance transform (AAEDT), previously presented for two-dimensional images, uses the gray-level information in, for example, area sampled images to calculate distances with sub-pixel precision. Here, we extend the studies of AAEDT to three dimensions, and to the Body-Centered Cubic (BCC) and Face-Centered Cubic (FCC) lattices, which are, in many respects, considered the optimal three-dimensional sampling lattices. We compare different ways of converting gray-level information to distance values, and find that the lesser directional dependencies of optimal sampling lattices lead to better approximations of the true Euclidean distance.

1 Introduction

1.1 Supersampling and Coverage

In a binary image, the spatial elements (spels) are classified either as part of an object or of the background. When imaging a continuous scene, this leads to jagged edges, and much of the information on the edge location and length is lost. With inspiration from anti-aliasing in computer graphics, it is shown in [13] that part of this information is preserved if the image is supersampled, and the intensity of a spel is proportional to the part of its Voronoi cell that is inside of an object, as in Figure 1. The intensity $c(\mathbf{p})$ of a spel $\mathbf{p}$ is then referred to as its coverage value. The theory is further developed in [12], and it is shown that some physical properties, such as volume and surface area, can be measured with higher accuracy in supersampled images. These qualities are important in, for example, cancer diagnostics [3,7].

In 2D imaging devices, the intensity of a spel is usually computed through integration over some environment of the spel center [2], and the result is more or less equivalent to a coverage value. 3D imaging techniques, such as computed tomography (CT), are often based on combining 2D images from different perspectives, and thus we feel comfortable to suggest and apply coverage based image processing methods in 3D as well.

E. Barcucci et al. (Eds.): DGCI 2014, LNCS 8668, pp. 88–98, 2014.

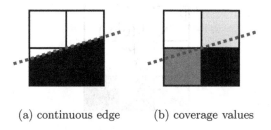

(a) continuous edge   (b) coverage values

Fig. 1. Assignment of intensity proportional to spel coverage

1.2   Euclidean Distance Transforms

The Euclidean distance d_E between two points $\mathbf{p}_1$ and $\mathbf{p}_2$ in n-dimensional space is defined as

$$d_E\left(\mathbf{p}_1, \mathbf{p}_2\right) = \sqrt{\sum_{i=1}^{n} \left(\mathbf{p}_1(i) - \mathbf{p}_2(i)\right)^2}.$$

The Euclidean distance transform (EDT) maps every point $\mathbf{p}$ to

$$d_{EDT}(\mathbf{p}) = \inf_{\mathbf{p}_\omega \in \Omega} \left(d_E(\mathbf{p}, \mathbf{p}_\omega)\right),$$

where Ω is some object(s) or a set of seed points. This transform is used in many essential operations in image processing, such as basic morphology, level sets, registration and path finding [1]. In 3D, it can be used for, among other things, visualization, modeling and animation [6].

Anti-Aliased Euclidean Distance Transform. The anti-aliased Euclidean distance transform (AAEDT) for two-dimensional images is presented in [5]. It uses coverage information to compute the Euclidean distance from an object with sub-spel precision. The distance d_{AAEDT} is defined as

$$d_{AAEDT}(\mathbf{p}) = \min_{\mathbf{p}_\omega \in \partial\Omega} \left(|\mathbf{d}_E(\mathbf{p}, \mathbf{p}_\omega)| + d_f(\mathbf{p}_\omega)\right), \tag{1}$$

where $|\mathbf{d}_E(\mathbf{p}, \mathbf{p}_\omega)|$ is the Euclidean distance between the centers of a background spel $\mathbf{p}$ and an edge spel $\mathbf{p}_\omega \in \partial\Omega$, $\partial\Omega$ being the edge of a binary and areasampled object Ω, and $d_f(\mathbf{p}_\omega)$ is the distance between the edge and the center of the edge spel. This is illustrated in Figure 2. The addition of the term $d_f(\mathbf{p}_\omega)$ is the source of the sub-pixel precision, which improves, for example, the accuracy of level sets and small-scale morphology.

1.3   Three-Dimensional Sampling Lattices

The spels in a digital image represent sample points, which are organized in a so-called sampling lattice. The most common sampling lattice is the Cartesian Cubic (CC) lattice, resulting in square spels in 2D, and cubic spels in 3D.

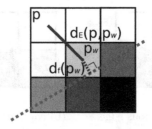

Fig. 2. Illustration of the anti-aliased Euclidean distance d_{AAEDT} between two spels p and p_ω

Unfortunately, the sampling properties of the CC lattice are strongly direction dependent [8, 14]. Consequently, to guarantee some minimum resolution in all directions within the image, some directions must be over-sampled, and redundant data must thus be stored and processed. An alternative is to use the Body-Centered Cubic (BCC) and Face-Centered Cubic (FCC) sampling lattices. Their direction dependence is much weaker, and, for band-limited signals, a minimum resolution in all directions can be obtained using $\sim 30\%$ fewer sample points than if a CC lattice were used [4, 8, 10, 14, 15].

1.4 Scope of This Paper

In AAEDT, $d_f(\mathbf{p}_\omega)$ is computed from the coverage value $c(\mathbf{p}_\omega)$ of the edge spel, usually under the assumption that the object surface intersecting the spel is locally flat, and that $d_f(\mathbf{p}_\omega)$ is measured along the surface normal [5]. However, as the orientation of the surface is unknown, the asymmetry of the Voronoi cell of the spel makes it difficult to construct a mapping between $c(\mathbf{p}_\omega)$ and $d_f(\mathbf{p}_\omega)$. Moreover, for spels not located on the edge, the vector propagation process may introduce a discrepancy between the direction to the nearest edge spel and that to the closest point on the surface. We propose that the lesser direction dependencies of the BCC and FCC sampling lattices, compared to the CC sampling lattice, lead to improved performance of AAEDT.

2 Method

2.1 Implementation

We use the graph-based AAEDT implementation presented in [9], which can be adapted to any dimensionality and sampling lattice by changing the definition of the spel neighborhood. In this way, we ensure fair comparison of the lattices.

2.2 Computation of $d_f(\mathbf{p}_\omega)$

We want to find an expression on the form $d_f(c)$ to approximate the value $d_f(\mathbf{p}_\omega)$ of an edge spel $\mathbf{p}_\omega$ from its coverage value $c(\mathbf{p}_\omega)$. As a reference for different approximation methods, we simulate $d_f(\mathbf{p}_\omega)$ and $c(\mathbf{p}_\omega)$ for the Voronoi cells of the CC, BCC and FCC lattices: We construct a set of planes, the normals of which are uniformly distributed within the symmetry regions of the Voronoi cells, intersecting the cell center. Using the Monte-Carlo sampling method [11], we approximate the portions of the cell that are located above and below the plane. We repeat the process while moving the plane away from the spel center in small steps, until the entire Voronoi cell is below the plane. The simulation output is shown in Figure 3, with plots of three different approximations $d_f(c)$ of $d_f(\mathbf{p}_\omega)$ from $c(\mathbf{p}_\omega)$, described below.

The implementation in [5] approximates $d_f(\mathbf{p}_\omega)$ within an edge spel using

$$d_f(c) = 0.5 - c(\mathbf{p}_\omega). \tag{2}$$

This is the exact relationship between $d_f(\mathbf{p}_\omega)$ and $c(\mathbf{p}_\omega)$ for for a spel of a CC lattice being intersected by a plane perpendicular to a lattice vector. Although $d_f(c)$ is only computed in the initialization of AAEDT, the computational simplicity of (2) is an attractive property. There is no equally simple formula for BCC and FCC lattices. However, linear regression on the simulation output in Figure 3 shows that (2) is actually an even better approximation of the relationship between $d_f(\mathbf{p}_\omega)$ and $c(\mathbf{p}_\omega)$ on these lattices, than on the CC lattice. Higher order regression on the data in Figure 3 leads to overfitting rather than improvement of the approximation.

As we do not calculate the orientation of the plane that intersects the spel, we want $d_f(c)$ to be orientation independent. As the ideal Voronoi cell of a sample point, with respect to sampling properties, is a ball, which is completely orientation independent, we derive $d_f(c)$ for a ball of unit volume. As our implementation uses 256 gray levels, we tabulate the relationship for 256 coverage values in the interval $[0, 1]$.

For the third approximation method plotted in Figure 3, we simply use the mean value of the Monte-Carlo simulation output. Again, we tabulate the relationship for 256 coverage values in the interval $[0, 1]$. This is the only approximation method in this study that is lattice dependent.

3   Experiments

3.1   Choice of Test Images

We study the behavior of AAEDT applied to images of supersampled binary balls, with 256 gray levels representing degrees of spel coverage, sampled on CC, BCC and FCC lattices. We use the exterior and interior distance from the ball surface as examples of convex and concave surfaces, respectively, of different orientations. By varying the ball radius r_s within some range $r_s \in [r_{min}, r_{max}], r_{min}, r_{max} \gg r_v$, where r_v is the average radius of a spel, we study the impact of surface curvature on the accuracy of AAEDT.

We use the distance from balls where $r_s \approx r_v$ to indicate the behavior of AAEDT for undersampled objects.

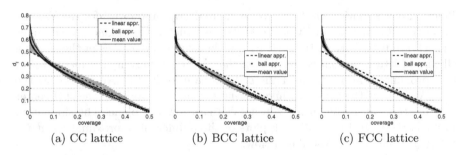

(a) CC lattice   (b) BCC lattice   (c) FCC lattice

Fig. 3. Approximative mappings between $d_f(\mathbf{p}_\omega)$, the distance between the intersecting plane and spel center, and the $c(\mathbf{p}_\omega)$, the spel coverage value, plotted on top of the output of a Monte-Carlo simulation of the relationship. As $d_f(c)$ behaves as an odd function centered at $c(\mathbf{p}_\omega) = 0.5$, it is only plotted for $c \in [0, 0.5]$.

For every lattice, we use a set of balls where the center points are evenly distributed within the symmetry region of the Voronoi cell of that lattice, so that the ball center is unlikely to coincide with a sample point. The sample density for all lattices is one spel per unit volume.

4   Results

The results are expressed in terms of

$$\epsilon(\mathbf{p}) = d_{EDT}(\mathbf{p}) - d_{AAEDT}(\mathbf{p}),$$

and its mean value $\epsilon(\mathbf{p})_m$, where $d_{EDT}(\mathbf{p})$ is the exact Euclidean distance transform, the unsigned relative error

$$|\epsilon_r(\mathbf{p})| = \frac{|d_{EDT}(\mathbf{p}) - d_{AAEDT}(\mathbf{p})|}{d_{EDT}(\mathbf{p})},$$

and the mean unsigned relative error $|\epsilon_r(\mathbf{p})|_m$. The error is evaluated within a Euclidean distance of 50 units from the ball surface, where the unit distance is defined in relation to the unit volume of the spels.

4.1   Bias Error

The mean error $\epsilon(\mathbf{p})_m$, computed on the CC, BCC and FCC lattices using linear, ball-based and mean value-based approximations of $d_f(c)$, is shown in Figures 4, 5 and 6.

For convex surfaces with $r_s \gg r_v$, and linear approximation of $d_f(c)$, the distance is underestimated on all lattices, although much more so on the CC lattice. The mean error on the CC lattice also exhibits a larger standard deviation than that on the BCC and FCC lattices. On the CC lattice, both the mean error and standard deviation are reduced by ball-based approximation of $d_f(c)$, and even further by mean value-based approximation. The error on the BCC lattice seems to be almost completely unbiased for ball-based approximation relative to the others, and tends somewhat towards underestimation for mean value-based approximation. On the FCC lattice, we see a tendency towards underestimation for all approximation methods, but it is the least apparent for ball-based approximation.

For concave surfaces with $r_s \gg r_v$, the mean error is very close to being unbiased for small r_s, and tends towards underestimation for less curved surfaces. As the bias increases, the standard deviation grows. As for convex surfaces, the result on the CC lattice is improved by ball-based and mean value-based approximation of $d_f(c)$. On the BCC and FCC lattices, it is very difficult to discern any bias for ball-based approximation, while there is a slight underestimation using mean value-based approximation.

For $r_s \approx r_v$, all lattices and approximations of $d_f(c)$ yield equivalent results, always underestimating the distance.

4.2 Error Range

Figures 7, 8 and 9 show the first (25th percentile), second (median) and third (75th percentile) quartiles of $|\epsilon_r(\mathbf{p})|$.

For convex surfaces with $r_s \gg r_v$, the CC lattice is clearly outperformed by the BCC and FCC lattices. The range of $|\epsilon_r(\mathbf{p})|$, indicated by the 25th and 75th percentiles, is highly concentrated around the median error on BCC and FCC, while it is notably larger on the CC lattice. However, the performance of the CC lattice is very much improved when ball-based or mean value-based approximation of $d_f(c)$ is used.

For concave surfaces, and for $r_s \approx r_v$, the performance is almost equivalent on all lattices. However, the growth of the range of $|\epsilon_r(\mathbf{p})|$, that occurs for small r_s, starts at an earlier stage on the CC lattice than on BCC and FCC.

In Figures 10, 11 and 12, we see $|\epsilon_r(\mathbf{p})|_m$ as a function of the distance d to the ball surface.

The approximation of $d_f(c)$ takes place at $d < 1$, at which state the difference between the lattices is small. In the cases of $r_s \gg r_v$, the BCC and FCC lattices seem to be at a small advantage in making this approximation, compared to the CC lattice, when ball-based or mean value-based approximation is used.

For all curvatures, lattices and approximations of $d_f(c)$, $|\epsilon_r(\mathbf{p})|_m$ decreases rapidly as the distance increases. For $r_s \gg r_v$, this decrease is more rapid on the BCC and FCC lattices than on the CC lattice.

5   Discussion

5.1   Bias Errors

The AAEDT has two possible sources of bias errors: The approximation of $d_f(c)$, and the risk that $|\mathbf{d}_{AAEDT}(\mathbf{p}, \mathbf{p}_\omega)|$ is incorrect due to omitted edge spels, as

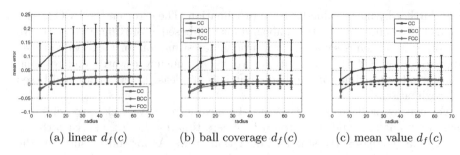

(a) linear $d_f(c)$ (b) ball coverage $d_f(c)$ (c) mean value $d_f(c)$

Fig. 4. Mean error with one standard deviation, convex surface, $r_s \gg r_v$

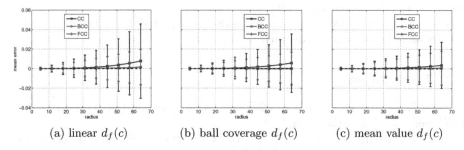

(a) linear $d_f(c)$ (b) ball coverage $d_f(c)$ (c) mean value $d_f(c)$

Fig. 5. Mean error with one standard deviation, concave surface, $r_s \gg r_v$

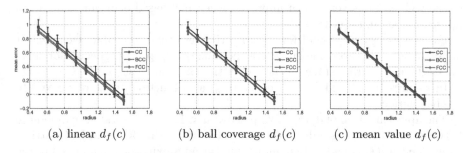

(a) linear $d_f(c)$ (b) ball coverage $d_f(c)$ (c) mean value $d_f(c)$

Fig. 6. Mean error with one standard deviation, convex surface, $r_s \approx r_v$

described in [9]. The former may cause both over- and underestimation, while the latter always causes overestimation of the distance.

Underestimation is likely a result from the fact that the distance from an edge to the center of an edge spel $\mathbf{p}_\omega$ depends not only on $c(\mathbf{p}_\omega)$, but also on the edge orientation. This is very prominent on the CC lattice, where the variation is a factor of $\sqrt{2}$, making it very difficult to make a representative model for mapping distance to coverage. In Figure 3, we can see that this causes problems for both linear and mean-value based approximation of $d_f(c)$, as the large variance close to $c = 0$ and $c = 1$ dulls the slope of the curve, leading to a large difference between

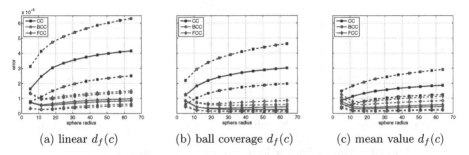

(a) linear $d_f(c)$ (b) ball coverage $d_f(c)$ (c) mean value $d_f(c)$

Fig. 7. First, second and third quartiles, convex surface, $r_s \gg r_v$

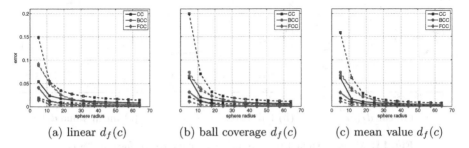

(a) linear $d_f(c)$ (b) ball coverage $d_f(c)$ (c) mean value $d_f(c)$

Fig. 8. First, second and third quartiles, concave surface, $r_s \gg r_v$

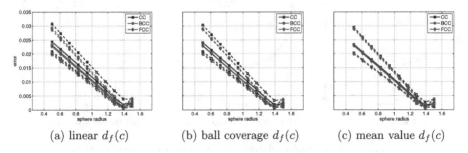

(a) linear $d_f(c)$ (b) ball coverage $d_f(c)$ (c) mean value $d_f(c)$

Fig. 9. First, second and third quartiles, convex surface, $r_s \approx r_v$

the largest $d_f(c)$ that can be assigned to an edge spel, and the smallest d_{AAEDT} that can be assigned to a background spel. For a ball, we have $\lim_{c_{ball} \to 0} d_f(c) = r_v$ and $\lim_{c_{ball} \to 1} d_f(c) = -r_v$, with a smooth transition from fully covered (or uncovered) to partly covered. It is possible that this, combined with the low directional dependencies of the BCC and FCC lattices, results in the low bias observed for the ball-based approximation of $d_f(c)$.

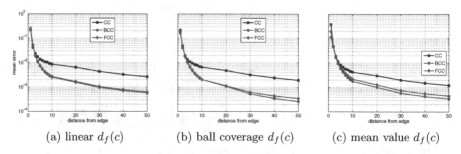

(a) linear $d_f(c)$ (b) ball coverage $d_f(c)$ (c) mean value $d_f(c)$

Fig. 10. Mean unsigned relative error vs. convex surface, $r_s \gg r_v$

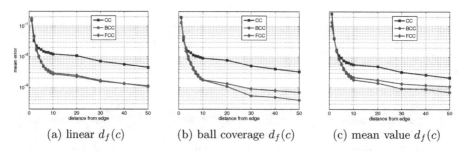

(a) linear $d_f(c)$ (b) ball coverage $d_f(c)$ (c) mean value $d_f(c)$

Fig. 11. Mean unsigned relative error vs. concave surface, $r_s \gg r_v$

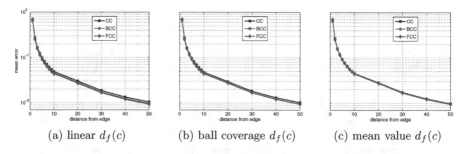

(a) linear $d_f(c)$ (b) ball coverage $d_f(c)$ (c) mean value $d_f(c)$

Fig. 12. Mean unsigned relative error vs. convex surface, $r_s \approx r_v$

5.2 Error Range

Figures 10, 11 and 12 show that $|\epsilon_r(\mathbf{p})|_m$ is smaller farther away from the surface. From this, we draw the conclusion that errors arise mainly from the approximation of $d_f(c)$, and not from the vector propagation.

It seems that the effect of improving the approximation of $d_f(c)$ is the most noticeable on the CC lattice. This is expected, as the cubic spels are more directionally dependent than the truncated octahedra and rhombic dodecahedra of the BCC and FCC lattices, respectively. Ironically, even the linear approximation used in [5] and [9] is much less accurate on the CC lattice than on BCC and FCC. Actually, although considerably improved when using the mean value-based approximation of $d_f(c)$, the performance of AAEDT on the CC lattice is still not as good as that on the BCC and FCC lattices, using only the linear approximation.

The best performance is achieved by using the ball-based approximation of $d_f(c)$ on the BCC and FCC lattices. As explained above, this behavior is likely to be a consequence of the treatment of edge spels with $c \approx 0$ and $c \approx 1$.

The increase of the median of $|\epsilon_r(\mathbf{p})|$ for large r_s in Figure 7 does not necessarily mean that AAEDT is less accurate for flat surfaces. As we compute the error for $d \le 50$ for all r_s, the ratio $n_{\text{close}}/n_{\text{far}}$, where n_{close} is the number of spels close to the surface and n_{far} the number of spels far away, is smaller for small balls. Hence, the ratio $n_{\text{large error}}/n_{\text{small error}}$ is also smaller. As the decrease in $|\epsilon_r(\mathbf{p})|$, visible in Figures 10 and 11, seems to be smooth, this affects both the mean value and median of $|\epsilon_r(\mathbf{p})|$.

The bad performance for $r_s \approx r_v$ shows that AAEDT is highly dependent on the sampling density being proportional to image scale, as $d_f(\mathbf{p}_\omega)$ cannot be accurately approximated from surfaces that are not locally flat.

5.3   Conclusions and Future Work

In this paper, we show how the performance of AAEDT in 3D can be significantly improved. We analyze the impact of the approximation of $d_f(c)$, and we explore the advantages of sampling lattices with lesser directional dependencies than the wide-spread CC lattice. Next, we hope to investigate the behavior of AAEDT in the presence of sharp corners and more complex surface curvature. In [5], it is suggested that gradient information is used to estimate the orientation of the surface, which may improve the performance on the BCC and FCC lattices even further.

Acknowledgements. The authors want to thank M.Sc. Max Morén for enabling the production of test images on the BCC and FCC lattices.

References

1. Borgefors, G.: Applications using distance transforms. In: Arcelli, C., Cordella, L.P., Sanniti di Baja, G. (eds.) Aspects of Visual Form Processing: Proceedings of the Second International Workshop on Visual Form, pp. 83–108. World Scientific Publishing (1994)
2. Chen, W., Li, M., Su, X.: Error analysis about ccd sampling in fourier transform profilometry. Optik - International Journal for Light and Electron Optics 120(13), 652–657 (2009)
3. Clifford Chao, K.S., Ozyigit, G., Blanco, A.I., Thorstad, W.L., O Deasy, J., Haughey, B.H., Spector, G.J., Sessions, D.G.: Intensity-modulated radiation therapy for oropharyngeal carcinoma: impact of tumor volume. International Journal of Radiation Oncology*Biology*Physics 59(1), 43–50 (2004)

4. Entezari, A.: Towards computing on non-cartesian lattices. Tech. rep. (2006)
5. Gustavson, S., Strand, R.: Anti-aliased Euclidean distance transform. Pattern Recognition Letters 32(2), 252–257 (2011)
6. Jones, M.W., Bærentsen, J.A., Sramek, M.: 3D distance fields: A survey of techniques and applications. IEEE Transactions on Visualization and Computer Graphics 12(4), 581–599 (2006)
7. Lebioda, A., Żyromska, A., Makarewicz, R., Furtak, J.: Tumour surface area as a prognostic factor in primary and recurrent glioblastoma irradiated with 192ir implantation. Reports of Practical Oncology & Radiotherapy 13(1), 15–22 (2008)
8. Linnér, E., Strand, R.: Aliasing properties of voxels in three-dimensional sampling lattices. In: Lirkov, I., Margenov, S., Waśniewski, J. (eds.) LSSC 2011. LNCS, vol. 7116, pp. 507–514. Springer, Heidelberg (2012)
9. Linnér, E., Strand, R.: A graph-based implementation of the anti-aliased Euclidean distance transform. In: International Conference on Pattern Recognition (August 2014)
10. Meng, T., Smith, B., Entezari, A., Kirkpatrick, A.E., Weiskopf, D., Kalantari, L., Möller, T.: On visual quality of optimal 3D sampling and reconstruction. In: Proceedings of Graphics Interface (2007)
11. Metropolis, N., Ulam, S.: The monte carlo method. Journal of the American Statistical Association 44(247), 335–341 (1949)
12. Sladoje, N., Lindblad, J.: High-precision boundary length estimation by utilizing gray-level information. IEEE Transactions on Pattern Analysis and Machine Intelligence 31(2), 357–363 (2009)
13. Sladoje, N., Nyström, I., Saha, P.K.: Measurements of digitized objects with fuzzy borders in 2D and 3D. Image and Vision Computing 23(2), 123–132 (2005)
14. Strand, R.: Sampling and aliasing properties of three-dimensional point-lattices (2010), http://www.diva-portal.org/smash/record.jsf?searchId=1&pid=diva2:392445&rvn=3
15. Theußl, T., Möller, T., Gröller, M.E.: Optimal regular volume sampling. In: Proceedings of the Conference on Visualization (2001)

Efficient Neighbourhood Computing for Discrete Rigid Transformation Graph Search[*]

Yukiko Kenmochi[1], Phuc Ngo[2], Hugues Talbot[1], and Nicolas Passat[3]

[1] Université Paris-Est, LIGM, CNRS, France
[2] CEA LIST – DIGITEO Labs, France
[3] Université de Reims Champagne-Ardenne, CReSTIC, France

Abstract. Rigid transformations are involved in a wide variety of image processing applications, including image registration. In this context, we recently proposed to deal with the associated optimization problem from a purely discrete point of view, using the notion of discrete rigid transformation (DRT) graph. In particular, a local search scheme within the DRT graph to compute a locally optimal solution without any numerical approximation was formerly proposed. In this article, we extend this study, with the purpose to reduce the algorithmic complexity of the proposed optimization scheme. To this end, we propose a novel algorithmic framework for just-in-time computation of sub-graphs of interest within the DRT graph. Experimental results illustrate the potential usefulness of our approach for image registration.

Keywords: image registration, discrete rigid transformation, discrete optimization, DRT graph.

1 Introduction

1.1 Discrete Rotations and Discrete Rigid Transformations

In continuous spaces (*i.e.*, $\mathbb{R}^n$), rotations are some of the simplest geometric transformations. However, in the discrete spaces (*i.e.*, $\mathbb{Z}^n$), their analogues, namely discrete rotations, are more complex. The induced challenges are not simply due to high-dimensionality: indeed, even in $\mathbb{Z}^2$, discrete rotations raise many difficulties, deriving mainly from their non-necessary bijectivity [1]. In this context, discrete rotations – and the closely related discrete rigid tansformations – have been widely investigated.

From a combinatorial point of view, discrete rotations have been carefully studied [2–4], in particular to shed light on remarkable configurations induced by the periodicity of rotations with respect to the discrete grid. At the frontier between combinatorics and algorithmics, the problem of 2D pattern matching under discrete rotations has also been explored [5, 6].

From an algorithmic point of view, efforts have been devoted to effectively compute discrete rotations. In particular, the quasi-shear rotations [7, 8] were introduced to preserve bijectivity, by decomposing rotations into successive quasi-shears.

---

[*] The research leading to these results has received funding from the French *Agence Nationale de la Recherche* (Grant Agreement ANR-2010-BLAN-0205).

E. Barcucci et al. (Eds.): DGCI 2014, LNCS 8668, pp. 99–110, 2014.

Finally, from an applicative point of view, discrete rotations have been used for image/signal processing purposes [9, 10]. Other strategies have also been proposed to pre-process 2D images in order to guarantee the preservation of topological properties under discrete rigid transformations [11].

Recently, we proposed a new paradigm to deal with discrete rotations, and more generally rigid transformations. This paradigm relies on a combinatorial structure, called discrete rigid transformation graph (DRT graph, for short) [12]. This structure describes the quantification of the parameter space of rigid transformations, in the framework of hinge angles, pioneered in [13–15].

The DRT graph has already allowed us to contribute to the state of the art on rigid transformations from a combinatorial point of view, by establishing the complexity of "free" [12] and "constrained" [16] discrete rigid transformations. From an algorithmic point of view, it has been used to characterise topological defects in transformed images [17]. Finally, we recently started to explore the applicative possibilities offered by the DRT graph. In particular, we have considered its potential usefulness in the context of image registration [18].

1.2   Registration Issues

In the context of image processing, geometric transformations are often considered for registration purposes [19]. Registration is indeed a complex, often ill-posed problem, that consists of defining the transformation that is required to correctly map a source image onto a target image.

Registration is mandatory in various application fields, from remote sensing [20] to medical imaging [21]. According to the specificities of these fields, registration can implicate different types of images (2D, 3D) and transformations, both rigid and non-rigid. However, the problem remains almost the same in all applications. Given two images A and B, we aim at finding a transformation T^* within a given transformation space $\mathbb{T}$. This transformation minimizes a given distance d between the image A and the transformed image $T(B)$ of the image B by T, $i.e.$

$$T^* = \arg \min_{T \in \mathbb{T}} d(A, T(B)) \tag{1}$$

In recent works [18], we investigated how to use the DRT graph in order to solve this problem in the case of rigid registration of 2D images. The novelty of this approach, with respect to the state of the art, was to provide exact transformation fields, so as to avoid any interpolation process and numerical approximations.

In this context, a preliminary algorithm was proposed for computing a local minimum for Eq. (1), thus providing a solution in a neighbourhood of depth $k \geq 1$, to the above registration problem. This algorithm strongly relies on the DRT graph, and consists of exploring a sub-graph defined around a given vertex, modeling an initial transformation. Its time complexity was $O(m^k N^2)$, which is linear with respect to the image size, but exponential with respect to the neighbourhood depth (with m the size of the 1-depth neighbourhood).

1.3 Contribution

We propose an improved algorithm (Sec. 3), which dramatically reduces the exponential complexity of that developed in [18]. Indeed, we show that the k-depth neighbourhood of a DRT graph can be computed with a time complexity $O(kN^2)$ (Sec. 4). Experiments emphasise the methodological interest of the proposed approach (Sec. 5).

2 Introduction to Discrete Rigid Transformation Graphs

2.1 Rigid Transformation Space

In the continuous space $\mathbb{R}^2$, a rigid transformation is a bijection $\mathcal{T} : \mathbb{R}^2 \to \mathbb{R}^2$, defined, for any $x = (x, y) \in \mathbb{R}^2$, by

$$\mathcal{T}(x) = \begin{pmatrix} \cos\theta & -\sin\theta \\ \sin\theta & \cos\theta \end{pmatrix} \begin{pmatrix} x \\ y \end{pmatrix} + \begin{pmatrix} a_1 \\ a_2 \end{pmatrix} \tag{2}$$

where $a_1, a_2 \in \mathbb{R}$ and $\theta \in [0, 2\pi[$ ($\mathcal{T}$ is sometimes noted $\mathcal{T}_{a_1 a_2 \theta}$). In order to apply such rigid transformations on $\mathbb{Z}^2$, a post-processing digitization is required. More precisely, a digitized rigid transformation $T : \mathbb{Z}^2 \to \mathbb{Z}^2$ is defined as $T = D \circ \mathcal{T}$ where $D : \mathbb{R}^2 \to \mathbb{Z}^2$ is a rounding function. In other words, for any $p = (p, q) \in \mathbb{Z}^2$, we have

$$T(p) = \begin{pmatrix} p' \\ q' \end{pmatrix} = D \circ \mathcal{T}(p) = \begin{pmatrix} [p\cos\theta - q\sin\theta + a_1] \\ [p\sin\theta + q\cos\theta + a_2] \end{pmatrix} \tag{3}$$

The use of the rounding function D implies that digitized rigid transformations are not continuous within the 3D parameter space induced by a_1, a_2 and θ. The transformations leading to such discontinuities are called critical transformations. In the space (a_1, a_2, θ), the subspace of critical transformations is composed of 2D surfaces $\Phi_{pqp'}$ and $\Psi_{pqq'}$, analytically defined, for any $p = (p, q) \in \mathbb{Z}^2$ and any vertical (resp. horizontal) pixel boundary $x = p' + \frac{1}{2}$ (resp. $y = q' + \frac{1}{2}$) with $p' \in \mathbb{Z}$ (resp. $q' \in \mathbb{Z}$), by

$$\Phi_{pqp'} : p\cos\theta - q\sin\theta + a_1 = p' + \frac{1}{2} \tag{4}$$

$$\Psi_{pqq'} : p\sin\theta + q\cos\theta + a_2 = q' + \frac{1}{2} \tag{5}$$

For a given triplet (p, q, p') (resp. (p, q, q')), $\Phi_{pqp'}$ (resp. $\Psi_{pqq'}$) is called a vertical (resp. horizontal) tipping surface in the parameter space (a_1, a_2, θ), and a vertical (resp. horizontal) tipping curve in the 2D plane (a_1, θ) (resp. (a_2, θ)).

For an image of size $N \times N$, $\Phi_{pqp'}$ and $\Psi_{pqq'}$ verify $p, q \in [\![0, N-1]\!]$ and $p', q' \in [\![0, N]\!]$. Examples of tipping surfaces and curves are illustrated in Fig. 1.

2.2 Discrete Rigid Transformation Graph

A set of tipping surfaces induces a subdivision of the (a_1, a_2, θ) space into classes, each consisting of transformations $\mathcal{T}_{a_1 a_2 \theta}$ such that $(a_1, a_2, \theta) \mapsto T = D \circ \mathcal{T}_{a_1 a_2 \theta}$ is constant. These classes – called discrete rigid transformations (DRTs) – indeed form 3D

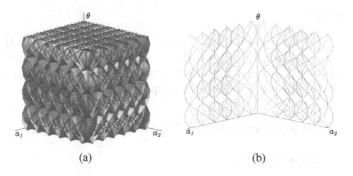

Fig. 1. (a) Tipping surfaces in the space (a_1, a_2, θ), and (b) their tipping curves [16]

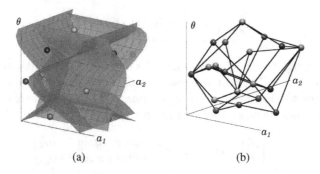

Fig. 2. (a) Subdivision of the (a_1, a_2, θ) parameter space into 3D cells by tipping surfaces, and (b) the associated DRT graph [17]

cells, bounded by tipping surfaces that correspond to discontinuities. By mapping each cell onto a vertex, and each tipping surface piece onto an edge, in a Voronoi/Delaunay paradigm, we can model this subdivided parameter space as a graph, called the DRT graph, as illustrated in Fig. 2.

Definition 1 ([12]). *A DRT graph $G = (V, E)$ is defined such that (i) each vertex $v \in V$ models a DRT; and (ii) each labelled edge $e = (v, w, f) \in E$, where f is either $\Phi_{pqp'}$ or $\Psi_{pqq'}$, connects two vertices $v, w \in V$ sharing the tipping surface f as boundary.*

For a given image I, each vertex is associated with a unique transformed image, induced by the DRT corresponding to the vertex. The existence of an edge between two vertices indicates a "neighbouring" relation between the two associated DRTs. More precisely the two transformed images differ by at most one over the N^2 pixels of I; the edge label f provides this information. This allows us to use the DRT graph to produce all the transformed images via successive elementary (*i.e.*, single-pixel) modifications.

2.3 Discrete Rigid Transformation Graph and Image Registration

The registration problem formalised in Eq. (1) consists of finding the transformation that best maps a source image onto a target image, with respect to a given distance.

In the discrete framework, the number of transformations is actually finite. In particular, in the case of rigid registration, the solution(s) to Eq. (1) can be found within the DRTs exhaustively modeled by the DRT graph. In other words, by considering a brute-force search, a solution, *i.e.*, a global optimum, can be determined. However, the DRT graph G of an image of size $N \times N$, has a high space complexity $O(N^9)$ [12] that induces the same time complexity both for its construction and exhaustive search.

This limits exploration of the whole structure to relatively small images. Nevertheless, as already discussed in [18], it is possible to perform a local search on G in order to determine a local optimum.

2.4   Local Search on a Discrete Rigid Transformation Graph

To find such an optimum, a local search begins at a given transformation, *i.e.*, a chosen vertex v of G. Then, it moves towards a better solution in its neighbourhood – following a gradient descent – as long as an improved solution can be found. Beyond the choice of the initial vertex – often guided by the application context – the most critical issue is the choice of a "good" search area around this vertex, *i.e.*, a depth of its neighbourhood. In particular, the trade-off is time efficiency versus exhaustiveness.

The neighbourhood of depth 1, noted $\mathcal{N}^1(v)$, actually corresponds to the set $\mathcal{N}(v)$ of vertices adjacent to v in G. More generally, neighbourhoods of depth $k \geq 1$, also called k-neighbourhoods, are then recursively obtained as

$$\mathcal{N}^k(v) = \mathcal{N}^{k-1}(v) \cup \bigcup_{u \in \mathcal{N}^{k-1}(v)} \mathcal{N}(u) \tag{6}$$

where $\mathcal{N}^0(v) = \{v\}$.

Our initial algorithm [18] was directly mapped on this recursive definition. As a consequence, this approach led to a high time complexity $O(m^k N^2)$, that is exponential with respect to the depth k of the neighbourhood with vertex degree m, which is supposed to be constant in average (Sec. 4.2). In the next section, we propose a more efficient algorithm, that removes this exponential cost.

3 k-Neighbourhood Construction Algorithm

We now propose an algorithm that efficiently computes the part of a DRT graph that models the neighbourhood of depth k around a given vertex. To this end, we need to handle the analytical representation of the cells associated to the DRT graph vertices, inside the subdivided parameter space of (a_1, a_2, θ) (Sec. 3.1). Then, we develop a construction strategy that relies on a sweeping plane technique introduced in [12] (Sec. 3.2). The final algorithm is described and formalized in Sec. 3.3.

3.1   Tipping Surfaces Associated to a Discrete Rigid Transformation

A vertex v of a DRT graph G corresponds to one discrete rigid transformation, that induces a unique transformed image I_v obtained by applying this transformation on an initial image I. In other words, for each pixel (p_i, q_i) of I_v, we know which pixel (p'_i, q'_i) of I transfers its value to (p_i, q_i). This correspondence is modeled by the following inequalities deriving from Eq. (3)

$$p'_i - \frac{1}{2} < p_i \cos \theta - q_i \sin \theta + a_1 < p'_i + \frac{1}{2} \tag{7}$$

$$q'_i - \frac{1}{2} < p_i \sin \theta + q_i \cos \theta + a_2 < q'_i + \frac{1}{2} \tag{8}$$

For an image I of size $N \times N$, each of the N^2 pixels generates 4 such inequalities. In the parameter space of (a_1, a_2, θ), the obtained $4N^2$ inequalities then define a 3D cell, denoted by $\mathcal{R}_v$, which gathers all the parameter triplets associated to the discrete rigid transformation corresponding to the vertex v.

When interpreting these inequalities in terms of tipping surfaces/curves (see Eqs. (4–5)), it appears that for each pixel of I_v, Eqs. (7–8) define a region of the parameter space that is bounded by two offset vertical (resp. horizontal) tipping surfaces/curves $\Phi_{p_i q_i p'_i}$ and $\Phi_{p_i q_i p'_i - 1}$ (resp. $\Psi_{p_i q_i q'_i}$ and $\Psi_{p_i q_i q'_i - 1}$). For any $i \in [\![1, N^2]\!]$, $\Phi_{p_i q_i p'_i}$ (resp. $\Psi_{p_i q_i q'_i}$) is called an upper tipping surface/curve, while $\Phi_{p_i q_i p'_i - 1}$ (resp. $\Psi_{p_i q_i q'_i - 1}$) is called a lower tipping surface/curve. The sets composed by these surfaces/curves, for all $i \in [\![1, N^2]\!]$, are denoted $\mathbf{S}_1^+(I_v)$ and $\mathbf{S}_1^-(I_v)$ (resp. $\mathbf{S}_2^+(I_v)$ and $\mathbf{S}_2^-(I_v)$).

We derive from Eqs. (7–8), that any cell $\mathcal{R}_v$ is directionally convex along the $a_\star$-axes [16]. This implies that for any θ value where it is defined, $\mathcal{R}_v$ is bounded by at least one upper (resp. lower) tipping surface, which constitutes the upper (resp. lower) part of its boundary in each $a_\star$-direction. This property can be used for constructing a DRT graph locally, or for obtaining topological information from a DRT graph such as a neighbourhood. One may notice that it is sufficient to consider only tipping surfaces of $\bigcup (\mathbf{S}_\star^+(I_v) \cup \mathbf{S}_\star^-(I_v))$ in order to obtain the k-neighbourhood of v, if $k < N$.

3.2 Sweeping Plane Algorithm for DRT Sub-graph Construction

In our new algorithm, the purpose is to build a k-neighbourhood "similarly" to the construction of a 1-neighbourhood in our previous version [18], that is by using a sweeping plane technique from one value θ_v within $\mathcal{R}_v$, to both the left-hand and right-hand sides along the θ-axis in the space (a_1, a_2, θ).

The differences between this new algorithm and the former are twofold. On the one hand, the range of the considered θ values is wider. Indeed, the sweep must be carried out inside $\mathcal{R}_v$ but also outside. On the other hand, a larger number of tipping surfaces are considered around $\mathcal{R}_v$, while only immediate neighbours were previously involved.

To ease the understanding of this algorithm, we first recall the general idea of the sweeping plane technique.

Given a set $\mathbf{S}$ of s_1 vertical and s_2 horizontal tipping surfaces, we aim to construct the DRT sub-graph G corresponding to a given range $[\theta_{start}, \theta_{end}]$. By comparison to [12], the plane is then swept from θ_{start} to θ_{end}, instead of 0 to 2π. From the very definition of tipping surfaces, this plane is subdivided into $(s_1 + 1) \times (s_2 + 1)$ 2D rectangular cells, generated by its intersection with the tipping surfaces of $\mathbf{S}$. More precisely, we have $(s_\star + 1)$ divisions in each $a_\star$-direction, except at the intersection of tipping

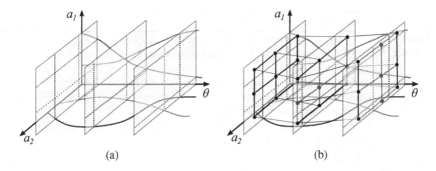

Fig. 3. DRT graph construction by the sweeping plane algorithm, with 2 vertical (blue, cyan) and 2 horizontal (red, magenta) tipping surfaces. (a) 3 × 3 rectangular cells generated by the tipping surfaces in each sweeping plane. (b) The associated DRT graph in each plane (in green: new vertices and edges in the second and third planes).

surfaces, where a rectangle disappears while a new appears. By observing these rectangle updates during the plane sweeping from θ_{start} to θ_{end}, we can construct the DRT sub-graph, where each rectangle corresponds to a vertex while each tipping surface between two rectangles corresponds to an edge. In other words, at each intersection of tipping surfaces, $s_\star$ new vertices and their associated $(3s_\star + 2)$ edges are generated, as illustrated in Fig. 3. (The reader is referred to [12] for more details.)

Our modified algorithm consists of using a topological sweep [22] in order to find the next closest intersection of tipping surfaces for updating the planar division. We consider at most $|S| - 2$ intersections at each update, by considering only the intersections of consecutive tipping surfaces in their ordered structure in the sweeping plane along each $a_\star$-axis, and find the closest one among them. After each update, the modifications of such intersections can be performed in constant time. We can also ignore the intersections that are not in the range between θ_{start} and θ_{end}. In particular, since we have $|\theta_{end} - \theta_{start}| \ll 2\pi$, the number of intersections can be considered a small constant.

Hereafter, we denote this specific procedure by $Sweep(\mathbf{S}, \theta_{start}, \theta_{end})$, and we write $G = Sweep(\mathbf{S}, \theta_{start}, \theta_{end})$ for the output DRT sub-graph.

3.3 k-Neighbourhood Construction

Finding the neighbouring vertices and edges of a given vertex v with depth k, is actually equivalent to constructing the DRT sub-graph containing those vertices and edges around v. Here, we assume to know a value θ_v lying into $\mathcal{R}_v$, and we use it as initial value of the sweeping algorithm. The plane is thus swept twice, in the space (a_1, a_2, θ), along the two directions of the θ-axis.

The key-point is how to limit the construction of the DRT sub-graph. For this purpose we verify, for each generated vertex u, its neighbourhood depth $t_v(u)$ with respect to v. If $t_v(u) > k$ for all vertices in the current sweeping plane, the process ends.

Algorithm 1. k-neighbourhood computation (in the left-hand side along the θ-axis)

Input: A DRT v (or its associated image I_v); a positive integer k.
Output: The DRT sub-graph $F = (V, E)$ containing the k-neighbours of v.

1   **for** $\star = 1, 2$ **do**
2   Determine the tipping surfaces associated to v: $\mathbf{S}_\star^+(I_v)$, $\mathbf{S}_\star^-(I_v)$ (Sec. 3.1).
3   In $\mathbf{S}_\star^+(I_v)$ (resp. $\mathbf{S}_\star^-(I_v)$), find the $(k + 1)$-th lowest (resp. uppermost) tipping surface $f_\star^+$ (resp. $f_\star^-$), that intersects the initial plane at θ_v.
4   Find and sort the $k + 1$ tipping surfaces that are lower (resp. upper) or equal to $f_\star^+$ (resp. $f_\star^-$), and put them in an ordered set $\mathbf{S}_\star$.

5   Initialize $V = \emptyset$, $E = \emptyset$
6   Initialize $\theta_{prev} = \theta_v$
7   **repeat**
8   **for** $\star = 1, 2$ **do**
9   Find the next left intersection $\theta_\star^+$ of $f_\star^+$ (resp. $\theta_\star^-$ of $f_\star^-$) with the other surface in $\mathbf{S}_\star$ for $\theta < \theta_{prev}$.
10     $\theta_{next} = \min\{\theta_1^+, \theta_1^-, \theta_2^+, \theta_2^-\} - \epsilon$ with $\epsilon \ll 1$
11     $\Delta F = Sweep(\mathbf{S}_1 \cup \mathbf{S}_2, \theta_{prev}, \theta_{next})$
12   **if** $\exists u \in \Delta V, t_v(u) \leq k$ **then**
13        $F = F \cup \Delta F$, $\theta_{prev} = \theta_{next}$
14   **if** *the next intersecting surface with $f_\star^+$ (or $f_\star^-$) is in $\mathbf{S}_\star$* **then**
15   Exchange their order in $\mathbf{S}_\star$.
16   **else**
17   Replace $f_\star^+$ (or $f_\star^-$) in $\mathbf{S}_\star$ with the new intersecting surface.

18   **until** $\forall u \in \Delta V, t_v(u) > k$;

When a vertex u is created, its depth $t_v(u)$ depends on that of the two vertices w_1 and w_2 to which it is adjacent in the $a_\star$-direction of the tipping surface intersection. We then have $t_v(u) = 1 + \min\{t_v(w_1), t_v(w_2)\}$. (An iterative backtracking process is also needed to update the depth of $w_\star$ and its successive neighbours, whenever $t_v(w_\star) > t_v(u) + 1$.)

In each $a_\star$-direction, by considering the $(k + 1)$ closest tipping surfaces around $\mathcal{R}_v$, we can obtain all the vertices u such that $t_v(u) \leq k$. In the θ-direction, we need to check if $t_v(u) > k$ for all vertices u in the current sweeping plane; if so, the sweeping ends.

The global process is described in Alg. 1. (Note that the algorithm describes only the k-neighbourhood construction in the left-hand side along the θ-axis, but the right-hand side can be constructed similarly.) The first loop (Lines 1–4) initializes the set of tipping surfaces that are needed to generate the k-neighbours of a given DRT v. We obtain $2(k + 1)$ vertical (resp. horizontal) tipping surfaces close to $\mathcal{R}_v$ at $\theta = \theta_v$, and sort and store them in the lists $\mathbf{S}_\star$. In the second loop (Line 7), we first verify how long we can keep the same tipping surface sets $\mathbf{S}_\star$ (Lines 9–10), and then build a DRT sub-graph by using the $Sweep$ algorithm for this verified θ interval (Line 11). After verifying if there still exists a generated vertex whose neighbourhood depth is $\leq k$ (Line 12), we update the tipping surface sets $\mathbf{S}_\star$ for the next interval (Lines 14–17).

Obviously, F is not the smallest sub-graph G including the k-neighbours of v. To obtain G from F, we simply keep vertices whose neighbourhood depth is $\leq k$.

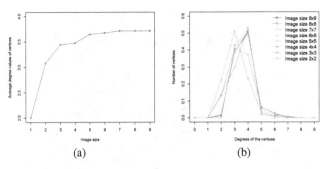

(a)   (b)

Fig. 4. (a) Average vertex degree in a 2D DRT graph. (b) Normalised vertex degree distribution in a 2D DRT graph.

4   Complexity Analysis

4.1   Time Complexity of k-Neighbourhood Construction Algorithm

In order to obtain the initial $2(k + 1)$ vertical (resp. horizontal) tipping surfaces of $\mathbf{S}_\star$, the time complexity is $O(N^2)$ for Line 2; $O(N^2)$ for Line 3 on the average case if we use Hoare's FIND algorithm [23]; and $O(N^2)$ and $O(k \log k)$ for finding and sorting the tipping surfaces in Line 4, respectively. Then, we carry out the plane sweep for each updated $\mathbf{S}_1 \cup \mathbf{S}_2$. For each iteration in the loop, the most costly parts are Lines 9 and 11, which require $O(N^2)$ and $O(k^2)$, respectively.

The next question concerns the number of updates for $\mathbf{S}_1 \cup \mathbf{S}_2$. If m is the degree of any vertex u of a DRT graph, this update number can be estimated as $m(2k + 1)$, since the union of $\mathcal{R}_u$ for all u in the k-neighbourhood of a given vertex v contains at most $2k + 1$ adjacent $\mathcal{R}_u$ in the θ-direction. Therefore, the time complexity is $O(mkN^2)$ for this iterative plane sweep loop.

The time complexity of Alg. 1 is thus $O(mkN^2)$, which is significantly lower than that of our previous algorithm [18], namely $O(m^k N^2)$. We observe, in the next section, that m can be estimated as a low constant value, leading to a final complexity of $O(kN^2)$.

4.2   Average Degree of DRT Graphs

The DRT graph space complexity for an image of size $N \times N$ is $O(N^9)$, both for vertices and edges [12]. In other words, the number of vertices and that of edges grow at the same rate. We can then infer that m is actually bounded, independently of N.

By analogy, let us imagine that we divide a 2D plane with straight lines defined randomly. Three lines will almost never intersect at a same point, and for a number of lines sufficiently large, the cells of the induced subdivision will be mostly triangles.

Following this analogy, we may infer that the degree of the vertices of the 2D DRT graphs in the planes (a_1, θ) and (a_2, θ) is close to 3, in average. However, this analogy has some limits. Indeed, the considered tipping curves are not straight lines, while their very regular structure implies that many curves often intersect at a same point.

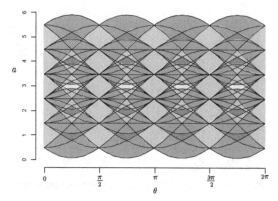

Fig. 5. Degree distribution in a 2D DRT graph, viewed in the dual subdivision of the parameter space. Each colour represents a given degree, that corresponds here to the number of curves bounding each cell (3: red, 4: green, 5: blue; 6: yellow).

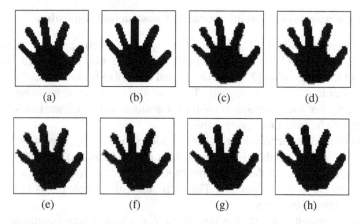

Fig. 6. Input images and results of the iterated local search for image registration. (a) Reference image, (b) target image, and (c) the initial transformed image of (b) with $(a_1, a_2, \theta) = (0.365, -0.045, 0.1423)$. (d–h) Local optima obtained from (c) by using k-neighbours for $k = 1, 3, 5, 10, 15$ respectively. Note that in (c–h), pixels are coloured if they are different from those in (a); yellow (resp. red) pixels are white (resp. black) in (c–h) and black (resp. white) in (a).

Nevertheless, we can assimilate a 2D DRT graph (which is the projection of a 3D DRT graph onto the $(a_\star, \theta)$ plane) to a planar graph whenever N is sufficiently large. Under such assumption, the Euler formula is valid, *i.e.*, we have $v - e + f = 2$, where v, e and f are the number of (0D) vertices, (1D) edges and induced (2D) cells, respectively. From the very definition of the DRT graph, we have $4f \leq 2e$. It then comes that $2e/v \leq 4 - 8/v$. As $v \gg 8$ in DRT graphs, we have $2e/v < 4$, where $2e/v$ is indeed the average degree of the 2D DRT graph. It follows that the average degree m of the 3D DRT graph (obtained by Cartesian product of two 2D DRT graphs) is lower than $2 \times 4 = 8$. This is confirmed by the experimental analysis, illustrated in Fig. 4(a).

In practice, the maximal degree of the vertices within a DRT graph also remains close to this average value. Indeed, the histograms depicted in Fig. 4(b) show that the

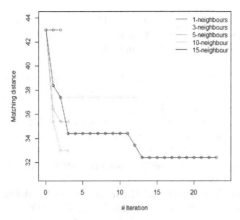

Fig. 7. Distance evolution during iterations of local search for the inputs in Fig. 6 (a) and (b), from the initial transformation in Fig. 6 (c), with respect to different depths k

2D DRT vertex degrees converge rapidly to a stable distribution that contains mainly degrees of value 3 and 4 (with a maximal value experimentally identified at 8). More qualitatively, Fig. 5 illustrates the distribution of these degrees of a 2D DRT graph.

5 Experiments

Iterated local search was applied to image registration. In this section we validate Alg. 1 in practice, and we observe its local behaviour when varying k. For simplicity, we use the same experimental settings as in [18], *i.e.,* two input binary images and a signed distance [24] for Eq. (1). In order to find an initial transformation, we use the first and second order central moments of a binary shape [25]. Experiments are carried out with different neighbourhood sizes, $k = 1, 3, 5, 10, 15$ on binary images of size 53×53 from the initial transformation, as illustrated in Fig. 6. We can observe in Fig. 7 that the locally optimal distance improves when we use a larger neighborhood, which is coherent in a gradient descent paradigm.

6 Conclusion

We have significantly improved the time complexity of the process of computing a neighbourhood of given depth within a DRT graph, without requiring the computation of the whole graph. This time complexity may be reduced in some cases, in particular if the image is binary by dealing only with the pixels that compose the binary object border. The proposed applications only validate our approach as a proof of concept. Nevertheless, an exact – *i.e.,* numerical error-free – strategy is novel in the field of image registration and may open the way to new image processing paradigms. In future work we will explore the notion of DRT graph in $\mathbb{Z}^3$.

References

1. Nouvel, B., Rémila, É.: Characterization of bijective discretized rotations. In: Klette, R., Žunić, J. (eds.) IWCIA 2004. LNCS, vol. 3322, pp. 248–259. Springer, Heidelberg (2004)

2. Nouvel, B., Rémila, E.: Configurations induced by discrete rotations: Periodicity and quasi-periodicity properties. Discrete Appl. Math. 147, 325–343 (2005)
3. Berthé, V., Nouvel, B.: Discrete rotations and symbolic dynamics. Theor. Comput. Sci. 380, 276–285 (2007)
4. Nouvel, B.: Self-similar discrete rotation configurations and interlaced Sturmian words. In: Coeurjolly, D., Sivignon, I., Tougne, L., Dupont, F. (eds.) DGCI 2008. LNCS, vol. 4992, pp. 250–261. Springer, Heidelberg (2008)
5. Jacob, M.A., Andres, E.: On discrete rotations. In: Proc. DGCI, pp. 161–174 (1995)
6. Amir, A., Kapah, O., Tsur, D.: Faster two-dimensional pattern matching with rotations. Theor. Comput. Sci. 368, 196–204 (2006)
7. Reveillès, J.P.: Géométrie discrète, calcul en nombres entiers et algorithmique. Thèse d'État, Université Strasbourg 1 (1991)
8. Andres, E.: The quasi-shear rotation. In: Miguet, S., Ubéda, S., Montanvert, A. (eds.) DGCI 1996. LNCS, vol. 1176, pp. 307–314. Springer, Heidelberg (1996)
9. Richman, M.S.: Understanding discrete rotations. In: Proc. ICASSP, vol. 3, pp. 2057–2060. IEEE (1997)
10. Andres, E., Fernandez-Maloigne, C.: Discrete rotation for directional orthogonal wavelet packets. In: Proc. ICIP, vol. 2, pp. 257–260. IEEE (2001)
11. Ngo, P., Passat, N., Kenmochi, Y., Talbot, H.: Topology-preserving rigid transformation of 2D digital images. IEEE T. Image Process. 23, 885–897 (2014)
12. Ngo, P., Kenmochi, Y., Passat, N., Talbot, H.: Combinatorial structure of rigid transformations in 2D digital images. Comput. Vis. Image Und. 117, 393–408 (2013)
13. Nouvel, B.: Rotations discrètes et automates cellulaires. PhD thesis, École Normale Supérieure de Lyon (2006)
14. Nouvel, B., Rémila, É.: Incremental and transitive discrete rotations. In: Reulke, R., Eckardt, U., Flach, B., Knauer, U., Polthier, K. (eds.) IWCIA 2006. LNCS, vol. 4040, pp. 199–213. Springer, Heidelberg (2006)
15. Thibault, Y., Kenmochi, Y., Sugimoto, A.: Computing upper and lower bounds of rotation angles from digital images. Pattern Recogn. 42, 1708–1717 (2009)
16. Ngo, P., Kenmochi, Y., Passat, N., Talbot, H.: On 2D constrained discrete rigid transformations. Ann. Math. Artif. Intell. (in press), doi:10.1007/s10472-014-9406-x
17. Ngo, P., Kenmochi, Y., Passat, N., Talbot, H.: Topology-preserving conditions for 2D digital images under rigid transformations. J. Math. Imaging Vis. 49, 418–433 (2014)
18. Ngo, P., Sugimoto, A., Kenmochi, Y., Passat, N., Talbot, H.: Discrete rigid transformation graph search for 2D image registration. In: Huang, F., Sugimoto, A. (eds.) PSIVT 2013. LNCS, vol. 8334, pp. 228–239. Springer, Heidelberg (2014)
19. Zitová, B., Flusser, J.: Image registration methods: A survey. Image Vision Comput. 21, 977–1000 (2003)
20. Schowengerdt, R.A.: Remote Sensing: Models and Methods for Image Processing, 3rd edn. Elsevier Academic Press (2007)
21. Noblet, V., Heinrich, C., Heitz, F., Armspach, J.P.: Recalage d'images médicales. Tech Ing (MED910) (2014)
22. Edelsbrunner, H., Guibas, L.J.: Topologically sweeping an arrangement. Journal Comput. Syst. Sci. 38, 165–194 (1989); Corrig. 42, 249–251 (1991)
23. Hoare, C.A.R.: Algorithm 65: find. Commun. ACM 4, 321–322 (1961)
24. Boykov, Y., Kolmogorov, V., Cremers, D., Delong, A.: An integral solution to surface evolution PDEs via geo-cuts. In: Leonardis, A., Bischof, H., Pinz, A. (eds.) ECCV 2006. LNCS, vol. 3953, pp. 409–422. Springer, Heidelberg (2006)
25. Flusser, J., Zitová, B., Suk, T.: Moments and Moment Invariants in Pattern Recognition. Wiley (2009)

The Minimum Barrier
Distance – Stability to Seed Point Position

Robin Strand[1], Filip Malmberg[1], Punam K. Saha[2], and Elisabeth Linnér[1]

[1] Centre for Image Analysis, Division of Visual Information and Interaction,
Uppsala University, Sweden
[2] Dept. of Electrical and Computer Engineering and Dept. of Radiology, University of Iowa,
Iowa City, IA 52242, United States

Abstract. Distance and path-cost functions have been used for image segmentation at various forms, e.g., region growing or live-wire boundary tracing using interactive user input. Different approaches are associated with different fundamental advantages as well as difficulties. In this paper, we investigate the stability of segmentation with respect to perturbations in seed point position for a recently introduced pseudo-distance method referred to as the minimum barrier distance. Conditions are sought for which segmentation results are invariant with respect to the position of seed points and a proof of their correctness is presented. A notion of δ-interface is introduced defining the object-background interface at various gradations and its relation to stability of segmentation is examined. Finally, experimental results are presented examining different aspects of stability of segmentation results to seed point position.

1 Introduction

Distance transforms and functions are widely used in image processing [1–8]. Intensity-weighted distance transforms take the pixel intensity values into consideration [6, 9–11]. This way, the homogeneity of intensity values in regions are quantified. This property makes intensity-weighted distances well-suited for image segmentation, where the goal is to group pixels in homogenous regions.

Here, image segmentation by intensity-weighted distances is achieved by assigning to each pixel the distance to, and the label of, the closest labeled *seed point*. The seed point can be given by a user or utilizing some domain knowledge. Stability to seed point position is a very important aspect of these segmentation methods; small perturbations in the seed point position should ideally not lead to significantly changed segmentation result. Different aspects on stability to seed point position for intensity-weighted distances have been examined [1, 12, 13].

Many distance- or cost functions in image segmentation by region growing have a *locality property*, which makes it possible to efficiently compute distance (or cost) defined as optimal paths by propagating values from adjacent points. The locality property is in one sense a deficiency, since global properties can not easily be included in the propagation. On the other hand, the locality property is essential for efficient computation, typically by wave-front propagation starting from points with zero cost or distance.

E. Barcucci et al. (Eds.): DGCI 2014, LNCS 8668, pp. 111–121, 2014.

A well-known and often used method, watershed, can intuitively be described as a flooding process of a topographic representation of gray scale images, where low intensities correspond to low altitude in the topographic representation. A recently developed method, the *minimum barrier distance*, is given by the minimum barrier that has to be passed to go from one point to another in the topographic representation of image data. The minimum barrier distance is a pseudo-metric, meaning that the properties identity, symmetry and triangle inequality, but not positivity, are obeyed [1]. We have showed that the minimum barrier distance has many properties that make it beneficial for image segmentation, e.g., stability to seed point position (using simple Dice's coefficient on segmentation results), noise, smoothing etc. [1, 13].

However the minimum barrier path cost function is not local in the above sense, and therefore standard wave-front propagation algorithms are not sufficient for computing the exact minimum barrier distance map [1, 13]. We have developed efficient algorithms for computing the minimum barrier distance [13] and approximations thereof [1]. A vectorial minimum barrier distance that takes multi-band image data, e.g., color or other multispectral images, as input has also been developed [14].

In this paper, we examine the stability to seed point position for the minimum barrier distance by introducing conditions under which the distance between points p and q equals the distance between p and another point q'. Also, we introduce the δ-interface which essentially is a region of uncertainty of border position in a segmentation result.

2   The Minimum Barrier Distance

We will consider $D = \{p = (x_1, x_2, \ldots, x_n) \in \mathbb{Z}^n : L_i \leq x_i \leq U_i\}$ as the image domain. The intensity at a point p is denoted by $f(p)$. A *path*, $\pi = \langle p_0, p_1, \ldots, p_n \rangle$, is a sequence of points $p_0, p_1, \ldots, p_n$, where each pair of consecutive points are adjacent given some adjacency. The maximum and minimum values along a path $\pi = \langle p_0, p_1, \ldots, p_n \rangle$ are

$$\max(\pi) = \max_{0 \leq i \leq n} f(p_i) \text{ and } \min(\pi) = \min_{0 \leq i \leq n} f(p_i),$$

respectively. The *minimum barrier* along a path π is defined as

$$\Phi(\pi) = \max(\pi) - \min(\pi).$$

The minimum barrier *distance* between two points p and q is defined as

$$\Phi(p, q) = \min_{\pi \in \Pi} \Phi(\pi),$$

where Π is the set of all paths between p and q. A path in Π that attains the minimum barrier distance is called an MBD-optimal path. The minimax and maximin distances between points are defined as

$$\Phi_{\max}(p, q) = \min_{\pi \in \Pi} \max(\pi) \text{ and } \Phi_{\min}(p, q) = \max_{\pi \in \Pi} \min(\pi),$$

respectively. The concatenation of two paths π and τ is denoted $\pi \cdot \tau$.

In [1], we gave an approximation of the minimum barrier distance, namely $\Phi(p, q) \approx \Phi_{\max}(p, q) - \Phi_{\min}(p, q)$. We showed that $\Phi_{\max} - \Phi_{\min}$ equals Φ in the continuous case and converges to Φ as the sampling density increases in the discrete case [1].

Distance Transform Algorithms and Segmentation

As previously mentioned, the key to seeded segmentation by intensity weighted distance transforms is the distance transform, where each pixel is assigned the distance to (and label of) the closest seed point. We have developed different algorithms that compute approximations of the minimum barrier distance-transform in $O(n \log n)$ time, where n is the number of pixels [1, 13]. In this manuscript, we will use our recently developed efficient algorithm for computing the exact minimum barrier distance [13]. The worst case time complexity is $O(mn \log n)$ (or $O(m(n + m))$, depending on the data structure), where m is the range of the weight function (here the number of intensitites that can be attained).

3 Stability of the Minimum Barrier Distance with Respect to Seed Point Position

In this section, we introduce this paper's theoretical results on conditions for which the minimum barrier distance between p and q equals the distance between p and q' and the δ-interface. The latter will also be used in the experiments in Section 4.

3.1 Invariance under Seed Point Position

In interactive segmentation, there is an uncertainty in the exact positions of seed points due to, for example, inter- and intra-user variability. Given a point p (for example in the object) and a user added seed point q (for example in the background). The used distance function has a high invariance under seed point position if there is a large set of points that can be used instead of q without altering the distance to the point p. Following this idea, conditions for which the minimum barrier distance value between a point p and seed point q equals the minimum barrier distance value between p and another seed point q' are given in the following theorem.

Theorem 1. *Let q and q' be points, and let $I = [I_{\min}, I_{\max}]$ be an interval such that any MBD-optimal paths π between q and q' are such that $\max(\pi) \leq I_{\max}$ and $\min(\pi) \geq I_{\min}$. Let p be a point such that*

- $\Phi_{\max}(q, p) > I_{\max}$. (†)
- $\Phi_{\min}(q, p) < I_{\min}$. (‡)

Then $\Phi(q', p) = \Phi(q, p)$.

Proof. Let π be an MBD-optimal path between p and q', and let π' be an MBD-optimal path between q and q'. Then $\Phi(\pi \cdot \pi') = \max(\max(\pi), \max(\pi')) - \min(\min(\pi), \min(\pi')) = \max(\pi') - \min(\pi') = \Phi(\pi')$, and so $\Phi(q', p) \leq \Phi(q, p)$.

Next, assume (*) that there exists a path π'' between q' and p such that $\Phi(\pi'') < \Phi(q, p)$. Then the following properties hold:

- $\max(\pi'') \geq \Phi_{\max}(q, p)$. (Otherwise $\max(\pi \cdot \pi'') < \Phi_{\max}(q, p)$, contradicting (†).)
- $\min(\pi'') \leq \Phi_{\min}(q, p)$. (Otherwise $\min(\pi \cdot \pi'') < \Phi_{\min}(q, p)$, contradicting (‡).)

From these properties, it follows that $\Phi(\pi \cdot \pi'') = \Phi(\pi'')$. But $\pi \cdot \pi''$ is a path between q and p and therefore $\Phi(\pi'') \not< \Phi(q, p)$, contradicting (*). □

Note that the strong conditions in Theorem 1 are not often satisfied for real images. The theorem and conditions are presented here mainly to give a deeper understanding of the minimum barrier distance.

3.2 The δ-Interface

In region growing segmentation among multiple objects [15], different objects are defined with respect to their seed points where a given point is assigned to an object whose seeds are closest to that point under a cost- or distance-function. Here, we formulate a notion of δ-interface that defines the region of uncertainty at the interface between two or more objects. Ideally, small perturbation in seed point positions will only effect the segmentation region within the uncertain region and, therefore, segmented core objects regions will be unaltered under such small perturbation in seed positions. The δ-interface will be used as a tool to quantify this stability.

The δ-interface is a region with 'thickness' δ where two regions meet. A δ-interface with $\delta = 0$ is the set of points located at the exact same distance to two seed points. Given a value of $\delta \geq 0$, the δ-interface between points p and q is the set $I(p, q, \delta) = \{r : |\Phi(p, r) - \Phi(q, r)| \leq \delta\}$.

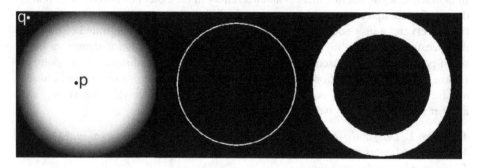

Fig. 1. Illustration of δ-interfaces. Left: example image and points p and q. Middle: $I(p, q, \delta)$ (shown in white), with small δ. Right: $I(p, q, \delta)$, with large δ.

The following Theorem gives a correspondence between the delta-interface between p, q and p, q'.

Theorem 2. $I(p, q, \delta) \subset I(p, q', \delta + \Phi(q, q'))$

Proof. Let $r \in I(p, q, \delta)$. We want to show that $r \in I(p, q', \delta + \Phi(q, q'))$, i.e. that

$$|\Phi(q', r) - \Phi(p, r)| \leq \delta + \Phi(q, q'). \tag{1}$$

(i) First of all, by the triangular inequality, $\Phi(q', r) - \Phi(p, r) \leq \Phi(q, r) + \Phi(q, q') - \Phi(p, r) \leq \delta + \Phi(q, q')$.

(ii) Secondly, by the triangle inequality, $|\Phi(r,p) - \Phi(q,p)| \leq \Phi(q,r)$ for any given points p, q, r. Therefore, $\Phi(q',r) - \Phi(p,r) \geq \Phi(q,r) - \Phi(q,q') - \Phi(p,r) \geq -\delta - \Phi(q,q')$. By combining (i) and (ii), we get exactly (1).   □

The following Corollary gives a set containing borders between objects defined by seed points p, q and p, q', respectively. The smaller the set, the more robust the minimum barrier distance is to seed point positioning.

Corollary 1. $I(p,q,\delta), I(p,q',\delta) \subset (I(p,q',\delta + \Phi(q,q')) \cap I(p,q,\delta + \Phi(q,q')))$

The sets $I(p,q',\Phi(q,q'))$ and $I(p,q,\Phi(q,q'))$, i.e. when $\delta = 0$, are illustrated in Figure 2.

Fig. 2. Illustration of Corollary 1 and the experiment in Section 4.1. Left: A gray-level image with three seed-points, p, q, and q'. Middle: The set $I(p,q',\Phi(q,q'))$ is shown in gray. The sets $I(p,q,0)$ and $I(p,q',0)$ are shown in white. Right: The set $I(p,q,\Phi(q,q'))$ is shown in gray. The sets $I(p,q,0)$ and $I(p,q',0)$ are shown in white. $E(p,q,q',0)$ is the intersection of $I(p,q',\Phi(q,q'))$ and $I(p,q,\Phi(q,q'))$.

4 Experiments and Results

Seventeen images from the grabcut dataset [16] are used for the experiments, see Figure 3. The images come with a reference segmentation, which we eroded/dilated to get object and background regions as shown for an example image in Figure 4. The images, converted to gray scale by using the mean of the three color band values, are used. The intensity range of the images is [0, 255].

4.1 Stability to Seed Point Position

Seed points added by a user with low accuracy or precision will differ in spatial distance. Ideally, a small perturbation in the seed point position gives a small difference in the segmentation result. In this section, we will evaluate the effect of a small change in position of the seed point q. This will be done by comparing the δ-interface of p and q with that of p and q' (such that q and q' are at a fixed spatial distance). We will see how

Fig. 3. Images from the grabcut database used for the experiments. Bottom: image numbering.

Fig. 4. Image 8 from the grabcut database used for the experiments. Left: original image. Right: Object region (white) and background region (gray).

the spatial distance between seed points q and q' affect the δ-interface. The set metric we will use in the evaluation is

$$E(p, q, q', \delta) = |I(p, q', \delta + \Phi(q, q')) \cap I(p, q, \delta + \Phi(q, q'))|$$

with $\delta = 0$. The set $I(p, q, 0)$ corresponds exactly to the set of pixels with equal distance between p and q. In other words, intuitively, this is the border between the regions that correspond to p and q, respectively in a segmentation. By comparing this set and the set obtained by $I(p, q', 0)$, where q' is a point close to q, the difference between a

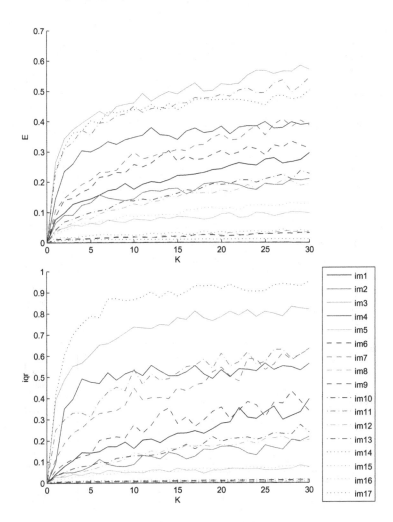

Fig. 5. Cardinality of the set $E(p, q, q', 0)$, for random points p (in the object), q (in the background), and q' (in the background such that $\|q - q'\| = K$) as a function of K. The mean values (top) and the interquartile range (iqr, the difference of the 75:th and 25:th percentiles, bottom) of 1000 executions for each image and value of K are plotted. The values are normalized with the size (number of pixels) of the images.

segmentation result obtained by seed points p and q compared to p and q' can be quantified. The quantification we will use is $E(p, q, q', 0)$, for random points p (in the object), q (in the background), and q' (in the background and such that $\|q - q'\|$ approximately equals a constant K). See the illustration in Figure 2.

Note that the quantification E used in the experiment only measures the difference between the interface $I(p, q, 0)$ and the interface $I(p, q', 0)$. Clear distinction by the distance function between object and background gives a small interface and the lower E is, the smaller the interface is. See Figure 1 and Figure 2 for illustrations.

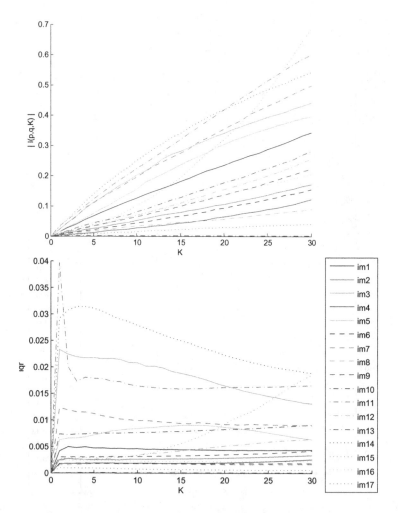

Fig. 6. Cardinality of the set $I(p, q, K)$ as a function of K. For each of the 17 images in Figure 3 $I(p, q, K)$, where q is a random object seed point and p is a random background seed point, is generated for $K = 1 \ldots 30$. The mean values (top) and the interquartile range (iqr, bottom) of 1000 executions for each image and value of K are plotted. The values are normalized with the size (number of pixels) of the images.

Since the cardinality of the set $\{q' \in \mathbb{Z}^2 : \|q - q'\| = K\}$ usually is very small in the digital space, we can not use points q and q' at *exactly* a Euclidean distance K. Instead, points q' are derived as follows: a random angle $\theta \in [0, 2\pi[$ is extracted and the point $q' = q + (\lfloor K \cos \theta \rceil, \lfloor K \sin \theta \rceil)$ is used. Here, $\lfloor \cdot \rceil$ is the rounding off function which gives the nearest integer.

A plot showing the mean overlap region from 1000 iterations per value of K for the 17 images in Figure 3 is shown in Figure 5.

4.2 Uncertainty of Border Position

The idea behind this experiment is to see how the cardinality of the δ-interface increases with increasing values of δ. By iteratively and randomly selecting an object seed point p and a background seed point q and computing the cardinality of the set $I(p, q, K)$ for increasing values of K, quantitative measures on the region where the uncertainty of the border position is high can be obtained. A plot on the mean cardinality of the uncertainty region from 1000 iterations per value of K for the 17 images in Figure 3 is shown in Figure 6.

5 Conclusions and Future Work

In Theorem 1, we gave conditions that guarantee that the minimum barrier distance between p and q equals the distance between p and a third point r. The conditions are correct by the proof. However, the practical implications of Theorem 1 are limited since the conditions are not often satisfied in real world images. Therefore, no evaluation of Theorem 1 is presented here.

In the first experiment, we evaluated how small perturbations in the seed point position changed the segmentation result by a measure on the difference between object region border when p and q are used as seed points, compared to when p and q', such that $\|q - q'\| \approx K$, are used. From the plots in Figure 5, we can see that there is a big difference between the images with lowest error (image 10, 17, 6, 8, 15, 5 in Figure 3) and the images with highest error (image 3, 11, 14, 4, 7, 9). The low-error images all have a distinct border between object and background, whereas this border is not as distinct in the images with high error. Also, images with an apparent texture all give high error. This is expected since only barriers based on intensity are taken into account in the minimum barrier distance given here. The vectorial minimum barrier distance, presented in [14], with appropriate point-wise texture features is expected to handle these situations better.

Statistical dispersions were computed by the interquartile range (the difference of the 75:th and 25:th percentiles) for the experiments, see Figure 5 and Figure 6. For experiment 1, the general trend was that the higher mean value, the higher interquartile distance. The interquartile distance values were roughly in the same range as the mean values. For experiment 2, the dispersion measures were lower, suggesting that these results are more reliable.

In the second experiment, the region with pixels such that the distances to the two seed points p and q are similar (K) is quantified. In most cases, a linear relationship between K and the cardinality of the δ-interface was observed. This relationship seems to not hold for image 15, probably due to the weak gradient in the border between object and background. Judging from the images with the lowest error (image 10, 17, 8, 4, 6, 2) and the images with highest error at $K = 30$ (image 15, 11, 14, 7, 3, 5), we conclude that this error measure seems to be less sensitive to texture.

In our future work, we plan to combine the minimum barrier distance with a spatial distance term to avoid problems with object 'leakage' due to the fact that after passing a large 'barrier', distance values are constant. Experiment 1 in this paper evaluates how small perturbations in the seed point positions change the segmentation result. As a result of the insightful reviews of this paper, we plan to define an uncertainty model from a probability displacement distribution and then to compute statistics while sampling q' in this distribution. This approach will more realistically model the distribution of seed points added by a large number of users. We also plan to do extensive evaluation of stability of several intensity-weighted distance functions, and methods that can be expressed as intensity-weighted distance- or cost-functions such as [17], with respect to seed point position, blur, inhomogeneity, noise, etc.

References

1. Strand, R., Ciesielski, K.C., Malmberg, F., Saha, P.K.: The minimum barrier distance. Computer Vision and Image Understanding 117(4), 429–437 (2013), special issue on Discrete Geometry for Computer Imagery
2. Rosenfeld, A., Pfaltz, J.L.: Distance functions on digital pictures. Pattern Recognition 1, 33–61 (1968)
3. Borgefors, G.: Distance transformations in arbitrary dimensions. Computer Vision, Graphics, and Image Processing 27, 321–345 (1984)
4. Borgefors, G.: On digital distance transforms in three dimensions. Computer Vision and Image Understanding 64(3), 368–376 (1996)
5. Danielsson, P.-E.: Euclidean distance mapping. Computer Graphics and Image Processing 14, 227–248 (1980)
6. Saha, P.K., Wehrli, F.W., Gomberg, B.R.: Fuzzy distance transform: theory, algorithms, and applications. Computer Vision and Image Understanding 86, 171–190 (2002)
7. Strand, R.: Distance functions and image processing on point-lattices: with focus on the 3D face- and body-centered cubic grids, Ph.D. thesis, Uppsala University, Sweden (2008), http://urn.kb.se/resolve?urn=urn:nbn:se:uu:diva-9312
8. Strand, R., Nagy, B., Borgefors, G.: Digital distance functions on three-dimensional grids. Theoretical Computer Science 412(15), 1350–1363 (2011)
9. Sethian, J.A.: Level Set Methods and Fast Marching Methods. Cambridge University Press (1999)
10. Fouard, C., Gedda, M.: An objective comparison between gray weighted distance transforms and weighted distance transforms on curved spaces. In: Kuba, A., Nyúl, L.G., Palágyi, K. (eds.) DGCI 2006. LNCS, vol. 4245, pp. 259–270. Springer, Heidelberg (2006)
11. Falcão, A.X., Stolfi, J., Lotufo, R.A.: The image foresting transform: Theory, algorithms, and applications. IEEE Transactions on Pattern Analysis and Machine Intelligence 26(1), 19–29 (2004)
12. Audigier, R., Lotufo, R.A.: Seed-relative robustness of watershed and fuzzy connectedness approaches. In: Brazilian Symposium on Computer Graphics and Image Processing. IEEE (2007)

13. Ciesielski, K.C., Strand, R., Malmberg, F., Saha, P.K.: Efficient algorithm for finding the exact minimum barrier distance. Computer Vision and Image Understanding 123, 53–64 (2014)
14. Kårsnäs, A., Strand, R., Saha, P.K.: The vectorial minimum barrier distance. In: 2012 21st International Conference on Pattern Recognition (ICPR), pp. 792–795 (2012)
15. Ciesielski, K.C., Udupa, J.K., Saha, P.K., Zhuge, Y.: Iterative relative fuzzy connectedness for multiple objects with multiple seeds. Computer Vision and Image Understanding 107(3), 160–182 (2007)
16. Rother, C., Kolmogorov, V., Blake, A.: GrabCut: Interactive Foreground Extraction using Iterated Graph Cuts. ACM Transactions on Graphics, SIGGRAPH 2004 (2004)
17. Bertrand, G.: On topological watersheds. Journal of Mathematical Image and Vision 22(2-3), 217–230 (2005)

Efficient Computation of the Outer Hull of a Discrete Path*

Srecko Brlek, Hugo Tremblay, Jérôme Tremblay, and Romaine Weber

Laboratoire de Combinatoire et d'Informatique Mathématique,
Université du Québec à Montréal,
CP 8888 Succ. Centre-ville, Montréal (QC) Canada H3C 3P8
{brlek.srecko,jerome.tremblay}@uqam.ca, hugo.tremblay@lacim.ca,
weberomaine@gmail.com

Abstract. We present here a linear time and space algorithm for computing the outer hull of any discrete path encoded by its Freeman chain code. The basic data structure uses an enriched version of the data structure introduced by Brlek, Koskas and Provençal: using quadtrees for representing points in the discrete plane $\mathbb{Z} \times \mathbb{Z}$ with neighborhood links, deciding path intersection is achievable in linear time and space. By combining the well-known wall follower algorithm for traversing mazes, we obtain the desired result with two passes resulting in a global linear time and space algorithm. As a byproduct, the convex hull is obtained as well.

Keywords: Freeman code, lattice paths, radix tree, discrete sets, outer hull, convex hull.

1 Introduction

The ever-growing use of digital screens in industrial, military and civil applications gave rise to a new branch of study of discrete objects: digital geometry, where objects are sets of pixels. In particular, their various geometric properties play an essential role, for allowing the design of efficient algorithms for recognizing patterns and extracting features: these are mandatory steps for an accurate interpretation of acquired images.

Convex objects play a prominent role in several branches of mathematics, namely functional analysis, optimization, probability and mathematical physics (see [1] for a detailed account of convex geometry and applications). In Euclidean geometry, given a finite set of points, the problem of finding the smallest convex set containing all of them led to the introduction of the geometric notion of *convex hull*. On the practical side, the computation of the convex hull proved to be one of the most fundamental algorithm in computational geometry as it has many applications ranging from operational research [2] to design automation [3]. It is also widely used in computer graphics, and particularly in image processing [4]. For example, the Delaunay triangulation of a d-dimensional set of points

---

* With the support of NSERC (Canada).

E. Barcucci et al. (Eds.): DGCI 2014, LNCS 8668, pp. 122–133, 2014.
© Springer International Publishing Switzerland 2014

in Euclidean space is equivalent to finding the convex hull of a set of $d + 1$-dimensional points [5]. It is well known that for the Euclidean case, algorithms for computing the convex hull of a set $S \subset \mathbb{R}^2$ run in $\mathcal{O}(n \log n)$ time where $n = |S|$ (see [6,7]). One can also show that such algorithms are optimal up to a linear constant (see [8,9,10] for the general case).

Nevertheless, by confining the problem to computing the convex hull of simple polygons, linear asymptotic bounds are achieved (see [11,12]). The digital version of this problem is a little more involved. For instance, one can compute the convex hull of a set of pixels S by first computing the Euclidean convex hull of S and then digitalizing the result [13]. This automatically yields $\mathcal{O}(n \log n)$ asymptotical bounds in the worst case. In the discrete case, the situation is surprisingly easier with the help of combinatorics on words, a field which recently led to the development of efficient tools to study digital geometry (see [17,18]). For instance, linear asymptotic bounds are obtained when considering discrete paths encoded by elementary steps. Indeed, Brlek et al. designed a linear time algorithm for computing the discrete convex hull of non self-intersecting closed paths in the square grid [14]. It is based on an optimal linear time and space algorithm for factorizing a word in Lyndon words designed by Duval [15]. The situation is more complicated for intersecting paths.

Here, we describe a linear algorithm for computing the outer hull of any discrete path using the data structure described in [16] where the authors designed a linear time and space algorithm for detecting path intersection. It rests on the encoding of points in the discrete plane $\mathbb{Z} \times \mathbb{Z}$ by quadrees deduced from the radix order representation of binary coordinate points. Then, each path is dynamically encoded by adding a pointer for each step of the discrete path encoded on the four letter alphabet {0,1,2,3}. Starting from that, the wall follower algorithm used for maze solutions allows to take at each intersection the rightmost available step. The resulting two-passes algorithm is linear in space and time. As a byproduct, the convex hull of any discrete path is computed in linear time.

2 Preliminaries

Given a finite alphabet Σ, a *word* w is a function $w : [1, 2, \ldots, n] \longrightarrow \Sigma$ denoted by its sequence of letters $w = w_1 w_2 \cdots w_n$, and $|w| = n$ is its *length*. For $a \in \Sigma$, $|w|_a$ is the number of letters a in w. The set of all words of length k is denoted by Σ^k. Consequently, $\Sigma^* = \bigcup_{i=0}^{\infty} \Sigma^i$ is the set of all finite words on Σ where $\Sigma^0 = \{\varepsilon\}$, the set consisting of the empty word. Σ^* together with the operation of concatenation form a monoid called *the free monoid on Σ*.

There is a bijection between the set of pixels and $\mathbb{Z}^2$ obtained by mapping $(a, b) \in \mathbb{Z}^2$ to the unitary square whose bottom left vertex coordinate is (a, b). Therefore, we may consider pixels as elements of $\mathbb{Z}^2$. By definition, a *discrete set* S is a set of pixels, i.e. $S \subset \mathbb{Z}^2$. Also, S is called *4-connected* if each pair of pixels share a common edge and *8-connected* if each pair of pixels share a common edge or vertex. Since any discrete set is a disjoint collection of 8-connected sets, we consider from now on that discrete sets are 4 or 8-connected.

A convenient way of representing discrete sets without hole is to use a word describing its contour. In 1961, Herbert Freeman proposed an encoding of discrete objects by specifying their contour using the four elementary steps ($\rightarrow$, $\uparrow$, $\leftarrow$, $\downarrow$) $\simeq$ $(0, 1, 2, 3)$ [19]. This encoding provides a convenient representation of discrete paths in $\mathbb{Z}^2$. By definition, a *discrete path* P is a sequence of points $P = \{p_1, p_2, \ldots, p_n\}$ where p_i and p_{i+1} are neighbors for $1 \leq i < n$. Intuitively, two points u and v are neighbors if and only if $u = v \pm e$ where $e \in \{(1, 0), (0, 1), (-1, 0), (0, -1)\}$.

It is clear from these definitions that any discrete path P is represented by a word $w \in \mathcal{F}^*$ where $\mathcal{F} = \{0, 1, 2, 3\}$ is the Freeman alphabet. It is worth mentioning that in the case of a closed discrete path, w is unique up to a circular permutation of its letters. For example, any circular permutation of the word $w = 001100322223$ represents the discrete path shown in Figure 1(a). One says that a word $w \in \mathcal{F}^*$ is *closed* if and only if $|w|_0 = |w|_2$ and $|w|_1 = |w|_3$. Further, w is called *simple* if it codes a non self-intersecting discrete path. For instance, $w = 001100322223$ is non-simple and closed.

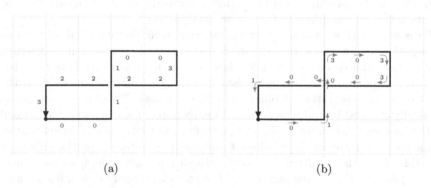

(a)   (b)

Fig. 1. (a) A discrete path coded by the word $w = 001100322223$; (b) and its first difference word $\Delta(w) = 01030330001$

It is sometimes useful to consider encoding of paths with turns instead of elementary steps. Such encoding is obtained from the contour word $w = w_1 \cdots w_n$ by setting

$$\Delta(w) = (w_2 - w_1)(w_3 - w_2) \cdots (w_n - w_{n-1})$$

where subtraction is computed modulo 4. $\Delta(w)$ is called the *first differences word of* w. Letters of $\Delta(w) \in \mathcal{F}^*$ are interpreted via the bijection $(0, 1, 2, 3) \simeq$ (forward, left turn, u-turn, right turn). For example, one can verify in Figure 1(b) that $\Delta(w) = 01030330001$ and that it codes the turns of w.

Now, every path w is contained in a smallest rectangle, or bounding box such that we can define the point W as in Figure 2(a). W is easily obtained in linear time by keeping track of the extremum coordinates while reading the word. It is worth mentioning that in the case of a closed simple path u, this coordinate corresponds to the point W of the standard decomposition of u obtained by considering the following four extremal points of the bounding box: W (lowest on the left side), N (leftmost on the top side), E (highest on the right side) and S (rightmost on the bottom side) (see Figure 2(b)).

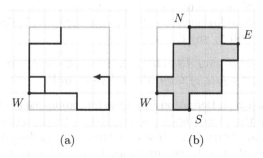

(a)   (b)

Fig. 2. (a) Smallest rectangle containing a discrete path and the point W; (b) Standard decomposition of a self-avoiding closed path

We close this section by recalling some notions about topological graph theory (see [20] for a thorough exposition of the subject). Let P be a discrete path (i.e. a sequence of integer points) coded by the word w. The image of P as a subset of $\mathbb{R}^2$ is noted G_P while the graph of its image embed in the plane $\mathbb{R}^2$ is noted $\mathcal{G}(P)$. Such embeddings in the plane are completely determined by associating a cyclic order on the edges around each vertex in the following way: Begin by fixing an orientation at each point (e.g. counterclockwise). Then, for each vertex v in G_P, define the cyclic permutation on incident edges of v. This defines a rotation scheme on G_P. One can then show that such a scheme is equivalent to an oriented embedding of G_P on a surface. For example, Figure 3 illustrates the path P coded by $w = 001233$, its graph G_P and its associated counterclockwise embedding $\mathcal{G}(P)$ in $\mathbb{R}^2$.

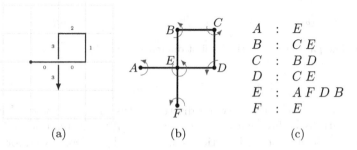

(a)   (b)   (c)

Fig. 3. (a) The graph G_P associated to the path coded by $w = 001233$; (b) The counterclockwise embedding $\mathcal{G}(P)$ in $\mathbb{R}^2$; (c) And its associated rotation scheme

3   Outer and Convex Hull

We recall from topology that given a set S, the boundary ∂S is the set of points in the closure of S, not belonging to the interior of S. Now, let S be a 8-connected discrete set. The *outer hull of S*, denoted Hull(S) is the boundary of the intersection of all discrete sets without hole containing S, i.e. the non self-intersecting path following the exterior contour of S. Definition 1 extends the notion of outer hull to any discrete path.

Definition 1. *Let P be any discrete path. Then, the outer hull of P, denoted by Hull(P) is the outer face of the embedded graph $\mathcal{G}(P)$.*

The difference between Definition 1 and the preceding one lies in the use of the embedding of P in the plane instead of a discrete set to describe the outer hull. This choice is not arbitrary as it allows the treatment of discrete line segments (i.e. Euclidean sets of area 0). For example, Figure 4 illustrates the outer hull of the path coded by $w = 021$. Remark that using Definition 1, the boundary of discrete line segments are coded by closed words, e.g. the outer hull of the horizontal line segment coded by 0 is coded by 02 (see Figure 4). This ensures that Definition 1 is a convenient generalization of the outer hull to discrete paths. Indeed, if P codes the boundary of a discrete set S, then P is simple and closed by definition. This gives $P = $ Hull(P) and since Hull(S) is the boundary $\partial(S)$ of S by definition, we have

$$\text{Hull}(S) = \partial(S) = P = \text{Hull}(P).$$

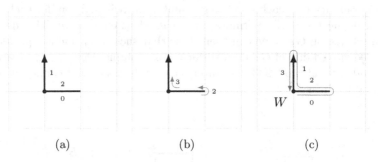

(a)   (b)   (c)

Fig. 4. (a) The path $w = 021$, (b) its first diference word $\Delta(w) = 23$ and (c) its outer hull Hull$(w) = 0213$

Since there is a bijection between discrete paths in $\mathbb{Z}^2$ and words on $\mathcal{F}$, we identify P with its coding word w and we write Hull(w) instead of Hull(P).

Finally, we recall some basic notions concerning digital convexity, for which a detailed exposure appears in [14,18]. Let S be an 8-connected discrete set. S is *digitally convex* if it is the Gauss digitalization of a convex subset R of $\mathbb{R}^2$, i.e. $S = \text{Conv}(R) \cap \mathbb{Z}^2$. The *convex hull of S*, denoted Conv(S) is the intersection of all convex sets containing S. In the case of a closed simple path w, Conv(w) is given by the Spitzer factorization of w (see [21,14]). Given $w = w_1 w_2 \cdots w_n \in \{0,1\}^*$, one can compute the NW part of this factorization as follows: Start with the list $(b_1, b_2, \ldots, b_n) = (w_1, w_2, \ldots, w_n)$. If the slope $\rho(b_i) = |b_i|_1/|b_i|_0$ of b_i is strictly smaller than that of b_{i+1} for some i, then

$$(b_1, b_2, \ldots, b_k) = (b_1, \ldots, b_{i-1}, b_i b_{i+1}, b_{i+2}, \ldots, b_k).$$

By repeating this process until it is no longer possible to concatenate any words, one obtains the Spitzer factorization of w. The NE, SE and SW parts of the factorization are obtained by rotations.

4   Algorithm

Let $w \in \mathcal{F}^*$ be a discrete path and G_w its graph representation. Remark that the application $g : w \mapsto G_w$ is not bijective since it is not injective (for example, $u = 0$ and $v = 02$ admits the same graph). Now, recall from Section 2 that the embedding $\mathcal{G}(w)$ of G_w in $\mathbb{R}^2$ gives rise to a rotation scheme (provided we fix an orientation). We use this embedding to compute the outer hull of w (i.e. the outer face of $\mathcal{G}(w)$): Fix an orientation $\mathcal{O}$ of the surface $\mathbb{R}^2$ and let $e_i = (u, v)$ be an arc from vertex u to v in $\mathcal{G}(w)$ such that e_i is an edge of the outer face of the embedding $\mathcal{G}(w)$ and such that e follows the fixed orientation $\mathcal{O}$. Next, compute $e_{i+1} = (v, \sigma_v(u))$ where σ_v is the cyclic permutation associated to v in $\mathcal{G}(w)$. By letting $\mathcal{O}$ be the counterclockwise orientation, one can iterate this process to obtain the outer face of $\mathcal{G}(w)$. For example, using the rotation scheme defined for $w = 001233$ in Figure 3(c) and starting with the arc (A, E), one computes the sequence of arcs

$$(A, E), \ (E, F), \ (F, E), \ (E, D), \ (D, C), \ (C, B), \ (B, E), \ (E, A)$$

which corresponds to the outer hull of w (see Figure 5).

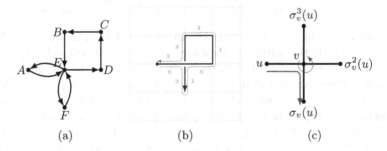

(a)   (b)   (c)

Fig. 5. (a) The sequence of arcs obtained by using the rotation scheme of Figure 3(c); (b) The outer hull of $w = 001233$; (c) The sequence (u, v), $(v, \sigma_v(u))$ corresponds to a right turn in the graph of a path, provided the orientation is counterclockwise

The correctness of this method follows from the so-called "right-hand rule" or "wall follower algorithm" for traversing mazes. Indeed, given an arc (u, v), taking the adjacent arc $(v, \sigma_v(u))$ amounts to "turning right" at vertex v (see Figure 5(c)). The underlying principle of our algorithm is thus to walk along the path, starting at an origin point on the outer hull and turning systematically right at each intersection and returning to the origin point. The preceding discussion guarantees that the resulting walk is then precisely the outer hull of w.

To efficiently implement this procedure, several problems must be addressed. First, as stated before, the walk needs to start on a coordinate of the outer hull, otherwise the resulting path may not describe the correct object. This can be solved by choosing the point W associated with the contour word w as the starting point.

Secondly, whenever a path returns to W (the simplest of which is the path coded by $w = 021$, see Figure 4), before continuing on, one must make sure that the algorithm does not stop until every such sub-path has been explored. An easy solution for managing that situation is to keep a list of all neighbors of W that are in the path P associated with w. This list has at most two elements since no vertex in P is located below or left of W.

Finally, one needs to recognize intersections and decide of the rightmost turn. We solve this problem by using a quadtree structure keeping information on neighborhood relations. This so-called radix quadtree structure was first introduced by Brlek, Koskas and Provençal in [16] for detecting path intersections. Given a discrete path w starting at $(x, y) \in \mathbb{N}^2$ and staying in the first quadrant, the *quadtree structure associated to* w (see [18] and [16] for the generalization to all four quadrants) is described as follows. $G = (N, R, T)$ is a quadtree where:

N is the set of vertices associated to points in the plane;

R is a set of edges representing the fatherhood relation: $r \in R$ is an edge from (x, y) to (x', y') $\iff$ (x', y') is a child of (x, y), that is if $(x', y') = (2x + \alpha, 2y + \beta)$ where $(\alpha, \beta) \in \{0, 1\}^2$;

T is a set of edges representing the neighborhood relation: $t \in T$ is an edge from (x, y) to (x', y') $\iff$ (x', y') is a neighbor of (x, y), that is if $(x', y') = (x, y) + e$ where $e \in \{(1, 0), (0, 1), (-1, 0), (0, -1)\}$ (see Example 2).

One should note that the quadtree structure is described in [16] with unidirectional edges while in this paper all edges are considered bidirectional. Further, by following the procedure described in [16] to build the quadtree, one adds neighborhood links between non-visited nodes during the recursion process. This is easily fixed by adding a boolean label to each neighborhood edge indicating if that specific edge is part of the discrete path or if it has been added by a recursive call. This ensures that the points u and v are neighbors if and only if there is a non-labeled neighborhood edge between these two vertices in the quadtree structure.

It is worth mentioning that this structure is computed in linear time and space. Moreover, it can be generalized to any discrete path as opposed to paths staying in the first quadrant.

Example 2. *Let* $w = 001100322223$ *be the word coding the discrete path in Figure 1 translated to the origin. The quadtree structure associated to w is represented in Figure 6. Parenthood and neighborhood relations are respectively represented by black and red edges. Visited nodes are marked by red squares.*

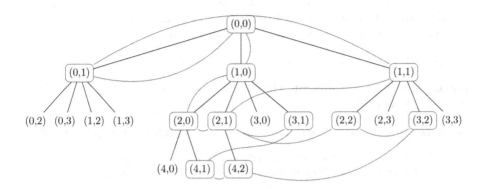

Fig. 6. Quadtree corresponding to the word $w = 001100322223$. Neighborhood edges added by recursive calls are omitted.

This gives rise to the following Algorithm 1 to compute the outer hull of a discrete path w, which proceeds as follows.

Algorithm 1. Outer hull

Require: A word $w \in \mathcal{F}^*$ coding a discrete path
Ensure: A simple word $w' \in \mathcal{F}^*$ describing Hull(w)
 1: Construct the quadtree G associated to w rooted in W
 2: Let W be the leftmost lowest coordinate on the bounding box of w
 3: Let N be the set of all visited neighbors of W
 4: c $\leftarrow$ W $+ (1,0)$ if it is in N or W $+ (0,1)$ otherwise
 5: $w' = $ Step(c $-$ W)
 6: **while** c $\neq$ W or N $\neq \varnothing$ **do**
 7:     turn $= 2 \mod 4$
 8:     **for** each neighbor v of c **do**
 9:         **if** $[\text{Step}(\text{v} - \text{c}) - \text{Lst}(w')] + 1 \mod 4 \leq [\text{turn}] + 1 \mod 4$ **then**
10:             turn $\leftarrow$ Step(v $-$ c) $-$ Lst(w')
11:             next $\leftarrow$ v
12:         **end if**
13:     **end for**
14:     $w' = w' \cdot$ Step(next $-$ c)
15:     remove c from N
16:     c $\leftarrow$ next
17: **end while**
18: **return** w'

It is assumed that the point W, that is the leftmost lower point of the bounding box is known. First, the quadtree G associated to w is built starting from W. Then, the graph G is traversed from its root W, following the path represented by w. At every intersection c, we need to:

(a) extract the letter α associated to the vector $\vec{cv}$ for each neighbor v of c;
(b) determine the turn associated to each v, that is $\Delta(w_c \cdot \alpha)$;
(c) choose the rightmost one, that is the closest to 3.

This procedure ends when returning to the point W.

Theorem 3 (Correctness of Algorithm 1). *For any word $w \in \mathcal{F}^*$, Algorithm 1 returns* Hull(w).

Proof. Let Hull(w) be of length $k \in \mathbb{N}^+$. We use the following loop invariant:

At the start of the ith iteration of the while loop in Line 6, w' is a prefix of length i of the contour word associated to Hull(w).

The invariant holds the first time Line 6 is executed, since at that time, w' is the first step of the outer hull of w computed at Line 5. Now, assume the invariant holds before the ith iteration of the loop. Then, Lines 8 to 13 find the rightmost turn at the current coordinate c. Then in Line 14, w' is concatenated with the step of this turn. By the right-hand rule for solving simply connected maze, considering rightmost turns yields coordinates on the outer hull of w. Consequently, at the end of the iteration, w' is a prefix of the contour word associated to Hull(w) of length $i + 1$. Finally, at the end of the loop, w' is a prefix of the contour word associated to Hull(w) of length k, that is $w' = $ Hull(w). Note that since any neighbor of W is on Hull(w), Line 15 clearly removes every element from N yielding, at termination, an empty set. $\qquad \square$

We end this section by showing that Algorithm 1 is linear in time and space. First, the quadtree structure is constructed in linear time (see [16]). Also, as stated before, the point W is easily computed in linear time. Consequently, computations in Line 1 are performed in linear time. Next, Line 2, 4 and 5 each take constant time. Moreover, the set N is constructed in linear time by accessing neighborhood informations of the root in the quadtree structure, so Line 3 takes linear time. Now, since any coordinate has at most four neighbors, the **for** loop in Line 8 is executed at most four time per iteration of the **while** loop. Line 15 takes constant time. This is due to the fact that N contains at most two elements. Since instructions in Line 7, 9, 10, 11, 14 and 16 all are computed in constant time, at most $k(4c_1 + c_2)$ computations occur during the execution of the **while** loop where $k \in \mathbb{N}^+$ is the length of Hull(w) and $c_1, c_2 \in \mathbb{R}$ some constants, thus making Algorithm 1 linear in time. Finally, the quadtree structure needs space linear in the length of the path, so that our algorithm is also linear in space.

Example 4. *Consider the word* $w = 001100322223$ *of Example 2. Then, Algorithm 1 yields* $w' = 001001223223$ *(see Figure 7). One can easily verify that* w' *is a simple path describing the outer hull of* w, *so* $\text{Hull}(w) = w'$.

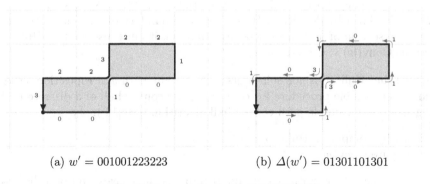

(a) $w' = 001001223223$ (b) $\Delta(w') = 01301101301$

Fig. 7. Outer hull of $w = 001100322223$

Our algorithm was implemented using the C++ programming language and tested with numerous examples (see Figure 8). The source code is available at `http://bitbucket.org/htremblay/outer_hull`.

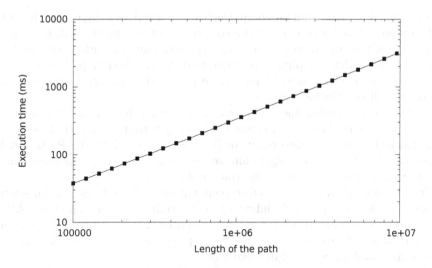

Fig. 8. Running time of Algorithm 1 for random discrete paths of length 10^5 to 10^7, with each point representing the mean running time of 100 random discrete paths of same length

Finally, we show how Algorithm 1 can be used to compute in linear time and space the convex hull of any discrete path. It relies on the following rather obvious result:

Proposition 5. *Let $w \in \mathcal{F}^*$ be a boundary word coding a discrete path. Then,*

$$\mathrm{Conv}(w) = \mathrm{Conv}(\mathrm{Hull}(w)).$$

Proof. If w is simple, then $\mathrm{Hull}(w) = w$ so the claim holds. Now, suppose w is non-simple. Then by definition, $\mathrm{Hull}(w)$ is the boundary of w. Since, $\mathrm{Conv}(w)$ is the intersection of all convex sets containing w, it must also contain $\mathrm{Hull}(w)$ and thus $\mathrm{Conv}(w) = \mathrm{Conv}(\mathrm{Hull}(w))$.   □

Recall that $\mathrm{Hull}(w)$ is non self-intersecting for any path w. Proposition 5 then yields a very simple procedure for computing the convex hull of a discrete path using Brlek et al. simple path convex hull algorithm (see [14]):

1. Start by computing $\mathrm{Hull}(w) = w'$;
2. Compute $\mathrm{Conv}(w')$.

It is clear that the preceding procedure computes the convex hull of a discrete path in linear time and space. Indeed, we showed in Section 4 that the first step is computed in linear time and space. Furthermore, it is shown in [14] that the second step is computed in a similar fashion.

5   Concluding Remarks

We presented an algorithm for computing the outer hull of a discrete path. This led to a procedure for computing the convex hull of any discrete set. Our algorithm is a significant improvement over the convex hull algorithm presented in [14] in the sense that computations can be made on any discrete path as opposed to non self-intersecting ones. Moreover, we proved that such computations can be made in linear time and space.

Instead of computing the outer hull of a discrete path P as described in this paper, one could want to compute the largest simply connected isothetic polygon such that all integers points on its boundary are visited by P. Although some modifications to our algorithm are necessary in order to perform such computations, the time complexity would not change.

In addition, this research begs to be generalized to three dimensional discrete spaces, that is geometry in Euclidean space $\mathbb{R}^3$ studying sets of unit cubes. Also, applications of our algorithm is not limited to convex hull problems. We plan on using it to study various path intersections problems such as primality, union, intersection and difference of discrete sets.

Acknowledgement. Martin Lavoie helped in implementing Algorithm 1. Thanks Martin. We are also grateful to the reviewers for the accurate comments which substantially improved the theoretical background of our work.

References

1. Gruber, P.M.: Convex and discrete geometry. Springer (2007)
2. Sherali, H., Adams, W.: A hierarchy of relaxations between the continuous and convex hull representations for zero-one programming problems. SIAM Journal on Discrete Mathematics 3(3), 411–430 (1990)
3. Kim, Y.S.: Recognition of form features using convex decomposition. Computer-Aided Design 24(9), 461–476 (1992)
4. Kim, M.A., Lee, E.J., Cho, H.G., Park, K.J.: A visualization technique for DNA walk plot using k-convex hull. In: Proceedings of the Fifth International Conference in Central Europe in Computer Graphics and Visualization, Plzeň, Czech Republic, Západočeská univerzita, pp. 212–221 (1997)
5. Okabe, A., Boots, B., Sugihara, K.: Spacial tesselations: Concepts and applications of Voronoi diagrams. Wiley (1992)
6. Graham, R.A.: An efficient algorithm for determining the convex hull of a finite planar set. Information Processing Letters 1(4), 132–133 (1972)
7. Chan, T.M.: Optimal output-sensitive convex hull algorithms in two and three dimensions. Discrete & Computational Geometry 16, 361–368 (1996)
8. Yao, A.C.C.: A lower bound to finding the convex hulls. PhD thesis, Stanford University (April 1979)
9. Chazelle, B.: An optimal convex hull algorithm in any fixed dimension. Discrete & Computational Geometry 10, 377–409 (1993)
10. Goodman, J.E., O'Rourke, J.: Handbook of discrete and computational geometry, 2nd edn. CRC Press (2004)
11. McCallum, D., Avis, D.: A linear algorithm for finding the convex hull of a simple polygon. Information Processing Letters 9(5), 201–206 (1979)
12. Melkman, A.: On-line construction of the convex hull of a simple polyline. Information Processing Letters 25, 11–12 (1987)
13. Chaudhuri, B.B., Rosenfeld, A.: On the computation of the digital convex hull and circular hull of a digital region. Pattern Recognition 31(12), 2007–2016 (1998)
14. Brlek, S., Lachaud, J.O., Provençal, X., Reutenauer, C.: Lyndon + Christoffel = digitally convex. Pattern Recognition 42, 2239–2246 (2009)
15. Duval, J.P.: Factorizing words over an ordered alphabet. J. Algorithms 4(4), 363–381 (1983)
16. Brlek, S., Koskas, M., Provençal, X.: A linear time and space algorithm for detecting path intersection. Theoretical Computer Science 412, 4841–4850 (2011)
17. Blondin Massé, A.: À l'intersection de la combinatoire des mots et de la géométrie discrète: Palindromes, symétries et pavages. PhD thesis, Université du Québec à Montréal (February 2012)
18. Provençal, X.: Combinatoire des mots, géométrie discrète et pavages. PhD thesis, Université du Québec à Montréal (September 2008)
19. Freeman, H.: On the encoding of arbitrary geometric configurations. IRE Transactions on Electronic Computers EC-10(2), 260–268 (1961)
20. Gross, J.L., Tucker, T.W.: Topological graph theory. Wiley (1987)
21. Spitzer, F.: A combinatorial lemma and its application to probability theory. Transactions of the American Mathematical Society 82, 323–339 (1956)

Voronoi-Based Geometry Estimator
for 3D Digital Surfaces[*]

Louis Cuel[1,2], Jacques-Olivier Lachaud[1], and Boris Thibert[2]

[1] Université de Savoie, Laboratoire LAMA,
Le bourget du lac, France
[2] Université de Grenoble, Laboratoire Jean Kuntzmann,
Grenoble, France

Abstract. We propose a robust estimator of geometric quantities such
as normals, curvature directions and sharp features for 3D digital
surfaces. This estimator only depends on the digitisation gridstep and is
defined using a digital version of the Voronoi Covariance Measure, which
exploits the robust geometric information contained in the Voronoi cells.
It has been proved in [1] that the Voronoi Covariance Measure is resilient
to Hausdorff noise. Our main theorem explicits the conditions under
which this estimator is multigrid convergent for digital data. Moreover,
we determine what are the parameters which maximise the convergence
speed of this estimator, when the normal vector is sought. Numerical
experiments show that the digital VCM estimator reliably estimates normals, curvature directions and sharp features of 3D noisy digital shapes.

1 Introduction

Differential quantities estimation, surface reconstruction and sharp feature detection are motivated by a large number of applications in computer graphics, geometry processing or digital geometry.

Digital geometry estimators. The commun way to link the estimated
differential quantities to the expect Euclidean one is the multigrid convergence principle: when the shape is digitized on a grid with *gridstep h*
tending to zero, the estimated quantity should converge to the expected
one. In dimension 2, several multigrid convergent estimators have been
introduced to approach normals [2, 3] and curvatures [3–5]. In 3D, empirical methods for normal and curvature estimation have been introduced in
[6]. More recently, a convergent curvature estimator based on covariance
matrix was presented in [7].

[*] This research has been supported in part by the ANR grants DigitalSnow ANR-11-
BS02-009, KIDICO ANR-2010-BLAN-0205 and TopData ANR-13-BS01-0008.

E. Barcucci et al. (Eds.): DGCI 2014, LNCS 8668, pp. 134–149, 2014.

Voronoi-based geometry estimation. Classical principal component analysis methods try to estimate normals by fitting a tangent plane or a higher-order polynomial (e.g. see [8]). In contrast, Voronoi-based methods try to fit the normal cones to the underlying shape, either geometrically [9] or more recently using the covariance of the Voronoi cells [1, 10]. Authors of [1] have improved the method of [10] by changing the domain of integration and the averaging process. The authors define the *Voronoi Covariance Measure* (VCM) of any compact sets, and show that this notion is stable under Hausdorff perturbation. Moreover, the VCM of a smooth surface encodes a part of its differential information, such as its normals and curvatures. With the stability result, one can therefore use the VCM to estimate differential quantities of a surface from a Hausdorff approximation such as a point cloud or a digital contour.

Voronoi Covariance measure background. The *Voronoi covariance measure* (VCM) has been introduced in [1] for normals and curvature estimations. Let K be a compact subset of $\mathbb{R}^3$ and d_K the *distance function* to K, i.e. the map $d_K(x) := \min_{p \in K} \|p - x\|$. A point p where the previous minimum is reached is called a *projection* of x on K. Almost every point admits a single projection on K, thus definining a map $p_K : \mathbb{R}^3 \to K$ almost everywhere. The *R-offset* of K is the R-sublevel set of d_K, i.e. the set $K^R := d_K^{-1}(]-\infty, R[)$. The VCM maps any integrable function $\chi : \mathbb{R}^3 \to \mathbb{R}^+$ to the matrix

$$\mathcal{V}_{K,R}(\chi) := \int_{K^R} (x - p_K(x))(x - p_K(x))^t \chi(p_K(x)) dx.$$

Remark that this definition matches the definition introduced in [1]: when χ is the indicatrix of a ball, one recovers a notion similar to the convolved VCM : $\mathcal{V}_{K,R}(\chi) := \int_{K^R \cap p_K^{-1}(\mathcal{B}_y(r))} (x - p_K(x))(x - p_K(x))^t dx$. The domain of integration $K^R \cap p_K^{-1}(\mathcal{B}_y(r))$ is the offset of K intersected with a union of Voronoi cells (cf. Figure 1). The stability result of [1] implies that

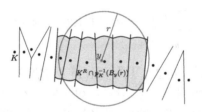

Fig. 1. VCM domain of integration

information extracted from the covariance matrix such as normals or principal directions are stable with respect to Hausdorff perturbation.

Contributions. The contributions of the paper can be sketched as follows. First, we define the estimator of the VCM in the case of digital data, for which we prove the multigrid convergence (Sect. 2, Theorem 1). We then show that the normal direction estimator, defined as the first eigenvector of the VCM estimator, is also convergent with a speed in $O(h^{\frac{1}{8}})$ (Sect. 3, Corollary 1). Furthermore, Theorem 2 specifies how to choose parameters r and R as functions of h to get the convergence. Finally, we present an experimental evaluation showing that this convergence speed is closer to $O(h)$ in practice (Sect. 4). Moreover, experiments indicate that the VCM estimator can be used to estimate curvature information and sharp features in the case of digital data perturbated by Hausdorff noise.

2   VCM on Digital Sets

In this section, we define an estimator of the VCM in the case of 3D digital input. Theorem 1 explicits the conditions under which this estimator is multigrid convergent for digital data.

2.1   Definition

Let X be a compact domain of $\mathbb{R}^3$ whose boundary is a surface of class C^2. We denote ∂X the boundary of X, $X_h := Dig_h(X) = X \cap (h\mathbb{Z})^3$ the Gauss digitisation of X, and $\partial_h X \subset \mathbb{R}^3$ the set of boundary surfels of X_h. We define a digital approximation of the VCM on a subset of the point cloud : $Z_h = \partial_h X \cap h(\mathbb{Z} + \frac{1}{2})^3$. For each point $x \in h(\mathbb{Z} + \frac{1}{2})^3$ with $x = (x_1, x_2, x_3)$, we can define the voxel of center x by $\mathrm{vox}(x) = [x_1 - \frac{1}{2}h, x_1 + \frac{1}{2}h] \times [x_2 - \frac{1}{2}h, x_2 + \frac{1}{2}h] \times [x_3 - \frac{1}{2}h, x_3 + \frac{1}{2}h]$. We then define the *digital VCM estimator* as

$$\widehat{\mathcal{V}}_{Z_h, R}(\chi) := \sum_{x \in \Omega_h^R} h^3 (x - p_{Z_h}(x))(x - p_{Z_h}(x))^t \chi(p_{Z_h}(x)),$$

where $\Omega_h^R = \{x \in Z_h^R \cap h(\mathbb{Z} + \frac{1}{2})^3, \mathrm{vox}(x) \subset Z_h^R\}$ is the set of centers of voxels entirely contained in Z_h^R, the R-offset of Z_h (see Fig. 2). Remark that the Hausdorff distance between ∂X and the point cloud Z_h used in the definition is less than h.

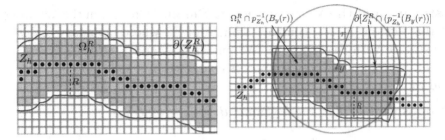

Fig. 2. Digitisation of the offset and its localisation

2.2 Multigrid Convergence of the VCM-Estimator

The main theoretical result of the paper is the following theorem. Roughly speaking, it quantifies the approximation of the VCM of a smooth surface by the digital VCM of its Gauss digitisation. We denote by $\|.\|_{op}$ the matrix norm induced by the Euclidean metric. Given a function $f : \mathbb{R}^n \to \mathbb{R}$, we let $\|f\|_\infty = \max_{x \in \mathbb{R}^n} |f(x)|$ and denote $\mathrm{Lip}(f) = \max_{x \neq y} |f(x) - f(y)| / \|x - y\|$ its Lipschitz constant.

Theorem 1. *Let X be a compact domain of $\mathbb{R}^3$ whose boundary ∂X is a C^2 surface with reach $\rho > 0$. Let $R < \frac{\rho}{2}$ and $\chi : \mathbb{R}^3 \to \mathbb{R}^+$ be an integrable function whose support is contained in a ball of radius r. Then for any $h > 0$ such that $h \leq \min\left(R, \frac{r}{2}, \frac{r^2}{32\rho}\right)$, one has*

$$\left\|\mathcal{V}_{\partial X, R}(\chi) - \widehat{\mathcal{V}}_{Z_h, R}(\chi)\right\|_{op} = O\Big(\mathrm{Lip}(\chi) \times [(r^3 R^{\frac{5}{2}} + r^2 R^3 + r R^{\frac{9}{2}})h^{\frac{1}{2}}]$$
$$+ \|\chi\|_\infty \times [(r^3 R^{\frac{3}{2}} + r^2 R^2 + r R^{\frac{7}{2}})h^{\frac{1}{2}} + r^2 R h]\Big).$$

In the theorem and in the following of the text, the constant involved in the notation $O(.)$ only depends on the reach of ∂X and on the dimension (which is three here).

For the proof of Theorem 1, we introduce the VCM of the point cloud Z_h, namely $\mathcal{V}_{Z_h, R}(\chi)$. By the triangle inequality, one has

$$\left\|\mathcal{V}_{\partial X, R}(\chi) - \widehat{\mathcal{V}}_{Z_h, R}(\chi)\right\|_{op} \leq \|\mathcal{V}_{\partial X, R}(\chi) - \mathcal{V}_{Z_h, R}(\chi)\|_{op} + \left\|\mathcal{V}_{Z_h, R}(\chi) - \widehat{\mathcal{V}}_{Z_h, R}(\chi)\right\|_{op}.$$

In Proposition 1, we bound the second term and in Proposition 2, we bound the first term.

Estimation of the VCM of a Point Cloud. Here and in the following of this section, X is a compact domain of $\mathbb{R}^3$ whose boundary ∂X is a C^2 surface with reach $\rho > 0$. We put $R < \frac{\rho}{2}$ and $\chi : \mathbb{R}^3 \to \mathbb{R}^+$ is an integrable function whose support is contained in a ball $\mathcal{B}_y(r)$ of center y and radius r.

Proposition 1. *For any $h \leq \min\left(R, \frac{r}{2}, \frac{r^2}{32\rho}\right)$, one has*

$$\left\| \mathcal{V}_{Z_h, R}(\chi) - \widehat{\mathcal{V}}_{Z_h, R}(\chi) \right\|_{op} = O\left[r^2 R^2 (\mathrm{Lip}(\chi) R + \|\chi\|_\infty) \, h^{\frac{1}{2}} + r^2 R \|\chi\|_\infty h \right].$$

Proof. **Step 1:** The aim of the first step is to prove that

$$\mathcal{V}_{Z_h, R}(\chi) = \int_{\mathrm{vox}(\Omega_h^R)} (x - p_{Z_h}(x))(x - p_{Z_h}(x))^{\mathbf{t}} \chi(p_{Z_h}(x)) dx + R^2 \|\chi\|_\infty O(hr^2).$$

Since $\mathrm{vox}(\Omega_h^R) \subset Z_h^R$, one has

$$
\begin{aligned}
\mathcal{V}_{Z_h, R}(\chi) = &\int_{\mathrm{vox}(\Omega_h^R)} (x - p_{Z_h}(x))(x - p_{Z_h}(x))^{\mathbf{t}} \chi(p_{Z_h}(x)) dx \\
&+ \int_{Z_h^R \setminus \mathrm{vox}(\Omega_h^R)} (x - p_{Z_h}(x))(x - p_{Z_h}(x))^{\mathbf{t}} \chi(p_{Z_h}(x)) dx
\end{aligned}
$$

By using the facts that $\|x - p_{Z_h}(x)\| \leq R$, χ is bounded by $\|\chi\|_\infty$, and the support of χ is contained in the ball $\mathcal{B}_y(r)$ (see Figure 2), the second term of the previous equation is bounded by

$$R^2 \times \|\chi\|_\infty \times \mathcal{H}^3 \left(\left[Z_h^R \setminus \mathrm{vox}(\Omega_h^R) \right] \cap p_{Z_h}^{-1}(\mathcal{B}_y(r)) \right).$$

Now, we claim that $Z_h^R \cap p_{Z_h}^{-1}(\mathcal{B}_y(r)) \subset p_{\partial X}^{-1}(\mathcal{B}_y(2r))$. Indeed, let $x \in Z_h^R \cap p_{Z_h}^{-1}(\mathcal{B}_y(r))$. The fact that the Hausdorff distance between Z_h and ∂X is less than h implies that $x \in \partial X^{R+h}$. Now, since $h \leq R$, Lemma 3 implies that $\|p_{\partial X}(x) - p_{Z_h}(x)\| \leq \sqrt{8h\rho} + h$, which leads to

$$\|p_{\partial X}(x) - y\| \leq \|p_{\partial X}(x) - p_{Z_h}(x)\| + \|p_{Z_h}(x) - y\| \leq \sqrt{8h\rho} + h + r \leq 2r.$$

Now, we show that $Z_h^R \setminus \mathrm{vox}(\Omega_h^R) \subset \partial X^{R+h} \setminus \partial X^{R-(\sqrt{3}+1)h}$. Indeed, as said just before, one has $Z_h^R \subset \partial X^{R+h}$. Furthermore, if $x \in \partial X^{R-(\sqrt{3}+1)h}$, then the fact that the Hausdorff distance between Z_h and ∂X is less than h implies that $x \in Z_h^{R-\sqrt{3}h}$. Let $c \in h(\mathbb{Z} + \frac{1}{2})^3$ be the center of a voxel containing x. The fact that $\mathrm{diam}(\mathrm{vox}(c)) = \sqrt{3}h$ implies that $\mathrm{vox}(c) \subset Z_h^R$, and thus $x \in Z_h^R$. We then get $Z_h^R \setminus \mathrm{vox}(\Omega_h^R) \subset \partial X^{R+h} \setminus \partial X^{R-(\sqrt{3}+1)h}$. We finally deduce that

$$\left[Z_h^R \setminus \mathrm{vox}(\Omega_h^R) \right] \cap p_Z^{-1}(\mathcal{B}_y(r)) \subset \left[\partial X^{R+3h} \setminus \partial X^{R-3h} \right] \cap p_{\partial X}^{-1}(\mathcal{B}_y(2r)), \quad (1)$$

whose volume is bounded by $O(hr^2)$ by Proposition 3, which allows us to conclude.

Step 2: We then have to bound the remaining term

$$\Delta = \int_{\text{vox}(\Omega_h^R)} (x - p_{Z_h}(x))(x - p_{Z_h}(x))^t \chi(p_{Z_h}(x)) \mathrm{d}x - \widehat{\mathcal{V}}_{Z_h,R}(\chi).$$

By decomposing Δ over all the voxels of $\text{vox}(\Omega_h^R)$, one has

$$\Delta = \sum_{c \in \Omega_h^R} \int_{\text{vox}(c)} \left[(x - p_{Z_h}(x))(x - p_{Z_h}(x))^t \chi(p_{Z_h}(x)) \right.$$
$$\left. - (c - p_{Z_h}(c))(c - p_{Z_h}(c))^t \chi(p_{Z_h}(c)) \right] \mathrm{d}x$$

As in Step 1, we can localise the calculation around the support of χ and we introduce the set of centers $D = \Omega_h^R \cap p_{\partial X}^{-1}(\mathcal{B}_y(2r))$. Using the relation $\chi(p_{Z_h}(c)) = \chi(p_{Z_h}(c)) + \chi(p_{Z_h}(x)) - \chi(p_{Z_h}(x))$, one gets $\Delta = \Delta_1 + \Delta_2$, where

$$\Delta_1 = \sum_{c \in D} \int_{\text{vox}(c)} (x - p_{Z_h}(x))(x - p_{Z_h}(x))^t [\chi(p_{Z_h}(x)) - \chi(p_{Z_h}(c))] \mathrm{d}x$$
$$\Delta_2 = \sum_{c \in D} \int_{\text{vox}(c)} [(x - p_{Z_h}(x))(x - p_{Z_h}(x))^t - (c - p_{Z_h}(c))(c - p_{Z_h}(c))^t] \chi(p_{Z_h}(c))$$

We are now going to bound Δ_1 and Δ_2. One has

$$\|\Delta_1\|_{\text{op}} \le \sum_{c \in D} \int_{\text{vox}(c)} \|x - p_{Z_h}(x)\| \, \|x - p_{Z_h}(x)^t\| \, \|\chi(p_{Z_h}(x)) - \chi(p_{Z_h}(c))\| \, \mathrm{d}x.$$

For all $c \in D$ and $x \in \text{vox}(c)$, we have $\|x - c\| \le \frac{\sqrt{3}}{2} h$. Furthermore, by definition of Ω_h^R, we have that x and c belong to $Z_h^R \subset \partial X^{R+h}$. Then, since $h \le R \le \frac{\rho}{2}$, Proposition 4 implies $\|p_{Z_h}(x) - p_{Z_h}(c)\| = O(h^{\frac{1}{2}})$ and then $\|\chi(p_{Z_h}(x)) - \chi(p_{Z_h}(c))\| = \text{Lip}(\chi)O(h^{\frac{1}{2}})$. Using the fact that $\|x - p_{Z_h}(x)\| \le R$, one has

$$\|\Delta_1\|_{\text{op}} = \text{Vol}(\text{vox}(D)) \times R^2 \times \text{Lip}(\chi) \times O(h^{\frac{1}{2}}).$$

Since $\text{vox}(D) \subset Z_h^R \cap p_{\partial X}^{-1}(\mathcal{B}_y(2r)) \subset \partial X^{R+h} \cap p_{\partial X}^{-1}(\mathcal{B}_y(2r))$ and $h \le R$, Proposition 3 implies that $\text{Vol}(\text{vox}(D)) = O(r^2 R)$. Finally $\|\Delta_1\|_{\text{op}} = \text{Lip}(\chi) \times O(r^2 R^3 h^{\frac{1}{2}})$.

Similarly, let us bound $\|\Delta_2\|_{\text{op}}$. We put $u = (x - c)$, $v = c - p_{Z_h}(c)$ and $w = p_{Z_h}(c) - p_{Z_h}(x)$. We can write $x - p_{Z_h}(x) = u + v + w$, and we get

$$\Delta_2 = \sum_{c \in D} \left[\int_{\text{vox}(c)} [(u + v + w)(u + v + w)^t - vv^t] \chi(p_{Z_h}(c)) \right].$$

From $\|u\| \leq h$, $\|v\| \leq R$ and $\|w\| = O(h^{\frac{1}{2}})$, we bound the integrand by $O(\|\chi\|_\infty (R\, h^{\frac{1}{2}} + h))$. From $\mathrm{Vol}(\mathrm{vox}(D)) = O(r^2 R)$, one has $\|\Delta_2\|_{\mathrm{op}} = O(\|\chi\|_\infty\, (R^2 r^2 h^{\frac{1}{2}} + r^2 R h))$.

Stability of the VCM. It is known that the VCM is stable. More precisely, Theorem 5.1 of [1] states that $\|\mathcal{V}_{\partial X, R}(\chi_r) - \mathcal{V}_{Z_h, R}(\chi_r)\|_{\mathrm{op}} = O(h^{\frac{1}{2}})$. However, the constant involved in $O(h^{\frac{1}{2}})$ depends on the whole surface ∂X. We provide here a more precise constant involving only local estimations, r and R. The proof is very similar to the one of [1], except that we localise the calculation of the integral. It is given in Appendix.

Proposition 2. *For any $h \leq R$ such that $\sqrt{8h\rho} + h \leq r$, one has*

$$\|\mathcal{V}_{\partial X, R}(\chi_r) - \mathcal{V}_{Z_h, R}(\chi_r)\|_{\mathrm{op}}$$
$$= O\Big(Lip(\chi) \times [(r^3 R^{\frac{5}{2}} + r^2 R^{\frac{7}{2}} + r R^{\frac{9}{2}}) h^{\frac{1}{2}}] + \|\chi\|_\infty \times [(r^3 R^{\frac{3}{2}} + r^2 R^{\frac{5}{2}} + r R^{\frac{7}{2}}) h^{\frac{1}{2}}] \Big).$$

End of proof of Theorem 1. Let $h \leq \min\left(R, \frac{r}{2}, \frac{r^2}{32\rho}\right)$. The assumption $h \leq \frac{r^2}{32\rho}$ implies that $\sqrt{8h\rho} + h \leq r$. Thus we can apply Proposition 1 and Proposition 2.

3   Multigrid Convergence of the Normal Estimator

Let X be a compact domain of $\mathbb{R}^3$ whose boundary ∂X is a surface of class C^2. We now want to estimate the normal, denoted by $n(p_0)$, of ∂X at a point p_0 from its Gauss digitisation. We define the normal estimator by applying the digital VCM on a Lipschitz function that approaches the indicatrix of the ball $\mathcal{B}_{p_0}(r)$.

Definition 1. *The normal estimator $\widehat{n}_{r,R}(p_0)$ is the unit eigenvector associated to the largest eigenvalue of $\widehat{\mathcal{V}}_{Z_h, R}(\chi_r)$, where χ_r is a Lipschitz function that is: equal to 1 on $\mathcal{B}_{p_0}(r)$, equal to $1 - (\|x - p_0\| - r)/r^{\frac{3}{2}}$ on $\mathcal{B}_{p_0}(r + r^{\frac{3}{2}}) \setminus \mathcal{B}_{p_0}(r)$, and equal to 0 elsewhere.*

Remark that the normal estimator is defined only up to the sign. The following theorem gives an error estimation between $\pm \widehat{n}_{r,R}(p_0)$ and $n(p_0)$.

Theorem 2. *Let X be a compact domain of $\mathbb{R}^3$ whose boundary ∂X is a C^2 surface with reach $\rho > 0$. Let $R < \frac{\rho}{2}$. Then for any $h > 0$ such that $h \leq \min\left(R, \frac{r}{2}, \frac{r^2}{32\rho}\right)$, the angle between the lines spanned by $\widehat{n}_{r,R}(p_0)$ and $n(p_0)$ satisfies*

$$\langle \hat{n}_{r,R}(p_0), n(p_0) \rangle = O\left((rR^{-\frac{3}{2}} + R^{-1} + r^{-\frac{1}{2}}R^{-\frac{1}{2}} + r^{-\frac{3}{2}} + r^{-\frac{5}{2}}R^{\frac{3}{2}})h^{\frac{1}{2}} + R^{-2}h + r^{\frac{1}{2}} + R^2 \right).$$

The following corollary is a direct consequence.

Corollary 1. *Let X be a compact domain of $\mathbb{R}^3$ whose boundary ∂X is a C^2 surface with reach $\rho > 0$. Let $a, b \in \mathbb{R}^+$, $r = ah^{\frac{1}{4}}$ and $R = bh^{\frac{1}{4}}$. Then for any $h > 0$ small enough, one has*

$$\langle \hat{n}_{r,R}(p_0), n(p_0) \rangle = O\left(h^{\frac{1}{8}} \right).$$

Proof of Theorem 2. We introduce the normalized VCM $\hat{N}_{r,R}(p_0) = \frac{3}{2\pi r^2 R^3} \hat{V}_{Z_h,R}(\chi_r)$. From Davis-Kahan $\sin(\theta)$ Theorem [11], up to the sign of $\pm \hat{n}_{r,R}(p_0)$, one has

$$\| \hat{n}_{r,R}(p_0) - n(p_0) \| \leq 2 \left\| \hat{N}_{r,R}(p_0) - n(p_0)n(p_0)^t \right\|_{op}.$$

It is therefore sufficient to bound the right hand side. The triangle inequality gives

$$\left\| \hat{N}_{r,R}(p_0) - n(p_0)n(p_0)^t \right\|_{op} \leq \frac{3}{2\pi R^3 r^2} \left\| \hat{V}_{Z_h,R}(\chi_r) - V_{\partial X,R}(\chi_r) \right\|_{op}$$

$$+ \frac{3}{2\pi R^3 r^2} \left\| V_{\partial X,R}(\chi_r) - V_{\partial X,R}(1_{\mathcal{B}_{p_0}(r)}) \right\|_{op}$$

$$+ \left\| \frac{3}{2\pi R^3 r^2} V_{\partial X,R}(1_{\mathcal{B}_{p_0}(r)}) - n(p_0)n(p_0)^t \right\|_{op}.$$

The proof of the theorem relies on Theorem 1, that controls the first term, and on the two following lemmas.

Lemma 1. *Under the assumption of Theorem 2, we have*

$$\frac{3}{2\pi r^2 R^3} \left\| V_{\partial X,R}(\chi_r) - V_{\partial X,R}(1_{\mathcal{B}_{p_0}(r)}) \right\|_{op} = O(r^{\frac{1}{2}}).$$

Proof. Since $\chi_r = 1_{\mathcal{B}_{p_0}(r)}$ on the ball $\mathcal{B}_{p_0}(r)$, by using similar arguments as previously, one has

$$\left\| V_{\partial X,R}(\chi_r) - V_{\partial X,R}(1_{\mathcal{B}_{p_0}(r)}) \right\|_{op} \leq \mathrm{Vol}\left(\partial X^R \cap \left[p_{\partial X}^{-1}(\mathcal{B}_y(r + r^{\frac{3}{2}})) \backslash p_{\partial X}^{-1}(\mathcal{B}_y(r)) \right] \right) \times R^2.$$

Proposition 3 implies that the volume $\mathrm{Vol}\left(\partial X^R \cap \left[p_{\partial X}^{-1}(\mathcal{B}_y(r + r^{\frac{3}{2}})) \backslash p_{\partial X}^{-1}(\mathcal{B}_y(r)) \right] \right)$ is less than $4R \times \mathrm{Area}\left(\mathcal{B}_y(r + r^{\frac{3}{2}}) \backslash \mathcal{B}_y(r) \right)$. The fact that this area is bounded by $O(r^{\frac{5}{2}})$ allows to conclude.

Lemma 2. *Under the assumption of Theorem 2, we have*

$$\left\|\frac{3}{2\pi R^3 r^2}\mathcal{V}_{\partial X,R}(1_{\mathcal{B}_{p_0}(r)}) - n(p_0)n(p_0)^t\right\|_{op} = O(r + R^2)$$

Proof. We have the following relation (see Theorem 1 of [12])

$$\mathcal{V}_{\partial X,R}(1_{\mathcal{B}_{p_0}(r)}) = \frac{2}{3}R^3\left[1 + O(R^2)\right]\int_{p\in\mathcal{B}_{p_0}(r)\cap S} n(p)n(p)^t \, dp. \qquad (2)$$

By the mean value theorem applied to the normal to ∂X, one has

$$\|n(p) - n(p_0)\| \leq \sup_{q\in S}\|Dn(q)\|_{op} \, l_{p,p_0},$$

where l_{p,p_0} is the length of a geodesic joining p and p_0. Since the chord (pp_0) belongs to the offset ∂X^R, where $R < \rho$, we have $l_{p,p_0} = O(\|p - p_0\|)$ (see [13] for example). Therefore $\|n(p) - n(p_0)\| = O(r)$ and thus $n(p)n(p)^t - n(p_0)n(p_0)^t = O(r)$. Consequently

$$\int_{p\in\mathcal{B}_{p_0}(r)\cap S} n(p)n(p)^t \, dp = \text{Area}(\mathcal{B}_{p_0}(r) \cap S)n(p_0)n(p_0)^t + \text{Area}(\mathcal{B}_{p_0}(r) \cap S) \, O(r).$$

Combining with Eq. (2), we have

$$\frac{3}{2R^3\text{Area}(\mathcal{B}_{p_0}(r) \cap S)}\mathcal{V}_{\partial X,R}(1_{\mathcal{B}_{p_0}(r)}) = \left[1 + O(R^2)\right] \times \left(n(p_0)n(p_0)^t + O(r)\right).$$

We conclude by using the fact that $\text{Area}(\mathcal{B}_{p_0}(r)\cap S)$ is equivalent to πr^2.

4   Experiments

We evaluate experimentally the multigrid convergence, the accuracy and robustness to Hausdorff noise of our normal estimator, and also its ability to detect features.

The first series of experiments analyzes the convergence of the normal estimation by VCM toward the true normal of the shape boundary ∂X. The shape "torus" is a torus of great radius 6 and small radius 2, and the shape "ellipsoid" is an ellipsoid of half-axes $\sqrt{90}$, $\sqrt{45}$ and $\sqrt{45}$. We measure the absolute angle error with $\epsilon(p) = \frac{180}{\pi}\cos^{-1}(\hat{n}(p) \cdot n(p))$ for every pointel $p \in Z_h$ of the digitized shape with several normalized norms:

$$l_1(\epsilon) \stackrel{def}{=} \frac{1}{\text{Card}(Z_h)}\sum_{p\in Z_h}\epsilon(p), \qquad l_\infty(\epsilon) \stackrel{def}{=} \sup_{p\in Z_h}\epsilon(p). \qquad (3)$$

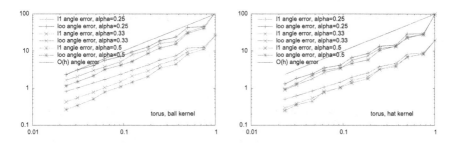

Fig. 3. Multigrid convergence of angle error of normal estimator (in degree). Abscissa is the gridstep h. Tests are run on torus shape for three kernel radii ($R = r = 3h^\alpha$ for $\alpha \in \{\frac{1}{4}, \frac{1}{3}, \frac{1}{2}\}$), two norms ($l_1$, l_∞): (left) kernel ball function χ_r^0, (right) kernel hat function χ_r^1.

In experiments we tried several kernel functions χ_r and we display results for two of them: the "ball" kernel $\chi_{p_0,r}^0(x) = 1$ if $\|x - p_0\| \leq r$, 0 otherwise; the "hat" kernel $\chi_{p_0,r}^1(x) = 1 - \|x - p_0\|/r$ if $\|x - p_0\| \leq r$, 0 otherwise. Figure 3 displays the norms of the estimation angle error in degrees, for finer and finer digitization steps. Corollary 1 predicts the multigrid convergence of the estimator when $r = ah^{\frac{1}{4}}$ and $R = bh^{\frac{1}{4}}$ at a rate in $O(h^{\frac{1}{8}})$. We observe the convergence of the estimator for parameters $R = r = 3h^{\frac{1}{4}}$, $R = r = 3h^{\frac{1}{3}}$, $R = r = 3h^{\frac{1}{2}}$, at an almost linear rate $O(h)$, for all norms. More experiments show that the most accurate results are obtained for $\alpha \in [\frac{1}{3}, \frac{1}{2}]$ if $R = r = ah^\alpha$. Note that the kernel function has not a great impact on normal estimates, as long as it has a measure comparable to the ball kernel.

We perturbate the shape "torus" with a Kanungo noise model of parameter $p = 0.25$ (the number p^d is the probability that a voxel at digital distance d from the boundary ∂X is flipped inside/out). This is not exactly a Hausdorff perturbation but most perturbations lie in a band of size $2h/(1 - p)$. Figure 4 shows that the normal is still convergent for all norms. Again convergence speed is experimentally closer to $O(h^{\frac{2}{3}})$, much better than the proved $O(h^{\frac{1}{8}})$.

We then assess the visual quality of the estimators on several shapes, by rendering the digital surfels according to their estimated normals. First of all, Figure 5 displays normal estimation results on a noisy "torus" shape perturbated with a strong Kanungo noise of parameter $p = 0.5$. Then, Figure 6 displays the visual improvement of using normals computed by the VCM estimator. In particular, comparing Fig.6b and Fig.6c shows that convolving Voronoi cell geometry is much more precise than convolving only surfel geometry. Furthermore, we have tested our estimator on many classical digital geometry shapes (see Figure 7).

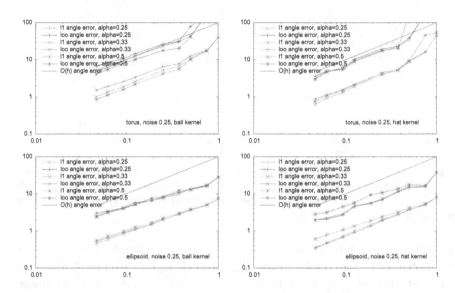

Fig. 4. Multigrid convergence of angle error of normal estimator (in degree) on a *noisy* shape. Abscissa is the gridstep h. Tests are run on "torus" shape (upper row) and on "ellipsoid" shape (lower row), perturbated by a Kanungo noise of parameter 0.25, for three kernel radii ($R = r = 3h^\alpha$ for $\alpha \in \{\frac{1}{4}, \frac{1}{3}, \frac{1}{2}\}$), two norms ($l_1$, l_∞): (left) kernel ball function χ_r^0, (right) kernel hat function χ_r^1.

Our VCM estimator is a matrix and carries also curvature information along other eigendirections. Mérigot *et al.* [1] proposed to detect sharp features by using the three eigenvalues l_1, l_2, l_3 of the VCM as follows: if $l_1 \geq l_2 \geq l_3$, compute $l_2/(l_1 + l_2 + l_3)$ and mark the point as *sharp* if this value exceeds a threshold T. Figure 8 shows such sharp features detection on the "bunny" dataset at many different scales, with $T = 0.1$ for all datasets (it corresponds to an angle of $\approx 25°$). This shows that the VCM information is geometrically stable and essentially scale-invariant. To conclude, we list below some information on computation times. This estimator has been implemented using the DGtal library [14], and will soon be freely available in it.

Image	size	#surfels	(R, r)	χ_r-VCM comput.	Orienting normals
"Al"	150^3	48017	$(30, 3)$	0.73 s	0.88 s
"rcruiser"	250^3	66543	$(30, 3)$	1.26 s	0.99 s
"bunny"	516^3	933886	$(30, 5)$	30.1 s	15.9 s
"Dig. Snow"	512^3	3035307	$(30, 5)$	82.1 s	53.6 s

Fig. 5. Visual result of the normal estimation on the "torus" shape perturbated with a strong Kanungo noise ($p = 0.5$) for gridsteps from left to right $h = 0.5, 0.25, 0.125, 0.0626$

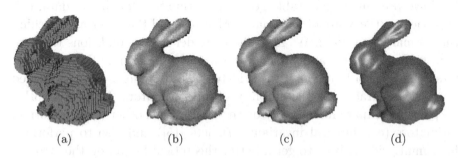

(a)   (b)   (c)   (d)

Fig. 6. Visual aspect of normal estimation on "bunny66" for $r = 3$: (a) trivial normals, (b) normals by χ_r^1 convolution of trivial normals with flat shading, (c) χ_r^1-VCM normals with flat shading, (d) χ_r^1-VCM normals with Gouraud shading

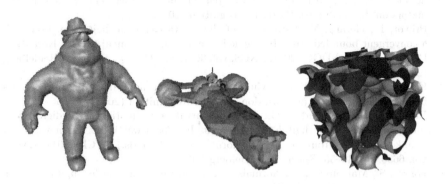

Fig. 7. Visual aspect of normal estimation on classical digital data structures: "Al" 150^3, "Republic cruiser" 250^3, "Digital snow" 512^3

Fig. 8. Sharp feature detection on "bunny" dataset at increasing resolutions ($R = 30$, $T = 0.1$): color is metallic blue when value is in $[0, \frac{2}{3}T]$, then goes through cyan and yellow in $]\frac{2}{3}T, T[$, till red in $[T, +\infty[$

5   Conclusion

We have presented new stable geometry estimators for digital data, one approaching the Voronoi Covariance Measure and the other approaching the normal vector field. We have shown under which conditions they are multigrid convergent and provided formulas to determine their parameters R and r as a function of the gridstep h. Experiments have confirmed both the accuracy and the stability of our estimators. In future works, we plan to compare numerically our estimator with other discrete normal estimators (e.g. integral invariants [7], jets [15]) and also to perform a finer multigrid analysis to get a better theoretical bound on the error.

References

1. Mérigot, Q., Ovsjanikov, M., Guibas, L.: Voronoi-based curvature and feature estimation from point clouds. IEEE Transactions on Visualization and Computer Graphics 17(6), 743–756 (2011)
2. de Vieilleville, F., Lachaud, J.O.: Comparison and improvement of tangent estimators on digital curves. Pattern Recognition (2008)
3. Provot, L., Gérard, Y.: Estimation of the derivatives of a digital function with a convergent bounded error. In: Debled-Rennesson, I., Domenjoud, E., Kerautret, B., Even, P. (eds.) DGCI 2011. LNCS, vol. 6607, pp. 284–295. Springer, Heidelberg (2011)
4. Kerautret, B., Lachaud, J.O.: Curvature estimation along noisy digital contours by approximate global optimization. Pattern Recognition (2009)
5. Roussillon, T., Lachaud, J.-O.: Accurate curvature estimation along digital contours with maximal digital circular arcs. In: Aggarwal, J.K., Barneva, R.P., Brimkov, V.E., Koroutchev, K.N., Korutcheva, E.R. (eds.) IWCIA 2011. LNCS, vol. 6636, pp. 43–55. Springer, Heidelberg (2011)
6. Fourey, S., Malgouyres, R.: Normals and curvature estimation for digital surfaces based on convolutions. In: Coeurjolly, D., Sivignon, I., Tougne, L., Dupont, F. (eds.) DGCI 2008. LNCS, vol. 4992, pp. 287–298. Springer, Heidelberg (2008)
7. Coeurjolly, D., Lachaud, J.-O., Levallois, J.: Integral based curvature estimators in digital geometry. In: Gonzalez-Diaz, R., Jimenez, M.-J., Medrano, B. (eds.) DGCI 2013. LNCS, vol. 7749, pp. 215–227. Springer, Heidelberg (2013)

8. Cazals, F., Pouget, M.: Estimating differential quantities using polynomial fitting of osculating jets. Computer Aided Geometric Design 22(2), 121–146 (2005)
9. Amenta, N., Bern, M.: Surface reconstruction by voronoi filtering. Discrete & Computational Geometry 22(4), 481–504 (1999)
10. Alliez, P., Cohen-Steiner, D., Tong, Y., Desbrun, M.: Voronoi-based variational reconstruction of unoriented point sets. In: Symposium on Geometry Processing (2007)
11. Davis, C.: The rotation of eigenvectors by a perturbation. Journal of Mathematical Analysis and Applications (1963)
12. Mérigot, Q.: Détection de structure géométrique dans les nuages de points. PhD thesis, Université Nice Sophia Antipolis (December 2009)
13. Morvan, J.M., Thibert, B.: Approximation of the normal vector field and the area of a smooth surface. Discrete & Computational Geometry 32(3), 383–400 (2004)
14. DGtal: Digital geometry tools and algorithms library, http://libdgtal.org
15. Cazals, F., Pouget, M.: Estimating differential quantities using polynomial fitting of osculating jets. Computer Aided Geometric Design 22(2), 121–146 (2005)
16. Weyl, H.: On the volume of tubes. American Journal of Mathematics, 461–472 (1939)
17. Federer, H.: Curvature measures. Trans. Amer. Math. Soc. 93(3), 418–491 (1959)
18. Cuel, L., Lachaud, J.-O., Thibert, B.: Voronoi-based geometry estimator for 3D digital surfaces. Technical report hal-00990169 (May 2014)

Appendix

We give here all the sketchs of the proofs to be self-content. Proposition 3 is classical and follows from the well-known tube formula for smooth surfaces [16]. Proposition 4 states that the projection map p_K onto a set K that is close to a smooth surface S behaves like the projection map p_S. It relies on classical properties of the projection map onto a set with positive reach. The proof of Proposition 2 is similar to the proof of Theorem 5.1 of [1], except that the calculations are done locally.

5.1   Hausdorff Measure of Offsets

Proposition 3. *Let $S \subset \mathbb{R}^3$ be a surface of class C^2 with reach $\rho > 0$. Let $R > 0$ and $\varepsilon > 0$ be such that $R + \varepsilon < \frac{\rho}{2}$. Then for any ball $\mathcal{B}$ of radius r, one has:*

a) $\mathrm{Vol}\left(S^R \cap p_S^{-1}(\mathcal{B} \cap S)\right) = O(Rr^2)$.
b) $\mathrm{Vol}\left((S^{R+\varepsilon} \backslash S^{R-\varepsilon}) \cap p_S^{-1}(\mathcal{B} \cap S)\right) = O(\varepsilon r^2)$.
c) $\mathrm{Area}\left(\partial[S^R \cap p_S^{-1}(\mathcal{B} \cap S)]\right) = O(Rr + r^2)$,

where the notation O involves a constant that only depends on the reach ρ.

Proof. The proof is based on the tube formula for surfaces of class C^2 [16, 17]. One has

$$\text{Vol}(S^R \cap p_S^{-1}(\mathcal{B} \cap S)) = \int_{\mathcal{B} \cap S} \int_{-R}^{R} (1 - t\lambda_1(x))(1 - t\lambda_2(x))dtdx,$$

where $\lambda_1(x)$ and $\lambda_2(x)$ are the principal curvatures of S at the point x. Now, since $|\lambda_1(x)|$ and $|\lambda_2(x)|$ are smaller than $\frac{1}{\rho}$, one has

$$\text{Vol}(S^R \cap p_S^{-1}(\mathcal{B} \cap S)) \le \int_{\mathcal{B} \cap S} dx \times \int_{-R}^{R} \left(1 + \frac{t}{\rho}\right)^2 dt \le \text{Area}(\mathcal{B} \cap S) \times 2R \left(1 + \frac{R^2}{3\rho^2}\right).$$

Point a) follow from the fact that $\text{Area}(\mathcal{B} \cap S) = O(r^2)$. We use the same kind of argument for Points b) and c).

5.2   Stability of the Projection on a Compact Set

Proposition 4. *Let S be a surface of $\mathbb{R}^3$ of class C^2 whose reach is greater than $\rho > 0$. Let K be a compact set such that $d_H(S, K) = \varepsilon < 2\rho$, and $R < \rho$ a positive number. If x and x' are points of S^R such that $d(x, x') \le \eta$, then :*

$$\left\| p_K(x) - p_K(x') \right\| \le 2\sqrt{8\varepsilon\rho} + 2\varepsilon + \frac{1}{1 - \frac{R}{\rho}}\eta$$

The proof of the proposition relies on Lemma 3 whose proof is given in [18].

Lemma 3. *Let S be a surface of $\mathbb{R}^3$ with a reach $\rho > 0$. Let K be a compact set such that $d_H(S, K) = \varepsilon$ with $\varepsilon \le 2\rho$. Let R be a number such that $R < \rho$. For every $x \in S^R$, one has*

$$p_K(x) \in B(p_S(x), \sqrt{8\varepsilon\rho} + \varepsilon)$$

Proof (of Proposition 4). By the triangle inequality, we have

$$\left\| p_K(x) - p_K(x') \right\| \le \left\| p_K(x) - p_S(x) \right\| + \left\| p_S(x) - p_S(x') \right\| + \left\| p_S(x') - p_K(x') \right\|.$$

It is well-known that the projection map p_S is $\frac{1}{1 - \frac{R}{\rho}}$-Lipschitz in S^R (Theorem 4.8 of [17]). We then have $\left\| p_S(x) - p_S(x') \right\| \le \frac{1}{1 - \frac{R}{\rho}}\eta$. The two other terms are bounded with Lemma 3.

5.3   Proof of Proposition 2

Similarly, as for equation (1) and using the hypothesis $h \leq \frac{\rho}{2}$ and $\sqrt{8h\rho} + h \leq r$, we have $p_{Z_h}^{-1}(\text{supp}(\chi)) \subset p_{\partial X}^{-1}(\mathcal{B}_y(2r))$. We then introduce the common set $E = \partial X^{R-h} \cap p_{\partial X}^{-1}(\mathcal{B}_y(2r))$, on which we are going to integrate. We have :

$$\mathcal{V}_{\partial X, R}(\chi_r) = \int_{\partial X^R \cap p_{\partial X}^{-1}(\mathcal{B}_y(2r))} (x - p_{\partial X}(x))(x - p_{\partial X}(x))^t \chi(p_{\partial X}(x))$$
$$= \int_E (x - p_{\partial X}(x))(x - p_{\partial X}(x))^t \chi(p_{\partial X}(x)) + Err_1,$$

where the error Err_1 satisfies

$$\|Err_1\|_{\text{op}} \leq R^2 \times \|\chi\|_\infty \times \text{Vol}(\partial X^R \cap p_{\partial X}^{-1}(\mathcal{B}_y(2r)) \setminus E).$$

Furthermore, one has $\partial X^R \cap p_{\partial X}^{-1}(\mathcal{B}_y(2r)) \setminus E = [\partial X^R \backslash \partial X^{R-h}] \cap p_{\partial X}^{-1}(\mathcal{B}_y(2r))$, whose volume is bounded by Proposition 3 by $O(r^2 h)$. Then

$$\|Err_1\|_{\text{op}} = \|\chi\|_\infty \times O(R^2 r^2 h).$$

Similarly, one has

$$\mathcal{V}_{\partial X, R}(\chi_r) = \int_E (x - p_{Z_h}(x))(x - p_{Z_h}(x))^t \chi(p_{Z_h}(x)) + Err_2,$$

where the error Err_2 satisfies $\|Err_2\|_{\text{op}} = (R + h)^2 \times \|\chi\|_\infty \times O(r^2 h)$. We now have to compare the two integrals on the common set E

$$\Delta = \int_E \left[(x - p_{\partial X}(x))(x - p_{\partial X}(x))^t \chi(p_{\partial X}(x)) - (x - p_{Z_h}(x))(x - p_{Z_h}(x))^t \chi(p_{Z_h}(x)) \right].$$

Following now the proof of Theorem 5.1 of [1], one has

$$\|\Delta\|_{\text{op}} \leq (R^2 \text{Lip}(\chi) + 2R\|\chi\|_\infty) \times [\text{Vol}(E) + (diam(E) + R + \sqrt{Rh}) \times \text{Area}(\partial E)] \times \sqrt{Rh}.$$

Proposition 3 gives that $\text{Vol}(E)$ is bounded by $O(r^2 R)$ and $\text{Area}(\partial E)$ is bounded by $O(rR + r^2)$. We then have

$$\|\Delta\|_{\text{op}} = O\left(\text{Lip}(\chi) \times [(r^3 R^{\frac{5}{2}} + r^2 R^{\frac{7}{2}} + rR^{\frac{9}{2}})h^{\frac{1}{2}}] + \|\chi\|_\infty \times [(r^3 R^{\frac{3}{2}} + r^2 R^{\frac{5}{2}} + rR^{\frac{7}{2}})h^{\frac{1}{2}}] \right).$$

Adding the bounds of $\|Err_1\|_{\text{op}}$, $\|Err_2\|_{\text{op}}$ and $\|\Delta\|_{\text{op}}$, we find the same bound :

$$\|\mathcal{V}_{\partial X, R}(\chi_r) - \mathcal{V}_{Z_h, R}(\chi_r)\|_{\text{op}}$$
$$= O\left(\text{Lip}(\chi) \times [(r^3 R^{\frac{5}{2}} + r^2 R^{\frac{7}{2}} + rR^{\frac{9}{2}})h^{\frac{1}{2}}] + \|\chi\|_\infty \times [(r^3 R^{\frac{3}{2}} + r^2 R^{\frac{5}{2}} + rR^{\frac{7}{2}})h^{\frac{1}{2}}] \right).$$

An Arithmetical Characterization of the Convex Hull of Digital Straight Segments*

Tristan Roussillon

Université de Lyon, CNRS
INSA-Lyon, LIRIS, UMR5205, F-69622, France
tristan.roussillon@liris.cnrs.fr

Abstract. In this paper, we arithmetically describe the convex hull of a digital straight segment by three recurrence relations. This characterization gives new insights into the combinatorial structure of digital straight segments of arbitrary length and intercept. It also leads to two on-line algorithms that computes a part of the convex hull of a given digital straight segment. They both run in constant space and constant time per vertex. Due to symmetries, they are enough to reconstruct the whole convex hull. Moreover, these two algorithms provide efficient solutions to the subsegment problem, which consists in computing the minimal parameters of a segment of a digital straight line of known parameters.

1 Introduction

The connection between continued fractions and the convex hull of lattice points lying above and below a straight segment whose endpoints are lattice points was already observed by Klein in the nineteenth century as mentioned in [7].

Based on this connection, many papers introduce output-sensitive algorithms to compute the convex hull of analytical point sets, such as the intersection of the fundamental lattice and an arbitrary half-plane [3,6,9,10], convex quadrics [3] or convex bodies [9]. In these papers, the authors propose a *geometrical* extension of the result of Klein, while in this paper, the connection between arithmetic and discrete ray casting, which is briefly described by Har-Peled in [9], is used to propose an arithmetical interpretation of the geometrical algorithm of Charrier and Buzer [6]. This new point of view leads to a simple *arithmetical* extension of the result of Klein to straight segments of arbitrary rational slope and arbitrary rational intercept.

More precisely, we introduce three recurrence relations, defining three sequences of integer pairs, viewed as points or vectors in the fundamental lattice $\mathbb{Z}^2$. The first two sequences, denoted by $\{L_k\}_{0...n}$ and $\{U_k\}_{0...n}$, both contain vertices of the convex hull of some lattice points lying on each side of a straight line (see fig. 1.a). There exists a close link between a separating line and a digital straight line (DSL). We refer the reader that is not familiar with digital

---

* This work has been mainly funded by DigitalSnow ANR-11-BS02-009 research grants.

E. Barcucci et al. (Eds.): DGCI 2014, LNCS 8668, pp. 150–161, 2014.

straightness to [11] and we use below the arithmetical framework introduced in [8,15]. For each $0 \leq k \leq n$, $L_k - (0,1)$ (resp. $L_k - (-1,1)$) is a vertex of the lower convex hull of the associated naive (resp. standard) DSL. Fig. 1.b is an illustration of the naive case. For the sake of clarity, we focus on the naive case in the rest of the paper. The last sequence, denoted by $\{v_k\}_{0...n}$, has also a simple geometrical interpretation. Indeed, we prove in section 2.3 that for each $0 \leq k \leq n$, $(L_k - U_k)$ and v_k are a pair of unimodular vectors. In other words, v_k is the direction vector of a digital straight segment (DSS) whose first lower and upper leaning points are respectively $L_k - (0,1)$ and U_k.

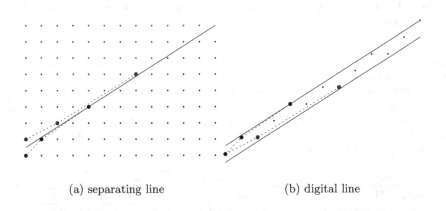

(a) separating line   (b) digital line

Fig. 1. Upper and lower convex hulls of lattice points of positive x-coordinate lying on each side of the straight line $\{(\alpha, \beta) \in \mathbb{R}^2 | 5\alpha - 8\beta = -4\}$ (Point $(0,0)$ is on the bottom left) (a). They are closely related to the upper and lower convex hulls of a naive and 8-connected digital straight segment of slope $5/8$, intercept -4 and first point $(0,0)$ (b).

Our arithmetical characterization goes beyond the scope of convex hull computation, because the convex hull of a DSS provides a substantial part of its combinatorial structure. Let us define a upper (resp. lower) *digital edge* as a DSS whose first and last point are upper (resp. lower) leaning points. The combinatorial structure of a digital edge has been studied since the seventies [4] and is well-known since the early nineties [5,15,18]. However, these works focus on digital edges or DSLs, which are infinite repetitions of digital edges, because the intercept of a DSL has no effect on its shape and can be assumed to be null without any loss of generality.

To the best of our knowledge, there are few works that extend such results to DSSs of arbitrary intercept and length. In [16], Yaacoub and Reveillès provide an algorithm to retrieve the convex hull of a naive DSS of slope a/b, intercept $\mu \in [0; b[$ and length $|b|$. But the presented algorithm fails to reach the claimed logarithmic complexity, because it takes as input the set of additive convergents of the continued fraction expansion of a/b.

Moreover, it is known for a long time that computing the convex hull of a DSS is a way of computing its parameters [2]. Several authors have recently investigated the problem of computing the minimal parameters of a subsegment of a DSL of known parameters [6,12,14,17]. Minimality is required to have a unique and compact representation of the segment slope and intercept. Due to the prior knowledge of one of its bounding digital straight line, all proposed algorithms outperform classical recognition algorithms [8,13], where each point must be considered at least once. Our simple arithmetical characterization leads to two algorithms, called **smartCH** and **reversedSmartCH**, which not only retrieve the vertices of a part of the DSS convex hull, but also compute its minimal parameters. They both runs in constant space and constant time per vertex. Their overall time complexity are among the best ones (see tab. 1 for a comparison).

Table 1. Theoretical comparison of **smartCH** and **reversedSmartCH** with convex hull algorithms (upper block) and subsegment algorithms (lower block). We consider a naive DSS Σ starting from $(0,0)$ and of slope $a/b = [u_1, \ldots, u_n]$ such that $0 \le a < b$. For the sake of clarity, we consider its left subsegment Σ' of slope $a'/b' = [u'_1, \ldots, u'_n]$ and of length $l \le b$ such that $\Sigma' = \{(x,y) \in \Sigma | 0 \le x \le l\}$. Time complexities depend on a, b, and l. If $l \ll b$, bounds depending on l are better. However, if l is close to b, bounds depending on the difference $b - l$ are better.

Algorithms	Time complexity	Remarks
smartCH	$O(\log l)$	on-line: $O(1)$ per vertex
reversedSmartCH	$O(\log(b - l))$	on-line: $O(1)$ per vertex, leaning points must be known
Reveillès et. al. [16]	$O(\sum_i^n u_i)$	$l = b$, $\{u_i\}$ must be known
Har-Peled [9]	$O(\log^2 l)$	on-line: $O(\log l)$ per vertex
Harvey [10]	$O(\log b)$	
Balza-Gomez et. al. [3]	$O(\log l)$	post-processing required
Charrier et. al. [6]	$O(\log l)$	on-line: $O(1)$ per vertex
smartDSS, Lachaud et. al. [12]	$O(\sum_i^n u'_i)$	
reversedSmartDSS, ibid.	$O(\log(b - l))$	$\{u_i\}$ must be known
Sivignon [17]	$O(\log l)$	
Ouattara et. al. [14]	$O(\log l)$	

In section 2, we introduce our arithmetical characterization and discuss its theoretical properties. New algorithms are derived in section 3.

2   A Simple Arithmetical Characterization

Let $\mathcal{L}(a, b, \mu)$ (or simply $\mathcal{L}$) be a straight line of equation $\{(\alpha, \beta) \in \mathbb{R}^2 | a\beta - b\alpha = \mu\}$ with $a, b, \mu \in \mathbb{Z}$, $\gcd(a, b) = 1$. Due to symmetries, let us assume w.l.o.g. that $0 \le a < b$. In addition, due to invariance by integral translation, let us assume w.l.o.g. that $-b < \mu \le 0$.

Let Λ be the restriction of the fundamental lattice $\mathbb{Z}^2$ to the lattice points of positive x-coordinate, i.e. $\Lambda := \{(x, y) \in \mathbb{Z}^2 | x \geq 0\}$. The straight line $\mathcal{L}$ always divides Λ into a upper domain, $\Lambda^+ := \{(x, y) \in \Lambda | ax - by \leq \mu\}$, and a lower one, $\Lambda^- := \{(x, y) \in \Lambda | ax - by > \mu\}$.

Definition 1 (Left hull (see fig. 1)). *The lower (resp. upper) left hull of Λ^+ (resp. Λ^-) is the part of the lower (resp. upper) convex hull located between the vertex of minimal x-coordinate and the vertex the closest to $\mathcal{L}$.*

In this section, we provide a simple arithmetical characterization of the upper and lower left hull of the lower and upper domain.

2.1 Recurrence Relations

Due to the asymmetric definition of Λ^+ (in which there is a large inequality) and Λ^- (in which there is a strict one), we introduce the two following notations: $\forall x \in \mathbb{R}$ (resp. $\forall x \in \mathbb{R} \setminus 0$), $[x]$ (resp. $\lfloor x \rfloor$) returns the integer $i \in \mathbb{Z}$ farthest to 0 such that $|i| \leq |x|$ (resp. $|i| < |x|$), i and x having same sign. We assume in this paper that these two floor functions run in $O(1)$. Moreover, the restriction of the *strict* floor function $\lfloor \cdot \rfloor$ to $\mathbb{R} \setminus 0$ does not cause any problem in our framework.

On the other hand, we recall that the *remainder* with respect to the straight line of slope a/b is a function $r_{(b,a)} : \mathbb{Z}^2 \to \mathbb{Z}$ such that $r_{(b,a)}(x, y) := (b, a) \wedge (x, y) = ax - by$. This value corresponds to the z-component of the cross-product $(b, a) \wedge (x, y)$ and is equal to the signed area of a parallelogram generated by (b, a) and (x, y). Note that the $\wedge$ operator is linear and antisymmetric.

In the sequel, $r_{(b,a)}(\cdot)$ is simplified into $r(\cdot)$ when the remainder refers to $\mathcal{L}$. Since a and b are given and constant, the difference $r(Q) - \mu$ of a point Q measures how far Q is from $\mathcal{L}$.

Let us consider the following set of recurrence relations (see fig. 2.1 for a numerical example):

$$L_0 = (0, 1), \ U_0 = (0, 0), \ v_0 = (1, 0) + \left[\frac{\mu - a}{-b} \right](0, 1) \tag{1}$$

$$\forall k \geq 1, \\ r(v_{k-1}) \neq 0 \quad \begin{cases} \text{if } r(v_{k-1}) > 0, & \begin{cases} L_k = L_{k-1} + \left\lfloor \frac{\mu - r(L_{k-1})}{r(v_{k-1})} \right\rfloor v_{k-1} \\ U_k = U_{k-1} \\ v_k = v_{k-1} + \left[\frac{\mu - (r(U_k) + r(v_{k-1}))}{(r(L_k) - r(U_k))} \right](L_k - U_k) \end{cases} \\ \text{if } r(v_{k-1}) < 0, & \begin{cases} L_k = L_{k-1} \\ U_k = U_{k-1} + \left[\frac{\mu - r(U_{k-1})}{r(v_{k-1})} \right] v_{k-1} \\ v_k = v_{k-1} + \left\lfloor \frac{\mu - (r(L_k) + r(v_{k-1}))}{(r(U_k) - r(L_k))} \right\rfloor (U_k - L_k) \end{cases} \end{cases} \tag{2}$$

The goal of this section is to prove the following theorem:

Theorem 1. *The sequence $\{L_k\}_{0...n}$ (resp. $\{U_k\}_{0...n}$) corresponds to the vertices of the lower (resp. upper) left hull of Λ^+ (resp. Λ^-).*

k	0	1	2	3	4
L_k	$(0,1)$	$(0,1)$	$(2,2)$	$(2,2)$	$(7,5)$
U_k	$(0,0)$	$(1,1)$	$(1,1)$	$(4,3)$	$(4,3)$
v_k	$(1,1)$	$(2,1)$	$(3,2)$	$(5,3)$	$(8,5)$

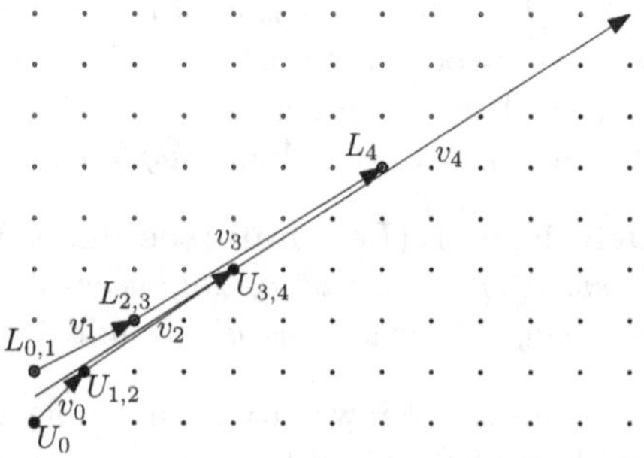

Fig. 2. We apply (1) and (2) for $a = 5$, $b = 8$ and $\mu = -4$. The first two sequences are respectively depicted with white and black disks, whereas the third sequence is depicted with arrows.

The proof of theorem 1 will be derived in section 3 from properties proved in section 2.3. The proof of some of these properties requires the following useful geometrical interpretation of integer divisions (see also [9]).

2.2   Integer Division and Ray Casting

Let us consider a point Q such that $r(Q) \geq \mu$ and a direction vector v whose coordinates are relatively prime and such that $r(v) < 0$.[1] Since $r(Q) \geq \mu$ and $r(v) < 0$, the ray emanating from Q in direction v intersects $\mathcal{L}$ at a point I (see fig. 3.a). The discrete ray casting procedure consists in computing the lattice point farthest from Q and lying on the line segment $[QI]$. Let I be equal to $Q + \tau v$ for some $\tau \in \mathbb{R}^+$. Since I belongs to $\mathcal{L}$ by definition, $r(I) = \mu$ and the linearity of the $\wedge$ product gives:

$$\mu = r(I) = r(Q) + \tau r(v) \Leftrightarrow \tau = \frac{\mu - r(Q)}{r(v)}.$$

Since the components of v are relatively prime and since τ is positive, the greatest integer $t \in \mathbb{Z}$ that is less than or equal to τ, i.e. $t = [\tau]$, leads to the lattice point $Q + tv$, which is the farthest from Q among those lying on $[QI]$. In the example illustrated by fig. 3.a, $t = 2$. Note that if we consider the half-open line segment $[QI[$ instead of the closed line segment $[QI]$, i.e. I is not included, we must use the strict floor function $\lfloor \cdot \rfloor$ instead of the large floor function $[\cdot]$.

Moreover, note that we can reverse a ray casting under some conditions. We will use this property to propose a dual characterization that leads to a reversed algorithm in section 3. Let us recall that the *position* of a point with respect to a direction vector s is a function $p_s : \mathbb{Z}^2 \to \mathbb{Z}$ such that $p_s(x, y) := (x, y) \wedge s$. In the sequel, we assume that $s = (0, 1)$ because we focus on the naive case and $p_s(x, y)$, which is merely denoted by $p(x, y)$, returns x.

---

[1] Note that points and vectors are both viewed as integer pairs $(x, y) \in \mathbb{Z}^2$.

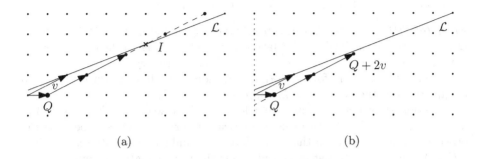

(a)   (b)

Fig. 3. In (a), the ray emanating from Q in direction v intersects $\mathcal{L} = \{(\alpha, \beta) \in \mathbb{R}^2 | 3\alpha - 8\beta = -2\}$ at I, because $Q = (1, 0)$ and $v = (2, 1)$ are such that $r(Q) = 3 \geq \mu = -2$ and $r(v) = -2 < 0$. The lattice point lying on the ray segment $[QI]$ and farthest from Q is $Q + 2v = (5, 2)$. Indeed, $\lfloor \frac{\mu - r(Q)}{r(v)} \rfloor = \lfloor \frac{5}{2} \rfloor = 2$. In (b), the point Q may be retrieved from $Q + 2v$ and v by a reversed discrete ray casting procedure, because $p(Q) = 1$ is strictly less than $p(v) = 2$.

It is easy to see that

$$p(Q) < p(v) \Rightarrow \left\lceil \frac{p(Q + tv)}{p(v)} \right\rceil = t. \tag{3}$$

Fig. 3.b shows the reversed version of the ray casting depicted in fig. 3.a.

2.3 A Unimodular Basis

We now prove several properties of (1) and (2). Let n be the index such that $r(v_n) = 0$. For the sake of clarity, we postpone the demonstration of the existence of such index to the end of the subsection. We first show that for each $0 \leq k \leq n$, points L_k and U_k lie on each side of $\mathcal{L}$, i.e.

$$\forall 0 \leq k \leq n, \; r(L_k) < \mu \text{ and } r(U_k) \geq \mu. \tag{4}$$

It is easy to see that (4) is true for $k = 0$ by (1) and that for all $0 \leq k \leq n$, the constructions of L_k from L_{k-1} (when $r(v_{k-1}) > 0$) and U_k from U_{k-1} (when $r(v_{k-1}) < 0$) are such that $r(L_k) < \mu$ and $r(U_k) \geq \mu$.

Moreover, for each $0 \leq k \leq n$, we show that there is a strong link between L_k, U_k and v_k:

Lemma 1. $\forall 0 \leq k \leq n$, v_k is the unique negative and valid Bezout vector[2] of $(L_k - U_k)$, i.e.

$$\forall 0 \leq k \leq n, \; r(L_k) < \mu - r(v_k) \leq r(U_k), \tag{5}$$

$$\forall 0 \leq k \leq n, \; v_k \wedge (L_k - U_k) = -1. \tag{6}$$

---

[2] The notion of valid Bezout vector is introduced in [6].

To prove lemma 1, we prove successively (5) and (6).

Proof (of (5)). Base case: Let us consider the ray emanating from $(1,0)$ in direction $L_0 = (0,1)$. The lattice point farthest from $(1,0)$ lying on the ray and below $\mathcal{L}$ is v_0 by (1). We have thus $r(v_0) \geq \mu$ and as a corollary, $r(v_0) + r(L_0) < \mu$, because the lattice point following v_0 in direction L_0 is above $\mathcal{L}$. Putting the two inequalities together we have $r(L_0) < \mu - r(v_0) \leq r(U_0)$.

Induction step: Let us assume that for some k between 1 and n, $r(L_{k-1}) < \mu - r(v_{k-1}) \leq r(U_{k-1})$. We focus on the case where $r(v_{k-1}) > 0$, because the other case is symmetric. In that case, due to the induction hypothesis, $r(U_k) = r(U_{k-1}) \geq \mu - r(v_{k-1})$. Let us consider now the ray emanating from $U_k + v_{k-1}$ in direction $L_k - U_k$. It intersects $\mathcal{L}$ because (i) $r(U_k + v_{k-1}) \geq \mu$ and (ii) $r(L_k - U_k) < 0$ by (4). The lattice point farthest from $U_k + v_{k-1}$ lying on the ray and below $\mathcal{L}$ is $U_k + v_k$ by (2). We have thus $r(U_k) + r(v_k) \geq \mu$ and as a corollary, $r(U_k) + r(v_k) + r(L_k - U_k) < \mu$, which is equivalent to $r(L_k) + r(v_k) < \mu$. As a consequence, we have $r(L_k) < \mu - r(v_k) \leq r(U_k)$, which concludes the proof. $\square$

To show (6), we use induction and the properties of the $\wedge$ operator (linearity and anticommutativity).

Proof (of (6)). Base case: $v_0 = (1,0) - c(0,1)$ for some constant $c \in \mathbb{Z}$ and $(L_0 - U_0) = (0,1)$ by (1). Therefore $v_0 \wedge (L_0 - U_0) = (1,0) \wedge (0,1) = -1$.

Induction step: let us assume that for some k between 1 and n, $v_{k-1} \wedge (L_{k-1} - U_{k-1}) = -1$. By (2), $v_k = v_{k-1} - c(L_k - U_k)$ and $(L_k - U_k) = (L_{k-1} - U_{k-1}) - c'v_{k-1}$ for some constants $c, c' \in \mathbb{Z}$. We conclude that $v_k \wedge (L_k - U_k) = v_{k-1} \wedge (L_{k-1} - U_{k-1})$, which is equal to -1 due to the induction hypothesis. $\square$

Note that (6) implies that $\forall 0 \leq k \leq n$, v_k and $(L_k - U_k)$ are *irreducible*, i.e. their coordinates are relatively prime. Indeed, v_0 and $(L_0 - U_0)$ are irreducible and for all $k \geq 1$, the greatest common divisor of their coordinates divides $v_k \wedge (L_k - U_k)$, which is equal to -1.

Geometrically, (6) implies that at each step $0 \leq k \leq n$, v_k and $(L_k - U_k)$ are a pair of unimodular vectors, while (4) and (5) guarantee that in such a basis the line segment $[L_k U_k]$ is always intersected by $\mathcal{L}$, whereas the line segment of endpoints L_k (resp. U_k) and $L_k + v_k$ (resp. $U_k + v_k$) is never intersected by $\mathcal{L}$.

We now end the subsection with the following lemma:

Lemma 2. *There exists a step $n \geq 1$ such that:*

$$r(L_n - U_n) = -1, \quad r(v_n) = 0. \tag{7}$$

$$U_n = \mu \text{ and } L_n = \mu - 1. \tag{8}$$

To prove lemma 2, it is enough to notice that $r(L_k - U_k)$ is always strictly negative by (4) and that $\forall 0 \leq k \leq n$, $r(L_{k-1} - U_{k-1}) < r(L_k - U_k)$ by (2). These inequalities and (6) guarantee that $r(L_n - U_n) = -1$ and $r(v_n) = 0$ at some step $n \geq 1$. Note that $v_n = (b, a)$, because v_n is irreducible. Then, by (4), the only possible values of $r(L_n)$ and $r(U_n)$ must be respectively $\mu - 1$ and μ.

3   Convex Hull Algorithms

The proof of theorem 1 is now straightforward:

- Ray casting is equivalent to integer division on remainders (section 2.2),
- (1) is equivalent to the initialization of Charrier and Buzer's algorithm [6].
- By lemma 1, (1) and (2) maintain the same invariant as Charrier and Buzer's algorithm [6], i.e. $\forall 0 \leq k \leq n$, v_k is the negative and valid bezout vector of $(L_k - U_k)$.
- Lemma 2 guarantees that the whole lower and upper left hulls are computed.

As a consequence, (1) and (2) provide a simple way to compute the left hull of the lower and upper domains. In the following sections, we show how to translate (1) and (2) first into a forward algorithm and then into a backward algorithm, based on (3).

3.1   A Forward Approach

We first propose a forward algorithm that computes L_k, U_k and v_k with increasing k, starting from L_0, U_0 and v_0. This algorithm is called **smartCH**, because it is on-line and runs in $O(1)$ per vertex and is thus optimal (see algorithm 1).

It is a rather straightforward translation of (1) and (2). There is a difference though: we add an extra constraint, which modifies the stopping criterion. Algorithm **smartCH** takes as input not only the slope a/b and the intercept μ of a DSL, but also the length l of the subsegment Σ' starting from $(0,0)$. If b is minimal for Σ' or, which is equivalent, if $l \gg b$, the algorithm iteratively computes L_k, U_k, and v_k from L_{k-1}, U_{k-1} and v_{k-1} until $r(v_k) = 0$. Otherwise, the algorithm stops as soon as it detects that a new lower or upper leaning point would lie outside Σ' (lines 3 and 7 of algorithm 2). In this case, we correct the last ray casting in order to get the last leaning point (lines 14 and 17) or the last direction vector of Σ' (lines 9 to 11 and 18 to 19 of algorithm 2). The components of the last direction vector gives the rational slope whose denominator is bounded by l and that best approximates a/b. This final step is computed in $O(1)$ instead of the $O(\log(l))$ steps required to compute the critical supporting lines of the lower and upper convex hulls as proposed in [6].

3.2   A Backward Approach

If all partial quotients are known, by using (1) and (2), it is obviously possible to compute U_k, L_k, and v_k with decreasing k from a given step $n \geq 1$. In this section, we show that these partial quotients can be computed from the positions of U_n, L_n and v_n.

As seen in section 2.2, ray casting can be reversed under some conditions. These conditions are actually fulfilled in our framework:

Algorithm 1. smartCH(a, b, μ, l)

Input: a, b, μ, l
Output: V, LHull and UHull, lower and upper left hull
`// initialisation`
1 stop := FALSE ;
2 U := (0,0) ; add U to UHull ;
3 L := (0,1) ; add L to LHull ;
4 $V := (1,0) + \left\lceil \frac{\mu - a}{-b} \right\rceil (0,1)$;
`// main loop`
5 **while** $r(v_k) \neq 0$ *and not stop* **do**
6   **if** $r(v_k) > 0$ **then**
7   stop := nextVertex(a, b, μ, l, L, U, V, LHull, $[.]$, $\lfloor . \rfloor$) ;
8   **if** $r(v_k) < 0$ **then**
9   stop := nextVertex(a, b, μ, l, U, L, V, UHull, $\lfloor . \rfloor$, $[.]$) ;

Algorithm 2. nextVertex(a, b, μ, l, X, Y, V, XHull, $floor_1$, $floor_2$)

Input: a, b, μ, l, X, Y, V, XHull, $floor_1$, $floor_2$
Output: X, V, XHull
1 $q := floor_1\left(\frac{\mu - r(X)}{r(V)}\right)$; `// first ray casting`
2 $X := X + q\,V$;
3 **if** $(p(X) \leq l)$ **then**
4   add X to XHull ;
5     $q := floor_2\left(\frac{\mu - (r(Y) + r(V))}{(r(X) - r(Y))}\right)$; `// second ray casting`
6     $V := V + q\,(X - Y)$;
7   **if** $(p(Y) + p(V) \leq l)$ **then return** *TRUE* ;
8   **else**
9        $V := V - q\,(X - Y)$;
10        $q := \left\lceil \frac{l - (p(Y) + p(V))}{p(X) - p(Y)} \right\rceil$;
11   **if** $q > 0$ **then** $V := V + q\,(X - Y)$;
12   **return** *FALSE* ;
13 **else**
14     $X := X - q\,V$;
15     $q := \left\lceil \frac{l - p(X)}{p(V)} \right\rceil$;
16   **if** $q > 0$ **then**
17        $X := X + q\,V$; add X to XHull ;
18     $q := \left\lceil \frac{l - (p(Y) + p(V))}{p(X) - p(Y)} \right\rceil$;
19   **if** $q > 0$ **then** $V := V + q\,(X - Y)$;
20   **return** *FALSE* ;

Theorem 2. *For each $0 \leq k \leq n$, the positions of L_k, U_k, v_k are ordered as follows:*

$$\forall 0 \leq k < n, \begin{cases} p(v_k) < p(L_{k+1}), & \text{if } r(v_k) > 0 \\ p(v_k) < p(U_{k+1}), & \text{if } r(v_k) < 0. \end{cases} \tag{9}$$

$$\forall 0 \leq k \leq n, \ p(L_k) < p(v_k) \ \text{and} \ p(U_k) < p(v_k). \tag{10}$$

Inequalities (9) are obvious and provide the necessary and sufficient condition to reverse the second ray casting (lines 5-6 of algorithm 2). Indeed, (3) requires that

$$p(Y) + p(V) < p(X - Y) \Leftrightarrow p(V) < p(X)$$

and according to the notation used in algorithm 2, $X = L_{k+1}$ if $V = v_k$ has a positive remainder, but $X = U_{k+1}$ otherwise.

Inequalities (10) provide the necessary and sufficient condition to reverse the first ray casting (lines 1-2 of algorithm 2). Indeed, (3) requires that $p(X) < p(V)$ and according to the notation used in algorithm 2, $X = L_k$ if $V = v_k$ has a positive remainder, but $X = U_k$ otherwise.

To complete the proof of theorem 2, we prove (10) by induction.

Proof (of (10)). Base case: Since $p(v_0) = 1$ while $p(L_0) = p(U_0) = 0$, (10) is obviously true for $k = 0$.

Induction step: Let us assume that $p(L_{k-1}) < p(v_{k-1})$ and that $p(U_{k-1}) < p(v_{k-1})$ for some k between 1 and n. Let us assume that $r(v_{k-1}) > 0$, the other case being symmetric. By (2), it is easy to see that $p(v_k) \geq p(v_{k-1}) + p(L_k - U_k)$. Since $U_k = U_{k-1}$ and $p(v_{k-1}) - p(U_{k-1}) > 0$ due to the induction hypothesis, we have $p(v_k) > p(L_k)$. Since $p(L_k) > p(U_k)$, we obviously have also $p(v_k) > p(v_{k-1}) > p(U_{k-1}) = p(U_k)$. □

Theorem 2 and (3) lead to a new set of recurrence relations, which is dual to (1) and (2):

$$\forall k \leq n, \atop p(U_k) \neq p(L_k)} \begin{cases} \text{if } p(L_k) < p(U_k), \begin{cases} L_{k-1} = L_k + \left\lceil \frac{p(L_k)}{p(v_{k-1})} \right\rceil v_{k-1} \\ U_{k-1} = U_k \\ v_{k-1} = v_k + \left\lfloor \frac{p(v_k) - p(U_k)}{p(U_k) - p(L_k)} \right\rfloor (L_k - U_k) \end{cases} \\ \text{if } p(L_k) > p(U_k), \begin{cases} L_{k-1} = L_k \\ U_{k-1} = U_k + \left\lceil \frac{p(U_k)}{p(v_{k-1})} \right\rceil v_{k-1} \\ v_{k-1} = v_k + \left\lfloor \frac{p(v_k) - p(L_k)}{p(U_k) - p(L_k)} \right\rfloor (U_k - L_k) \end{cases} \end{cases} \tag{11}$$

This set of recurrence relations has properties similar to (2) and straightforwardly leads to a backward algorithm, called **reversedSmartCH**, which is on-line and runs in $O(1)$ per vertex. As done with **smartCH**, we can add a length constraint to stop the algorithm sooner and solve the subsegment problem. We do not provide more details due to lack of space. However, we compare below our implementation of **smartCH** and **reversedSmartCH** to the algorithms whose implementation is available in DGtal [1].

3.3 Experiments

We generate random DSSs, starting from $(0,0)$, whose slope a/b has a continued fraction expansion of constant depth n equal to 15.[3] For each DSS, we consider one of its left subsegment starting from $(0,0)$. Its length l is determined by the denominator of the k-th convergent of the bounding DSS slope, so that the subsegment slope of minimal denominator has a continued fraction expansion of depth n' equal to k. In fig. 4, we plot the running times (in seconds) of 100.000 calls of several algorithms against parameter k, which is ranging from 1 to 14. As expected in tab. 1, we observe that the running time of the forward methods (a) is linear in n' (and thus logarithmic in l), whereas the running time of the backward methods (b) is linear in $n - n'$ (and thus logarithmic in $b - l$). Our generic implementation of **smartCH** outperforms [6], smartDSS [12] and is comparable to [17]. Moreover, our implementation of **reversedSmartCH** outperforms reversedSmartDSS [12], which is much more space consuming because continued fraction expansions of all DSS slopes are stored in a shared data structure that grows at each call.

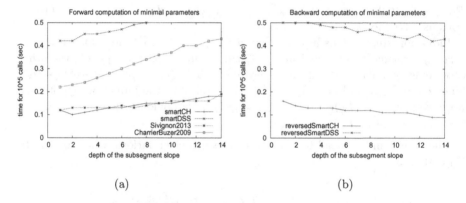

(a)   (b)

Fig. 4. We plot the running times of the algorithms whose implementation is available in DGtal [1], i.e. [6,12,17], against parameter k, which is equal to the depth of the subsegment slope

4 Conclusion

In this paper, we propose a simple arithmetical characterization of the convex hull of DSSs, which gives new insights into the combinatorial structure of DSSs of arbitrary intercept and length. This characterization and its dual, lead to two on-line algorithms that computes the left hull of a given DSS. The first one, called **smartCH**, returns vertices of decreasing remainders, but increasing positions, while the second one, called **reversedSmartCH**, returns vertices of

---

[3] Note that the depth is set to 15 and all partial quotients are randomly chosen in $\{1, \ldots, 4\}$ so that numerators and denominators are not greater than $2^{31} - 1$.

increasing remainders, but decreasing positions. They both run in constant space and constant time per vertex. They also provide a logarithmic-time and efficient solution to the subsegment problem.

References

1. DGtal: Digital geometry tools and algorithms library, `http://libdgtal.org`
2. Anderson, T.A., Kim, C.E.: Representation of digital line segments and their preimages. Computer Vision, Graphics, and Image Processing 30(3), 279–288 (1985)
3. Balza-Gomez, H., Moreau, J.-M., Michelucci, D.: Convex hull of grid points below a line or a convex curve. In: Bertrand, G., Couprie, M., Perroton, L. (eds.) DGCI 1999. LNCS, vol. 1568, pp. 361–374. Springer, Heidelberg (1999)
4. Brons, R.: Linguistic Methods for the Description of a Straight Line on a Grid. Computer Graphics and Image Processing 3(1), 48–62 (1974)
5. Bruckstein, A.M.: Self-Similarity Properties of Digitized Straight Lines. Contemporary Mathematics 119, 1–20 (1991)
6. Charrier, E., Buzer, L.: Approximating a real number by a rational number with a limited denominator: A geometric approach. Discrete Applied Mathematics 157(16), 3473–3484 (2009)
7. Davenport, H.: The Higher Arithmetic: Introduction to the Theory of Numbers. Oxford University Press, Oxford (1983)
8. Debled-Rennesson, I., Reveillès, J.P.: A linear algorithm for segmentation of digital curves. International Journal of Pattern Recognition and Artificial Intelligence 9(4), 635–662 (1995)
9. Har-Peled, S.: An output sensitive algorithm for discrete convex hulls. Computational Geometry 10(2), 125–138 (1998)
10. Harvey, W.: Computing Two-Dimensional Integer Hulls. SIAM Journal on Computing 28(6), 2285–2299 (1999)
11. Klette, R., Rosenfeld, A.: Digital straitghness – a review. Discrete Applied Mathematics 139(1-3), 197–230 (2004)
12. Lachaud, J.O., Said, M.: Two efficient algorithms for computing the characteristics of a subsegment of a digital straight line. Discrete Applied Mathematics 161(15), 2293–2315 (2013)
13. Lindenbaum, M., Bruckstein, A.: On recursive, $o(n)$ partitioning of a digitized curve into digital straight segments. IEEE Transactions on Pattern Analysis and Machine Intelligence 15(9), 949–953 (1993)
14. Ouattara, J.S.D., Andres, E., Largeteau-Skapin, G., Zrour, R., Tapsob, T.M.Y.: Remainder Approach for the Computation of Digital Straight Line Subsegment Characteristics. Submitted to Discrete Applied Mathematics (2014), doi:10.1016/j.dam.2014.06.006
15. Reveillès, J.P.: Géométrie Discrète, calculs en nombres entiers et algorithmique. Thèse d'etat, Université Louis Pasteur (1991)
16. Reveillès, J.P., Yaacoub, G.: A sublinear 3D convex hull algorithm for lattices. In: DGCI 1995, pp. 219–230 (1995)
17. Sivignon, I.: Walking in the Farey Fan to Compute the Characteristics of a Discrete Straight Line Subsegment. In: Gonzalez-Diaz, R., Jimenez, M.-J., Medrano, B. (eds.) DGCI 2013. LNCS, vol. 7749, pp. 23–34. Springer, Heidelberg (2013)
18. Voss, K.: Coding of digital straight lines by continued fractions. Computers and Artificial Intelligence 10(1), 75–80 (1991)

Parameter-Free and Multigrid Convergent Digital Curvature Estimators[*]

Jérémy Levallois[1,2], David Coeurjolly[1], and Jacques-Olivier Lachaud[2]

[1] Université de Lyon, CNRS,
INSA-Lyon, LIRIS, UMR5205, F-69621, France
[2] Université de Savoie, CNRS,
LAMA, UMR5127, F-73776, France

Abstract. In many geometry processing applications, the estimation of differential geometric quantities such as curvature or normal vector field is an essential step. Focusing on multigrid convergent estimators, most of them require a user specified parameter to define the scale at which the analysis is performed (size of a convolution kernel, size of local patches for polynomial fitting, etc). In a previous work, we have proposed a new class of estimators on digital shape boundaries based on Integral Invariants. In this paper, we propose new variants of these estimators which are parameter-free and ensure multigrid convergence in 2D. As far as we know, these are the first parameter-free multigrid convergent curvature estimators.

Keywords: Curvature estimation, multigrid convergence, integral invariants, digital straight segments, parameter-free estimators.

1 Introduction

Estimating differential quantities like curvatures on discrete data is a tricky task and generally relies on some user supervision to specify some computation window. Indeed, the user has to balance between a small window which preserves most likely sharp features and a big window which offer a better accuracy in flatter smooth zones. Even worse, there may not exist a window size that is appropriate to the whole data. For digital data, another fundamental issue is related to the *multigrid convergence* property of geometric estimators: this property holds whenever the geometric estimation on a digitized shape is more and more accurate as the digitization step gets finer and finer. It is clear that a user supervision cannot be considered when multigrid convergence is involved. The question is then: can we design *parameter-free* curvature(s) estimator on digital data ? Furthermore, can this estimator be adaptive to the local data characteristics ?

For 2D digital curves, *tangent* estimation from maximal digital straight segments answers these two questions in a nice way. Indeed, it requires no parameter

---

[*] This work has been mainly funded by DIGITALSNOW ANR-11-BS02-009 and KIDICO ANR-2010-BLAN-0205 research grants.

E. Barcucci et al. (Eds.): DGCI 2014, LNCS 8668, pp. 162–175, 2014.
© Springer International Publishing Switzerland 2014

and is proven to be multigrid convergent [7,13]. For *curvature* estimation, the *digitization grid step h* is required to get the scale of the shape but is not compulsory to get relative estimations. Two accurate curvature estimators are parameter-free and adaptive: one based on maximal digital circular arcs (MDCA) [12], one based on squared curvature minimization [5]. However, their convergence is not proven. To get convergence, authors generally define the computation window as a function of h. Binomial or Gaussian convolution estimators [8,4] use some $1/h^{\alpha}$ window size, digital integral invariants [2] use some $h^{\frac{1}{3}}$ radius size. The polynomial fitting of [11] also requires a thickness parameter. For 3D digital surfaces, the only multigrid convergent estimator of the curvature tensor that we are aware of is the digital integral invariant (II for short) of [2,3]; it relies on a $h^{\frac{1}{3}}$ radius size.

This paper proposes a theoretically sound method to get rid of user specified parameters. The idea is to use length properties of maximal segments as functions of h in order to determine automatically a correct computation window. We show how this approach can set automatically the digital radius size required for the multigrid convergence of the II curvature estimator, without any knowledge of the grid step h. By this way, we obtain the first parameter-free multigrid convergent curvature estimator for 2D contours. This approach is also extensible to 3D by a careful use of axis-aligned slices in the shape. Although we have hints of multigrid convergence, still some work is required to prove it fully.

2   Preliminaries

We denote by Z any subset of $\mathbb{Z}^2$ (or $\mathbb{Z}^3$, depending on the context). In dimension 2, $Bd(Z)$ denotes the topological boundary of Z, seen as a cellular cartesian complex. It is thus composed of $0-$ and $1-$cells (resp. *pointels* and *linels*). By convention, we decide to map pointels coordinates to $\mathbb{Z}^2$.

Definition 1 (Standard Line and Digital Straight Segment). *The set of points* $(x,y) \in \mathbb{Z}^2$ *satisfying* $\mu \le ax - by < \mu + |a| + |b|$, *with a, b and μ integer numbers, is called the* standard digital line *with slope a/b and shift μ. Any connected subset of pixels of a standard digital line is a* digital straight segment *(DSS for short).*

Definition 2 (Maximal Segment and Maximal Segment Pencil [7]). *A sequence of pointels* $\{p_i, \ldots, p_j\} \subset Bd(Z)$ *is a* maximal segment *iff* $\{p_i, \ldots, p_j\}$ *is a DSS which cannot be extended neither to its front nor to its back while still being a DSS. At a given pointel $p \in Bd(Z)$, the* pencil of maximal segment *at p is the set of maximal segments on $Bd(Z)$ containing p.*

In the multigrid convergence framework, digital objects Z are given by the digitization of a continuous object for a given scale factor h. More formally, given a family of Euclidean shapes $\mathbb{X}$, we denote by $\mathbf{D}_h(X)$ the *Gauss digitization* of $X \in \mathbb{X}$ in a $d-$dimensional grid of grid step h, i.e.

$$\mathbf{D}_h(X) \stackrel{def}{=} \left(\frac{1}{h} \cdot X\right) \cap \mathbb{Z}^d \qquad (1)$$

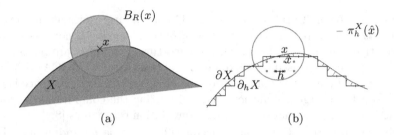

Fig. 1. Integral invariant computation (*left*) and notations (*right*) in dimension 2

Similarly to [3] we denote $\partial_h X$ the h−boundary of X, *i.e.* a $d − 1$-dimensional subset of $\mathbb{R}^d$ corresponding to the geometrical embedding of the boundary of the Gauss digitization of X at grid step h. In our multigrid convergence framework, quantities are estimated on $\partial_h X$ and then compared to associated expected values on ∂X (see Fig. 1). Note that a discrete/combinatorial view of $\partial_h X$ is given by $Bd(Z)$ with $Z = \mathsf{D}_h(X)$. In many situations, maximal segments and maximal segment pencils play a very important role in multigrid digital contour geometry processing [6,13]. For the purpose of this paper, let us focus on the asymptotic properties of lengths of maximal segment:

Lemma 1 (Asymptotic Laws of Maximal Segments [6,13]). *Let X be some convex shape of $\mathbb{R}^2$, with at least C^3-boundary and bounded curvature. The discrete length of maximal segments in $Bd(Z)$ for $Z = \mathsf{D}_h(X)$ follows:*

- *the shortest is lower bounded by $\Omega(h^{-\frac{1}{3}})$;*
- *the longest is upper bounded by $O(h^{-\frac{1}{2}})$;*
- *their average length, denoted $L_D(Z)$, is such that:*

$$\Theta(h^{-\frac{1}{3}}) \le L_D(Z) \le \Theta\left(h^{-\frac{1}{3}} \log\left(\frac{1}{h}\right)\right). \tag{2}$$

In [2,3], we have proposed convergent curvature estimators based on Integral Invariants [10,9]. For short, the idea is to move a geometrical kernel (ball with radius R) at each surface point and to compute integrals on the intersection between the ball and the digital shape. In dimension 2, the estimator $\hat{\kappa}_R(Z, \hat{x}, h)$ is defined as a function of the number of grid points in $(h \cdot Z) \cap B_R(\hat{x})$, $B_R(\hat{x})$ being the ball with radius R centered at $\hat{x}$. In dimension 3, mean curvature estimation can be obtained from the number of points in $(h \cdot Z) \cap B_R(\hat{x})$. Instead of simply estimating the measure (volume) of $X \cap B_R(x)$ by discrete summation, we estimate in 3D the covariance matrix of $(h \cdot Z) \cap B_R(\hat{x})$. Its eigenvalues/eigenvectors give us quantitative and directional information from which we can design estimators $\hat{\kappa}_R^1(Z, \hat{x}, h)$ and $\hat{\kappa}_R^2(Z, \hat{x}, h)$ of principal curvatures κ^1 and κ^2 for $x \in \partial X$. Their multigrid convergence properties are summed up below.

Theorem 1 (Uniform multigrid convergence of $\hat{\kappa}_R$, $\hat{\kappa}_R^1$, and $\hat{\kappa}_R^2$ [2,3]). *Let X be some convex shape in $\mathbb{R}^2$ or $\mathbb{R}^3$, with at least C^3-boundary and bounded* *curvature. Then, there exist positive constants h_0, k and K, for any $h \leq h_0$, setting $R = kh^{\frac{1}{3}}$, we have: $\forall x \in \partial X, \forall \hat{x} \in \partial_h X, \|\hat{x} - x\|_\infty \leq h$,*

$$|\hat{\kappa}_R(\mathsf{D}_h(X), \hat{x}, h) - \kappa(X, x)| \leq Kh^{\frac{1}{3}} \tag{3}$$

and more specifically for X in $\mathbb{R}^3$, $\forall i \in \{1, 2\}$:

$$|\hat{\kappa}_R^i(\mathsf{D}_h(X), \hat{x}, h) - \kappa^i(X, x)| \leq Kh^{\frac{1}{3}} \tag{4}$$

A key point in Theorem 1 is that the radius R of the ball has to be in $\Theta(h^{\frac{1}{3}})$ to get the convergence result. In the following, we use geometrical characteristics of $Bd(Z)$ (its maximal segment length distribution) to automatically select the appropriate local or global radius R while keeping the multigrid convergence property.

3   Multigrid Convergence of 2D Parameter-Free Curvature

Let us first define our new curvature estimator on digital objects $Z \subset \mathbb{Z}^2$:

Definition 3. *Given $Z \subset \mathbb{Z}^2$, the parameter-free digital curvature estimator $\hat{\kappa}^*$ at a pointel $p \in Bd(Z)$ is defined as:*

$$\hat{\kappa}^*(Z, p) \overset{def}{=} \frac{3\pi}{2\rho(Z)} - \frac{3A(Z, p)}{\rho(Z)^3} \tag{5}$$

where $\rho(Z) = L_D^2(Z)$ and $A(Z, p) = Card(B_{\rho(Z)}(p) \cap Z)$.

To rephrase the definition, we first compute the average discrete length of all maximal segments of $Bd(Z)$. Then the symbol ρ is the square of this length. The estimation $\hat{\kappa}^*(Z, p)$ is a function of the number of digital points in Z intersected with the ball of radius ρ centered at p. In the following, we fill the gap between the parameter-free estimator $\hat{\kappa}^*$ and 2D II curvature estimator as described in [2,3].

First of all, the multigrid convergence framework implies we have a scale factor h between maximal segment lengths on $Bd(Z)$ and distances in the Euclidean space on which $\partial_h X$ is defined. Hence, for $Z = \mathsf{D}_h(X)$, inserting $\rho(Z) = L_D^2(Z)$ into Lemma 1 implies

$$\Theta(h^{\frac{1}{3}}) \leq h\rho(Z) \leq \Theta(h^{\frac{1}{3}} \log^2(\frac{1}{h})) . \tag{6}$$

In [2,3], we have define a 2D II digital curvature estimator $\hat{\kappa}_R$ that depends on a ball radius R. Its multigrid convergence is guaranteed whenever the radius is in $\Theta(h^{\frac{1}{3}})$. The quantity $\rho(Z)$ relies only on the digital contour geometry of Z. Except for the $\log^2(\cdot)$ term, $h\rho(Z)$ is thus an excellent candidate for parameter R, since it follows approximately $\Theta(h^{\frac{1}{3}})$, to design a parameter free curvature estimator.

Theorem 2 (Uniform convergence of curvature estimator $\hat{\kappa}^*$). *Let X be some convex shape of $\mathbb{R}^2$, with at least C^3-boundary and bounded curvature. Let $Z = D_h(X)$. Then, there exist a positive constant h_0, for any $0 < h \leq h_0$, we have, $\forall x \in \partial X$ and $\forall p \in Bd(Z)$*

$$\|hp - x\|_\infty \leq h \Rightarrow \left| \frac{1}{h}\hat{\kappa}^*(Z, p) - \kappa(X, x) \right| \leq O(h^{\frac{1}{3}} \log^2(\frac{1}{h})) . \tag{7}$$

Note that $p \in Bd(Z)$ implies $hp \in \partial_h X$. The parameter-free curvature is rescaled by h in order to compare comparable shapes.

Proof. First, we expand $\frac{1}{h}\hat{\kappa}^*(Z, p)$ as

$$\frac{1}{h}\hat{\kappa}^*(Z, p) = \frac{3\pi}{2h\rho(Z)} - \frac{3A(Z, p)}{h\rho(Z)^3} = \frac{3\pi}{2h\rho(Z)} - \frac{3\,Card(B_{\rho(Z)}(p) \cap D_h(X))}{h\rho(Z)^3}$$

$$= \frac{3\pi}{2h\rho(Z)} - \frac{3\,Card(B_{(h\rho(Z))/h}(\frac{1}{h} \cdot (hp)) \cap D_h(X), h)}{h\rho(Z)^3}$$

$$= \hat{\kappa}_R(D_h(X), \hat{x}, h) \quad (\text{with } R \overset{def}{=} h\rho(Z) \text{ and } \hat{x} \overset{def}{=} hp \text{ and } [2]) .$$

It suffices now to bound $|\hat{\kappa}_R(D_h(X), \hat{x}, h) - \kappa(X, x)|$ according to the asymptotic behavior of $R \overset{def}{=} h\rho(Z)$. According to Eq.(6), R is contained between two bounds:

If $R = \Theta(h^{\frac{1}{3}})$, we are in the hypothesis of Theorem 2, so the error term is in $O(h^{\frac{1}{3}})$.

If $R = \Theta(h^{\frac{1}{3}} \log^2(\frac{1}{h}))$, we expand the error term in Theorem 2 using Eq. (18) of [2] (with $\alpha' = 1$ and $\beta = 1$ in general case):

$$|\hat{\kappa}_R(D_h(X), \hat{x}, h) - \kappa(X, x)|$$

$$\leq O(h\rho(Z)) + O\left(\frac{h}{(h\rho(Z))^2}\right) + O\left(\frac{h}{(h\rho(Z))^2}\right)(1 + O((h\rho(Z))^2) + O(h)) \tag{8}$$

$$\leq O(h^{\frac{1}{3}} \log^2(\frac{1}{h})) + O\left(\frac{h^{\frac{1}{3}}}{\log^4(\frac{1}{h})}\right) + O(h) + O\left(\frac{h^{\frac{4}{3}}}{\log^4(\frac{1}{h})}\right) \tag{9}$$

$O(h^{\frac{1}{3}} \log^2(\frac{1}{h}))$ is the dominant error term in the latter expression. Gathering the two cases and recalling that $\frac{1}{h}\hat{\kappa}^*(Z, p) = \hat{\kappa}_R(D_h(X), \hat{x}, h)$, we conclude that the error in Eq.(7) is exactly $O(h^{\frac{1}{3}} \log^2(\frac{1}{h}))$. □

The previous curvature estimator is thus convergent. It requires the scale parameter h only to determine the unity used for measuring curvatures. But all curvatures are correct relatively. A possible drawback of the previous estimator is that the ball radius is not adaptive to shape features (for instance sharp features). Instead of using the same ball radius for the whole shape, we can use the maximal segment pencil at each pointel to detect a local size for the radius. We will denote by $\rho(Z, p)$ the square of average length of the maximal segment pencil at pointel p.

Definition 4. *Given* $Z \subset \mathbb{Z}^2$, *the* local parameter-free curvature estimator $\hat{\kappa}_l^*$ *at a pointel* $p \in Bd(Z)$ *is given by:*

$$\hat{\kappa}_l^*(Z,p) \stackrel{def}{=} \frac{3\pi}{2\rho(Z,p)} - \frac{3A'(Z,p)}{\rho(Z,p)^3} \tag{10}$$

where $A'(Z,p) = Card(B_{\rho(Z,p)}(p) \cap Z)$.

In this local version, some maximal segments may have a too long length which prevents us to have multigrid convergence proof. Indeed, if, in the maximal segment pencil, lengths are in the global range of Eq. (2), good multigrid behavior of this local estimator can be expected. Issues arise for maximal segments with longest length in $O(h^{-\frac{1}{2}})$ in Lemma 1. In this pathological case, no convergence can be expected. In Sect. 5, we experimentally show very good convergence properties on this estimator.

4   3D Parameter-Free Curvature Tensor Estimators

In this section, we present parameter-free curvature tensor estimators in 3D. To sum up, we use the lengths of maximal segments in object slices to automatically set the integral invariant radius parameter. Let us first start with a proposition on smooth manifolds. Let X be any object in $\mathbb{R}^3$ with C^2-smooth boundary ∂X whose absolute principal curvatures are bounded by some constant K. The normal to ∂X at x is denoted by $\boldsymbol{n}(x)$. The principal curvatures at x are denoted by $\kappa_1(x)$ and $\kappa_2(x)$.

Proposition 1. *For any* $x \in \partial X$, *let* $\pi_e(x)$ *be the plane containing* x *and orthogonal to vector* $e \in \{\boldsymbol{x}, \boldsymbol{y}, \boldsymbol{z}\}$. *Let* $\partial X_e(x)$ *be the set that is the intersection* $X \cap \pi_e(x)$. *Then at least two of the sets* $\partial X_{\boldsymbol{x}}(x)$, $\partial X_{\boldsymbol{y}}(x)$ *and* $\partial X_{\boldsymbol{z}}(x)$ *are locally curves whose curvatures are bounded by* $\sqrt{2}K$ *in absolute value.*

The proof is available in Appendix A.1. The radius $\rho'(Z)$ of integral invariant computation will be defined as the square of some *average of lengths of maximal segments for a digital object Z in $\mathbb{Z}^3$*, a more formal definition will be given just after. We may define now our parameter-free curvature estimators in 3D:

Definition 5. *Given* $Z \subset \mathbb{Z}^3$, *the* parameter-free mean curvature estimator $\hat{H}^*$ *at a pointel* $p \in Bd(Z)$ *is defined as:*

$$\hat{H}^*(Z,p) \stackrel{def}{=} \frac{8}{3\rho'(Z)} - \frac{4V_{\rho'(Z)}(Z,p)}{\pi\rho'(Z)^4} , \tag{11}$$

where $V_{\rho'(Z)}(Z,p) = Card(B_{\rho'(Z)}(p) \cap Z)$.

As discussed in [9,3], directional curvature information and thus curvature tensor can be estimated from the eigenvalues of the covariance matrix[1] of $Card(B_{\rho'(Z)}(p) \cap Z)$.

---

[1] The covariance matrix of $Y \subset \mathbb{R}^3$ is defined by $J(Y) \stackrel{def}{=} \int_Y (p - \overline{Y})(p - \overline{Y})^T dp$ where $\overline{Y}$ is the centroid of Y.

Definition 6. *Let Z be a digital shape in $\mathbb{Z}^3$, we define the* parameter-free prin-cipal curvature estimators $\hat{\kappa}^{1*}$ *and* $\hat{\kappa}^{2*}$ *of Z at point $p \in Bd(Z)$ as*

$$\hat{\kappa}^{1*}(Z,p) = \frac{6(\hat{\lambda}_2 - 3\hat{\lambda}_1)}{\pi\rho'(Z)^6} + \frac{8}{5\rho'(Z)} \;, \quad \hat{\kappa}^{2*}(Z,p) = \frac{6(\hat{\lambda}_1 - 3\hat{\lambda}_2)}{\pi\rho'(Z)^6} + \frac{8}{5\rho'(Z)}, \quad (12)$$

where $\hat{\lambda}_1$ and $\hat{\lambda}_2$ are the two greatest eigenvalues of the covariance matrix of $B_{\rho'(Z)}(p) \cap Z$.

Let us now precise what is the $\rho'(Z)$ parameter. We provide one global definition $\rho'(Z)$ and one local definition $\rho'(Z,p)$ for $p \in Bd(Z)$:

Definition 7. *Given a digital object Z, each surfel $p \in Bd(Z)$ is orthogonal to two slices $\pi_{e_1(p)}$ and $\pi_{e_2(p)}$. For each slice $\pi_{e_i(p)} \cap Z$, the pencil of maximal segments covering p determines a set of integers $l_i(p)$, formed by the lengths of these maximal segments. Finally, we number by $M(p)$ the slice containing the longest maximal segment (i.e. the slice i whose set $l_i(p)$ contains the biggest integer). Then, we define*

- *$\rho'(Z)$ is the square of the average of maximal segment lengths for all slices $\pi_{e_i(p)} \cap Z$ of Z;*
- *$\rho'(Z,p)$ is the square of the average value of $l_{M(p)}(p)$.*

As in 2D, some pathological cases may appear leading to the fact that $h\rho'(Z,p) \in \Theta(1)$. In which case, nothing could be expected in terms of multigrid convergence. Again, experimental analysis shows that this bad behavior is not observed. Un-like the 2D case, we do not have a complete knowledge about the multigrid behavior of $h\rho'(Z)$. Let use first express it as a conjecture.

Conjecture 1. *Let X be some convex shape of $\mathbb{R}^3$, with C^3-boundary and bounded curvature. Let $Z \overset{def}{=} D_h(X)$, then there exists a positive constant h_0,*

$$\forall 0 < h \leq h_0, \quad \Theta(h^{\frac{1}{3}}) \leq h\rho'(Z) \leq \Theta(h^{\frac{1}{3}} \log^2(\frac{1}{h})) . \quad (13)$$

The rationale behind this conjecture can be sketched as follows. Slicing the objects in all directions, from Proposition 1, we know that at least two third of the slices define convex curves with bounded local curvature information. Since two slices go through one surfel, at least one slice per surfel provides a convex curve with bounded curvature. We thus expect that the lengths of more than half of the maximal segments follow Eq.(13) bounds. Hence, computing the mean of all these lengths provides a stable and consistent quantity which would also follow Eq. (13). In Sect 5, we provide a complete experimental evaluation which supports this conjecture. Assuming Conjecture 1, we can prove the two following observations:

Observation 1 (Uniform convergence of $\hat{H}^*$.). *Let X be some convex shape of $\mathbb{R}^3$, with at least C^3-boundary and bounded curvature. Let $Z \overset{def}{=} D_h(X)$, then* *there exists a positive constant h_0, for any $0 < h \leq h_0$, we have $\forall x \in \partial X$ and $\forall p \in Bd(Z)$,*

$$\|hp - x\|_\infty \leq h \Rightarrow \left| \frac{1}{h}\hat{H}^*(Z,p) - H(X,x) \right| \leq O(h^{\frac{1}{3}}\log^2(\frac{1}{h})). \qquad (14)$$

Assuming Conjecture 1, the proof is similar to the proof of Theorem 2.

Observation 2 (Uniform convergence of $\hat{\kappa}^{1*}$ and $\hat{\kappa}^{2*}$). *Let X be some convex shape of $\mathbb{R}^3$, with at least C^3-boundary and bounded curvature. For $i \in \{1,2\}$, recall that $\kappa^i(X,x)$ is the i-th principal curvature of ∂X at boundary point x. Let $Z \overset{def}{=} D_h(X)$, then, there exists a positive constant h_0, for any $0 < h \leq h_0$, we have $\forall x \in \partial X$ and $\forall p \in Bd(Z)$,*

$$\|hp - x\|_\infty \leq h \Rightarrow \left| \frac{1}{h}\hat{\kappa}^{i*}(Z,p) - \kappa^i(X,x) \right| \leq O(h^{\frac{1}{3}}log^2(\frac{1}{h})). \qquad (15)$$

Proof is available in Appendix A.2.

5   Experimental Evaluation

We present an experimental evaluation of our parameter-free curvature estimators described before, in 2D and 3D (mean and principal curvatures). All these estimators are implemented in the open-source C++ library DGtal [1]. DGtal provides us a way to construct parametric and implicit 2D and 3D shapes for a given grid step h. Furthermore, DGtal holds a collection of estimators and several tools to facilitate the comparison between estimators. In dimension 2, we compare our estimators with a parameter-free curvature estimator called *Most-centered Digital Circular Arc curvature estimator* (MDCA) [12], which gives good results but whose multigrid convergence — although observed — is unfortunately not proven. In dimension 3, there is no parameter-free estimator which provides some multigrid convergence. Therefore, considering an implicit or parametric shape on which the exact curvature is known, we present two different global curvature error measurement for a shape at a given grid step h: the l_∞ norm measures the average of the absolute error between estimated and true curvature (it corresponds to the uniform convergence in previous theorems), and the l_2 norm is the square root of the average of squared errors (it better reflects an average behavior of the estimator).

As described in Sect. 3, we build our estimators by moving a geometrical kernel (an Euclidean ball in dD) of radius $h\rho$ in 2D and $h\rho'$ in 3D, and centering it on each surface elements (*surfels*). The volume or the covariance matrix of the intersection of the kernel and the digital object is then estimated by simple pixel or voxel enumeration. Since the radius of the kernel is $h\rho$ or $h\rho'$, we first need to estimate them.

In Fig. 2, we study $h\rho'$. We see that experimentally it follows the expected asymptotic behavior of Conjecture 1, i.e. they are bounded between $\Theta(h^{\frac{1}{3}})$ and $\Theta(h^{\frac{1}{3}}\log^2(\frac{1}{h}))$. Hence, they define a consistent kernel radius for curvature estimators. In Fig. 3 we present asymptotic error measurements for the proposed

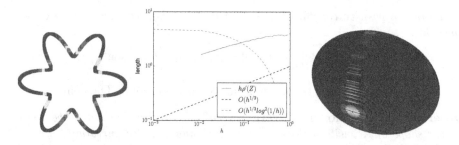

Fig. 2. Clustering of squared length statistics using $K-$means mapped to "flower" contour points (left). Comparison of asymptotic behavior (in log-space) of $h\rho'(Z)$ (center) and mapping of length of $h\rho'(Z, p)$ on an ellipsoid (right).

parameter-free curvature estimators $\hat{\kappa}^*$ and $\hat{\kappa}_l^*$. These graphs also display error measurements for MDCA [12] and our former non parameter-free version of this estimator $\hat{\kappa}_R$ (setting $R = kh^{1/3}$ for some constant k) [2]. We observe that all estimators are convergent with convergence speed at least in $O(h^{\frac{1}{3}})$ except for the local parameter-free estimator on the multigrid ellipse.

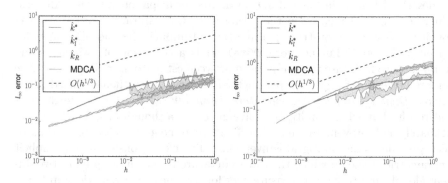

Fig. 3. Comparison in log-space of l_∞ curvature error on multigrid ellipses (left) and flowers (right)

In Fig. 4, we present an experimental evaluation of local estimator $\hat{\kappa}_l^*$ which adapts the kernel size for each point of the digital contour. The main argument for this estimator is to offer an adaptive estimation, for instance to better handle sharp features. As depicted in Fig. 4 for a "flower" shape, we first observe that the local estimator $\hat{\kappa}_l^*$ has a similar behavior to the global estimator $\hat{\kappa}^*$ for the l_∞ norm. We also note that, thanks to the adaptive nature of $\hat{\kappa}_l^*$, the l_2 error is lower for $\hat{\kappa}_l^*$ than $\hat{\kappa}^*$. For high resolution objects (*i.e.* small h), $\hat{\kappa}_l^*$ is very time consuming (we need to create a new kernel at each point and we cannot use differential masks as described in [2]). To speed up its computation, we introduce $K-$means variants: we distribute the lengths of maximal segments into K bins by K-means clustering, in order to have a limited number of kernels. Hence

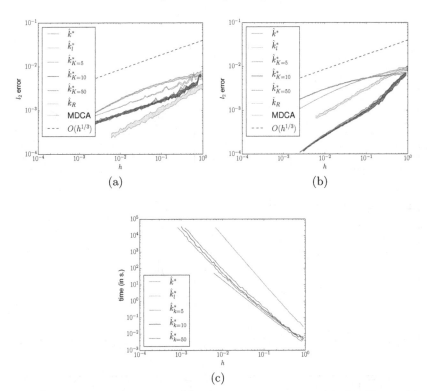

(a)   (b)

(c)

Fig. 4. Comparison in log-space of l_2 curvature error on multigrid ellipses (a) and multigrid flowers (b) with the local $\hat{\kappa}_l^*$ estimator and with different number of precomputed kernels. (c) Computational efficiency of local estimators.

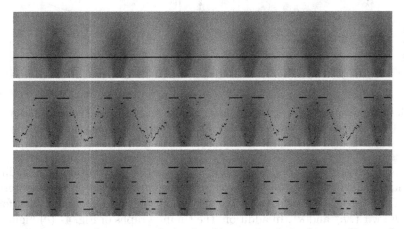

Fig. 5. Curvature scale-space analysis of a flower: x−axis is the curvilinear abscissa, y−axis is the kernel radius, curvature values are mapped between the blue (lowest curvature) and the yellow color (highest curvature). In black are drawn the radius $\rho(Z)$ for global estimator $\hat{\kappa}^*$ (first row), radii $\rho(Z,p)$ for local estimator $\hat{\kappa}_l^*$ (second row), and radii $\rho(Z,p)$ after K−mean clustering for local estimator $\hat{\kappa}_{K=5}^*$. (last row)

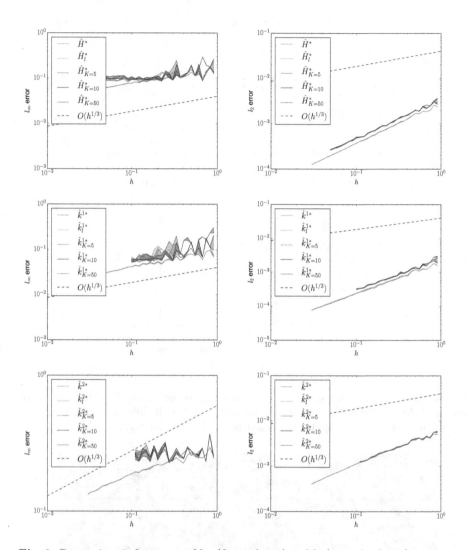

Fig. 6. Comparison in log-space of l_∞ (first column) and l_2 (second column) mean curvature (top), principal curvatures (middle and bottom) errors on a multigrid ellipsoid

mask precomputations are possible. Fig. 2 shows the distribution of K-radii in 2D and 3D. In Fig. 4, we tested curvature estimators based on this clustering with $K \in \{5, 10, 20\}$. We first observe very good multigrid accuracy, even for a small K, w.r.t. $\hat{\kappa}_l^*$ and $\hat{\kappa}^*$. In addition, the timing graphs of Fig. 4−(c) highlight the interest of considering $K-$means clustering to get an efficient and accurate local estimator. To better understand the local and global length properties of maximal segments, we display on Fig 5 a scale-space view of curvature estimation on the classical "flower" shape and the ball radii respectively used by $\hat{\kappa}^*$, $\hat{\kappa}_l^*$ and $\hat{\kappa}_{K=5}^*$.

Fig. 7. (Left) Mean curvature mapped on "bunny" at different resolution using $\hat{H}_l^*$ (yellow color is the highest curvature, blue the lowest). (Right) First principal direction on "bunny" using $\hat{k}_l^*$ estimator.

Fig. 6 presents 3D result on parameter-free curvature tensor estimators $\hat{H}^*$ and $\hat{\kappa}^{i*}$. We also observe the expected $O(h^{\frac{1}{3}})$ convergence speed. Similarly to 2D, local estimators (with or without K-means clustering) on Fig. 6 (second column) show good multigrid convergence.

6 Conclusion

In this article, we have proposed variants of integral invariant estimator to obtain parameter-free curvature estimators in 2D and 3D. In dimension 2, we have demonstrated that the parameter-free curvature estimator is also multigrid convergent. As far as we know, this is the first parameter-free curvature estimator with this multigrid property. In dimension 3, we have defined several parameter-free curvature tensor estimators with very good multigrid convergence behaviors. However, convergence proofs rely on an interesting open conjecture on the length distribution of maximal segment in object slices.

References

1. DGTAL: Digital geometry tools and algorithms library, http://dgtal.org
2. Coeurjolly, D., Lachaud, J.-O., Levallois, J.: Integral based curvature estimators in digital geometry. In: Gonzalez-Diaz, R., Jimenez, M.-J., Medrano, B. (eds.) DGCI 2013. LNCS, vol. 7749, pp. 215–227. Springer, Heidelberg (2013)
3. Coeurjolly, D., Lachaud, J.O., Levallois, J.: Multigrid Convergent Principal Curvature Estimators in Digital Geometry. Computer Vision and Image Understanding (June 2014), http://liris.cnrs.fr/publis/?id=6625
4. Esbelin, H.A., Malgouyres, R., Cartade, C.: Convergence of binomial-based derivative estimation for 2 noisy discretized curves. Theoretical Computer Science 412(36), 4805–4813 (2011)
5. Kerautret, B., Lachaud, J.O.: Curvature estimation along noisy digital contours by approximate global optimization. Pattern Recognition 42(10), 2265–2278 (2009)
6. Lachaud, J.O.: Espaces non-euclidiens et analyse d'image: modèles déformables riemanniens et discrets, topologie et géométrie discrète. Habilitation à diriger des recherches, Université Bordeaux 1, Talence, France (2006)

7. Lachaud, J.O., Vialard, A., de Vieilleville, F.: Fast, accurate and convergent tangent estimation on digital contours. Image and Vision Computing 25(10), 1572–1587 (2007)
8. Malgouyres, R., Brunet, F., Fourey, S.: Binomial convolutions and derivatives estimation from noisy discretizations. In: Coeurjolly, D., Sivignon, I., Tougne, L., Dupont, F. (eds.) DGCI 2008. LNCS, vol. 4992, pp. 370–379. Springer, Heidelberg (2008)
9. Pottmann, H., Wallner, J., Huang, Q., Yang, Y.: Integral invariants for robust geometry processing. Computer Aided Geometric Design 26(1), 37–60 (2009)
10. Pottmann, H., Wallner, J., Yang, Y., Lai, Y., Hu, S.: Principal curvatures from the integral invariant viewpoint. Computer Aided Geometric Design 24(8-9), 428–442 (2007)
11. Provot, L., Gérard, Y.: Estimation of the derivatives of a digital function with a convergent bounded error. In: Debled-Rennesson, I., Domenjoud, E., Kerautret, B., Even, P. (eds.) DGCI 2011. LNCS, vol. 6607, pp. 284–295. Springer, Heidelberg (2011)
12. Roussillon, T., Lachaud, J.-O.: Accurate curvature estimation along digital contours with maximal digital circular arcs. In: Aggarwal, J.K., Barneva, R.P., Brimkov, V.E., Koroutchev, K.N., Korutcheva, E.R. (eds.) IWCIA 2011. LNCS, vol. 6636, pp. 43–55. Springer, Heidelberg (2011)
13. de Vieilleville, F., Lachaud, J.O., Feschet, F.: Maximal digital straight segments and convergence of discrete geometric estimators. Journal of Mathematical Image and Vision 27(2), 471–502 (2007)

A   Proofs

A.1   Proof of Proposition 1

Proof. Since no ambiguity may arise, we remove (x) from all notations. Please also consider Fig. 8 for illustrations. First of all, if one of the π_x, π_y and π_z is the tangent plane at x, then the two other planes are normal planes to ∂X at x (they contains the normal n). In this case, Euler's theorem tells that any curve defined by the intersection of a normal plane and the surface ∂X has a curvature

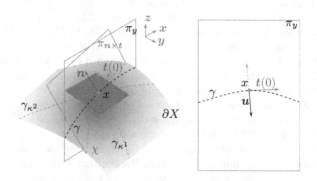

Fig. 8. Notations for Proposition 1

κ equals to $\kappa_1 \cos^2 \theta + \kappa_2 \sin^2 \theta$ for some angle θ. It is then immediate that $|\kappa|$ lies in-between $[\min(|\kappa_1|, |\kappa_2|), \max(|\kappa_1|, |\kappa_2|)]$, and is therefore bounded by K on these two planes.

Otherwise, for each $e \in \{x, y, z\}$, the set ∂X_e is locally a 3D curve that crosses x on the surface ∂X. First remark that there is at most one $e \in \{x, y, z\}$ such that $n \cdot e \geq \frac{\sqrt{2}}{2}$. Indeed, let $n = ax + by + cz$ and for instance $n \cdot x \geq \frac{\sqrt{2}}{2}$, then $b^2 + c^2 = 1 - a^2 = 1 - (n \cdot x)^2 < \frac{1}{2}$. Hence both $b = n \cdot y$ and $c = n \cdot z$ are smaller than $\frac{\sqrt{2}}{2}$. We only consider a vector $e \in \{x, y, z\}$ such that $n \cdot e \leq \frac{\sqrt{2}}{2}$.

Let χ be the curve defined by the intersection of ∂X and the plane $\pi_{n \times t}$ containing n and the tangent t at x of ∂X_e. From Meusnier's theorem, we have the following relationship between the curvature of κ_χ of χ and the curvature κ of ∂X_e at x: $\kappa_\chi = \kappa \cdot \cos \alpha$, α being the angle between planes π_e and $\pi_{n \times t}$. Since $\cos \alpha = n \cdot e$ and $\frac{\sqrt{2}}{2} \leq \cos \alpha \leq 1$, we finally have:

$$|\kappa_\chi| \leq |\kappa| \leq \sqrt{2}|\kappa_\chi|. \tag{16}$$

Again, since $|\kappa_\chi|$ lies in-between $[\min(|\kappa_1|, |\kappa_2|), \max(|\kappa_1|, |\kappa_2|)]$, it is bounded by K and we have the final result.

A.2   Proof of Uniform Multigrid Convergence of $\hat{\kappa}^{1*}$ and $\hat{\kappa}^{2*}$

Proof. As in 2D, we need to check the convergence of both error bounds of ρ'. Assuming that Conjecture 1 is true, we have: If $h\rho'(Z) = \Theta(h^{\frac{1}{3}})$, we are in the hypothesis of Theorem 1, the error term is in $O(h^{\frac{1}{3}})$. If $h\rho'(Z) = \Theta(h^{\frac{1}{3}} \log^2(\frac{1}{h}))$, we can decompose the error term using Equation (28) of [3] (setting $\mu_i = 1$ in Eq.(28)):

$$|\frac{1}{h}\hat{\kappa}^{1*}(Z) - \kappa^1(X, x)| = |\hat{\kappa}_R^1(\mathsf{D}_h(X), \hat{x}, h) - \kappa^1(X, x)|$$

$$\leq O(R) + O(\frac{h}{R^2}) \tag{17}$$

$$\leq O(h^{\frac{1}{3}} \log^2(\frac{1}{h})) + O(\frac{h^{\frac{1}{3}}}{\log^4(\frac{1}{h})}) \tag{18}$$

The upper bound error term is $O(h^{\frac{1}{3}} \log^2(\frac{1}{h}))$. Proof for $\hat{\kappa}^{2*}$ follows similarly. Finally, $\Theta(h^{\frac{1}{3}}) \leq h\rho' \leq \Theta(h^{\frac{1}{3}} \log^2(\frac{1}{h}))$ implies $|\frac{1}{h}\hat{\kappa}^{i*}(Z) - \kappa^i(X, x)| \leq O(h^{\frac{1}{3}} log^2(\frac{1}{h}))$.   □

Freeman Digitization and Tangent Word Based Estimators

Thierry Monteil

CNRS – Université Montpellier 2, France
http://www.lirmm.fr/~monteil

Abstract. This paper deals with the digitization of smooth or regular curves (beyond algebraic, analytic or locally convex ones). The first part explains why the Freeman square box quantization is not well-defined for such curves, and discuss possible workarounds to deal with them. In the second part, we prove that first-order differential estimators (tangent, normal, length) based on tangent words are multi-grid convergent, for any (C^1) regular curve, without assuming any form of convexity.

Keywords: Freeman digitization, symbolic coding, cutting sequence, smooth curve, tangent word, tangent estimation, multigrid convergence.

1 Freeman Square Box Quantization ...

In his survey paper [5], Freeman defines the *square box quantization* of tracings as follows: given a continuous curve $\gamma : [0,1] \to \mathbb{R}^2$ and a (square) grid G, we associate the (ordered) sequence of pixels that are intersected by γ.

There is a slight ambiguity when the curve crosses the grid since an edge belongs to two pixels and a vertex belongs to four pixels. Freeman solves this ambiguity by breaking the rotational symmetry and defines half-open pixels in a way that the pixels form a partition of the plane $\mathbb{R}^2$. The pixel (m, n) is defined as the set $\{(x, y) \in \mathbb{R}^2 \mid (m - 1/2)h < x \leq (m + 1/2)h$ and $(n - 1/2)h < y \leq (n+1/2)h\}$ (here, h denotes the mesh of the grid, m and n are integers, Freeman does not consider any non-integer shift and places $(0, 0)$ at the center of some pixel). A vertex now belongs to its lower left pixel, a horizontal edge belongs to its bottom pixel and a vertical edge belongs to its left pixel (see Figure 1).

Fig. 1. Two ambiguous discretizations of curves, solved by redefining pixel shape

E. Barcucci et al. (Eds.): DGCI 2014, LNCS 8668, pp. 176–189, 2014.

Then, noticing that two consecutive pixels in the sequence are "essentially" 1-connected (they share an edge, unless the curves goes to the bottom left vertex of a pixel), Freeman codes the sequence of pixels by a word on the alphabet $\{0, 1, 2, 3\}$, depending on whether a pixel is located right, above, left, or below the previous one. We denote by $F(\gamma, G)$ the Freeman *chain code* of the square box quantization of the curve γ through the grid G (see Figure 2).

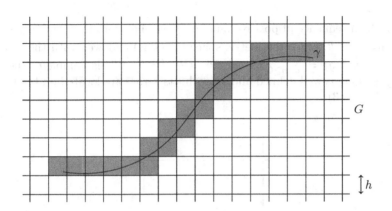

Fig. 2. $F(\gamma, G) = 00000101010101001000$

The hypothesis about the regularity of the curve is very weak (the curve is only assumed to be continuous), and allows a lot of pathological constructions such as plane filling Peano curves that will prevent Freeman quantization to be well defined.

There is a huge literature about the estimation of differential operators applied to a smooth curve through the knowledge of its Freeman discretizations at arbitrary small scales. Most of the proven methods require, in addition to some level of regularity (γ should be C^2 or C^3), the curve to be strictly convex. The aim of this paper, in the sequel of [11], is to understand Freeman discretization of smooth or regular curves, beyond convex of analytic ones.

Let I be the unit interval $[0, 1]$. A curve $\gamma : I \rightarrow \mathbb{R}^2$ is said to be *regular* if it is C^1 and if $||\gamma'(t)|| > 0$ for any $t \in I$. It is said to be *smooth* if it is moreover of class C^∞.

2   ... Is Not Well Defined for Smooth Curves

Another problem appears here, which is not addressed in Freeman's survey: the "sequence" of pixels may not be well defined. Let us construct a typical example of bad behaviour. The map $s = \begin{pmatrix} \mathbb{R} \rightarrow & \mathbb{R} \\ x \mapsto & 0 & \text{if } x \leq 0 \\ x \mapsto e^{-1/x^2} & \text{otherwise} \end{pmatrix}$ is smooth and positive on positive numbers. Hence, the bell map defined by $b(x) = s(x - 1/3)s(2/3 - x)$ is smooth and positive on $(1/3, 2/3)$. Now let us define recursively a sequence of maps b_n by $b_0 = b$ and

$$b_{n+1} = \begin{pmatrix} \mathbb{R} \to & \mathbb{R} \\ x \mapsto & b_n(3x) & \text{if } 0 < x < 1/3 \\ x \mapsto b_n(3(x - 2/3)) & \text{if } 2/3 < x < 1 \\ x \mapsto & 0 & \text{otherwise} \end{pmatrix}.$$

Let (a_n) be a sequence of positive numbers such that, for any n and any $k \leq n$, $\|a_n b_n^{(k)}\|_\infty \leq 2^{-n}$. Since the metric space $C^\infty(I, \mathbb{R})$ endowed with the distance defined by $d(f, g) = \sum_{n \geq 0} 2^{-\min(1, \sup_{t \in I} |f^{(n)}(t) - g^{(n)}(t)|)}$ is complete, the map $\phi = \sum_{n \geq 0} a_n b_n$ is well defined and smooth. Its zeroes are located on the Cantor set (see Figure 3).

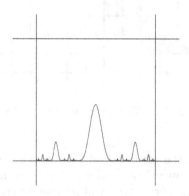

Fig. 3. The curve defined as the graph of ϕ

The Freeman digitization of the smooth curve defined by $\gamma(t) = (t, \phi(t))$ is not well defined with respect to the unit grid (or any grid containing the horizontal axis): since the Cantor set has uncountably many connected components, we created an uncountable sequence of pixels ! But this construction is sensitive to the choice that Freeman did along the edges. For example, replacing $\gamma = (x, y)$ by $(x, -y)$ leads to a single pixel, and it can be opposed that such intersections with the horizontal edge are irrelevant and could be considered as trivial.

To deal with such an objection, let us assume that vertices and edges do not belong to any pixel, that is, a pixel is selected only when the curve pass through its interior. Unfortunately, this is not sufficient to solve our problem: it is possible to construct an oscillating smooth curve that intersects the interior of the pixels infinitely many times. For this purpose, let us define the smooth map $\psi = \sum_{n \geq 0} (-1)^n a_n b_n$ (see Figure 4).

The Freeman digitization of the smooth curve defined by $\eta(t) = (t, \psi(t))$ is not well defined with respect to the unit grid (or any grid containing the horizontal

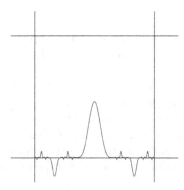

Fig. 4. The curve defined as the graph of ψ

axis): η passes through the interior of different pixels infinitely many times, and no two of them are consecutive (letting the chain coding from the sequence of pixels hard to define)!

Hence, we have a serious problem with the coding of smooth curves that cannot be fixed with a convention on choosing how to code a curve that crosses the vertex of the grid or turns around along an edge. Let us discuss some possible workarounds, leading to different research directions.

3   Some Workarounds

3.1   Restrict

A first possibility to deal with such situations is to forbid them. For example, in [6], the authors impose that the curve "passes only once between two neighboring nodes of the grid". Such a condition impacts both the curve and the grid.

Another possibility is to restrict to a class of curves that can not present such oscillations. This is the case for analytic curves. A curve $\gamma = (x, y) : [0, 1] \to \mathbb{R}^2$ is said to be *analytic* if on a neighbourhood of any time $t_0 \in [0, 1]$, its coordinates $x(t)$ and $y(t)$ can be written as $\sum_{n \geq 0} a_n (t - t_0)^n$. If such an analytic curve meets an (say) horizontal edge infinitely many times, then $x(t)$ takes the same value for infinitely many $t \in [0, 1]$: such an accumulation forces $x(t)$ to be constant and the curve is a horizontal straight line.

A similar representation of curves by words was introduced by dynamicists and are called *cutting sequences*. They have been introduced by Hadamard [7], and are used in the symbolic coding of geodesics in hyperbolic or (piecewise) euclidean spaces. It should be noticed that the problems we encountered are avoided for a similar reason: geodesics are locally straight (while their long-range behaviour may be intricate).

3.2   Extend

Conversely, we can face the problem and extend Freeman chain code to generalized sequences indexed by linear orders (instead of sequences on finite linear orders (words)).

Let us describe the possible orders appearing in a generalized Freeman chain code. Such orders are countable. Indeed, since we count intersections with interiors of pixels and since the curve is assumed to be continuous, the set of times t that γ spends in a pixel has non-empty interior. Since the real line is separable, there are only countably many non-trivial pixel intersections in the sequence. There are no other obstructions on the order type of the generalized sequence of pixels.

Theorem 1. *Any countable linear order can be obtained from the Freeman square box quantization of a smooth curve.*

Proof. The complement of the Cantor set in I (around which we built the map ψ) is a disjoint union of open intervals. This countable set (which we denote by $\mathcal{I}$) inherits from the linear order of I (given two distinct intervals $A = (a, b)$ and $B = (c, d)$ in $\mathcal{I}$, either $a < b < c < d$ or $c < d < a < b$). Since it is dense (for any $A < B$ in $\mathcal{I}$, there exists a $C \in \mathcal{I}$ such that $A < C < B$), and has no maximum nor minimum, it is (order) isomorphic to the chain $\mathbb{Q}$ of rationals. The chain of rationals has the following universal property [2]: any countable linear order is isomorphic to a subset of the chain of the rationals.

Given a countable linear order $\mathcal{L}$, we can see it as a subset of $\mathcal{I}$. Unfortunately, it is not sufficient to keep only the bells that are defined on the elements of $\mathcal{L}$. Indeed, some consecutive elements may appear in $\mathcal{L}$, and the related bells may not have opposite signs (see Figure 5).

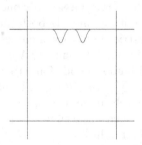

Fig. 5. Two consecutive bells in the same direction select only one pixel

So, we have to ensure local sign alternation for consecutive elements of $\mathcal{L}$. Let $\sim$ be the binary relation defined on $\mathcal{L}$ by $A \sim B$ if, and only if, the set $\{X \in \mathcal{L} \mid A \leq X \leq B \text{ or } B \leq X \leq A\}$ is finite. This defines an equivalence relation whose classes are either (order) isomorphic to a finite linear order $\{0, \ldots, n-1\}$, to the chain $\mathbb{N}$ of non-negative integers, to the chain $\mathbb{Z}\setminus\mathbb{N}$ of negative integers, or to the chain $\mathbb{Z}$ of integers, depending on the existence of a minimum or a maximum. Two consecutive elements of $\mathcal{L}$ belong to the same class.

For each class $\mathcal{C} \subseteq \mathcal{L}$, we can define an oscillating smooth map $\chi_{\mathcal{C}} : I \to \mathbb{R}$ whose graph has a Freeman sequence that is order isomorphic to $\mathcal{C}$. Let us do it for the most complex case where $\mathcal{C}$ is isomorphic to $\mathbb{Z}$. The sequence $b_n = (1+n/(1+|n|))/2$ is an increasing sequence from $\mathbb{Z}$ to I. If $(a_n)_{n\in\mathbb{Z}}$ is a sufficiently fast decreasing sequence of positive real numbers, the map defined by $\chi_{\mathcal{C}}(x) = \sum_{n\in\mathbb{Z}}(-1)^n a_n s(x - b_n)s(b_{n+1} - x)$ is convenient (see Figure 6).

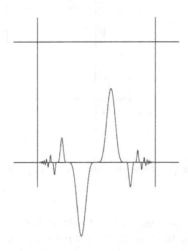

Fig. 6. An oscillating smooth map whose Freeman sequence is order-isomorphic to $\mathbb{Z}$

Now, since equivalent classes form intervals in $\mathcal{L}$, the quotient $\mathcal{J} = \mathcal{L}/\sim$ inherits a countable linear order from the one of $\mathcal{L}$. We can see it as a subset of $\mathcal{I}$. For each element $\mathcal{C}$ of $\mathcal{J}$, which is identified with an open interval (a, b) of I, we can define the smooth map $\lambda_{\mathcal{C}}$ by $\lambda_{\mathcal{C}}(x) = \chi_{\mathcal{C}}((x - a)/(b - a))$, which vanishes out of (a, b), and whose Freeman coding is isomorphic to $\mathcal{C}$.

Again, we can sum the family of maps $(\lambda_{\mathcal{C}})_{\mathcal{C}\in\mathcal{J}}$ in a way that it converges in $C^\infty(I, \mathbb{R})$. We constructed a smooth map $I \to \mathbb{R}$ whose graph is a curve whose Freeman sequence is order isomorphic to $\mathcal{L}$. □

3.3   Blur

The mesh of the grid somehow corresponds to the scale of precision of the optical device. But the constructions above play with the sharpness of the interpixel edges, as if the optical device is infinitely precise there. A possible workaround is to consider that the optical device does not have an infinite precision between consecutive cells. This corresponds to thicken the width of both edges and vertices between pixels in the mathematical model (see Figure 7). Hence, a regular curve can not oscillate between two plain pixels anymore. The blurred edge belongs to both pixels nondeterministically: if a regular curve oscillates between a pixel and a blurred edge, the device may detect it or not, as if the edge belongs to this plain pixel. If the curve passes from a plain pixel to another plain pixel, or if the curve threads its way through blurred edges, the device must detect it. Among all possible outputs, one is a finite word.

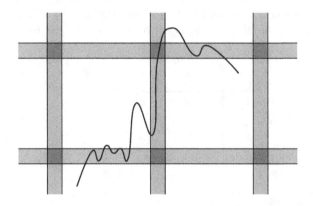

Fig. 7. Thicken the interpixel vertices and edges (but keep the long-range information)

Multigrid convergence (corresponding to using better and better optical devices) will therefore have to deal with two parameters (corresponding to the sources of imprecision) : the mesh of the grid and the width of the blurred interpixel zone. Their ratio may not be constant along the increase of precision, and depending on the speed in which the width becomes small compared to the mesh of the grid, some smoothed fractal details (as defined before) may appear or not.

3.4   Look Almost Everywhere

The set of grids of given mesh $h > 0$, which are translate to each other, can be identified with the torus $\mathbb{R}^2/h\mathbb{Z}^2$ of area h^2 (corresponding to possible shifts of the grid). It therefore inherits a natural finite Lebesgue measure.

The aim of this workaround is to prove that bad phenomena are very rare with respect to this measure.

Theorem 2. *Let γ be a regular curve and h be a positive mesh. For almost every grid G of mesh h, γ does not cross any vertex of G and intersects the edges of G transversally. For such generic grids G, the Freeman chain code $F(\gamma, G)$ is a well defined finite word.*

Proof. Since the image of a regular curve has Lebesgue measure zero in $\mathbb{R}^2$, and since the set of grid vertices is countable, we have that for almost every grid G with mesh h, the curve γ does not hit a vertex of G.

Sard's lemma asserts that given a C^k map $f : A \subseteq \mathbb{R}^m \to \mathbb{R}^n$ (for $k \geq \max(1, m - n + 1)$), the image of critical points $f(\{t \in A \mid \text{rank}(\text{Diff}_t(f)) < n\})$ has (Lebesgue) measure zero in $\mathbb{R}^n$. Let us apply this lemma to the first coordinate $x : I \to \mathbb{R}$ of γ. The set $\{X \in \mathbb{R} \mid (\exists t \in I)(X = x(t) \text{ and } x'(t) = 0)\}$ has Lebesgue measure zero. By countable union, the set $\{\kappa \in [0, h] \mid (\exists k \in \mathbb{Z})(\exists t \in I)(x(t) = k + \kappa \text{ and } x'(t) = 0)\}$ also has measure zero. The same holds for the second coordinate y, hence the set $\{(\kappa, \lambda) \in [0, h]^2 \mid ((\exists k \in \mathbb{Z})(\exists t \in I)(x(t) = k + \kappa \text{ and } x'(t) = 0)) \text{ or } (\exists t \in I)(y(t) = l + \lambda \text{ and } y'(t) = 0))\}$ has measure zero. This set corresponds to the set of grids that are not transversally intersected by γ.

Now, let G be a generic grid. At each time t when γ intersects G, the intersection is transverse: there is an open interval O containing t such that for any $t' \neq t$ in O, $\gamma(t') \notin G$. Moreover, $\gamma(O \cap [0, t))$ and $\gamma(O \cap (t, 1])$ are included in two distinct adjacent connected components of $\mathbb{R}^2 \setminus G$ (pixel interiors). Since the set $\gamma^{-1}(G)$ is a closed subset of the compact interval I and is made of isolated points, it is finite. Hence, there is a finite set of times $\leq 0 < t_0 < \cdots < t_n \leq 1$ such that for each i, $\gamma(t_i)$ is on the grid and $\gamma((t_{i-1}, t_i))$ and $\gamma((t_i, t_{i+1}))$ are included in two distinct edge-adjacent pixel interiors. $\qquad\square$

That said, note that the set of grids that are not transversally intersected by γ may not be countable (an antiderivative f of ϕ admits uncountably many singular values, hence the curve defined by $\gamma(t) = (t, f(t))$ intersects uncountably many horizontal lines non-transversally).

With this workaround, by looking modulo almost everywhere, the arbitrary choices made by Freeman in the boundaries of the pixels become irrelevant. Hence, we get a more symmetric and intrinsic discretization scheme that does not have to specify non-canonical choices (since those happen only on a set of zero measure).

This workaround has another big advantage: the action of $SL(2, \mathbb{Z})$ on $\mathbb{R}^2$ that appears in the continued fractions algorithm is central in the study of discrete straight segments (and tangent words), and inherent to the lattice structure of the grid. It sends lines with rational slopes to the vertical and horizontal axes. So, applying such a map to a well-coded curve that oscillates around a rational slope may lead to an ill-coded one. By looking almost everywhere, we avoid such a situation since the set of rational slopes is countable.

This workaround was the one we used in [11] to define tangent words and we will stick to that framework for the remaining of the paper.

4   First-Order Differential Operators via Tangent Words

Given a regular curve, the limit object we get while zooming into a point is a straight line. Freeman chain codes of straight segments are known to be exactly the *balanced* words. Hence, it seems natural to decompose a discretized curve into maximal balanced words in order to approach the tangents of the real curve, a strategy which has been widely studied. But this is not what the multigrid discretization scheme does: it discretizes a curve, at various scales. Tangent words

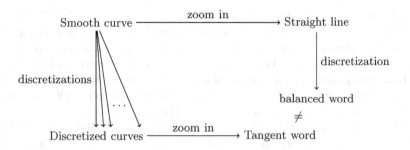

Fig. 8. Not even a noncommutative diagram

are the finite words that appear in the coding of a smooth or regular curve at arbitrary small scale.

More formally, a finite word u is said to be *tangent* if there exists a regular curve $\gamma : [0, 1] \to \mathbb{R}^2$ and a sequence of grids (G_n) whose meshes converge to zero, such that u is a factor of the Freeman chain code $F(\gamma, G_n)$ of the curve γ through the grid G_n for any integer n.

Theorem 3. *Let $\gamma : [0, 1] \to \mathbb{R}^2$ be a regular curve. For any sequence (G_n) of grids whose meshes converge to zero, the minimal length of a maximal tangent word in $F(\gamma, G_n)$ goes to ∞ with n.*

Proof. Assume by contradiction that there exists a sequence of grids (G_n) whose mesh converges to zero such that there exists an integer L such that, for any N, there exists $n \geq N$ and a maximal tangent word $w(n)$ in $W(n) = F(\gamma, G_n)$ whose length is less than L. Up to taking a subsequence (there are finitely many words of length $\leq L$), we can assume that there exists a finite word w that is a maximal tangent word of $W(n)$, for any n: $w = W(n)_{i_n \ldots j_n}$ for some indices i_n, j_n and neither $W(n)_{i_n-1 \ldots j_n}$ nor $w = W(n)_{i_n \ldots j_n+1}$ is tangent (when defined). Since γ is regular, it is not trivial, hence it has positive length. In particular, the length $|W(n)|$ of $W(n)$ goes to infinity with n. Hence, for n large enough, $i_n > 0$ or $j_n < |W(n)| - 1$. Up to symmetry and up to taking a subsequence, we can assume that $j_n < |W(n)| - 1$ for all n. Now, the letter $W(n)_{j_n+1}$ takes at most four values: one of them, which we call a appears infinitely often: wa is therefore a tangent word, that extends w in $W(n)$ for infinitely many n, which contradicts the maximality of w.   □

The simplicity of this result (a pigeonhole argument) should not be surprising: the definition of tangents words contains its adaptation to the behaviour of regular curves. What is lucky is that tangent words have a simple combinatorial characterization and can be recognized in linear time [11], so they are ready to serve as drop-in replacement where balanced words are not optimal.

Now, this feature allows us to easily ensure multigrid convergence of first-order differential operators by replacing decomposition of the curve into maximal balanced words by a decomposition into maximal tangent words. Due to a lack of space, the proofs of the following results are postponed in the appendix.

4.1   Length Estimation

Given a regular curve γ, and a sequence of grids (G_n) whose meshes h_n go to zero, we can estimate its length as follows: for each n, greedily decompose $F(\gamma, G_n)$ into maximal (to the left) tangent words $F(\gamma, G_n) = w_0^n \ldots w_{k_n-1}^n$, and compute the length of the associated polygonal line:

$$l_n = h_n \sum_{i=0}^{k_n-1} \sqrt{(|w_i^n|_0 - |w_i^n|_2)^2 + (|w_i^n|_1 - |w_i^n|_3)^2}$$

The sequence l_n converges to the length of γ when n goes to ∞.

4.2   Tangent Estimation

Let us first recall the definition of the directions of a tangent word u (see the definition of slope in [8]). If u is a tangent word, there exists a regular curve $\gamma : I \to \mathbb{R}^2$ and a sequence (G_n) of grids whose meshes go to 0 and such that for all n, u is a factor of $F(\gamma, G_n)$. In particular, for any integer n, there exist two sequences (t_n^1) and (t_n^2) in I such that u is the Freeman code of $\gamma|_{]t_n^1, t_n^2[}$ with respect to the grid G_n. Up to taking a subsequence (the segment I is compact), we can assume that (t_n^1) and (t_n^2) both converge to some $t \in I$: we say that the direction of $\gamma'(t)$ is a *direction* of u. The *set of directions* of u corresponds to all possible choices on the curve γ, the sequence of grids (G_n) and sequences of times (t_n^1) and (t_n^2). The set of directions of a tangent words can be computed from the continued fraction algorithm introduced to recognize it. The set of directions of a balanced word is a non-trivial interval, while the set of directions of a non-balanced tangent words is a singleton.

Now, we can obtain a tangent (or normal) estimator as follows: given a pixel of a discretization of a regular curve γ, we can choose any maximal tangent word that contains this pixel and take any of its directions as an estimation.

As for length estimation, this leads to uniformly multigrid convergent estimators. Convex combinations of estimations (corresponding to λ-MST [10]) also work. While existing convergence results for balanced word based estimators require the curve to be C^3 and piecewise strictly convex, the use of maximal tangent words in place of balanced words only require the curve to be C^1 (with the same algorithms).

4.3   Maximal Symmetric Tangent Words

Feschet and Tougne [4] defined a tangent estimator based on maximal symmetric balanced words around a given pixel (symmetric in the sense that the position of the pixel is in the middle of the word). Since, even under very strong hypotheses, the length of such words does not converges to ∞ around some pixels, [9] proved that it is not a multigrid convergent tangent estimator.

But, the same argument as in Theorem 3 works for maximal symmetric tangent words: any word w can be extended simultaneously in both directions as awb for $(a, b) \in \{0, 1, 2, 3\}^2$ in at most 16 ways. Since 16 is a finite number, the same pigeonhole argument applies. Hence, apart from the ends of the word (a case which can be dealt with by considering closed curves and circular words, or by simply looking away from the ends), the length of the smallest maximal tangent word which is symmetric around any letter converges to ∞ when the mesh of the grid goes to zero, letting the estimator to be uniformly convergent.

5   Why Does Convexity Matter for Maximal Segment Based Estimators?

A more detailed description of the tangent convex words was provided in [8]. In particular, we should notice that any tangent convex word is the concatenation of two balanced words, which should not impact the estimations so much. In particular, while the smallest length of maximal balanced words is not the same as for maximal tangent words, their average lengths are similar.

However, this is not the case for general tangent words, whose minimal factorization into balanced words can have arbitrary many factors, and even linear with respect to the length of the tangent word (think of the tangent words $(0011)^n$).

This difference between (piecewise) convex curves and more general smooth curves is highlighted by this counting argument [12]: the set of tangent convex words of length n has size $\Theta(n^3)$, which is the same as for balanced words, whereas the set of smooth tangent words of length n has exponential size.

Those facts tend to indicate that small oscillations are the main reason for maximal segment-based estimators not to work for general smooth curves.

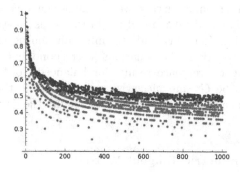

Fig. 9. Smallest maximal balanced (red) and tangent (blue) word in the coding of circles (abscissa=inverse of the mesh, ordinate = length of the word (in a loglog scale))

6   Conclusion

We tried to explain that the problems arising in the digitization of regular curves are related to soften fractal oscillations. We saw that the use of tangent words in place of balanced words has the effect of smoothing those irregularities, especially around points whose curvature vanishes. They are very natural objects, since their definition is adapted to the framework of regular (or smooth) curves, and qualitative convergence properties come for free. Actually, quantitative results on the speed of convergence can also be proven: *if* $\gamma : [0,1] \to \mathbb{R}^2$ *is any regular* C^2 *curve and* (G_n) *is a sequence of grids whose meshes* h_n *converge to zero, the minimal length of a maximal tangent word in* $F(\gamma, G_n)$ *belongs to* $\Omega(h_n^{-\frac{1}{3}})$, a result which requires much weaker hypotheses than in [1] and [3].

References

1. Balog, A., Bárány, I.: On the convex hull of the integer points in a disc. In: Proceedings of the Seventh Annual Symposium on Computational Geometry, SCG 1991, pp. 162–165. ACM, New York (1991)
2. Cantor, G.: Beiträge zur Begründung der transfiniten Mengenlehre. Mathematische Annalen 46(4), 481–512 (1895)
3. de Vieilleville, F., Lachaud, J.-O., Feschet, F.: Convex digital polygons, maximal digital straight segments and convergence of discrete geometric estimators. J. Math. Imaging Vision 27(2), 139–156 (2007)
4. Feschet, F., Tougne, L.: Optimal time computation of the tangent of a discrete curve: Application to the curvature. In: Bertrand, G., Couprie, M., Perroton, L. (eds.) DGCI 1999. LNCS, vol. 1568, pp. 31–40. Springer, Heidelberg (1999)
5. Freeman, H.: Computer processing of line-drawing images. Computing Surveys 6, 57–97 (1974)
6. Groen, F.C.A., Verbeek, P.W.: Freeman-code probabilities of object boundary quantized contours. Computer Graphics and Image Processing 7(3), 391–402 (1978)
7. Hadamard, J.: Les surfaces à courbures opposées et leurs lignes géodésiques. J. Math. Pures et Appl. 4, 27–73 (1898)
8. Hoarau, A., Monteil, T.: Persistent patterns in integer discrete circles. In: Gonzalez-Diaz, R., Jimenez, M.-J., Medrano, B. (eds.) DGCI 2013. LNCS, vol. 7749, pp. 35–46. Springer, Heidelberg (2013)
9. Lachaud, J.-O.: On the convergence of some local geometric estimators on digitized curves. Research Report, 1347-05 (2005)
10. Lachaud, J.-O., Vialard, A., de Vieilleville, F.: Fast, accurate and convergent tangent estimation on digital contours. Image Vision Comput. 25(10), 1572–1587 (2007)
11. Monteil, T.: Another definition for digital tangents. In: Debled-Rennesson, I., Domenjoud, E., Kerautret, B., Even, P. (eds.) DGCI 2011. LNCS, vol. 6607, pp. 95–103. Springer, Heidelberg (2011)
12. Monteil, T.: The complexity of tangent words. In: WORDS. Electronic Proceedings in Theoretical Computer Science, vol. 63, pp. 152–157 (2011)

7   Appendix: Postponed Proofs

7.1   Length Estimation (Section 4.1)

If w is a finite word over the alphabet $\{0, 1, 2, 3\}$, let us denote
$||w|| = \sqrt{(|w|_0 - |w|_2)^2 + (|w|_1 - |w|_3)^2}$.

Since the curve γ is of class C^1, it is rectifiable and its length can be approximated as follows: if $(\{0 = t_0^n \leq t_1^n \leq \cdots \leq t_{k_n-1}^n \leq t_{k_n}^n = 1\})_{n \in \mathbb{N}}$ is a sequence of subdivisions of the unit interval such that $\lim_{n \to \infty} \max_{0 \leq i \leq k_n - 1} |t_{i+1}^n - t_i^n| = 0$, then the length of γ is given by

$$l(\gamma) = \lim_{n \to \infty} \sum_{i=0}^{k_n-1} ||\gamma(t_{i+1}^n) - \gamma(t_i^n)||$$

Given a grid G_n and a decomposition of $F(\gamma, G_n)$ into maximal tangent words $w_0^n \ldots w_{k_n-1}^n$, each word w_i^n corresponds to a sequence $s_i^n = (p_0^{i,n}, \ldots, p_{|w_i^n|}^{i,n})$ of $|w_i^n| + 1$ pixels of G_n, where the last pixel of s_i^n is the first pixel of s_{i+1}^n.

The value

$$l_n = h_n \sum_{i=0}^{k_n-1} ||w_i^n||$$

corresponds to the length of the polygonal line $(x_0^n, \ldots, x_{k_n}^n)$, where x_i^n is the center of the pixel $p_i^n := p_{|w_{i-1}^n|}^{i-1,n} = p_0^{i,n}$.

We can pick, for any i, a time $t_i^n \in I$ where $\gamma(t_i^n)$ pass through the pixel p_i^n. Since for any i, $||x_i^n - \gamma(t_i^n)|| \leq h_n$, we have

$$\left| \sum_{i=0}^{k_n-1} ||\gamma(t_{i+1}^n) - \gamma(t_i^n)|| - h_n \sum_{i=0}^{k_n-1} ||w_i^n|| \right| \leq 2 h_n k_n$$

Theorem 3 asserts that the minimum length of the w_i^n (for $0 < i < k_n - 1$) goes to infinity, hence $2 h_n k_n$ goes to zero (the error we made in the extremities of the segments become negligible with respect to their length).

We are almost done, except that $\max_{0 \leq i \leq k_n - 1} |t_{i+1}^n - t_i^n|$ does not necessarily converge to zero when n goes to infinity (some w_i^n can be very long compared to $1/h_n$). To achieve this, let us decompose the long maximal tangent words into smaller ones $(F(\gamma, G_n) = \bar{w}_0^n \ldots \bar{w}_{\bar{k}_n-1}^n)$ in a way that their minimal length (except on the boundaries) still goes to infinity, but such that the maximal value of $h_n |\bar{w}_i^n|$ goes to zero. Hence, constructing the $\bar{t}_i^n$ accordingly, we now have

$$l(\gamma) = \lim_{n \to \infty} \sum_{i=0}^{\bar{k}_n-1} ||\gamma(\bar{t}_{i+1}^n) - \gamma(\bar{t}_i^n)|| = \lim_{n \to \infty} h_n \sum_{i=0}^{\bar{k}_n-1} ||\bar{w}_i^n||$$

Now, tangent words enjoy some balance property (they stay close to a segment): if w is a tangent word that is written as a concatenation of words $w = w_0 \ldots w_{l-1}$, then $\left| ||w|| - \sum_{i=0}^{l-1} ||w_i|| \right| \leq 4l$. This concludes the proof since it implies $\lim_{n \to \infty} \left| h_n \sum_{i=0}^{\bar{k}_n-1} ||\bar{w}_i^n|| - h_n \sum_{i=0}^{k_n-1} ||w_i^n|| \right| = 0$. $\square$

7.2   Tangent Estimation (Section 4.2)

The structure of this proof is similar as above. Let t be a point in I and (G_n) be a sequence of grids whose meshes go to zero. For any n, let u_n be a maximal tangent factor of $F(\gamma, G_n)$ *containing* $\gamma(t)$, in the sense that there exists an interval $I_{u_n} \subseteq I$ containing t such that $F(\gamma\lceil_{I_{u_n}}, G_n) = u_n$.

The first (resp. last) letter of u_n corresponds to an edge e_n (resp. e'_n) of G_n. The set $D(u_n)$ of directions of u_n is included in the set of directions of vectors $b - a$ for $(a, b) \in e_n \times e'_n$, which we denote by $\bar{D}(u_n)$. Since γ crosses e and e', the mean value theorem implies there exists a point t_n in I_{u_n} such that the direction of $\gamma'(t_n)$ belongs to $\bar{D}(u_n)$.

Theorem 3 asserts that the length of u_n goes to infinity, hence the distance between e_n and e'_n goes to infinity and the diameter of $\bar{D}(u_n)$ converges to zero (to be precise, we should deal separately with the special case along the axes where γ oscillates around a set of parallel edges).

Now, as in the previous proof, if the length of u_n grows too fast, it is possible that t_n does not converge to the point t. Again, we can shorten u_n to a word u'_n that still contains t, whose length still goes to infinity, but whose length is negligible with respect to the inverse of the mesh of G_n. This leads to an interval $I_{u'_n}$, a point t'_n in $I_{u'_n}$ and two sets of slopes $D(u'_n)$ and $\bar{D}(u'_n)$ as before, with the additional property that t'_n converges to t. Even if $\bar{D}(u'_n)$ is not necessarily included in $\bar{D}(u_n)$, the word u'_n is also tangent and $D(u'_n)$ is included in $D(u_n)$ (which is included in $\bar{D}(u_n)$). Since the diameter of $\bar{D}(u_n)$ goes to zero, and contains $\gamma'(t'_n)$ which converges to $\gamma'(t)$, any choice of direction in $\bar{D}(u_n)$ converges to $\gamma'(t)$.   □

Determination of Length and Width
of a Line-Segment by Using a Hough Transform

Zezhong Xu[1,2], Bok-Suk Shin[1], and Reinhard Klette[1]

[1] Department of Computer Science, The University of Auckland,
Auckland, New Zealand
[2] College of Computer Information Engineering, Changzhou Institute of Technology,
Changzhou, Jiangsu, China
`zezhongx@gmail.com`, `{b.shin,r.klette}@auckland.ac.nz`

Abstract. The standard Hough transform does not provide length and
width of a line-segment detected in an image; it just detects the normal parameters of the line. We present a novel method for determining
also length and width of a line segment by using the Hough transform.
Our method uses statistical analysis of voting cells around a peak in
the Hough space. In image space, the voting cells and voting values are
analysed. The functional relationship between voting variance and voting
angle is deduced. We approximate this relationship by a quadratic polynomial curve. In Hough space, the statistical variances of columns around
a peak are computed and used to fit a quadratic polynomial function.
The length and width of a line segment are determined simultaneously by
resolving the equations generated by comparing the corresponding coefficients of two functions. We tested and verified the proposed method on
simulated and real-world images. Obtained experimental results demonstrate the accuracy of our novel method for determining length and width
of detected line segments.

Keywords: Hough transform, length, width, curve fitting.

1   Introduction

Line segments are important when analyzing geometric shapes in images for
machine vision applications; see, for example, [16]. In particular this problem
also involves a need to extract parameters of line segments in images, such as
width and length.

A class of methods for line detection applies least-square fitting; see, for example, [15,17,19,22]. These methods are in general sensitive to outliers; they require
that feature points are clustered.

The *Hough transform* (HT) [1,8,11,23,24] defines an alternative class of methods. The basic HT does not provide length or width of a detected line segment;
it only provides the two *normal parameters* d and α of a line; see Eq. (1) below
for those two parameters. This paper contributes to the HT subject.

In principle, the HT is able to detect the length of a line segment. After having
the direction of a set of approximately collinear pixels detected, we can project

E. Barcucci et al. (Eds.): DGCI 2014, LNCS 8668, pp. 190–201, 2014.
© Springer International Publishing Switzerland 2014

the estimated collinear image features on the x- or y-axis in image space; see, for example, [5,18,25]; the length of the line-segment is then determined as the Euclidean distance between the estimated two endpoints.

There are also HT methods which use the *butterfly distribution* in the Hough space, as identified in [10]. These butterfly-techniques have origins in methods proposed earlier. Akhtar [2] calculates the length of a detected line segment based on the spreading of voting cells in a column around the peak. Ioannou [12] estimates the line-segment length by analyzing the total vote values of cells in the peak column. In [3,4,13,14], the endpoints are detected by resolving simultaneously equations obtained by the first and the last non-zero-value voting cells in any two columns around the peak; the length is then again calculated as the Euclidean distance between the estimated two endpoints.

These methods detect the length besides the standard HT output of normal parameters of a detected line segment. But, they do not contribute to the calculation of the width of the line segment.

Du et al. [6,7] consider the complete parameter description of a line segment, defined by direction, length, width, and position. Here, length is obtained by measuring the vertical width of a butterfly wing. The width of a line segment is computed by comparing the actual voting value and theoretical voting values in a specific column. Reliable length and width are obtained using a *Mean Square Error* (MSE) estimation by considering multiple columns. This method is affected by image noise. The detection accuracy relies on a very fine quantization of the Hough space.

This paper proposes an HT method for obtaining the length and width of a detected line segment. The voting variance is analyzed in image space, and a 2nd order functional relationship is deduced. In Hough space, the statistical variances of columns around a peak are computed and used to fit a quadratic polynomial function. Length and width of a line segment are determined by resolving the equations generated by comparing the corresponding coefficients of two functions.

The rest of the paper is organized as follows. Section 2 analysis the voting variance in image space. Section 3 introduces the voting distribution in Hough space, and calculates the length and width of a line segment. Section 4 provides experimental results. Section 5 concludes.

2   Voting Analysis in Image Space

Following [8], the standard Hough transform applies the following equation

$$d \;=\; x \cdot \cos\alpha \;+\; y \cdot \sin\alpha \tag{1}$$

for representing a straight line by normal parameters α and d. This representation was introduced in [20] when defining a transformation in continuous space, today known as the *Radon transform*; this transform is a generalization of the Hough transform.

All pixels on the line-segment in an image vote for all possible cells (α_i, d_j) in Hough space. For a pixel, given a voting angle $\alpha_i \in [0, \pi)$, the corresponding d_j-value is computed. The cell (α_i, d_j) is *voted for* by increasing the voting value at this cell by 1. Let H_{ij} be the voting value of cell (α_i, d_j) in Hough space.

For a voting angle α_i, the number of voting cells and voting values of each cell are analyzed first; then we deduce a functional relationship between voting variance and voting angle.

The actual normal parameters of a line segment are denoted by (α_0, d_0). Let L and T denote the length and the width of the line segment. For abbreviation, let S and C be short for the values of sine and cosine of $|\alpha_i - \alpha_0|$, respectively:

$$S = \sin|\alpha_i - \alpha_0| \quad \text{and} \quad C = \cos|\alpha_i - \alpha_0| \qquad (2)$$

2.1   Voting Cells and Voting Values

Regarding a voting angle α_i, the number of voting cells is proportional to the number of parallel bars intersected by the considered line-segment. The voting value H_{ij}, corresponding to the voting angle α_i and the distance d_j, is proportional to the length of the bar intersected by the line-segment.

For detecting line segments with different length and width, we consider two cases for estimating the number of voting cells and voting values.

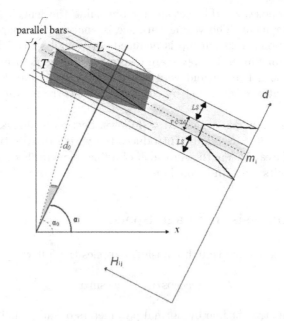

Fig. 1. The number of voting cells and voting values for $|\alpha_i - \alpha_0| < \arctan(T/L)$. Actual parameters are normal parameters d_0 and α_0, and width T and length L. For the remaining parameters in the figure see the text for explanations.

For a voting angle α_i, if $|\alpha_i - \alpha_0| < \arctan(T/L)$ then there are $T \cdot C + L \cdot S$ parallel bars crossing the considered line-segment in total. This is illustrated in Fig. 1. At the middle of the parallel bars, the number of voting cells equals

$$T \cdot C - L \cdot S \tag{3}$$

and the voting values are identical. On both outer sides of parallel bars, there are $L \cdot S$ voting cells for each side; and the voting values decrease to 0 gradually.

For a voting angle α_i, if $|\alpha_i - \alpha_0| > \arctan(T/L)$ then there are $L \cdot S + T \cdot C$ parallel bars crossing the considered line-segment in total. This is illustrated in Fig. 2. At the middle of the parallel bars, the number of voting cells equals

$$L \cdot S - T \cdot C \tag{4}$$

and the voting values are identical. On both outer sides of the parallel bars, there are $T \cdot C$ voting cells on each side; and the voting values decrease to 0 gradually.

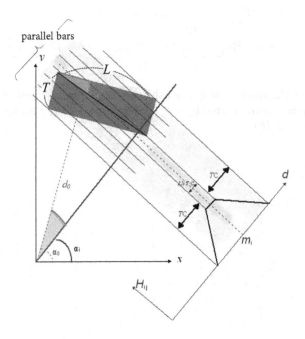

Fig. 2. The number of voting cells and voting values for $|\alpha_i - \alpha_0| > \arctan(T/L)$. See the text for explanations.

2.2   Voting Variances

In both cases for a voting angle α_i, we consider the voting cells along the axis d to be a random variable. The voting values of corresponding cells define a probabilistic density function. The voting variance σ_i^2, which corresponds to voting angle α_i, is calculated based on the corresponding probabilistic density function.

For both discussed cases, the voting variance σ_i^2 is calculated as follows:

$$\sigma_i^2 = \frac{L^2 \sin^2 |\alpha_i - \alpha_0| + T^2 \cos^2 |\alpha_i - \alpha_0|}{12}$$

$$= \frac{(L^2 - T^2) \sin^2 |\alpha_i - \alpha_0| + T^2}{12} \tag{5}$$

We only consider those voting cells around the peak in Hough space. It means that $|\alpha_i - \alpha_0|$ is small, and that we can approximate $\sin |\alpha_i - \alpha_0|$ by $|\alpha_i - \alpha_0|$. Thus we have the following:

$$\sigma_i^2 \approx \frac{(L^2 - T^2)(\alpha_i - \alpha_0)^2 + T^2}{12}$$

$$= \frac{(L^2 - T^2)(\alpha_i^2 + \alpha_0^2 - 2\alpha_i\alpha_0) + T^2}{12}$$

$$= \frac{(L^2 - T^2)\alpha_i^2 - 2\alpha_0(L^2 - T^2)\alpha_i + (L^2 - T^2)\alpha_0^2 + T^2}{12} \tag{6}$$

This shows that the functional relationship between voting variance σ^2 and voting angle α can be approximated by a 2nd order curve (called f for later reference) as expressed in Eq. (6).

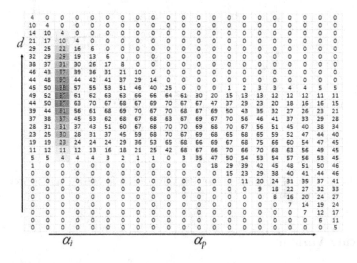

Fig. 3. Voting distribution in an α_i column for one "thick" line segment. Blue gradient cells illustrate that the voting values decrease gradually on both sides.

3   Statistical Distribution in Hough Space

For a line segment in an image, all collinear pixels vote for all possible cells in the Hough space. Due to various uncertainties, the voting in a column is considered as being a random variable. The voting value at each cell defines a probabilistic distribution. We compute the statistical variances in columns near the peak and use them to fit a quadratic polynomial curve, called g for later reference.

After voting, a peak is detected and represented by (α_p, d_p). This is just a coarse estimate for the actual normal parameters (α_0, d_0).

In that α_i-column which is close to the peak α_p, the middle cells have approximately identical voting values. Those voting values are larger than the voting values at outer cells. See Fig. 3 for an illustration.

3.1   Statistical Variances

For each column α_i in a peak region, the statistical mean m_i and statistical variance σ_i^2 are computed as follows:

$$m_i = \sum_{j \in W} [H_{ij} \cdot d_j] / \sum_{j \in W} H_{ij}$$

$$\sigma_i^2 = \sum_{j \in W} [H_{ij} \cdot (d_j - m_i)^2] / \sum_{j \in W} H_{ij} \tag{7}$$

where W defines the peak region in the Hough space.

3.2   Quadratic Polynomial Curve Fitting

Based on a voting analysis as discussed above, the functional relationship between statistical variance σ^2 and angle α can be approximated by a quadratic polynomial curve.

We fit a quadratic polynomial curve g to pairs (σ_i^2, α_i), all calculated in the peak region. Formally, this is denoted by

$$g : \sigma^2 = g(\alpha)$$
$$\triangleq e_2 \alpha^2 + e_1 \alpha + e_0 \tag{8}$$

3.3   Length and Width of Line-Segment

We compute length and width of a detected line segment based on the coefficients of the fitted function.

Following Eqs. (6) and (8), we obtain the following equational system:

$$(L^2 - T^2)/12 = e_2 \tag{9}$$
$$-2\alpha_0(L^2 - T^2)/12 = e_1 \tag{10}$$
$$((L^2 - T^2)\alpha_0^2 + T^2)/12 = e_0 \tag{11}$$

By solving simultaneously those equations, the length L and width T of the line segment are as follows:

$$L = \sqrt{12}\sqrt{e_2 + e_0 - \frac{e_1^2}{4e_2}} \tag{12}$$

$$T = \sqrt{12}\sqrt{e_0 - \frac{e_1^2}{4e_2}} \tag{13}$$

This defines our novel closed-form solution.

4   Experimental Results

We tested and verified the proposed method for determining the length and width of a detected line segment. We used a set of simulated image data as well as real-world images.

Used simulated binary images are of size 200×200. Each image contains a representation of one digitized line-segment as well as background image noise. A background pixel is called *noisy* if it is black due to the generated background noise. For the digitised line segments we have all their parameters available, including length and width, defining the *ground truth*. The direction, position, length, and width of a synthesised line segment are generated randomly in our test data.

Figure 4 illustrates an example for line segment detection. The length and width are accurately calculated when applying the proposed method.

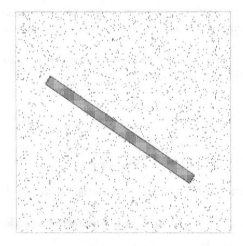

Fig. 4. Illustration of an example of our simulated binary images for determining the length and width for a line-segment with the proposed method. The blue box is drawn according to calculated length and width of the detected line-segment.

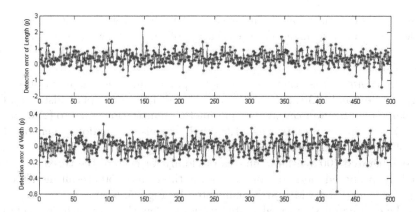

Fig. 5. Detection errors in the common case. *Top*: Length. *Bottom*: Width.

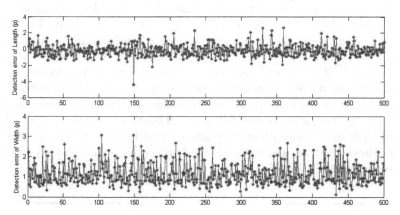

Fig. 6. Detection errors in the coarse-quantization case. *Top*: Length. *Bottom*: Width.

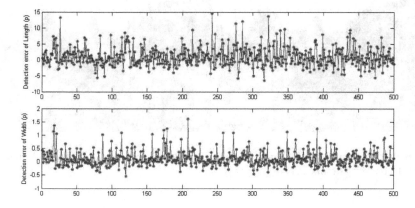

Fig. 7. Detection errors in the heavy-noise case. *Top*: Length. *Bottom*: Width.

Our method focuses on the accuracy of length and width calculation for a single detected line segment. For the accuracy of length and width determination, three cases have been considered in terms of different quantization steps and noise scales.

In the *common case* there are no noisy pixels, and the quantization of the Hough space equals $(\Delta\alpha, \Delta d) = (1°, 1p)$ (the unit for d is the pixel distance). For the *coarse-quantization case*, we set parameter quantization equal to $(\Delta\alpha, \Delta d) = (4°, 4p)$. For the *heavy-noise case*, 1,000 noisy pixels are randomly generated in each of the 200×200 images used.

We generated 500 synthetic images randomly for each of the three cases and tested the proposed method. For each of the three cases, 500 resulting detection errors for length and width are documented by Figs. 5, 6, and 7.

In the common case, the calculated values for length and width are accurate. The mean errors of length and width are equal to 0.4853 and 0.0781, respectively. When the Hough space is quantized at $(\Delta\alpha, \Delta d) = (4°, 4p)$, the mean errors of length and width are equal to 0.5796 and 1.1478, respectively. The length detection is accurate, while the width detection is sensitive to the quantization interval Δd. By adding 1,000 noisy pixels, the mean errors of length and width are equal to 3.0772 and 0.2613, respectively. The calculated width values are still accurate but length calculation is now effected by the given image noise.

For testing on recorded images, we use image sequences published in Set 5 of EISATS [9]. Images are of size 640×480. Those image sequences have been recorded for studying algorithms for vision-based driver assistance, in particular for algorithms detecting and tracking lane borders. (A review about visual lane analysis is given in [21]).

Fig. 8. Detection results for lane markers in real-world images. *Left:* Original images. *Right:* Detected lane markers.

Lane-detection results for one image of this data set is shown in Fig. 8. Only pixels in the lower half of the images are processed by supposing that lane borders are constrained to this image region. We are able to detect both frontiers (i.e. left and right border lines) of one lane marker as individual line segments

Fig. 9. Detection results for a building facade and road images. *Left:* Original images. *Right:* Detected result.

when using the accurate line detector reported in [23]. However, lane-border detection usually does not require such a fine and accurate line detection; it is more appropriate to detect one lane marker as a line segment of some width. This also supports the typically following step of lane-border tracking that only such line segments are accepted which do have a width within a given interval estimated for lane markers.

We also test the proposed method on building facade images and road images; see Fig. 9 for two examples, also showing detected lines. For the building facade image, all linear features with different length and width are detected. Two wide roads are detected in the shown aerial road view.

Line-segment features in images have varying lengths and widths. The proposed method calculates the length and width of these linear features using a Hough transform.

5   Conclusions

This paper proposes a novel method for line-segment length and width calculation using a Hough transform. We analyse the voting variance. We derive a functional relationship between the voting variance and the voting angle. This relation is approximated by a 2nd-order function f. Due to quantization errors and image noise, we consider voting in an α-column as being a random variable, and voting values define a probabilistic distribution. We compute the corresponding statistical variances and use them to fit a quadratic polynomial curve g.

We obtain three equations by comparing the coefficients of functions f and g. We calculate the length and width of a line segment by solving simultaneously these three equations. Various simulated and real-world images have been used for testing the proposed method, and also for illustrating new opportunities which are not yet available with previously specified line detection algorithms.

Experimental results verify the accuracy and feasibility of the proposed solution for line-segment length and width detection.

Acknowledgments. The first author thanks Jiangsu Overseas Research & Training Program for University Prominent Young & Middle-aged Teachers and Presidents for granting a scholarship to visit and undertake research at The University of Auckland.

References

1. Aggarwal, N., Karl, W.: Line detection in images through regularized Hough transform. IEEE Trans. Image Processing 15, 582–591 (2006)
2. Akhtar, M.W., Atiquzzaman, M.: Determination of line length using Hough transform. Electronics Letters 28, 94–96 (1992)
3. Atiquzzaman, M., Akhtar, M.W.: Complete line segment description using the Hough transform. Image and Vision Computing 12, 267–273 (1994)
4. Atiquzzaman, M., Akhtar, M.W.: A robust Hough transform technique for complete line segment description. Real-Time Imaging 1, 419–426 (1995)
5. Costa, L.F., Ben-Tzvi, B., Sandler, M.: Performance improvements to the Hough transform. In: UK IT 1990 Conference, pp. 98–103 (1990)
6. Du, S., Tu, C., van Wyk, B.J., Chen, Z.: Collinear segment detection using HT neighborhoods. IEEE Trans. Image Processing 20, 3912–3920 (2011)
7. Du, S., Tu, C., van Wyk, B.J., Ochola, E.O., Chen, Z.: Measuring straight line segments using HT butterflies. PLoS ONE 7(3), e33790 (2012)
8. Duda, R.O., Hart, P.E.: Use of the Hough transformation to detect lines and curves in pictures. Comm. ACM 15, 11–15 (1972)
9. EISATS: *.enpeda.*. image sequence analysis test site (2013), http://www.mi.auckland.ac.nz/EISATS
10. Furukawa, Y., Shinagawa, Y.: Accurate and robust line segment extraction by analyzing distribution around peaks in Hough space. Computer Vision Image Understanding 92, 1–25 (2003)
11. Hough, P.V.C.: Methods and means for recognizing complex patterns. U.S. Patent 3.069.654 (1962)
12. Ioannou, D.: Using the Hough transform for determining the length of a digital straight line segment. Electronics Letters 31, 782–784 (1995)
13. Kamat, V., Ganesan, S.: A robust Hough transform technique for description of multiple line segments in an image. In: Int. Conf. Image Processing, pp. 216–220 (1998)
14. Kamat, V., Ganesan, S.: Complete description of multiple line segments using the Hough transform. Image Vision Computing 16, 597–613 (1998)

15. Kiryati, N., Bruckstein, A.M.: What's in a Set of Points? IEEE Trans. Pattern Analysis Machine Intelligence 14, 496–500 (1992)
16. Klette, R.: Concise Computer Vision. Springer, London (2014)
17. Netanyahu, N.S., Weiss, I.: Analytic line fitting in the presence of uniform random noise. Pattern Recognition 34, 703–710 (2001)
18. Nguyen, T.T., Pham, X.D., Jeon, J.: An improvement of the standard Hough transform to detect line segments. In: IEEE Int. Conf. Industrial Technology, pp. 1–6 (2008)
19. Qjidaa, H., Radouane, L.: Robust line fitting in a noisy image by the method of moments. IEEE Trans. Pattern Analysis Machine Intelligence 21, 1216–1223 (1999)
20. Radon, J.: Über die Bestimmung von Funktionen durch ihre Integralwerte längs gewisser Mannigfaltigkeiten. Berichte Sächsische Akademie Wissenschaften, Math.-Phys. Kl. 69, 262–267 (1917)
21. Shin, B.-S., Xu, Z., Klette, R.: Visual lane analysis and higher-order tasks: A concise review. Machine Vision Applications (to appear, 2014)
22. Weiss, I.: Line fitting in a noisy image. IEEE Trans. Pattern Analysis Machine Intelligence 11, 325–329 (1989)
23. Xu, Z., Shin, B.-S.: Line segment detection with Hough transform based on minimum entropy. In: Klette, R., Rivera, M., Satoh, S. (eds.) PSIVT 2013. LNCS, vol. 8333, pp. 254–264. Springer, Heidelberg (2014)
24. Xu, Z., Shin, B.-S.: A statistical method for peak localization in Hough space by analysing butterflies. In: Klette, R., Rivera, M., Satoh, S. (eds.) PSIVT 2013. LNCS, vol. 8333, pp. 111–123. Springer, Heidelberg (2014)
25. Yamato, J., Ishii, I., Makino, H.: Highly accurate segment detection using Hough transformation. Systems and Computers in Japan 21, 68–77 (1990)

Stable Shape Comparison of Surfaces via Reeb Graphs

Barbara Di Fabio[1,3] and Claudia Landi[2,3]

[1] Dipartimento di Matematica, Università di Bologna, Italy
barbara.difabio@unibo.it
[2] DISMI, Università di Modena e Reggio Emilia, Italy
claudia.landi@unimore.it
[3] ARCES, Università di Bologna, Italy

Abstract. Reeb graphs are combinatorial signatures that capture shape properties from the perspective of a chosen function. One of the most important questions is whether Reeb graphs are robust against function perturbations that may occur because of noise and approximation errors in the data acquisition process. In this work we tackle the problem of stability by providing an editing distance between Reeb graphs of orientable surfaces in terms of the cost necessary to transform one graph into another by edit operations. Our main result is that the editing distance between two Reeb graphs is upper bounded by the extent of the difference of the associated functions, measured by the maximum norm. This yields the stability property under function perturbations.

Keywords: Shape similarity, editing distance, Morse function.

1 Introduction

In shape comparison, a widely used scheme is to measure the dissimilarity between signatures associated with each shape rather than match shapes directly [14,12,18].

Reeb graphs are signatures describing shapes from topological and geometrical perspectives. In this framework, shapes are modeled as spaces X endowed with scalar functions f. The role of f is to explore geometrical properties of the space X. The Reeb graph of $f : X \to \mathbb{R}$ is obtained by shrinking each connected component of a level set of f to a single point [15].

Reeb graphs have been used as an effective tool for shape analysis and description tasks since [17,16]. The Reeb graph has a number of characteristics that make it useful as a search key for 3D objects. First, a Reeb graph always consists of a one-dimensional graph structure and does not have any higher dimension components such as the degenerate surface that can occur in a medial axis. Second, by defining the function appropriately, it is possible to construct a Reeb graph that is invariant to translation and rotation, or even more complicated isometries of the shape.

E. Barcucci et al. (Eds.): DGCI 2014, LNCS 8668, pp. 202–213, 2014.

One of the most important questions is whether Reeb graphs are robust against perturbations that may occur because of noise and approximation errors in the data acquisition process. Heuristics have been developed so that the Reeb graph turns out to be resistant to connectivity changes caused by simplification, subdivision and remesh, and robust against noise and certain changes due to deformation [10,4].

In this paper we tackle the robustness problem for Reeb graphs from a theoretical point of view. The main idea is to generalize to the case of surfaces the techniques developed in [6] to prove the stability of Reeb graphs of curves against function perturbations. Indeed the case of surfaces appears as the most interesting area of applications of the Reeb graph as a shape descriptor.

To this end, we introduce a combinatorial dissimilarity measure, called an *editing distance*, between Reeb graphs of surfaces in terms of the cost necessary to transform one graph into another by edit operations. Thus our editing distance between Reeb graphs belongs to the family of Graph Edit Distances [9], widely used in pattern analysis. As shown in [9], some of these Graph Edit Distances are metrics, some other are only pseudo-metrics. Our editing distance turns out to have all the properties of a pseudo-metric. The main result we provide is that the editing distance between two Reeb graphs is never greater than the extent of the difference of the associated functions, measured by the maximum norm, yielding the stability property under function perturbations.

In the literature, some other comparison methodologies have been proposed to compare Reeb graphs and estimate the similarity of the shapes described by Reeb graph.

In [10] the authors propose a Multiresolutional Reeb Graph (MRG) based on geodesic distance. Similarity between 3D shapes is calculated using a coarse-to-fine strategy while preserving the topological consistency of the graph structures to provide a fast and efficient estimation of similarity and correspondence between shapes.

In [4] the authors discuss a method for measuring the similarity and recognizing sub-part correspondences of 3D shapes, based on the synergy of a structural descriptor, like the Extended Reeb Graph, with a geometric descriptor, like spherical harmonics.

Although the matching frameworks proposed in [10] and [4] are characterized by several computational advantages, the methods provided for Reeb graphs comparison have not been proved to be stable with respect to noise in the data, differently from the method proposed here.

Only recently the problem of Reeb graph stability has been investigated from the theoretical point of view.

In [6] an editing distance between Reeb graphs of curves endowed with Morse functions is introduced and shown to yield stability. Importantly, despite the combinatorial nature of this distance, it coincides with the natural pseudo-distance between shapes [8], thus showing the maximal discriminative power for this sort of distances.

The work in [2] about a stable distance for merge trees is also pertinent to the stability problem for Reeb graphs: merge trees are known to determine contour trees, that are Reeb graphs for simple domains.

Recently a functional distortion distance between Reeb graphs has been proposed in the preprint [1], with proven stable and discriminative properties. The functional distortion distance is intrinsically continuous, whereas the editing distance we propose is combinatorial.

In conclusion, the novelty of this paper is the announcement of a new combinatorial method to compare Reeb graphs in a stable way when shapes can be modeled as surfaces. An outline of the proof of this stability result is also given here, while full details and technicalities can be found in the technical report [7].

Outline. Section 2 reviews Reeb graphs. Section 3 introduces the admissible editing deformations to transform Reeb graphs into each other. In Section 4 the editing distance is defined. Section 5 illustrates the stability of Reeb graphs with respect to the editing distance. Section 6 concludes the paper.

2   Preliminaries on Reeb Graphs

An overview of the properties of Reeb graphs from the mathematical foundations to its history in the Computer Graphics context can be found in [3].

Since the main focus of this paper is on theoretical aspects, and computational issues being postponed to a future research, the appropriate setting for studying Reeb graphs is the following one.

$\mathcal{M}$ is a smooth (i.e. differentiable of class at least C^2) closed (i.e. compact and without boundary) orientable surface, and $f : \mathcal{M} \to \mathbb{R}$ is a simple Morse function on $\mathcal{M}$, i.e., a smooth function such that its Hessian matrix at each critical point is non-singular and, for every two distinct critical points p and q of f, the associated critical level sets $f^{-1}(f(p))$ and $f^{-1}(f(q))$ are disjoint.

Definition 1. *For every $p, q \in \mathcal{M}$, set $p \sim q$ whenever p, q belong to the same connected component of $f^{-1}(f(p))$. The quotient space $\mathcal{M}/ \sim$ is a finite and connected simplicial complex of dimension 1 known as the* Reeb graph *associated with f.*

Hence, the Reeb graph of a simple Morse function $f : \mathcal{M} \to \mathbb{R}$ is a graph whose vertices are the connected components of the critical levels of f that contain a critical point.

The Reeb graph associated with f will be denoted by Γ_f, its vertex set by $V(\Gamma_f)$, and its edge set by $E(\Gamma_f)$. Moreover, if $v_1, v_2 \in V(\Gamma_f)$ are adjacent vertices, i.e., connected by an edge, we will write $e(v_1, v_2) \in E(\Gamma_f)$.

The critical points of f correspond bijectively to the vertices of Γ_f. In particular, the assumption that $\mathcal{M}$ is orientable ensures that the vertices of Γ_f can be either of degree 1 (when corresponding to minima or maxima of f), or of degree 3 (when corresponding to saddles of f). Moreover, if $\mathcal{M}$ has genus $\mathfrak{g}$, Γ_f has exactly $\mathfrak{g}$ linearly independent cycles.

In what follows, we label each vertex of Γ_f by the value taken by f at the corresponding critical point. We denote such a labeled graph by (Γ_f, ℓ_f), where $\ell_f : V(\Gamma_f) \to \mathbb{R}$ is the restriction of $f : \mathcal{M} \to \mathbb{R}$ to the set of its critical points. In a labeled Reeb graph, each vertex v of degree 3 has at least two of its adjacent vertices, say w, w', such that $\ell_f(w) < \ell_f(v) < \ell_f(w')$. An example is displayed in Figure 1.

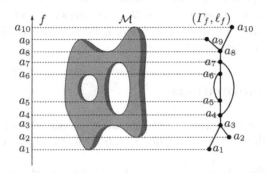

Fig. 1. Left: the height function $f : \mathcal{M} \to \mathbb{R}$; center: the surface $\mathcal{M}$ of genus $\mathfrak{g} = 2$; right: the associated labeled Reeb graph (Γ_f, ℓ_f)

Following [13], it can be seen that, given a graph on an even number of vertices, all of which are of degree 1 or 3, appropriately labeled, there is a simple Morse function whose labeled Reeb graph is the given one. This result requires the following definition.

Definition 2. *We shall say that two labeled Reeb graphs* $(\Gamma_f, \ell_f), (\Gamma_g, \ell_g)$ *are isomorphic, and we write* $(\Gamma_f, \ell_f) \cong (\Gamma_g, \ell_g)$, *if there exists a graph isomorphism* $\Phi : V(\Gamma_f) \to V(\Gamma_g)$ *such that, for every* $v \in V(\Gamma_f)$, $f(v) = g(\Phi(v))$ *(i.e.* Φ *preserves edges and vertices labels).*

Proposition 1 (Realization theorem). *Let* (G, ℓ) *be a labeled graph, where* G *is a graph with m linearly independent cycles, on an even number of vertices, all of which are of degree 1 or 3, and* $\ell : V(G) \to \mathbb{R}$ *is an injective function such that, for any vertex v of degree 3, at least two among its adjacent vertices, say w, w', are such that* $\ell(w) < \ell(v) < \ell(w')$. *Then an orientable closed surface* $\mathcal{M}$ *of genus* $\mathfrak{g} = m$, *and a simple Morse function* $f : \mathcal{M} \to \mathbb{R}$ *exist such that* $(\Gamma_f, \ell_f) \cong (G, \ell)$.

One may wonder if such surface and function are also unique, up to labeled graph isomorphism. Following [7], we answer to this question by considering two equivalence relations on the space of functions, and studying how they are mirrored by Reeb graphs isomorphisms.

Definition 3. *Let* $\mathcal{D}(\mathcal{M})$ *be the set of self-diffeomorphisms of* $\mathcal{M}$. *Two simple Morse functions* $f, g : \mathcal{M} \to \mathbb{R}$ *are called* right-equivalent *if there exists* $\xi \in$

$\mathcal{D}(\mathcal{M})$ *such that* $f = g \circ \xi$. *Moreover,* f, g *are called* right-left equivalent *if there exist* $\xi \in \mathcal{D}(\mathcal{M})$ *and an orientation preserving self-diffeomorphism* η *of* $\mathbb{R}$ *such that* $f = \eta \circ g \circ \xi$.

Proposition 2 (Uniqueness theorem). *If* f, g *are simple Morse functions on a closed surface, then*

1. *f and g are right-left equivalent if and only if their Reeb graphs Γ_f and Γ_g are isomorphic by an isomorphism that preserves the vertex order;*
2. *f and g are right-equivalent if and only if their labeled Reeb graphs (Γ_f, ℓ_f) and (Γ_g, ℓ_g) are isomorphic.*

3   Editing Deformations

In this section we present the moves that allow to edit Reeb graphs into each other. Basically, these moves amount to finite ordered sequences of elementary deformations.

Elementary deformations allow us to transform a Reeb graph into another with either a different number of vertices (*birth* (B) and *death* (D)), or with the same number of vertices endowed with different labels (*relabeling* (R) and *moves by Kudryavtseva* (K$_1$), (K$_2$), (K$_3$) [11]). We underline that the definition of the deformations of type (B), (D) and (R) is essentially different from the definition of analogous deformations in the case of Reeb graphs of curves as given in [6], even if the associated cost will be the same (see Section 4). This is because the degree of the involved vertices is 2 for Reeb graphs of closed curves, 1 and 3 for Reeb graphs of surfaces.

Definition 4. *With the convention of denoting the open interval with endpoints a, b by* $]a, b[$, *the elementary deformations* (B), (D), (R), (K$_i$), $i = 1, 2, 3$, *can be defined as follows.*

(B) For a fixed edge $e(v_1, v_2) \in E(\Gamma_f)$, with $\ell_f(v_1) < \ell_f(v_2)$, T is an *elementary deformation* of (Γ_f, ℓ_f) *of type* (B) if $T(\Gamma_f, \ell_f)$ is a labeled Reeb graph (Γ_g, ℓ_g) such that
   - $V(\Gamma_g) = V(\Gamma_f) \cup \{u_1, u_2\}$;
   - $E(\Gamma_g) = (E(\Gamma_f) - \{e(v_1, v_2)\}) \cup \{e(v_1, u_1), e(u_1, u_2), e(u_2, v_2)\}$;
   - $\ell_f(v_1) < \ell_g(u_i) < \ell_g(u_j) < \ell_f(v_2)$, with $\ell_g^{-1}(]\ell_g(u_i), \ell_g(u_j)[) = \emptyset$, $i, j \in \{1, 2\}, i \neq j$, and $\ell_{g|V(\Gamma_f)} = \ell_f$.

(D) For fixed edges $e(v_1, u_1), e(u_1, u_2), e(u_1, v_2) \in E(\Gamma_f)$, u_2 being of degree 1, such that $\ell_f(v_1) < \ell_f(u_i) < \ell_f(u_j) < \ell_f(v_2)$, with $\ell_f^{-1}(]\ell_f(u_i), \ell_f(u_j)[) = \emptyset$, $i, j \in \{1, 2\}, i \neq j$, T is an *elementary deformation* of (Γ_f, ℓ_f) *of type* (D) if $T(\Gamma_f, \ell_f)$ is a labeled Reeb graph (Γ_g, ℓ_g) such that
   - $V(\Gamma_g) = V(\Gamma_f) - \{u_1, u_2\}$;
   - $E(\Gamma_g) = (E(\Gamma_f) - \{e(v_1, u_1), e(u_1, u_2), e(u_2, v_2)\}) \cup \{e(v_1, v_2)\}$;
   - $\ell_g = \ell_{f|V(\Gamma_f) - \{u_1, u_2\}}$.

(R) T is an *elementary deformation* of (Γ_f, ℓ_f) *of type* (R) if $T(\Gamma_f, \ell_f)$ is a labeled Reeb graph (Γ_g, ℓ_g) such that
- $\Gamma_g = \Gamma_f$;
- $\ell_g : V(G) \to \mathbb{R}$ induces the same vertex-order as ℓ_f except for at most two non-adjacent vertices, say u_1, u_2, for which, if $\ell_f(u_1) < \ell_f(u_2)$ and $\ell_f^{-1}(]\ell_f(u_1), \ell_f(u_2)[) = \emptyset$, then $\ell_g(u_1) > \ell_g(u_2)$, and $\ell_g^{-1}(]\ell_g(u_2), \ell_g(u_1)[) = \emptyset$.

(K_1) For fixed edges $e(v_1, u_1), e(u_1, u_2), e(u_1, v_4), e(u_2, v_2), e(u_2, v_3) \in E(\Gamma_f)$, with two among v_2, v_3, v_4 possibly coincident, and either $\ell_f(v_1) < \ell_f(u_1) < \ell_f(u_2) < \ell_f(v_2), \ell_f(v_3), \ell_f(v_4)$, with $\ell_f^{-1}(]\ell_f(u_1), \ell_f(u_2)[) = \emptyset$, or $\ell_f(v_2), \ell_f(v_3), \ell_f(v_4) < \ell_f(u_2) < \ell_f(u_1) < \ell_f(v_1)$, with $\ell_f^{-1}(]\ell_f(u_2), \ell_f(u_1)[) = \emptyset$, T is an *elementary deformation* of (Γ_f, ℓ_f) *of type* (K_1) if $T(\Gamma_f, \ell_f)$ is a labeled Reeb graph (Γ_g, ℓ_g) such that:
- $V(\Gamma_g) = V(\Gamma_f)$;
- $E(\Gamma_g) = (E(\Gamma_f) - \{e(v_1, u_1), e(u_2, v_2)\}) \cup \{e(v_1, u_2), e(u_1, v_2)\}$;
- $\ell_{g|V(\Gamma_g) - \{u_1, u_2\}} = \ell_f$, and either $\ell_f(v_1) < \ell_g(u_2) < \ell_g(u_1) < \ell_f(v_2)$, $\ell_f(v_3), \ell_f(v_4)$, with $\ell_g^{-1}(]\ell_g(u_2), \ell_g(u_1)[) = \emptyset$, or $\ell_f(v_2), \ell_f(v_3), \ell_f(v_4) < \ell_g(u_1) < \ell_g(u_2) < \ell_f(v_1)$, with $\ell_g^{-1}(]\ell_g(u_1), \ell_g(u_2)[) = \emptyset$.

(K_2) For fixed edges $e(v_1, u_1), e(u_1, u_2), e(v_2, u_1), e(u_2, v_3), e(u_2, v_4) \in E(\Gamma_f)$, with u_1, u_2 of degree 3, v_2, v_3 possibly coincident with v_1, v_4, respectively, and $\ell_f(v_1), \ell_f(v_2) < \ell_f(u_1) < \ell_f(u_2) < \ell_f(v_3), \ell_f(v_4)$, with $\ell_f^{-1}(]\ell_f(u_1), \ell_f(u_2)[) = \emptyset$, T is an *elementary deformation* of (Γ_f, ℓ_f) *of type* (K_2) if $T(\Gamma_f, \ell_f)$ is a labeled Reeb graph (Γ_g, ℓ_g) such that:
- $V(\Gamma_g) = V(\Gamma_f)$;
- $E(\Gamma_g) = (E(\Gamma_f) - e(v_1, u_1), e(u_2, v_3)\}) \cup \{e(u_1, v_3), e(v_1, u_2)\}$;
- $\ell_{g|V(\Gamma_g) - \{u_1, u_2\}} = \ell_f$ and $\ell_f(v_1), \ell_f(v_2) < \ell_g(u_2) < \ell_g(u_1) < \ell_f(v_3)$, $\ell_f(v_4)$, with $\ell_g^{-1}(]\ell_g(u_2), \ell_g(u_1)[) = \emptyset$.

(K_3) For fixed edges $e(v_1, u_2), e(u_1, u_2), e(v_2, u_1), e(u_1, v_3), e(u_2, v_4) \in E(\Gamma_f)$, with u_1, u_2 of degree 3, v_2, v_3 possibly coincident with v_1, v_4, respectively, and $\ell_f(v_1), \ell_f(v_2) < \ell_f(u_2) < \ell_f(u_1) < \ell_f(v_3), \ell_f(v_4)$, with $\ell_f^{-1}(]\ell_f(u_2), \ell_f(u_1)[) = \emptyset$, T is an *elementary deformation* of (Γ_f, ℓ_f) *of type* (K_3) if $T(\Gamma_f, \ell_f)$ is a labeled Reeb graph (Γ_g, ℓ_g) such that:
- $V(\Gamma_g) = V(\Gamma_f)$;
- $E(\Gamma_g) = (E(\Gamma_f) - e(v_1, u_2), e(u_1, v_3)\}) \cup \{e(v_1, u_1), e(u_2, v_3)\}$;
- $\ell_{g|V(\Gamma_g) - \{u_1, u_2\}} = \ell_f$ and $\ell_f(v_1), \ell_f(v_2) < \ell_g(u_1) < \ell_g(u_2) < \ell_f(v_3)$, $\ell_f(v_4)$, with $\ell_g^{-1}(]\ell_g(u_1), \ell_g(u_2)[) = \emptyset$.

All the elementary deformations above defined are schematically displayed in Table 1.

We observe that, differently from the case of curves [6], it is not sufficient to consider only deformations of type (B), (D) and (R). The necessity to add those of type (K_i), $i = 1, 2, 3$, can be in fact deduced by observing the changes a Reeb graph undergoes when it is dynamically associated with Morse functions that at a some instant fail to be simple (Table 1). Only in some particular cases, such as when some of the vertices v_j are of degree 1, operations (K_i), $i = 1, 2, 3$, can be obtained by composition of operations (B), (D) and (R).

Table 1. Elementary deformations of a labeled Reeb graph

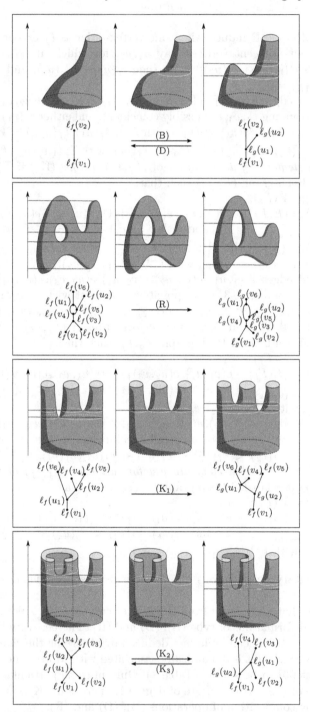

Since each type of elementary deformation transforms a labeled Reeb graph into another one, we can apply elementary deformations iteratively to transform labeled Reeb graphs into each other.

Definition 5. *We shall call* deformation *of* (Γ_f, ℓ_f) *any finite ordered sequence* $T = (T_1, T_2, \ldots, T_r)$ *of elementary deformations such that* T_1 *is an elementary deformation of* (Γ_f, ℓ_f), T_2 *is an elementary deformation of* $T_1(\Gamma_f, \ell_f)$, ..., T_r *is an elementary deformation of* $T_{r-1}T_{r-2} \cdots T_1(\Gamma_f, \ell_f)$. *We shall denote by* $T(\Gamma_f, \ell_f)$ *the result of the deformation* T *applied to* (Γ_f, ℓ_f). *Moreover, we shall call* identical deformation *any deformation such that* $T(\Gamma_f, \ell_f) \cong (\Gamma_f, \ell_f)$.

Proposition 3. *Let* (Γ_f, ℓ_f) *and* (Γ_g, ℓ_g) *be two labeled Reeb graphs associated with simple Morse functions* $f, g : \mathcal{M} \to \mathbb{R}$. *Then the set of all the deformations* T *such that* $T(\Gamma_f, \ell_f) \cong (\Gamma_g, \ell_g)$ *is non-empty.*

In other words, any two Reeb graphs of simple Morse functions on a given surface can be transformed into each other by a finite sequence of elementary deformations. This result is a consequence of the fact that, through a finite sequence of elementary deformations of type (B), (D),(R), (K$_i$), $i = 1, 2, 3$, every Reeb graph can be transformed into one having only one maximum, one minimum, and all the cycles, if any, of length 2 [7].

4   Editing Distance

Given two labeled Reeb graphs (Γ_f, ℓ_f) and (Γ_g, ℓ_g) associated with simple Morse functions $f, g : \mathcal{M} \to \mathbb{R}$, we denote by $\mathcal{T}((\Gamma_f, \ell_f), (\Gamma_g, \ell_g))$ the set of all possible deformations between (Γ_f, ℓ_f) and (Γ_g, ℓ_g). Let us associate a cost with each editing deformation in $\mathcal{T}((\Gamma_f, \ell_f), (\Gamma_g, \ell_g))$.

Definition 6. *Let* T *be an elementary deformation such that* $T(\Gamma_f, \ell_f) \cong (\Gamma_g, \ell_g)$.

- *If* T *is of type* (B) *inserting the vertices* $u_1, u_2 \in V(\Gamma_g)$, *then we define the associated cost as*
$$c(T) = \frac{|\ell_g(u_1) - \ell_g(u_2)|}{2}.$$

- *If* T *is of type* (D) *deleting the vertices* $u_1, u_2 \in V(\Gamma_f)$, *then we define the associated cost as*
$$c(T) = \frac{|\ell_f(u_1) - \ell_f(u_2)|}{2}.$$

- *If* T *is of type* (R) *relabeling the vertices* $v \in V(\Gamma_f) = V(\Gamma_g)$, *then we define the associated cost as*
$$c(T) = \max_{v \in V(\Gamma_f)} |\ell_f(v) - \ell_g(v)|.$$

- *If* T *is of type* (K$_i$), *with* $i = 1, 2, 3$, *relabeling the vertices* $u_1, u_2 \in V(\Gamma_f)$, *then we define the associated cost as*
$$c(T) = \max\{|\ell_f(u_1) - \ell_g(u_1)|, |\ell_f(u_2) - \ell_g(u_2)|\}.$$

Moreover, if $T = (T_1, \ldots, T_r)$ is a deformation such that $T_r \cdots T_1(\Gamma_f, \ell_f) \cong (\Gamma_g, \ell_g)$, we define the associated cost as $c(T) = \sum_{i=1}^{r} c(T_i)$.

Now we can define the editing distance between labeled Reeb graphs as the infimum cost we have to bear to transform one graph into the other [7, Thm. 3.3].

Theorem 1. *For every two labeled Reeb graphs (Γ_f, ℓ_f) and (Γ_g, ℓ_g), we set*

$$d((\Gamma_f, \ell_f), (\Gamma_g, \ell_g)) = \inf_{T \in \mathcal{T}((\Gamma_f, \ell_f), (\Gamma_g, \ell_g))} c(T).$$

Then d is a pseudo-metric on isomorphism classes of labeled Reeb graphs.

We recall that a pseudo-metric is non-negative, symmetric, and has the triangle inequality, but may be unable to distinguish different objects.

We do not exclude that our editing distance may have also the coincidence axiom as in the case of curves. If so, it would turn to be a metric. The main difficulty is that the linearization technique used in the case of curves does not work in the case of surfaces. We are currently investigating different techniques.

5   Stability Result

Let $\mathcal{F}(\mathcal{M}, \mathbb{R})$ be the set of smooth real valued functions on $\mathcal{M}$, endowed with the C^∞ topology, and let us stratify such a space, as done by Cerf in [5]. Let us denote by $\mathcal{F}^0$ the submanifold of $\mathcal{F}(\mathcal{M}, \mathbb{R})$ of co-dimension 0 that contains all the simple Morse functions $f : \mathcal{M} \to \mathbb{R}$. Then, let $\mathcal{F}^1 = \mathcal{F}_\alpha^1 \cup \mathcal{F}_\beta^1$ be the submanifold of $\mathcal{F}(\mathcal{M}, \mathbb{R})$ of co-dimension 1, where: $\mathcal{F}_\alpha^1$ represents the set of functions whose critical levels contain exactly one critical point, and the critical points are all non-degenerate, except exactly one; $\mathcal{F}_\beta^1$ the set of Morse functions whose critical levels contain at most one critical point, except for one level containing exactly two critical points.

The main result, proven in [7], is the following one.

Theorem 2 (Stability Theorem). *For every $f, g \in \mathcal{F}^0$,*

$$d((\Gamma_f, \ell_f), (\Gamma_g, \ell_g)) \leq \|f - g\|_{C^0},$$

where $\|f - g\|_{C^0} = \max_{p \in \mathcal{M}} |f(p) - g(p)|$.

The proof relies on two intermediate results. The first one states that, considering a linear path connecting two functions $f, g \in \mathcal{F}^0$ and not traversing strata of co-dimension greater than 0, the editing distance between the Reeb graphs associated with its end-points is upper bounded by the distance of f and g computed in the C^0-norm. In this case the graph (Γ_g, ℓ_g) can be obtained transforming (Γ_f, ℓ_f) with a sequence of elementary deformations of type (R).

– Let $f, g \in \mathcal{F}^0$ and let us consider the path $h : [0,1] \to \mathcal{F}(\mathcal{M}, \mathbb{R})$ defined by $h(\lambda) = (1 - \lambda)f + \lambda g$. If $h(\lambda) \in \mathcal{F}^0$ for every $\lambda \in [0,1]$, then $d((\Gamma_f, \ell_f), (\Gamma_g, \ell_g)) \leq \|f - g\|_{C^0}$.

The second result states that, if two functions $f, g \in \mathcal{F}^0$ can be connected by a linear path having only one point which belong to a stratum $\mathcal{F}^1$ and do not traverse strata of co-dimension greater than 1, the cost to transform (Γ_f, ℓ_f) into (Γ_g, ℓ_g) is again upper bounded by $\|f - g\|_{C^0}$. In particular, crossing a stratum $\mathcal{F}^1_\alpha$ ($\mathcal{F}^1_\beta$, resp.), means that the Reeb graph is undergoing an elementary deformation of type (B) or (D) ((R) or (K_i), $i = 1, 2, 3$, resp.).

– Let $f, g \in \mathcal{F}^0$ and let us consider the path $h : [0,1] \to \mathcal{F}(\mathcal{M}, \mathbb{R})$ defined by $h(\lambda) = (1 - \lambda)f + \lambda g$. If $h(\lambda) \in \mathcal{F}^0$ for every $\lambda \in [0,1] \setminus \{\overline{\lambda}\}$, with $0 < \overline{\lambda} < 1$, and h transversely intersects $\mathcal{F}^1$ at $\overline{\lambda}$, then $d((\Gamma_f, \ell_f), (\Gamma_g, \ell_g)) \leq \|f - g\|_{C^0}$.

As an example illustrating the stability property of the editing distance, consider $f, g : \mathcal{M} \to \mathbb{R}$ as in Figure 2. Let $f(q_i) - f(p_i) = a$, $i = 1, 2, 3$. It holds that $d((\Gamma_f, \ell_f), (\Gamma_g, \ell_g)) \leq \frac{a}{2}$, showing that the editing distance is bounded by the norm of the difference between the functions. Indeed, for every $0 < \epsilon < \frac{a}{2}$, we can apply to (Γ_f, ℓ_f) a deformation of type (R), that relabels the vertices p_i, q_i, $i = 1, 2, 3$, in such a way that $\ell_f(p_i)$ is increased by $\frac{a}{2} - \epsilon$, and $\ell_f(q_i)$ is decreased by $\frac{a}{2} - \epsilon$, composed with three deformations of type (D) that delete p_i with q_i, $i = 1, 2, 3$. Thus, since the total cost is equal to $\frac{a}{2} - \epsilon + 3\epsilon$, by the arbitrariness of ϵ, it holds that $d((\Gamma_f, \ell_f), (\Gamma_g, \ell_g)) \leq \frac{a}{2}$.

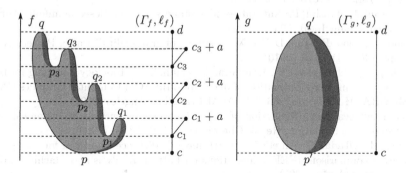

Fig. 2. For these two simple Morse functions f, g it is easy to see that $d((\Gamma_f, \ell_f), (\Gamma_g, \ell_g))$ is bounded from above by the norm of $f - g$

6   Discussion

Building on arguments similar to those given in [6] for curves, we have presented a combinatorial dissimilarity measure for Reeb graphs of surfaces. For a complete analogy with the case of curves, we still need to prove that the editing distance is not only a pseudo-metric but actually a metric. Also, it would be useful to prove that, as for curves, it discriminates shapes as well as the natural pseudo-distance.

From the computational viewpoint, it would be very interesting to find an analogue of the editing distance for the case when the considered surfaces are discrete models, e.g. triangular meshes, and the functions accordingly discrete.

References

1. Bauer, U., Ge, X., Wang, Y.: Measuring distance between Reeb graphs. No. arXiv:1307.2839v1 (2013)
2. Beketayev, K., Yeliussizov, D., Morozov, D., Weber, G.H., Hamann, B.: Measuring the distance between merge trees. In: Topological Methods in Data Analysis and Visualization V (TopoInVis 2013) (2013)
3. Biasotti, S., Giorgi, D., Spagnuolo, M., Falcidieno, B.: Reeb graphs for shape analysis and applications. Theoretical Computer Science 392, 5–22 (2008)
4. Biasotti, S., Marini, S., Spagnuolo, M., Falcidieno, B.: Sub-part correspondence by structural descriptors of 3D shapes. Computer-Aided Design 38(9), 1002–1019 (2006)
5. Cerf, J.: La stratification naturelle des espaces de fonctions différentiables réelles et le théorème de la pseudo-isotopie. Inst. Hautes Études Sci. Publ. Math. 39, 5–173 (1970)
6. Di Fabio, B., Landi, C.: Reeb graphs of curves are stable under function perturbations. Mathematical Methods in the Applied Sciences 35(12), 1456–1471 (2012)
7. Di Fabio, B., Landi, C.: Reeb graphs of surfaces are stable under function perturbations. Tech. Rep. 3956, Università di Bologna (February 2014), http://amsacta.cib.unibo.it/3956/
8. Donatini, P., Frosini, P.: Natural pseudodistances between closed manifolds. Forum Mathematicum 16(5), 695–715 (2004)
9. Gao, X., Xiao, B., Tao, D., Li, X.: A survey of graph edit distance. Pattern Anal. Appl. 13(1), 113–129 (2010)
10. Hilaga, M., Shinagawa, Y., Kohmura, T., Kunii, T.L.: Topology matching for fully automatic similarity estimation of 3D shapes. In: ACM Computer Graphics (Proc. SIGGRAPH 2001), pp. 203–212. ACM Press, Los Angeles (2001)
11. Kudryavtseva, E.A.: Reduction of Morse functions on surfaces to canonical form by smooth deformation. Regul. Chaotic Dyn. 4(3), 53–60 (1999)
12. Ling, H., Okada, K.: An efficient earth mover's distance algorithm for robust histogram comparison. IEEE Transactions on Pattern Analysis and Machine Intelligence 29, 840–853 (2007)
13. Masumoto, Y., Saeki, O.: A smooth function on a manifold with given reeb graph. Kyushu J. Math. 65(1), 75–84 (2011)
14. Ohbuchi, R., Takei, T.: Shape-similarity comparison of 3D models using alpha shapes. In: 11th Pacific Conference on Computer Graphics and Applications, pp. 293–302 (2003)
15. Reeb, G.: Sur les points singuliers d'une forme de Pfaff complétement intégrable ou d'une fonction numérique. Comptes Rendus de L'Académie ses Sciences 222, 847–849 (1946)

16. Shinagawa, Y., Kunii, T.L.: Constructing a Reeb Graph automatically from cross sections. IEEE Computer Graphics and Applications 11(6), 44–51 (1991)
17. Shinagawa, Y., Kunii, T.L., Kergosien, Y.L.: Surface coding based on morse theory. IEEE Computer Graphics and Applications 11(5), 66–78 (1991)
18. Wu, H.Y., Zha, H., Luo, T., Wang, X.L., Ma, S.: Global and local isometry-invariant descriptor for 3D shape comparison and partial matching. In: 2010 IEEE Conference on Computer Vision and Pattern Recognition (CVPR), pp. 438–445 (2010)

About Multigrid Convergence of Some Length Estimators*

Loïc Mazo and Étienne Baudrier

ICube, University of Strasbourg, CNRS,
300 Bd Sébastien Brant - CS 10413 - 67412 ILLKIRCH, France

Abstract. An interesting property for curve length digital estimators is the convergence toward the continuous length and the associate convergence speed when the digitization step h tends to 0. On the one hand, it has been proved that the local estimators do not verify this convergence. On the other hand, DSS and MLP based estimators have been proved to converge but only under some convexity and smoothness or polygonal assumptions. In this frame, a new estimator class, the so called *semi-local estimators*, has been introduced by Daurat *et al.* in [4]. For this class, the pattern size depends on the resolution but not on the digitized function. The semi-local estimator convergence has been proved for functions of class $\mathcal{C}^2$ with an optimal convergence speed that is a $\mathcal{O}(h^{\frac{1}{2}})$ without convexity assumption (here, optimal means with the best estimation parameter setting). A semi-local estimator subclass, that we call *sparse estimators*, is exhibited here. The sparse estimators are proved to have the same convergence speed as the semi-local estimators under the weaker assumptions. Besides, if the continuous function that is digitized is concave, the sparse estimators are proved to have an optimal convergence speed in h. Furthermore, assuming a sequence of functions $G_h \colon h\mathbb{Z} \to h\mathbb{Z}$ discretizing a given Euclidean function as h tends to 0, sparse length estimation computational complexity in the optimal setting is a $\mathcal{O}(h^{-\frac{1}{2}})$.

1 Introduction

The ability to perform the measurement of geometric features on digital representations of continuous objects is an important goal in a world becoming more and more digital. We focus in this paper on one classical digital problem: the length estimation. The problem is to estimate the length of a continuous curve S knowing a digitization of S. As information is lost during the digitization step, there is no reliable estimation without *a priori* knowledge and it is difficult to evaluate the estimator performances. In order to refine the evaluation of the estimators, a property, so called *convergence property* is desirable: the estimation convergence toward the true length of the curve S when the grid step h tends to 0. This property can be viewed as a robustness to digitization grid change.

---

* This work was supported by the Agence Nationale de la Recherche through contract ANR-2010-BLAN-0205-01.

E. Barcucci et al. (Eds.): DGCI 2014, LNCS 8668, pp. 214–225, 2014.

The local estimators based on a fixed pattern size do not satisfy the convergence property [13]. The adaptive estimators based on the Maximal Digital Straight Segment (MDSS) or the Minimum Length Polygon (MLP) satisfy the convergence property under assumptions of convexity, 4-connectivity for closed simple curves (also called Jordan curves) [3]. The semi-local estimators, introduced by Daurat et al [4] for function graphs, verifies the convergence property under smoothness assumption but without convexity hypothesis. We present here a subclass of the semi-local estimators, the *sparse estimators* that only need information on a small part of the function values and keep the convergence property.

The paper is organized as follows. In Section 2, some necessary notations and conventions are recalled, then existing estimators and their convergence properties are detailed. In Section 3, the sparse estimators are defined, their convergence properties are given in the general case and then in the concave cases (we make a distinction between the concavity of the continuous function and the concavity of the piecewise affine function related to the discretization). Due to the lack of place, no formal proofs can be provided in the present article. Nevertheless a series of lemmas outlines them and an experiment exemplifies the lemmas. The reader that want to dive more deeper in the proofs can find them in [15]. Section 4 concludes the article and gives directions for future works. In Appendix A two counterexamples about the concavity are exhibited. Appendix B presents a minimal error on the sparse estimation of the length of a segment of parabola.

2   Background

2.1   Digitization Model

In this work, we have restricted ourselves to the digitizations of function graphs. So, let us consider a continuous function $g : [a, b] \to \mathbb{R}$ ($a < b$), its graph $\mathcal{C}(g) = \{(x, g(x)) \mid x \in [a, b]\}$ and a positive real number r, the *resolution*. We assume to have an orthogonal grid in the Euclidean space $\mathbb{R}^2$ whose set of grid points is $h\mathbb{Z}^2$ where $h = 1/r$ is the *grid spacing*. We use the following notations: $\lfloor x \rfloor_h$ is the greatest multiple of h less than or equal to x, $\{x\}_h = x - \lfloor x \rfloor_h$. Finally, for any function f defined on an interval, $L(f)$ denotes the length of $\mathcal{C}(f)$, the graph of f ($L(f) \in [0, +\infty]$).

The common methods to model the digitization of the graph $\mathcal{C}(g)$ at the resolution r are closely related to each others. In this paper, we assume an *object boundary quantization* (OBQ). This method associates to the graph $\mathcal{C}(g)$ the h-*digitization* set $\mathcal{D}^O(g, h) = \{(kh, \lfloor g(kh) \rfloor_h) \mid k \in \mathbb{Z}\}$. The set $\mathcal{D}^O(g, h)$ contains the uppermost grid points which lie in the hypograph of g, hence it can be understood as a part of the boundary of a solid object. Provided the slope of g is limited by 1 in modulus, $\mathcal{D}^O(g, h)$ is an 8-connected digital curve. Observe that if g is a function of class C^1 such that the set $\{x \in [a, b] \mid |g'(x)| = 1\}$ is finite, then by symmetries on the graph $\mathcal{C}(g)$, it is possible to come down to the case where $|g'| \leq 1$. So, we assume that g is a Lipschitz function which Lipschitz constant 1. Hence, the set $\mathcal{D}^O(g, h)$ is 8-connected for any h and the curve $\mathcal{C}(g)$

is rectifiable $(L(g) < +\infty)$. Moreover, the h-digitization set $\mathcal{D}^O(g, h)$ can be described by its first point and its *Freeman code* [9], $\mathcal{F}(g, h)$, with the alphabet $\{0, 1, 7\}$. For any word $\omega \in \{0, 1, 7\}^k$ $(k \in \mathbb{N})$, we set $\|\omega\| = \sqrt{k^2 + j^2}$ where j is the number of letters 1 minus the number of letters 7 in the word ω.

2.2   Local Estimators

Local length estimators (see [10] for a short review) are based on parallel computations of the length of fixed size segments of a digital curve. For instance, an 8-connected curve can be split into 1-step segments. For each segment, the computation return 1 whenever the segment is parallel to the axes (Freeman's code is even) and $\sqrt{2}$ when the segment is diagonal (Freeman's code is odd). Then all the results are added to give the curve length estimation.

This kind of local computation is the oldest way to estimate the length of a curve and has been widely used in image analysis. Nevertheless, it has not the convergence property. In [13], the authors introduce a general definition of local length estimation with sliding segments and prove that such computations cannot give a convergent estimator for straight lines whose slope is small (less than the inverse of the size of the sliding segment). In [17], a similar definition of local length estimation is given with disjoint segments. Again, it is shown that the estimator failed to converge for straight lines (with irrational slopes). This behavior is experimentally confirmed in [3] on a test set of five closed curves. Moreover, the non-convergence is established in [5,18] for almost all parabolas.

2.3   Adaptive Estimators: MDSS and MLP

Adaptive length estimators gather estimators relying on a segmentation of the discrete curve that depends on each point of the curve: a move on a point can change the whole segmentation. Unlike local estimators, it is possible to prove the convergence property of adaptive length estimators under some assumptions. Adaptive length estimators include two families of length estimators, namely the Maximal Digital Straight Segment (MDSS) based length estimators and the Minimal Length Polygon (MLP) based length estimators.

Definition and properties of MDSS can be found in [12,7,3]. Efficient algorithms have been developed for segmenting curves or function graphs into MDSS and to compute their characteristics in a linear time [12,8,7]. The decomposition in MDSS is not unique and depends on the start-point of the segmentation and on the curve travel direction. The convergence property of MDSS estimators has been proved for convex polygons whose MDSS polygonal approximation[1] is also convex [11, Th. 13 and the proof]: given a convex polygon $\mathcal{C}$ and a grid spacing h (below some threshold), the error between the estimated length $L_{\text{est}}(\mathcal{C}, h)$ and the true length of the polygon $L(\mathcal{C})$ is such that

---

[1] Though the digitization of a convex set is digitally convex, it does not mean that a polygonal curve related to a convex polygonal curve via a MDSS segmentation process is also convex.

$$|L(S) - L_{\text{est}}(S, h)| \leq (2 + \sqrt{2})\pi h. \tag{1}$$

Empirical MDSS multigrid convergence has also been tested in [3,6] on smooth nonconvex planar curves. The obtained convergence speed is a $O(h)$ as in the convex polygonal case. Nevertheless it has not been proved under these assumptions. Another way to obtain an estimation of the length of a curve using MDSS is to take the slopes of the MDSSs to estimate the tangent directions and then to compute the length by numerical integration [2,3,14]. The estimation is unique and has been proved to be multigrid convergent for smooth curves (of class C^2 with bounded curvature in [2], of class C^3 with strictly positive curvature in [14]). The convergence speed is a $O(h^{\frac{1}{3}})$ [14] and thus, worse than (1).

Let $\mathcal{C}$ be a simple closed curve lying in-between two polygonal curves γ_1 and γ_2. Then there is a unique polygon, the MLP, whose length is minimal between γ_1 and γ_2. The length of the MLP can be used to estimate the length of the curve $\mathcal{C}$. At least two MLP based length estimators have been described and proved to be multigrid convergent for convex, smooth or polygonal, simple closed curves, the SB-MLP proposed by Sloboda $et~al.$ [16] and the AS-MLP, introduced by Asano $et~al.$ [1]. For both of them, and for a given grid spacing h, the error between the estimated length $L_{\text{est}}(\mathcal{C}, h)$ and the true length of the curve $L(\mathcal{C})$ is a $O(h)$:

$$|L(\mathcal{C}) - L_{\text{est}}(\mathcal{C}, h)| \leq Ah$$

where $A = 8$ for SB-MLP and $A \approx 5.844$ for AS-MLP.

On the one hand, as estimators described in this section are adaptive, the convergence theorems are difficult to establish and rely on strong hypotheses. On the other hand, the study of the MDSS in [6] shows that the MDSS size tends to 0 and their discrete length tends toward infinity as the grid step tends to 0. Thereby, one could ask whether combining a local estimation with an increasing window size as the resolution grows would give a convergent estimator under more general assumptions and/or with simpler proofs of convergence. The following sections explore this question.

2.4   Semi-local Length Estimators

The notion of semi-local estimator appears in [4]. At a given resolution, a semi-local estimator resembles a local estimator: it can be implemented via a parallel computation, each processor handling a fixed size segment of the curve. Nevertheless, in the framework of semi-local estimation, the processors must be aware of the resolution from which the size of the segments depends.

More formally, let $g : [a, b] \to \mathbb{R}$ be a 1-Lipschitz function[2]. Hence, at any resolution, the Freeman's code describing the discretization of g belongs to the set $\mathcal{P} = \bigcup_{n \in \mathbb{N}} \{0, 1, 7\}^n$.

---

[2] In [4], the hypothesis on g is not clear. On the one hand, the code $\mathcal{F}(g, h)$ is supposed to have $\{0, 1\}$ as alphabet. On the other hand, [4, Prop 1] does not retain any hypothesis on g but its class of differentiability. Indeed, in the proof, the derivative of g needs not be positive nor limited by 1.

A *semi-local estimator* is a pair (H, p) where

- $H :]0, \infty[\to \mathbb{N}^*$ gives the relative size of the segments given a grid spacing h and
- $p : \mathcal{P} \to [0, \infty[$ gives the estimated feature (here, the length) associated to a (finite) Freeman's code.

At a given grid spacing h, the Freeman's code describing the digitization of the curve $\mathcal{C}(g)$ is segmented in N_h codes ω_i of length $H(h)$ and a rest $\omega_* \in \{0, 1, 7\}^j$, $j < H(h)$. Then, the length of the curve $\mathcal{C}(g)$ is estimated by

$$L^{\mathrm{SL}}(g, h) = h \sum_{i=0}^{N_h - 1} p(\omega_i).$$

In [4], the authors give a proof of convergence for functions of class C^2.

Theorem 1 ([4, Prop. 1]). *Let (H, p) be a semi-local estimator such that:*

1. $\lim_{h \to 0} h H(h) = 0$,
2. $\lim_{h \to 0} H(h) = +\infty$,
3. $\max \{ p(\omega) - \|\omega\| \mid \omega \in \{0, 1, 7\}^k \} = o(k)$ *as* $k \to +\infty$.

Then, for any function $g \in C^2([a, b])$, the estimation $L^{\mathrm{SL}}(g, h)$ converges toward the length of the curve $\mathcal{C}(g)$. Furthermore, if the term $o(k)$ in the third hypothesis is a constant and $H(h) = \Theta(h^{-\frac{1}{2}})$, then $L(g) - L^{\mathrm{SL}}(g, h) = \mathcal{O}(h^{\frac{1}{2}})$.

$H(h)$ stands for the size of a Freeman's code ω while $h H(h)$ is the real length of the computation step. In the above theorem, the first hypothesis states that the real length $h H(h)$ tends to 0. If instead of diminishing the grid spacing, we keep it constant while doing a magnification of the curve with a factor $1/h$, the second hypothesis states that the size $H(h)$ of a code tends to infinity. Finally, and informally speaking, the last hypothesis states that the function p applied to a code ω must return a value close to the diameter[3] of the subset of $\mathcal{D}^O(g, h)$ associated to ω.

3   Sparse Estimators

In this section, we introduce a new notion, derived from semi-local estimators. Yet, on the contrary to semi-local estimators, we discard the information given by the codes ω_i but their extremities. It is as if we had two resolutions, one for the space (the abscissas), one for the calculus (the ordinates).

We have noted earlier that the hypotheses about semi-local estimators in [4] are ambiguous. May be for the same reasons than Daurat *et al.*, we are tempted to do so. Indeed, in all of our proofs, we do not need the "1" in the 1-Lipschitz hypothesis. But from a practical point of view, k-Lipschitz function for $k > 1$

---

[3] The maximal Euclidean distance between two points of the subset.

may give non 8-connected digitization and it does not make a lot of sense to measure the length of a set of disconnected points (though we could define a discrete curve as the curve, in the usual mathematical sense, of a function from $\mathbb{Z}$ to $\mathbb{Z}$). Hence, in the following definition, as in the statement of our theorems, we assume a 1-Lipschitz function while we intentionally forget the "1" in the statements of the lemmas.

3.1 Definition

Definition 1. *Let* $H \colon {]0, +\infty[} \to \mathbb{N}^*$ *such that* $\lim_{h \to 0} H(h) = +\infty$ *and* $\lim_{h \to 0} h H(h) = 0$. *We say that* H *is sparsity function. Let* $g : [a, b] \to \mathbb{R}$ *be a 1-Lipschitz function. The* H-*sparse estimator of the length of the curve* $\mathcal{C}(g)$ *is defined by*

$$L^{Sp}(g, h) = h \sum_{i=0}^{N_h} \|\omega_i\|$$

where $\omega_i \in \{0, 1, 7\}^{H(h)}$ *for* $i \neq N_h$, $\omega_{N_h} \in \{0, 1, 7\}^j$ *with* $j \in (0, H(h)]$ *and the concatenation of the words* ω_i *equals* $\mathcal{F}(g, h)$.

An Illustration is given Figure 1.

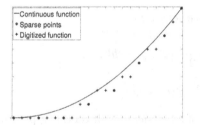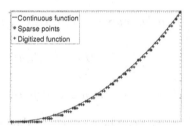

Fig. 1. Sparse estimation at two resolutions

3.2 Convergence

In this section, we establish that the sparse length estimators are convergent for 1-Lipschitz functions. Moreover, Theorem 2 gives a bound on the error at grid spacing h for functions of class C^2.

Let $m = h H(h)$ and A, B be resp. the minimum and the maximum of the integer interval $\{k \in \mathbb{N} \mid kh \in [a, b]\}$. The proof of Theorem 2 relies on two lemmas. The first one evaluates the difference between the length of the curve $\mathcal{C}(g)$ and the length of the curve of the piecewise affine function g_m defined on $[Ah, Bh]$ by $g_m(Ah + km) = g(Ah + km)$ $(k \in \mathbb{N})$ and $g_m(Bh) = g(Bh)$. The second lemma evaluates the difference between $\mathrm{L}(g_m)$ and the length of the piecewise affine function g_m^h defined on $[Ah, Bh]$ by $g_m^h(Ah + km) = \lfloor g_m(Ah + km) \rfloor_h = \lfloor g(Ah + km) \rfloor_h$ $(k \in \mathbb{N})$ and $g_m^h(Bh) = \lfloor g(Bh) \rfloor_h$. Figure 2 shows the three functions g, g_m, g_m^h on an interval $[Ah + km, Ah + (k + 1)m]$.

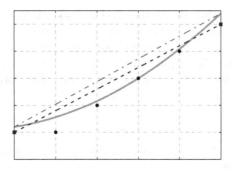

Fig. 2. The two parts of the estimation error: the curve g (in green, solid) to its chord g_m (in magenta, dotted-dashed) then the curve chord to the chord g_m^h (in blue, dashed) of the digitized curve $\mathcal{D}^O(g, h)$ (black points)

Lemma 1. *Let g be a Lipschitz function.*

– *For any sparsity function H, we have*

$$\lim_{h \to 0} L(g_m) = L(g).$$

– *If furthermore g is of class C^2, we have for any h*

$$|L(g_m) - L(g)| \leq m \frac{b - a}{2} \|\varphi'\|_\infty + 2h \|\varphi\|_\infty \tag{2}$$

where the function φ is defined on $\mathbb{R}$ by $\varphi(t) = \sqrt{1 + g'(t)^2}$.

Lemma 1 can be seen as an adaptation of a classical result on the approximation of a curve by its chords, the difficulty comes from the 1-Lipschitz hypothesis and the fixed sparsity step $H(h)$.

Lemma 2. *Let f_1 and f_2 be two piecewise affine functions defined on $[c, d] \subset \mathbb{R}$ with a common subdivision having p steps. Suppose that $f_1 \leq f_2$ and $\|f_1 - f_2\|_\infty \leq e$ for some $e \in \mathbb{R}$. Then*

$$|L(f_1) - L(f_2)| \leq pe.$$

Theorem 2 relies on Lemma 1 and Lemma 2 which is applied to the piecewise affine functions g_m and g_m^h, taking $e = h$.

Theorem 2. *Let H be a sparsity function and $g : [a, b] \to \mathbb{R}$ a 1-Lipschitz function. Then, the estimator L^{SP} converge toward the length of the curve $C(g)$. Furthermore, if g is of class C^2, we have*

$$L(g) - L^{\mathrm{SP}}(g, h) = \mathcal{O}(hH(h)) + \mathcal{O}\left(\frac{1}{H(h)}\right). \tag{3}$$

Formula 3 shows two opposite trends for the determination of the sparsity step $H(h)$: $\mathcal{O}(hH(h))$ – the discretization error – corresponds to the curve sampling error and tends to reduce the step $H(h)$ while $\mathcal{O}\left(\frac{1}{H(h)}\right)$ – the quantization error – corresponds to the error due to the quantization of the sample points and tends to extend the step. The optimal convergence speed in $h^{\frac{1}{2}}$ is then obtained taking $H(h) = \Theta(h^{-\frac{1}{2}})$. Thus, only one in about $h^{-\frac{1}{2}}$ value is needed to make a sparse estimation (which justifies the adjective sparse). Then, the complexity in the optimal case is a $\mathcal{O}(r^{\frac{1}{2}})$.

3.3   Concave Functions

In this section, we assume besides that the function g is differentiable and concave on $[a, b]$. Under these hypotheses, we can improve the bound on the convergence speed of the estimated length toward the true length of the curve $\mathcal{C}(g)$. The functions g_m and g_m^h are those defined in Section 3.2. Lemmas 3 and 4 are improvements of Lemmas 1 and 2 for concave curves. Figure 3 shows some experiments that illustrate the convergence rate obtained with Theorem 3.

Lemma 3. *If g is of class C^2 and $g'' \leq 0$ on $[a, b]$, then*

$$|L(g) - L(g_m)| \leq \frac{(b-a)\|g''\|_\infty^2}{8} m^2 + 2h\|\varphi\|_\infty \tag{4}$$

where the function φ is defined on $\mathbb{R}$ by $\varphi(t) = \sqrt{1 + g'(t)^2}$.

The right part of Inequality (4) contains two terms as in Inequality (2) but only the first term has been improved with m becoming m^2. The second term standing for the error on the edges remains the same.

Lemma 4.

$$\left|L(g_m) - L(g_m^h)\right| \leq (b-a)\,\frac{1}{H(h)^2} + h\|g'\|_\infty.$$

From Lemma 4 and Lemma 3, we derive the following bound on the speed of convergence when the function g is concave.

Theorem 3. *Let H be a sparsity function and $g : [a, b] \to \mathbb{R}$ a concave 1-Lipschitz function of class C^2. Then, we have*

$$L(g) - L^{Sp}(g, h) = \mathcal{O}(h^2 H(h)^2) + \mathcal{O}\left(\frac{1}{H(h)^2}\right).$$

Concavity allows squarring each term (compared to Theorem 2), which does not change the optimal size for $H(h)$ but improves the optimal convergence speed up to h.

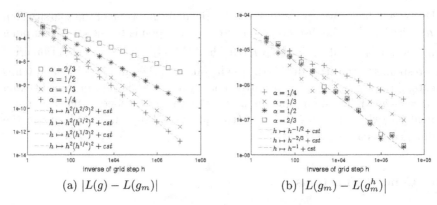

(a) $|L(g) - L(g_m)|$ (b) $|L(g_m) - L(g_m^h)|$

Fig. 3. Experimental convergence rates. We have computed the length of the curve $y = \ln(x)$, $x \in [1,2]$, using the sparse estimators defined by $H(h) = \lfloor h^{-\alpha} \rfloor$ where $\alpha \in \{\frac{1}{4}, \frac{1}{3}, \frac{1}{2}, \frac{2}{3}\}$, for the resolutions defined by $r = \lfloor 1.5^{1+3n} \rfloor$, $n \in [0,13]$. (a) Discretization error (the errors on the left and the right bounds of the interval have been withdrew). We observe the convergence in $\mathcal{O}(h^2 H(h)^2)$ which appears in Theorem 3. (b) Quantization error. For $\alpha \in \{\frac{1}{4}, \frac{1}{3}, \frac{1}{2}\}$, we observe the convergence is a $\mathcal{O}(1/H(h)^2)$, which appears in Theorem 3. For $\alpha = \frac{2}{3}$, the condition *(iii)* of Prop. 1 is satisfied and thus the piecewise affine function g_m^h is concave. Hence, we can observe that the convergence is a $\mathcal{O}(h)$ as deduced from Lemma 5.

3.4   Strong Concavity

When the function g is concave, the piecewise affine function g_m is clearly also concave. Nevertheless, the second piecewise function g_m^h is not necessary concave even on the sub-domain $J_h = [Ah, Ah + N_0 m]$ where N_0 is the greatest integer such that $J_h \subseteq [a,b]$. Indeed, we exhibit in Appendix A a function g that is concave and for which the function g_m^h is nonconcave for any h below some threshold. This section gives some sufficient conditions for g_m^h to be also concave and studies the consequences on the convergence speed of such an assumption.

Proposition 1. *Let H be a sparsity function and $g : [a,b] \to \mathbb{R}$ a concave function of class C^2. If one of the following condition holds, then there exists $h_0 > 0$ such that, for any $h < h_0$, the piecewise affine function g_m^h is concave on J_h.*

(i) $H(h) = h^{-\frac{1}{2}}$ and $\max(g'') < -1$.
(ii) $H(h) = h^{-\frac{1}{2}}$ and $g(x) = ax^2 + bx + c$ where $a \leq -\frac{1}{2}$.
(iii) $hH(h)^2 \to +\infty$ as $h \to 0$ and $\max(g'') < 0$.

The following lemma is an improvement of Lemma 4 for two concave piecewise affine functions.

Lemma 5. *Let f_1 and f_2 be two concave piecewise affine functions with the same monotonicity defined on $[c,d] \subset \mathbb{R}$ such that $f_1 \leq f_2$ and $\|f_1 - f_2\|_\infty \leq e$ for some $e \in \mathbb{R}$. Then*

$$|L(f_1) - L(f_2)| \leq B\,e.$$

where B is a constant related to the slope of f_1 and f_2 at c.

Corollary 1. *Let H be a sparsity function and $g : [a, b] \to \mathbb{R}$ a concave function of class C^2. If, for some $h_0 > 0$, the function g_m^h is concave on J_h for any $h < h_0$, then we have*

$$L(g) - L^{Sp}(g, h) = \mathcal{O}(h^2 H(h)^2) + \mathcal{O}(h).$$

From Corollary 1, it follows that, to speed up the convergence, we shall take the smallest sparsity step $H(h)$ provided the hypothesis about the concavity is satisfied. According to Proposition 1, this should lead us to choose the function H such that H dominates $h^{-\frac{1}{2}}$ as $h \to 0$. For instance, we can take $H(h) = h^{-\frac{1}{2}-\varepsilon}$ where $\varepsilon > 0$ and $\varepsilon \approx 0$. Then, the convergence speed is $h^{1-2\varepsilon}$. Note that h is a minimal error bound that cannot be improved in the general case since for the function g defined by $g(x) = (\frac{19}{48})^2 - x^2$, $x \in [\frac{1}{16}, \frac{19}{48}]$, we have shown that $L(g) - L^{Sp}(g, h) \geq 0.06h$ (see Appendix B).

4   Conclusion

In this article, we have studied some convergence properties of a class of semi-local length estimators in the concave and the general cases. These estimators need few information about the curve: the proportion of points of the curve used for the computation tends to 0 as the resolution tends toward infinity. That is why we propose to call them sparse estimators. In a future work, we plan to extend our estimators to the nD Euclidean space to compute k-volumes, $k < n$. We have also to study how the material presented in this article behave with Jordan curves obtained as boundary of solid objects through various discretization schemes. Furthermore, the definition of the sparse estimators relies on Jordan's definition for curve length. It would be interesting to keep the main idea from these estimators while relying on the more general definition of Minkowski (as in [2]). This could be more realistic in the framework of multigrid convergence, since physic objects cannot be considered as smooth (nor convex, etc.) at any resolution. Another extension of this work is to check whether the proofs of convergence obtained for sparse estimators can help to obtain new proofs for the convergence of adaptative length estimators as the MDSS. This could lead to the definition of a larger class of geometric feature estimators including sparse estimators and MDSS. Eventually, there is a need to find how to estimate the resolution of a given curve.

A   Strong Concavity: Counterexamples

In this appendix, we show that a piecewise affine function can be concave and its digitization, beyond some resolution, never concave (that is, the piecewise affine function g_m^h defined in Sec. 3.2 is not concave for grid spacing h below some threshold). The first counterexample uses a local estimator and the second one uses a sparse estimator. Both counterexamples rely on the following theorem proved in [18] (in fact, an extended version of the theorem is needed for the second counterexample). This theorem asserts that, given a function $x \mapsto ax^2 + bx + c$, the distribution in $[0, h]$ of the values of the expression $\{a(kh)^2 + b(kh) + c\}_h$, $k \in \mathbb{N}$, which are the errors resulting from the quantization, tends toward the equidistribution.

Theorem 4 ([18, Lemma 2 and Prop. 3]). *Let $a, b \in \mathbb{R}$, $a < b$. Let $g : [a, b] \to \mathbb{R}$ be a polynomial function of degre 2 with derivative in $[0, 1]$. Then, for all $[u, v] \subseteq [0, 1]$,*

$$\lim_{h \to 0} \frac{\text{card}\{x \in h\mathbb{N} \cap [a, b] \mid \{g(x)\}_h \in [hu, hv]\}}{\text{card}(h\mathbb{N} \cap [a, b])} = v - u.$$

For the first counterexample, we digitize the parabola associated to the function $g(x) = 2x - x^2$, $x \in [0, 1]$ and we split this parabola into segments of size $5h$. Thanks to Theorem 4, we prove that, for each grid spacing h below some threshold, we can choose an integer p for which the fractional part $\{g_m^h(ph)\}_h$ is such that the finite difference $g_m^h((p+5)h) - g_m^h(ph)$ is less than or equal to the grid spacing h while the finite difference $g_m^h((p+10)h) - g_m^h((p+5)h)$ is greater than or equal to twice the grid spacing h. Thus, the function g_m^h is not concave on $[0, 1]$.

For the second counterexample, we discretize the parabola $y = g(x) = \frac{1}{50}(2x - x^2)$, $x \in [0, 1]$ and we use segments of size $H(h) = \left\lfloor h^{-\frac{1}{2}} \right\rfloor$. Again, we have shown that there exists $h_0 > 0$ such that for any h, $0 < h < h_0$, there exists an integer p for which the slope of g_m^h on $[ph, ph + H(h)h]$ is greater than its slope on $[ph + H(h)h, ph + 2H(h)h]$.

B   Inferior Bound for the Method Error in the Concave Case

We give an inferior bound on the difference between the true length $L(g)$ of the parabola $y = g(x) = (\frac{19}{48})^2 - x^2$ for $x \in [\frac{1}{16}, \frac{19}{48}]$ and the length $L^{\text{Sp}}(g, h)$, obtained with the sparse estimator defined by the sparsity function $H(h) = \left\lfloor h^{-\frac{1}{2}} \right\rfloor$. Let g_m and g_m^h be the piecewise affine functions defined in Section 3.2. Then the lengths of their curves satisfy $L(g_m^h) + 0.06h \leq L(g_m) \leq L(g)$ for any $h = (12(8p + 1))^{-2}$ where $p \in \mathbb{N}$. Moreover, the bounds of the interval $[\frac{1}{16}, \frac{19}{48}]$ are multiples of h. Hence, there is no error due to the bounds. Eventually, for any $p \in \mathbb{N}$ and $h = (12(8p + 1))^{-2}$, we get $L(g) - L^{\text{Sp}}(g, h) \geq 0.06h$.

References

1. Asano, T., Kawamura, Y., Klette, R., Obokata, K.: Minimum-length polygons in approximation sausages. In: Arcelli, C., Cordella, L.P., Sanniti di Baja, G. (eds.) IWVF 2001. LNCS, vol. 2059, pp. 103–112. Springer, Heidelberg (2001)

2. Coeurjolly, D.: Algorithmique et géométrie discrète pour la caractérisation des courbes et des surfaces. Ph.D. thesis, Université Lyon 2 (2002)
3. Coeurjolly, D., Klette, R.: A comparative evaluation of length estimators of digital curves. IEEE Trans. Pattern Anal. Mach. Intell. 26(2), 252–257 (2004)
4. Daurat, A., Tajine, M., Zouaoui, M.: Les estimateurs semi-locaux de périmètre. Tech. rep., LSIIT CNRS, UMR 7005, Université de Strasbourg (2011), http://hal.inria.fr/hal-00576881
5. Daurat, A., Tajine, M., Zouaoui, M.: Patterns in discretized parabolas and length estimation. In: Brlek, S., Reutenauer, C., Provençal, X. (eds.) DGCI 2009. LNCS, vol. 5810, pp. 373–384. Springer, Heidelberg (2009)
6. De Vieilleville, F., Lachaud, J.O., Feschet, F.: Convex digital polygons, maximal digital straight segments and convergence of discrete geometric estimators. Journal of Mathematical Imaging and Vision 27(2), 139–156 (2007)
7. Debled-Rennesson, I., Reveillès, J.P.: A linear algorithm for segmentation of digital curves. International Journal of Pattern Recognition and Artificial Intelligence 09(04), 635–662 (1995)
8. Dorst, L., Smeulders, A.W.: Discrete straight line segments: parameters, primitives and properties. In: Vision Geometry: Proceedings of an AMS Special Session Held October 20-21, 1989,[at Hoboken, New Jersey], vol. 119, pp. 45–62 (1991)
9. Freeman, H.: On the encoding of arbitrary geometric configurations. IRE Transactions on Electronic Computers EC-10(2), 260–268 (1961)
10. Klette, R., Rosenfeld, A.: Digital Geometry. Morgan Kaufmann (2004)
11. Klette, R., Žunić, J.: Multigrid convergence of calculated features in image analysis. Journal of Mathematical Imaging and Vision 13(3), 173–191 (2000)
12. Kovalevsky, V.: New definition and fast recognition of digital straight segments and arcs. In: Proceedings of the 10th International Conference on Pattern Recognition, vol. ii, pp. 31–34 (1990)
13. Kulkarni, S.R., Mitter, S.K., Richardson, T.J., Tsitsiklis, J.N.: Local versus non-local computation of length of digitized curves. IEEE Trans. Pattern Anal. Mach. Intell. 16(7), 711–718 (1994)
14. Lachaud, J.O.: Espaces non-euclidiens et analyse d' image: modèles déformables riemanniens et discrets, topologie et géométrie discrète. Habilitation á diriger des recherches, Université de Bordeaux 1 (Décembre 2006)
15. Mazo, L., Baudrier, E.: About multigrid convergence of some length estimators (extended version). Tech. rep., ICube, University of Strasbourg, CNRS (2014), http://hal.inria.fr/hal-00990694
16. Sloboda, F., Zatko, B., Stoer, J.: On approximation of planar one-dimensional continua. In: Advances in Digital and Computational Geometry, pp. 113–160 (1998)
17. Tajine, M., Daurat, A.: On local definitions of length of digital curves. In: Nyström, I., Sanniti di Baja, G., Svensson, S. (eds.) DGCI 2003. LNCS, vol. 2886, pp. 114–123. Springer, Heidelberg (2003)
18. Tajine, M., Daurat, A.: Patterns for multigrid equidistributed functions: Application to general parabolas and length estimation. Theoretical Computer Science 412(36), 4824–4840 (2011)

Non-additive Bounded Sets of Uniqueness in $\mathbb{Z}^n$

Sara Brunetti[1], Paolo Dulio[2,*], and Carla Peri[3]

[1] Dipartimento di Ingegneria dell'Informazione e Scienze Matematiche,
Via Roma, 56, 53100 Siena, Italy
sara.brunetti@unisi.it

[2] Dipartimento di Matematica "F. Brioschi", Politecnico di Milano,
Piazza Leonardo da Vinci 32, I-20133 Milano, Italy
paolo.dulio@polimi.it

[3] Università Cattolica S. C.,
Via Emilia Parmense, 84, 29122 Piacenza, Italy
carla.peri@unicatt.it

Abstract. A main problem in discrete tomography consists in looking for theoretical models which ensure uniqueness of reconstruction. To
this, lattice sets of points, contained in a multidimensional grid $\mathcal{A} =
[m_1] \times [m_2] \times \cdots \times [m_n]$ (where for $p \in \mathbb{N}$, $[p] = \{0, 1, ..., p - 1\}$), are
investigated by means of X-rays in a given set S of lattice directions.
Without introducing any noise effect, one aims in finding the minimal
cardinality of S which guarantees solution to the uniqueness problem.

In a previous work the matter has been completely settled in dimension two, and later extended to higher dimension. It turns out that $d + 1$
represents the minimal number of directions one needs in $\mathbb{Z}^n$ ($n \geq d \geq 3$),
under the requirement that such directions span a d-dimensional subspace of $\mathbb{Z}^n$. Also, those sets of $d + 1$ directions have been explicitly
characterized.

However, in view of applications, it might be quite difficult to decide
whether the uniqueness problem has a solution, when X-rays are taken
in a set of more than two lattice directions. In order to get computational simpler approaches, some prior knowledge is usually required on
the object to be reconstructed. A powerful information is provided by
additivity, since additive sets are reconstructible in polynomial time by
using linear programming.

In this paper we compute the proportion of non-additive sets of unique-
ness with respect to additive sets in a given grid $\mathcal{A} \subset \mathbb{Z}^n$, in the important
case when d coordinate directions are employed.

Keywords: Additive set, bad configuration, discrete tomography, non-
additive set, uniqueness problem, X-ray.

1 Introduction

One of the main problems of discrete tomography is to determine finite subsets
of the integer lattice $\mathbb{Z}^n$ by means of their X-rays taken in a finite number of

---

* Corresponding author.

E. Barcucci et al. (Eds.): DGCI 2014, LNCS 8668, pp. 226–237, 2014.

lattice directions. Given a lattice direction u, the X-ray of a finite lattice set E in the direction u counts the number of points in E on each line parallel to u. The points in E can model the atoms in a crystal, and new techniques in high resolution transmission electron microscopy allow the X-rays of a crystal to be measured so that the main goal of discrete tomography is to use these X-rays to deduce the local atomic structure from the collected counting data, with a view to applications in the material sciences. The high energies required to produce the discrete X-rays of a crystal mean that only a small number of X-rays can be taken before the crystal is damaged. Therefore, discrete tomography focuses on the reconstruction of binary images from a small number of X-rays.

In general, it is hopeless to obtain uniqueness results unless the class of lattice sets is restricted. In fact, for any fixed set S of more than two lattice directions, to decide whether a lattice set is uniquely determined by its X-rays along S is NP-complete [9]. Thus, one has to use a priori information, such as convexity or connectedness, about the sets that have to be reconstructed (see for instance [8], where convex lattice sets are considered). An important class, which provides a computational simpler approach to the uniqueness and reconstruction problems, is that of additive sets introduced by P.C. Fishburn and L.A. Shepp in [7] (see the next section for all terminology). Additive sets with respect to a finite set of lattice directions are uniquely determined by their X-rays in the given directions, and they are also reconstructible in polynomial time by use of linear programming. The notions of additivity and uniqueness are equivalent when two directions are employed, whereas, for three or more directions, additivity is more demanding than uniqueness, as there are non-additive sets which are unique [6]. More recently, additivity have been reviewed and settled by a more general treatment in [10]. Thanks to this new approach, the authors showed that there are non-additive lattice sets in $\mathbb{Z}^3$ which are uniquely determined by their X-rays in the three standard coordinate directions by exhibiting a counter-example (see [10, Remark 2 and Figure 2]). This answers in the negative a question raised by Kuba at a conference on discrete tomography in Dagsthul (1997), that every subset E of $\mathbb{Z}^3$ might be uniquely determined by its X-rays in the three standard unit directions of $\mathbb{Z}^3$ if and only if E is additive.

In previous works we restricted our attention to bounded sets, i.e. lattice sets contained in a given grid $\mathcal{A} = [m_1] \times [m_2] \times \cdots \times [m_n]$. In particular, in [2] we addressed the problem in dimension two and we proved that for a given set S of four lattice directions, there exists a rectangular grid $\mathcal{A}$ such that all the subsets of $\mathcal{A}$ are uniquely determined by their X-rays in the directions in S. In [4] we extended the previous uniqueness results to higher dimensions, by showing that $d + 1$ represents the minimal number of directions one needs in $\mathbb{Z}^n$ ($n \geq d \geq 3$), under the requirement that such directions span a d-dimensional subspace of $\mathbb{Z}^n$. Also, those sets of $d + 1$ directions have been explicitly characterized.

We also recall that Fishburn et al. [7] noticed that an explicit construction of non-additive sets of uniqueness has proved rather difficult even though it might be true that non-additive uniqueness is the rule rather than exception. In particular they suggest that for some set of X-ray directions of cardinality larger than two, the proportion of lattice sets E of uniqueness that are not also additive approaches 1 as E gets large. They leave it as an open question in the discussion section. In [3] we presented a procedure for constructing non-additive sets in $\mathbb{Z}^2$ and we showed that when S contains the coordinate directions this proportion does not depend on the size of the lattice sets into consideration.

In the present paper we focus on non-additive sets in $\mathbb{Z}^n$ and estimate the proportion of non-additive sets of uniqueness with respect to additive sets in a given grid $\mathcal{A}$, when the set S contains d coordinate directions (see Theorem 1). It turns out that such proportion tends to zero as $\mathcal{A}$ gets large so that the probability to have an additive set is high. From the viewpoint of the applications, this suggest the use of linear programming for good quality solutions as the reconstruction problem for additive sets is polynomial.

2   Definitions and Preliminaries

The standard orthonormal basis for $\mathbb{Z}^n$ will be $\{e_1, \ldots, e_n\}$, and the coordinates with respect to this orthonormal basis $x_1, \ldots, x_n$. A vector $u = (a_1, \ldots, a_n) \in \mathbb{Z}^n$, where $a_1 \geq 0$, is said to be a *lattice direction*, if $\gcd(a_1, ..., a_n) = 1$. We refer to a finite subset E of $\mathbb{Z}^n$ as a *lattice set*, and we denote its cardinality by $|E|$. For a finite set $S = \{u_1, u_2, ..., u_m\}$ of directions in $\mathbb{Z}^n$, the *dimension* of S, denoted by $\dim S$, is the dimension of the vector space generated by the vectors $u_1, u_2, ..., u_m$. Moreover, for each $I \subseteq S$, we denote $u(I) = \sum_{u \in I} u$, with $u(\emptyset) = 0 \in \mathbb{Z}^n$. Any two lattice sets E and F are *tomographically equivalent* if they have the same X-rays along the directions in S. Conversely, a lattice set E is said to be *S-unique* if there is no lattice set F different from but tomographically equivalent to E.

An *S-weakly bad configuration* is a pair of lattice sets (Z, W) each consisting of k lattice points not necessarily distinct (counted with multiplicity), $z_1, ..., z_k \in Z$ and $w_1, ..., w_k \in W$ such that for each direction $u \in S$, and for each $z_r \in Z$, the line through z_r in direction u contains a point $w_r \in W$. If all the points in each set Z, W are distinct (multiplicity 1), then (Z, W) is called an *S-bad configuration*. If for some $k \geq 2$ an S-(weakly) bad configuration (Z, W) exists such that $Z \subseteq E$, $W \subseteq \mathbb{Z}^n \backslash E$, we then say that a lattice set E has an *S-(weakly) bad configuration*. This notion plays a crucial role in investigating uniqueness problems, since a lattice set E is S-unique if and only if E has no S-bad configurations [7].

For $p \in \mathbb{N}$, denote $\{0, 1, ..., p - 1\}$ by $[p]$. Let $\mathcal{A} = [m_1] \times [m_2] \times \cdots \times [m_n]$ be a fixed lattice grid in $\mathbb{Z}^n$. We shall restrict our considerations to lattice sets contained in a given lattice grid $\mathcal{A}$, referred to as *bounded sets*. We say that a set S is a *valid* set of directions for $\mathcal{A}$, if for all $i \in \{1, \ldots, n\}$, the sum h_i of the absolute values of the i-th coordinates of the directions in S satisfies the condition $h_i < m_i$. Notice that this definition excludes trivial cases when S contains a direction with so large (or so small) slope, with respect to $\mathcal{A}$, such that each line with this slope meets $\mathcal{A}$ in no more than a single point. If each subset $E \subset \mathcal{A}$ is S-unique in $\mathcal{A}$, we then say that S is a *set of uniqueness* for $\mathcal{A}$. For our purpose, we define additivity in terms of solutions of linear programs.

The reconstruction problem can be formulated as an integer linear program (ILP). Since the NP-hardness of the reconstruction problem for more than two directions reflects in the integrality constraint, relaxation of ILP are considered (see, for instance [1], [11],[14],[15]). In this setting, a lattice set is *S-additive* if it is the unique solution of the relaxed linear program (LP). Moreover, a set is *S*-additive if and only if it has no *S*-weakly bad configurations. Additivity is also fundamental for treating uniqueness problems, due to the following facts (see [5, Theorem 2]):

1. Every *S-additive* set is *S-unique*.
2. There exist *S-unique* sets which are not *S-additive*.

A set which is not *S*-additive will be simply said *non-additive*, when confusion is not possible.

In [2] we characterized all the minimal sets S of planar directions which are sets of uniqueness for $\mathcal{A}$, and in [4] we studied the problem in higher dimension. In particular, we stated the following necessary condition on minimal sets S of lattice directions to be sets of uniqueness for $\mathcal{A}$.

Proposition A ([4, Proposition 8]). *Let $S \subset \mathbb{Z}^n$ be a set of distinct lattice directions such that $|S| = d + 1$ and $\dim S = d \geq 3$ ($n \geq d \geq 3$). Suppose that S is a valid set of uniqueness for a finite grid $\mathcal{A} = [m_1] \times [m_2] \times \cdots \times [m_n] \subset \mathbb{Z}^n$. Then S is of the form*

$$S = \{u_1, ..., u_d, w = u(I) - u(J)\}, \tag{1}$$

where the vectors $u_1, ..., u_d$ are linearly independent, and I, J are disjoint subsets of $\{u_1, ..., u_d\}$ such that $|I| \equiv |\{w\} \cup J| \pmod 2$.

Examples of sets S of the form (1) which are contained in $\mathbb{Z}^3$ and $\mathbb{Z}^4$ are presented in Subsections 3.1 and 3.2, respectively.

Among the sets S of the form (1) we then specified which ones are sets of uniqueness for $\mathcal{A}$, by employing an algebraic approach introduced by Hajdu and Tijdeman in [12]. To illustrate the result we need some further definitions (see also [4]).

For a vector $u = (a_1, \ldots, a_n) \in \mathbb{Z}^n$, we simply write x^u in place of the monomial $x_1^{a_1} x_2^{a_2} \ldots x_n^{a_n}$. Consider now any lattice vector $u \in \mathbb{Z}^n$, where $u \neq 0$. Let $u_- \in \mathbb{Z}^n$ be the vector whose entries equal the corresponding entries of u if negative, and are 0 otherwise. Analogously, let $u_+ \in \mathbb{Z}^n$ be the vector whose entries equal the corresponding entries of u if positive, and are 0 otherwise.

For any finite set S of lattice directions in $\mathbb{Z}^n$, we define the polynomial

$$F_S(x_1, \ldots, x_n) = \prod_{u \in S} \left(x^{u_+} - x^{-u_-} \right). \tag{2}$$

For example, for $S = \{e_1, e_2, e_3, e_1 + e_2 - e_3\} \subset \mathbb{Z}^3$ we get

$F_S(x_1, x_2, x_3) = (x_1 - 1)(x_2 - 1)(x_3 - 1)(x_1 x_2 - x_3) = -x_1 x_2 x_3^2 + x_2 x_3^2 + x_1 x_3^2 - x_3^2 + \\ + x_1^2 x_2^2 x_3 - x_1 x_2^2 x_3 - x_1^2 x_2 x_3 + 2 x_1 x_2 x_3 - x_2 x_3 - x_1 x_3 + x_3 - x_1^2 x_2^2 + x_1 x_2^2 + x_1^2 x_2 - x_1 x_2.$

Given a function $f : \mathcal{A} \to \mathbb{Z}$, its *generating function* is the polynomial defined by

$$G_f(x_1, \ldots, x_n) = \sum_{(a_1,\ldots,a_n) \in \mathcal{A}} f(a_1, \ldots, a_n) x_1^{a_1} \ldots x_n^{a_n}.$$

Conversely, we say that the function f is *generated* by a polynomial $P(x_1, \ldots, x_n)$ if $P(x_1, \ldots, x_n) = G_f(x_1, \ldots, x_n)$. Notice that the function f generated by the polynomial $F_S(x_1, \ldots, x_n)$ vanishes outside $\mathcal{A}$ if and only if the set S is valid for $\mathcal{A}$.

Furthermore, to a monomial $kx_1^{a_1} x_2^{a_2} \ldots x_n^{a_n}$ we associate the lattice point $z = (a_1, \ldots, a_n) \in \mathbb{Z}^n$, together with its *weight* k. We say that a point $(a_1, \ldots, a_n) \in \mathcal{A}$ is a *multiple positive point* for f (or G_f) if $f(a_1, \ldots, a_n) > 1$. Analogously, $(a_1, \ldots, a_n) \in \mathcal{A}$ is said to be a *multiple negative point* for f if $f(a_1, \ldots, a_n) < -1$. Such points are simply referred to as *multiple points* when the signs are not relevant. For a polynomial $P(x_1, \ldots, x_n)$ we denote by P^+ (resp. P^-) the set of lattice points corresponding to the monomials of $P(x_1, \ldots, x_n)$ having positive (resp. negative) sign, referred to as *positive* (resp. *negative*) lattice points. We also write $P = P^+ \cup P^-$.

The *line sum* of a function $f : \mathcal{A} \to \mathbb{Z}$ along the line $x = x_0 + tu$, passing through the point $x_0 \in \mathbb{Z}^n$ and with direction u, is the sum $\sum_{x = x_0 + tu, x \in \mathcal{A}} f(x)$. Further, we denote $\|f\| = \max_{x \in \mathcal{A}} \{|f(x)|\}$. We can easily check that the function f generated by $F_S(x_1, \ldots, x_n)$ has zero line sums along the lines taken in the directions belonging to S.

Hajdu and Tijdeman proved that if $g : \mathcal{A} \to \mathbb{Z}$ has zero line sums along the lines taken in the directions of S, then $F_S(x_1, \ldots, x_n)$ divides $G_g(x_1, \ldots, x_n)$ over $\mathbb{Z}$ (see [12, Lemma 3.1] and [13]). We recall that two functions $f, g : \mathcal{A} \subset \mathbb{Z}^n \to \{0, 1\}$ are *tomographically equivalent* with respect to a given finite set S of lattice directions if they have equal line sums along the lines corresponding to the directions in S. Note that two non trivial functions $f, g : \mathcal{A} \to \{0, 1\}$ which are tomographically equivalent can be interpreted as characteristic functions of two lattice sets which are tomographically equivalent. Further, the difference $h = f - g$ of f and g has zero line sums. Hence there is a one-to-one correspondence between S-bad configurations contained in $\mathcal{A}$ and non-trivial functions $h : \mathcal{A} \to \mathbb{Z}$ having zero line sums along the lines corresponding to the directions in S and $\|h\| \le 1$.

Let us consider a set $S = \{u_1, \ldots, u_d, w = u(I) - u(J)\}$, where I, J are disjoint subsets of $\{u_1, \ldots, u_d\}$. We define

$$D = \{\pm v : v = u(X) - u(I) \ne 0, X \subseteq I \cup J \cup \{w\}\}. \tag{3}$$

In [4] we proved the following result.

Theorem B ([4, Theorem 12]). *Let $S \subset \mathbb{Z}^n$ be a set of distinct lattice directions such that $S = \{u_r = (a_{r1}, \ldots, a_{rn}) : r = 1, \ldots, d+1\}$ ($n \ge d \ge 3$), where $u_1, \ldots, u_d$ are linearly independent, $u_{d+1} = u(I) - u(J)$, and I, J are disjoint subsets of $\{u_1, \ldots, u_d\}$ such that $|I| \equiv |\{w\} \cup J| \pmod 2$. Suppose S is valid for the grid*

$\mathcal{A} = [m_1] \times [m_2] \times \cdots \times [m_n]$. *Denote* $\sum_{r=1}^{d+1} |a_{ri}| = h_i$, *for each* $i \in \{1, \ldots, n\}$. *Suppose that* $g : \mathcal{A} \to \mathbb{Z}$ *has zero line sums along the lines in the directions in* S, *and* $\|g\| \leq 1$. *Then* g *is identically zero if and only if for each* $v = (v_1, \ldots, v_n) \in D$, *there exists* $s \in \{1, \ldots, n\}$ *such that* $|v_s| \geq m_s - h_s$.

From the geometrical point of view, a set S of lattice directions is a set of uniqueness for a grid $\mathcal{A}$ if and only if S and $\mathcal{A}$ are chosen according to assumptions in Theorem B, and the resulting set D is such that its members satisfy the conditions of the theorem.

3   Non-additive Bounded Set of Uniqueness

In this section we study non-additive sets of uniqueness contained in a given grid $\mathcal{A}$, in the important case when S contains the coordinate directions. In [3] we showed that when $S \subset \mathbb{Z}^2$ contains the coordinate directions the proportion of lattice sets E of uniqueness that are not also additive does not depend on the size of the lattice sets into consideration and is given by

$$\frac{2}{2^{|F_S|} - 2},$$

where $|F_S|$ denotes the cardinality of the set of points corresponding to the polynomial $F_S(x_1, x_2)$.

In the present paper we aim to extend this estimate to higher dimension. Before presenting the general result, we wish to consider two preliminary cases, concerning $\mathbb{Z}^3$ and $\mathbb{Z}^4$ respectively, which motivate the general result presented below.

3.1   Non-additive Sets in $\mathbb{Z}^3$

Let us consider the case $n = d = 3$. Let $S = \{e_1, e_2, e_3, w = u(I) - u(J)\} \subset \mathbb{Z}^3$, where I, J are disjoint subsets of $\{e_1, e_2, e_3\}$ such that $|I| \equiv |\{w\} \cup J| \pmod 2$. Since w is a direction distinct from e_1, e_2, e_3, we have $e_1 \in I$ and, up to exchanging the role of e_2 and e_3, we have the following choices for w.

1. $w = e_1 + e_2 + e_3 = (1, 1, 1)$, where $I = \{e_1, e_2, e_3\}, J = \emptyset$.
2. $w = e_1 + e_2 - e_3 = (1, 1, -1)$, where $I = \{e_1, e_2, \}, J = \{e_3\}$.
3. $w = e_1 - e_2 - e_3 = (1, -1, -1)$, where $I = \{e_1\}, J = \{e_2, e_3\}$.

In order to apply Theorem B, we now evaluate the set D defined by (3), in all these cases.

1. If $I = \{e_1, e_2, e_3\}, J = \emptyset$, then by choosing $X = \{w, e_i\}$, for $i = 1, 2, 3$, we get $v = e_i \in D$.
2. If $I = \{e_1, e_2, \}, J = \{e_3\}$, then for $X = \{e_1, e_2, e_3\}$ we get $v = e_3 \in D$. For $X = \{e_1\}$ we get $v = -e_2 \in D$, and for $X = \{e_2\}$ we get $v = -e_1 \in D$.

3. If $I = \{e_1\}, J = \{e_2, e_3\}$, then for $X = \{e_1, e_i\}$ we get $v = e_i \in D$, where $i = 2, 3$. For $X = \{w, e_1, e_2, e_3\}$ we get $v = e_1 \in D$.

Since in all the previous cases we have $h_i = 2$, for $i = 1, 2, 3$, then the set S is a set of uniqueness for the grid $\mathcal{A} = [m_1] \times [m_2] \times [m_3]$ if and only if $m_i - h_i \leq 1$ for each $i = 1, 2, 3$, that is $m_i = 3$. This implies that $\mathcal{A}$ contains a unique S-weakly bad configuration given by F_S. The non-additive sets of uniqueness in F_S are precisely F_S^- and F_S^+. All the other subsets of F_S, are additive. Therefore, the proportion of bounded non-additive sets of uniqueness w.r.t. those additive is given by

$$\frac{2 \cdot 2^{|\mathcal{A} \setminus F_S|}}{2^{|\mathcal{A}|} - 2 \cdot 2^{|\mathcal{A} \setminus F_S|}} = \frac{2}{2^{|F_S|} - 2}. \tag{4}$$

3.2   Non-additive Sets in $\mathbb{Z}^4$

Let us now consider the case $n = 4$. We first note that the condition $|I| \equiv |\{w\} \cup J| \pmod 2$ in Theorem B implies that $|I \cup J|$ must be odd. Therefore, we have $|I \cup J| = 3$ and we can distinguish the following cases:

1. $n = 4 > d = |I \cup J| = 3$;
2. $n = 4 = d > |I \cup J| = 3$.

1. Suppose $n = 4 > d = |I \cup J| = 3$. Up to permutations of the standard orthonormal vectors, we can assume $S = \{e_1, e_2, e_3, w = u(I) - u(J)\} \subset \mathbb{Z}^4$, where I, J are disjoint subsets of $\{e_1, e_2, e_3\}$ such that $I \cup J = \{e_1, e_2, e_3\}$, and $|I| \equiv |\{w\} \cup J| \pmod 2$. By identifying $\mathbb{Z}^3$ with the subspace $H = \{(z_1, z_2, z_3, 0) : z_1, z_2, z_3 \in \mathbb{Z}\}$ in $\mathbb{Z}^4$, we can repeat the same considerations as in the previous subsection. Therefore, we have $S, D \subset H$ and the set S is a set of uniqueness for the grid $\mathcal{A} = [m_1] \times [m_2] \times [m_3] \times [m_4]$ if $m_i = 3$ for $i = 1, 2, 3$, and $m_4 \geq 1$. We shall simply write $m = m_4$. In this case the grid $\mathcal{A}$ can be arbitrary large in one direction and we shall compute the proportion of bounded non-additive sets of uniqueness w.r.t. those additive in $\mathcal{A}$ as a function of m. For the sake of simplicity, we assume $w = e_1 + e_2 + e_3 = (1, 1, 1, 0)$, since all the other cases are analogous. We have

$$F_S(x_1, x_2, x_3) = (x_1 - 1)(x_2 - 1)(x_3 - 1)(x_1 x_2 x_3 - 1),$$

and all the S-weakly bad configurations contained in $\mathcal{A}$ correspond to polynomials of the form $F_S(x_1, x_2, x_3)P(x_4)$, where $P(x_4)$ is a polynomial in x_4 with degree less than or equal to $m - 1$. Thus $P(x_4) = a_{m-1}x_4^{m-1} + \cdots + a_0$, where the coefficients $a_{m-1}, \cdots, a_0$ are not all zero.
Let us consider two polynomials

$$Q_1(x_1, x_2, x_3, x_4) = F_S(x_1, x_2, x_3)P_1(x_4) = F_S(x_1, x_2, x_3)\left(a_{m-1}x_4^{m-1} + \cdots + a_0\right),$$

$$Q_2(x_1, x_2, x_3, x_4) = F_S(x_1, x_2, x_3)P_2(x_4) = F_S(x_1, x_2, x_3)\left(b_{m-1}x_4^{m-1} + \cdots + b_0\right).$$

The corresponding sets of points Q_1, Q_2 are equal if and only if $a_i = 0$ implies $b_i = 0$, for all $i = 0, \cdots, m - 1$. Thus the number of S-weakly bad configurations contained in $\mathcal{A}$ equals the number of polynomials $P(x_4) = a_{m-1} x_4^{m-1} + \cdots + a_0$ whose coefficients belong to the set $\{0, 1\}$, except the null polynomial.

Let us denote by $\mathcal{F}$ the set of all points in $\mathcal{A}$ which belong to some S-weakly bad configuration. Notice that each S-weakly bad configuration contains two non-additive sets consisting of the set of positive (resp. negative) points. By multiplying the corresponding polynomial by -1, these two non-additive sets exchange each other. Therefore, the number of non-additive sets which are contained in $\mathcal{F}$ equals the number of polynomials $P(x_4) = a_{m-1} x_4^{m-1} + \cdots + a_0$ whose coefficients belong to the set $\{-1, 0, 1\}$, except the null polynomial. Thus we have $3^m - 1$ non-additive sets in $\mathcal{F}$. Any other non-additive set in $\mathcal{A}$ is obtained by adding some points of $\mathcal{A} \setminus \mathcal{F}$ to a non-additive set in $\mathcal{F}$. Therefore, the number of non-additive sets contained in $\mathcal{A}$ is given by $2^{|\mathcal{A} \setminus \mathcal{F}|} (3^m - 1)$.

The proportion of bounded non-additive sets of uniqueness in $\mathcal{A}$ with respect to those additive is given by

$$\frac{2^{|\mathcal{A} \setminus \mathcal{F}|} (3^m - 1)}{2^{|\mathcal{A}|} - 2^{|\mathcal{A} \setminus \mathcal{F}|} (3^m - 1)} = \frac{3^m - 1}{2^{|F_S| m} - 3^m + 1}. \tag{5}$$

For $m = 1$ we get

$$\frac{2}{2^{|F_S|} - 2},$$

as in (4). Moreover, as $\mathcal{A}$ gets large, we get

$$\lim_{m \to \infty} \frac{3^m - 1}{2^{|F_S| m} - 3^m + 1} = 0,$$

so that the set of non-additive sets is negligible.

2. Let us now suppose $n = 4 = d > |I \cup J| = 3$. Up to permutations of the standard orthonormal vectors, we can assume $S = \{e_1, e_2, e_3, e_4, w = u(I) - u(J)\} \subset \mathbb{Z}^4$, where I, J are disjoint subsets of $\{e_1, e_2, e_3\}$ such that $I \cup J = \{e_1, e_2, e_3\}$, and $|I| \equiv |\{w\} \cup J| \pmod{2}$. By identifying $\mathbb{Z}^3$ with the subspace $H = \{(z_1, z_2, z_3, 0) : z_1, z_2, z_3 \in \mathbb{Z}\}$ in $\mathbb{Z}^4$, we can repeat the same considerations as in the previous subsection. In particular, we have $h_i = 2$, for $i = 1, 2, 3$, $h_4 = 1$, and the set S is a set of uniqueness for the grid $\mathcal{A} = [m_1] \times [m_2] \times [m_3] \times [m_4]$ if and only if $m_i = 3$ for each $i = 1, 2, 3$, and $m_4 = m \geq 2$. Again we assume $w = e_1 + e_2 + e_3 = (1, 1, 1, 0)$, so that we have

$$F_S(x_1, x_2, x_3, x_4) = (x_1 - 1)(x_2 - 1)(x_3 - 1)(x_1 x_2 x_3 - 1)(x_4 - 1) = F_T(x_1, x_2, x_3)(x_4 - 1),$$

where $T = \{e_1, e_2, e_3, w\}$. All the S-weakly bad configurations contained in $\mathcal{A}$ correspond to polynomials of the form $F_S(x_1, x_2, x_3, x_4) P(x_4)$, where $P(x_4)$ is a polynomial in x_4 with degree less than or equal to $m - 2$. Thus

$P(x_4) = a_{m-2}x_4^{m-2} + \cdots + a_0$, where the coefficients $a_{m-2}, \cdots, a_0$ are not all zero. Denote by $\mathcal{F}$ the set of all points in $\mathcal{A}$ which belong to some S-weakly bad configuration. Then, by the same arguments as in the previous case we have that $\mathcal{F}$ contains $3^{m-1} - 1$ non-additive sets. Thus, we obtain the following estimate for the proportion of bounded non-additive sets of uniqueness in $\mathcal{A}$ with respect to those additive.

$$\frac{2^{|\mathcal{A}\setminus\mathcal{F}|}\left(3^{m-1} - 1\right)}{2^{|\mathcal{A}|} - 2^{|\mathcal{A}\setminus\mathcal{F}|}\left(3^{m-1} - 1\right)} = \frac{3^{m-1} - 1}{2^{|\mathcal{F}|} - (3^{m-1} - 1)} = \frac{3^{m-1} - 1}{2^{|F_T|m} - 3^{m-1} + 1}, \quad (6)$$

as $|\mathcal{F}| = m|F_T|$.
For $m = 2$ we get

$$\frac{2}{2^{|F_S|} - 2},$$

as $|F_S| = 2|F_T|$.
Again, as $\mathcal{A}$ gets large, we get

$$\lim_{m\to\infty} \frac{3^{m-1} - 1}{2^{15m} - 3^{m-1} + 1} = 0,$$

so that the set of non-additive sets is negligible.

3.3   Non-additive Sets in $\mathbb{Z}^n$

We now consider the general case. In the following, for $p, q \in \mathbb{N}$ with $1 \le p < q$, we denote $(p, q] = \{z \in \mathbb{N} : p < z \le q\}$. Further, to unify different cases, when $p = q$ we still adopt the notation $(p, q]$ with the convention that $\prod_{j\in(p,q]} z_j = 1$, for every $z_j \in \mathbb{Z}$.

Theorem 1. *Let $S = \{e_1, \ldots, e_d, w = u(I) - u(J)\}$ be a set of $d + 1$ distinct directions in $\mathbb{Z}^n$, where $n \ge d \ge 3$, $I \cup J = \{e_1, \ldots, e_k\}$ ($3 \le k \le d$), and $|I| \not\equiv |J| \pmod 2$. Let $\mathcal{A} = [m_1] \times [m_2] \times \cdots \times [m_n] \subset \mathbb{Z}^n$, where $m_i = 3$ for $i = 1, \ldots, k$, $m_i \ge 2$ for $i = k + 1, \ldots, d$, and $m_i \ge 1$ for $i = d + 1, \ldots, n$. Then the set S is a set of uniqueness for $\mathcal{A}$ and the proportion of non-additive sets of uniqueness in $\mathcal{A}$ with respect to those additive is given by*

$$\frac{3^{\prod_{i\in(k,d]}(m_i-1)\prod_{j\in(d,n]}m_j} - 1}{2^{(2^{k+1}-1)\prod_{i\in(k,n]}m_i} - \left(3^{\prod_{i\in(k,d]}(m_i-1)\prod_{j\in(d,n]}m_j} - 1\right)}. \quad (7)$$

Proof. By Theorem B, in order to prove that S is a set of uniqueness for $\mathcal{A}$, we have to show that for each $v = (v_1, \ldots, v_n) \in D$ there exists $i \in \{1, \ldots, n\}$ such that $|v_i| \ge m_i - h_i$.
We have $w = \sum_{i=1}^k \delta_i e_i$, where $\delta_i = 1$ if $e_i \in I$, and $\delta_i = -1$ if $e_i \in J$, so that $h_i = 2$ for $i = 1, \ldots, k$, $h_i = 1$ for $i = k+1, \ldots, d$, and $h_i = 0$ for $i = d+1, \ldots, n$. If $v = (v_1, \ldots, v_n) \in D$, then $v_j = 0$ for $k + 1 \le j \le n$, and $v_i \ne 0$ for some $i_o \in \{1, \ldots, k\}$, so that $|v_{i_o}| \ge m_{i_o} - h_{i_o} = 1$. This proves that S is a set of uniqueness for $\mathcal{A}$.

We have

$$F_S(x_1, \ldots, x_d) = \left(x^{w_+} - x^{-w_-}\right) \prod_{i=1}^{d} (x_i - 1).$$

Let us denote $F_T(x_1, \ldots, x_k) = (x^{w_+} - x^{-w_-}) \prod_{i=1}^{k} (x_i - 1)$. Then

$$F_S(x_1, \ldots, x_d) = F_T(x_1, \ldots, x_k) \prod_{i \in (k,d]} (x_i - 1).$$

All the S-weakly bad configurations contained in $\mathcal{A}$ correspond to polynomials of the form $F_S(x_1, \ldots, x_d)P(x_{k+1}, \ldots, x_n)$, where the degree $deg_i P(x_{k+1}, \ldots, x_n)$ of $P(x_{k+1}, \ldots, x_n)$ with respect to x_i, where $i = k+1, \ldots, n$, satisfies the conditions

$$\begin{aligned} deg_i P(x_{k+1}, \ldots, x_n) &< m_i - 1 \quad \text{for } i = k+1, \ldots, d, \\ deg_i P(x_{k+1}, \ldots, x_n) &< m_i \qquad \text{for } i = d+1, \ldots, n. \end{aligned} \tag{8}$$

Thus we have

$$P(x_{k+1}, \ldots, x_n) = \sum a_{r_{k+1}, \ldots, r_n} x_{k+1}^{r_{k+1}} \ldots x_n^{r_n}, \tag{9}$$

where $r_{k+1} \in [m_{k+1} - 1], \ldots, r_d \in [m_d - 1], r_{d+1} \in [m_{d+1}], \ldots, r_n \in [m_n]$. Each S-weakly bad configuration contained in $\mathcal{A}$ corresponds to a polynomial of the form

$$P(x_{k+1}, \ldots, x_n)F_T(x_1, \ldots, x_k) \prod_{i \in (k,d]} (x_i - 1),$$

where $P(x_{k+1}, \ldots, x_n)$ is given by (9).

Let us denote by $\mathcal{F}$ the set of points in $\mathcal{A}$ which belong to some S-weakly bad configuration. Notice that each S-weakly bad configuration contains two non-additive sets consisting of the set of positive (resp. negative) points. By multiplying the corresponding polynomial by -1, these two non-additive sets exchange each other. Therefore, the number of non-additive sets which are contained in $\mathcal{F}$ equals the number of polynomials $P(x_{k+1}, \ldots, x_n)$ given by (9), whose coefficients belong to the set $\{-1, 0, 1\}$, except the null polynomial. Such a number is given by

$$3^{\prod_{i \in (k,d]} (m_i - 1) \prod_{j \in (d,n]} m_j} - 1.$$

Any other non-additive set in $\mathcal{A}$ is obtained by adding some points of $\mathcal{A} \setminus \mathcal{F}$ to a non-additive set in $\mathcal{F}$. Thus, the proportion of non-additive sets of uniqueness in $\mathcal{A}$ with respect to those additive results

$$\begin{aligned} &\frac{2^{|\mathcal{A} \setminus \mathcal{F}|} \left(3^{\prod_{i \in (k,d]} (m_i - 1) \prod_{j \in (d,n]} m_j} - 1\right)}{2^{|\mathcal{A}|} - 2^{|\mathcal{A} \setminus \mathcal{F}|} \left(3^{\prod_{i \in (k,d]} (m_i - 1) \prod_{j \in (d,n]} m_j} - 1\right)} \\ &= \frac{3^{\prod_{i \in (k,d]} (m_i - 1) \prod_{j \in (d,n]} m_j} - 1}{2^{|\mathcal{F}|} - \left(3^{\prod_{i \in (k,d]} (m_i - 1) \prod_{j \in (d,n]} m_j} - 1\right)}. \end{aligned} \tag{10}$$

Since $|\mathcal{F}| = |F_T| \prod_{i \in (k,n]} m_i$ and $|F_T| = 2^{k+1} - 1$, we get

$$\frac{3^{\prod_{i \in (k,d]}(m_i-1)} \prod_{j \in (d,n]} m_j - 1}{2^{(2^{k+1}-1)} \prod_{i \in (k,n]} m_i - \left(3^{\prod_{i \in (k,d]}(m_i-1)} \prod_{j \in (d,n]} m_j - 1\right)},$$

as required. □

When $n = d = k = 3$, as in Subsection 3.1, we have

$$\prod_{i \in (k,d]} (m_i - 1) = \prod_{j \in (d,n]} m_j = \prod_{i \in (k,n]} m_i = 1, \quad 2^{k+1} - 1 = 2^4 - 1 = |F_S|,$$

so that formula (7) gives (4).

When $4 = n = d > k = 3$, as in Subsection 3.2 case 2, we have

$$\prod_{i \in (k,d]} (m_i-1) = m_4 - 1 = m - 1, \quad \prod_{j \in (d,n]} m_j = 1, \quad \prod_{i \in (k,n]} m_i = m_4 = m, \quad |F_T| = 2^{k+1} - 1 = 15,$$

so that formula (7) gives (6).

Moreover, if $3 \le k < n$, then for $(m_{k+1}, \ldots, m_n) \to (\infty, \ldots, \infty)$ we have

$$\frac{3^{\prod_{i \in (k,d]}(m_i-1)} \prod_{j \in (d,n]} m_j - 1}{2^{(2^{k+1}-1)} \prod_{i \in (k,n]} m_i - \left(3^{\prod_{i \in (k,d]}(m_i-1)} \prod_{j \in (d,n]} m_j - 1\right)} \to 0.$$

4   Conclusions

We have determined explicitly the proportion of bounded non-additive sets of uniqueness with respect to those additive. The resulting ratio has been computed as a function of the dimensions of the confining grid. This allows us to prove that, in the limit case when the grid gets large, the above proportion tends to zero, meaning that the probability that a random selected set is additive increases. Further improvements could be explored by considering more general sets S of directions. In this case the tomographic grid, obtained as intersections of lines parallel to the X-ray directions corresponding to nonzero X-ray, is not necessarily contained in the confining rectangular grid $\mathcal{A}$, and the computation of the proportion of bounded non-additive sets of uniqueness with respect to those additive, seems to be a more challenging problem to be investigated.

References

1. Aharoni, R., Herman, G.T., Kuba, A.: Binary vectors partially determined by linear equation systems. Discr. Math. 171, 1–16 (1997)
2. Brunetti, S., Dulio, P., Peri, C.: Discrete Tomography determination of bounded lattice sets from four X-rays. Discrete Applied Mathematics 161(15), 2281–2292 (2013), doi:10.1016/j.dam.2012.09.010

3. Brunetti, S., Dulio, P., Peri, C.: On the Non-additive Sets of Uniqueness in a Finite Grid. In: Gonzalez-Diaz, R., Jimenez, M.-J., Medrano, B. (eds.) DGCI 2013. LNCS, vol. 7749, pp. 288–299. Springer, Heidelberg (2013)
4. Brunetti, S., Dulio, P., Peri, C.: Discrete Tomography determination of bounded sets in $\mathbb{Z}^n$. Discrete Applied Mathematics, doi:10.1016/j.dam.2014.01.016
5. Fishburn, P.C., Lagarias, J.C., Reeds, J.A., Shepp, L.A.: Sets uniquely determined by projections on axes II. Discrete case. Discrete Math. 91, 149–159 (1991), doi:10.1016/0012-365X(91)90106-C
6. Fishburn, P.C., Schwander, P., Shepp, L., Vanderbei, R.: The discrete Radon transform and its approximate inversion via linear programming. Discrete Applied Math. 75, 39–61 (1997), doi:10.1016/S0166-218X(96)00083-2
7. Fishburn, P.C., Shepp, L.A.: Sets of uniqueness and additivity in integer lattices. In: Herman, G.T., Kuba, A. (eds.) Discrete Tomography: Foundations, Algorithms and Application, pp. 35–58. Birkhäuser, Boston (1999)
8. Gardner, R.J., Gritzmann, P.: Discrete tomography: Determination of finite sets by X-rays. Trans. Amer. Math. Soc. 349, 2271–2295 (1997), doi:10.1090/S0002-9947-97-01741-8
9. Gardner, R.J., Gritzmann, P., Prangenberg, D.: On the computational complexity of reconstructing lattice sets from their X-rays. Discrete Math. 202, 45–71 (1999), doi:10.1016/S0012-365X(98)00347-1
10. Gritzmann, P., Langfeld, B., Wiegelmann, M.: Uniquness in Discrete Tomography: three remarks and a corollary. SIAM J. Discrete Math. 25, 1589–1599 (2011), doi:10.1137/100803262
11. Gritzmann, P., Prangenberg, D., de Vries, S., Wiegelmann, M.: Success and failure of certain reconstruction and uniqueness algorithms in discrete tomography. Intern. J. of Imaging System and Techn. 9, 101–109 (1998), doi:10.1002/(SICI)1098-1098(1998)9:2/3<101::AID-IMA6>3.0.CO;2-F
12. Hajdu, L., Tijdeman, R.: Algebraic aspects of discrete tomography. J. Reine Angew. Math. 534, 119–128 (2001), doi:10.1515/crll.2001.037
13. Hajdu, L., Tijdeman, R.: Algebraic Discrete Tomography. In: Herman, G.T., Kuba, A. (eds.) Advances in Discrete Tomography and Its Applications, pp. 55–81. Birkhäuser, Boston (2007), doi:10.1007/978-0-8176-4543-4_4
14. Weber, S., Schnoerr, C., Hornegger, J.: A linear programming relaxation for binary tomography with smoothness priors. Electr. Notes Discr. Math. 12, 243–254 (2003)
15. Weber, S., Schüle, T., Hornegger, J., Schnörr, C.: Binary tomography by iterating linear programs from noisy projections. In: Klette, R., Žunić, J. (eds.) IWCIA 2004. LNCS, vol. 3322, pp. 38–51. Springer, Heidelberg (2004)

Back-Projection Filtration Inversion
of Discrete Projections

Imants Svalbe[1], Andrew Kingston[2], Nicolas Normand[3],
and Henri Der Sarkissian[3,4]

[1] School of Physics, Monash University, Melbourne, Australia
`imants.svalbe@monash.edu`
[2] Research School of Physical Sciences, Australian National University,
Canberra, Australia
`andrew.kingston@anu.edu.au`
[3] LUNAM Université, Université de Nantes, IRCCyN UMR CNRS 6597,
Nantes, France
`{nicolas.normand,henri.dersarkissian}@univ-nantes.fr`
[4] Keosys, Saint-Herblain, France

Abstract. We present a new, robust discrete back-projection filtration
algorithm to reconstruct digital images from close-to-minimal sets of arbitrarily oriented discrete projected views. The discrete projections are
in the Mojette format, with either Dirac or Haar pixel sampling. The
strong aliasing in the raw image reconstructed by direct back-projection
is corrected via a de-convolution using the Fourier transform of the discrete point-spread function (PSF) that was used for the forward projection. The de-convolution is regularised by applying an image-sized digital
weighting function to the raw PSF. These weights are obtained from the
set of back-projected points that partially tile the image area to be reconstructed. This algorithm produces high quality reconstructions at and
even below the Katz sufficiency limit, which defines a minimal criterion
for projection sets that permit a unique discrete reconstruction for noise-
free data. As the number of input discrete projected views increases, the
PSF more fully tiles the discrete region to be reconstructed, the de-
convolution and its weighting mask become progressively less important.
This algorithm then merges asymptotically with the perfect reconstruction method found by Servières et al in 2004. However the Servières
approach, for which the PSF must exactly tile the full area of the reconstructed image, requires $O(N^2)$ uniformly distributed projection angles
to reconstruct $N \times N$ data. The independence of each (back-) projected
view makes our algorithm robust to random, symmetrically distributed
noise. We present, as results, images reconstructed from sets of $O(N)$
projected view angles that are either uniformly distributed, randomly
selected, or clustered about orthogonal axes.

Keywords: Discrete tomography, image reconstruction from discrete
projections, inverse problems.

E. Barcucci et al. (Eds.): DGCI 2014, LNCS 8668, pp. 238–249, 2014.
© Springer International Publishing Switzerland 2014

1 Introduction

A Mojette projection of a 2D digital image $\mathcal{I}$ is comprised of the sums of image pixel intensities that are located along parallel lines, oriented at some set of angles defined by pairs of co-prime integers, (p_i, q_i) [3]. A set of Mojette projections can be used to reconstruct, either approximately or exactly, an image of the original data. A discrete point-spread function (PSF) is defined by back-projection of a single point by a set of projected views, (p_i, q_i). The PSF links the original image data to the Dirac Mojette back-projected image, $\mathcal{M}_{pq}$, through:

$$\mathcal{M}_{pq} = \mathcal{I} * PSF_{pq} \qquad (1)$$

where $*$ denotes spatial convolution.

The Mojette Transform (MT) [3], is one of several inherently discrete image projection techniques, like the (closely related) Finite Radon Transform (FRT) [9], where each image projection is defined explicitly by the discrete image data. We prefer to approach image reconstruction by first defining the discrete image that we want to display and then deciding what sets of projected views are sufficient to reconstruct that image.

Our aim is to do tomography this way, i.e. to transform real, noisy projection data into a form that is discretely Mojette-like as possible, and then use the Mojette inverse to reconstruct the image.

Inversion from Mojette projection sets may also shine some theoretical light on the Katz Criterion [6] used in discrete tomography. Katz showed that any $N \times N$ image can be reconstructed exactly from a set of discrete projections (p_i, q_i) provided $\max\left(\sum |p_i|, \sum |q_i|\right) \geq N$. Here we set $K = \frac{\max(\sum |p_i|, \sum |q_i|)}{N}$, so that $K = 1$ for a projection set at the Katz limit, whilst a set with $K < 1$ is below the Katz limit and cannot reconstruct an exact image. Katz specifies the spatial resolution and view angle requirements that permit exact digital inversion, but says nothing about approximate reconstructions, the effect of noise, the equivalence (or not) of different sets of angles, nor about the constraints imposed by the dynamic range of quantised image values.

There are other algorithms to invert Mojette projections, but these methods have severe limitations. The corner-based algorithm of Normand, and related geometric techniques [10], work only for noise-free projection sets that satisfy or exceed the Katz condition. Alternatively Mojette data can be mapped to the periodic form of the FRT, for which inversion by back-projection is exact, or else by applying the central slice theorem using Fourier [9] or number-theoretic transforms [1,2]. Direct inversion of the projection matrix is possible, but requires inverting very large matrices that are often ill-posed. Other methods, such as conjugate gradient [11] or partially ordered sets, require an iterative or statistical approach [4] that negate the advantages of using a direct reconstruction algorithm.

The intensity at each forward-projected Mojette bin is back-projected across the image reconstruction space, along the same discrete lines along which that

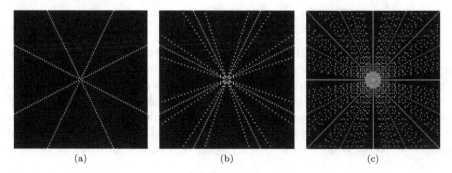

(a)   (b)   (c)

Fig. 1. Left to right: back-projected reconstruction of a single point using (a) 4, (b) 12 and (c) 60 discretely projected Dirac Mojette views. Image data confined to the area inside the red circle in (c) can be reconstructed exactly because the translated PSFs (green circles) intersect back-projected pixels that all have zero intensity.

data was projected (but with the mean projected sum, not the individual pixel values that make up each sum). The back-projection method is "blind" to other values in the projection data, it does not try to uncouple projection bins and pixel values. A back-projection algorithm is then tolerant of noise on the projections; it uses the same method for whatever data lies in each projection bin.

2   Image Reconstruction and the Discrete PSF

Fig. 1a shows four back-projected rays (at angles $(\pm 1, 2), (\pm 2, 1)$) for a single point located at the centre of an image. Back-projecting intensity values of 1.0 at the peak and $-1/(4 - 1)$ elsewhere, results in a normalised intensity of 1.0 at their intersection point, zero along the projected rays and -0.333 elsewhere. Back-projecting 12 symmetric projections $((1, 2), (1, 3)$ and $(2, 3)$ in Fig. 1b) with intensity 1.0 or $-1/(12 - 1)$ gives 1.0 at the intersection point, zero along each of the 12 projected rays, and -0.0909 elsewhere. Fig. 1c shows the normalised Mojette back-projection for 60 projected views (the symmetric set of shortest vectors $(p, q) = (0, 1)$ to $(\pm 3, 7)$), yielding a centre value of 1.0 (white), the central circular region has value zero (grey), other pixels have value $-1/59$.

Note how the zero pixels increasingly tile the region that surrounds the intersection point. All translations of this PSF inside the red circle in Fig. 1c can be reconstructed exactly (as a single 1 on a background of 0). A PSF can only be translated as far as the green circles in that figure, before the negative background value (here the black pixels) will cause reconstruction errors. For a 59×59 image, 3208 Mojette projections $((0, 1)$ to $(31, 49))$ are needed to uniformly tile a disc of radius $R = 58$ (since $31^2 + 49^2 = 3362$ and $58^2 = 3364$). Hence direct, unfiltered, back-projection of discrete Mojette data works exactly, but it requires approximately $\left(\frac{6}{\pi}\right)^2 2N(2N - 1)$ views to reconstruct an $N \times N$ image [13]. This result confirmed earlier work done by Servières [12]. The shape of the image region of interest (ROI) also matters, as it determines how many, and which, array pixels need to be exactly tiled by the discrete PSF.

If you have only $M \ll N^2$ views, you can try to interpolate the missing $N^2 - M$ views that you don't know, using the M views that you do know [13,12]. For example, we can synthesize the projected view for $(13, 14)$ from the known projection for $(1, 1)$. Interpolation of digital profiles is tough work, especially for Dirac rather than the smoother Haar or higher-order spline projections.

If the image array is assumed to be periodic, the continuation of each projected ray (other than for $(1, 0)$ and $(0, 1)$) will pass across the ROI several times, increasing the number of pixels that are filled by back-projection.

For the FRT [9], where the array size is prime, $p \times p$, and for composite $N \times N$ arrays [7,8], periodic back-projection can fully tile the ROI, the PSF is then "perfect", making exact inversion possible with $O(N)$ projections from an $N \times N$ image. A reconstruction method, for uniformly-distributed angle sets, that merges elements of the discrete Mojette, Fourier and compressed sensing techniques appeared recently [5].

2.1   Reconstruction of Images Using a Finite, Discrete PSF

Consider a finite PSF, such as our previous example composed of 60 views, which reconstructs, perfectly, any image inside the red circle of Fig. 2a. If we try to use this PSF to reconstruct a larger circular ROI (e.g. 60×60, the blue circle in Fig. 2a), we need to correct for each of the discrete un-corrected negative contributions that fall outside the flat zone of the blue PSF, i.e. all of the black points that lie inside the green circle shown in Fig. 2a.

A discrete, back-projected image, $\mathcal{M}_{pq}$, reconstructed from a set of view angles, (p, q), is exactly equivalent to convolution of the discrete PSF with the original image, $\mathcal{I}$. This follows from the definition of discrete back-projection for Dirac Mojette data [3]. This is easy to demonstrate; Fig. 3 shows an original image, $\mathcal{I}$, and the direct back-projected image, $\mathcal{M}_{pq}$, for the set of 60 (p, q) angles shown in Fig. 2. The back-projected image is identical to the convolution of the original image data by the PSF shown in Fig. 2 (the differences on subtraction are $O(10^{-15})$ and result from finite float computational precision). Then,

$$\mathcal{M}_{pq} * \text{PSF}_{pq}^{-1} = \mathcal{I} \tag{2}$$

$$\mathcal{I} = F^{-1} \left\{ \frac{F\{\mathcal{M}_{pq}\}}{F\{\text{PSF}_{pq}\}} \right\} \tag{3}$$

where $F\{\cdot\}$ denotes the 2D finite Fourier transform (FFT) of the zero-padded image data. Equation (3) is the basis for the de-convolution process presented in this work. Recovery of $\mathcal{I}$ from the direct, back-projected image $\mathcal{M}$, is contingent on the FFT of the PSF being well-conditioned or regularised. To ensure these conditions, a weighted version of the raw PSF, denoted as PSF^+ in equation (6), is then used in (3).

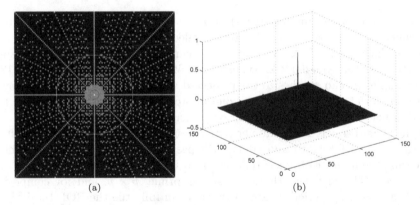

(a)   (b)

Fig. 2. Left: (a) Normalised, back-projected PSF reconstructed from 60 discrete projections at angles (p, q) using the shortest vector lengths (white = 1, grey = 0, black < 0). Images confined inside the red circle (with a diameter of about 23 pixels) can be reconstructed exactly by direct back-projection. Reconstruction of images lying inside the blue circle, with diameter 60 pixels, using the same 60 projected views, requires correction for all of the missing (black) back-projected points the lie inside the green circle with diameter 120 pixels. Right: (b) a 3D view of the PSF in (a).

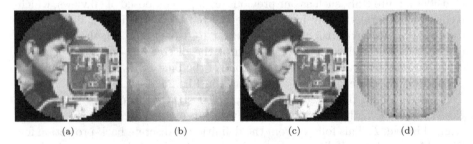

(a)   (b)   (c)   (d)

Fig. 3. (a): Original image $\mathcal{I}$, circular ROI diameter 60 pixels. (b): Raw back-projected reconstruction from 60 discrete Mojette views of the original data. Image (b) is exactly the same as the result obtained by convolving image (a) with the discrete PSF shown in Fig. 2b. (c): Reconstructed image from the 416 Mojette shortest (p, q) views (65×65 pixels, $K = 57 \gg 1$), PSNR = 46.62, MSE = 1.30. This is about 10% of the number of views required for exact, unfiltered Mojette back-projection. (d): reconstruction errors for this image are structural and arise from inversion of the PSF; they are not evidently strongly image-related.

2.2   Image Reconstruction Examples with $K \gg 1$

We verify this approach by applying (3) to reconstruct images from discrete projection sets comprised of many views, but far fewer than the $O(N^2)$ views required for exact $N \times N$ reconstruction. Fig. 3c shows an example image, reconstructed from 416 Mojette projections, that were obtained from a circular ROI in 65×65 discrete image data (a cropped portion of the "cameraman" image).

Adding normally distributed noise to each of the 416 projections only slightly reduced the reconstructed image PSNR (from 46.62 to 43.8 ± 0.2), confirming the robustness of the back-projection approach. The reconstructed images for $K \gg 1$ are of high quality, but our aim here is to reduce the number of discrete projections required, so that $K \simeq 1$, or preferably, $K < 1$. Using fewer projection directions makes the outer zones of the PSF increasingly sparse and the flat zone of zeros smaller, so that the PSF inverse becomes less well-conditioned and requires some regularisation.

3   Regularisation of the PSF

Our approach to recover images for projection sets where $K \simeq 1$, is to weight the image of the PSF, in the form as shown in Fig. 2, so that the outer regions of the PSF are made smoother and closer to zero, whilst preserving the central flat zone. These weights should be smooth and close to unity for $O(N^2)$ projected views, where no or little correction of the back-projected image is required. For increasingly sparse views, the weights should be closer to zero around the periphery of the PSF, and remain strongly discretised along those directions where more correction is needed.

For sets of projected views that exceed the Katz Criterion ($K \geq 1$), we construct a weight function, that we call W_{pn}, that reflects the correlation between the spatial distribution of points across the region of support in the PSF that are correctly back-projected and the distribution of points where the back-projected data is absent.

We generate first an image (p) of the PSF that contains the correctly back-projected points over the region of image support. These points are set to one, all other points are set to zero. We generate a second image (n) of the PSF that contains the (complementary) set of uncorrected back-projected points over the region of image support. These points are set to one, all other points are set to zero. We then cross-correlate these distributions, across the full extent of the back-projected space, as

$$W_{pn} = (p \star n) * (D \star D) \tag{4}$$

where $\star$ denotes the cross-correlation product and D has the uniform value of 1 over the image region of support and is otherwise zero. Pixels of W_{pn} that fall within the central flat zone of the PSF (the area of perfect reconstruction) should not be down-weighted, so the weight for these pixels is set to 1. The function W_{pn} is then normalised to have maximum value 1, as shown in Fig. 4a.

For sets of projected views that fall below the Katz Criterion ($K \leq 1$), the back-projected area becomes increasingly sparse outside the small flat central zone of exact reconstruction and the region of support, D, plays an increasingly smaller contribution. For these cases, the weight function, that we call T_{pn}, is constructed as a direct correlation of the spatial distributions of the correctly back-projected points in the PSF vs the non-back-projected points;

$$T_{pn} = p \star n \tag{5}$$

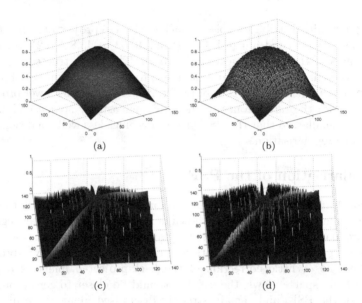

Fig. 4. Top row: (a-b) Left: (a) Weight function W_{pn} for the PSF for the set of 52 angles $((1,0)$ to $(\pm 2,7))$, 63×63 image, circular region of support. Right: (b) Weight function T_{pn} for the same PSF as in (a).
Bottom row: (c-d) Left: (c) The PSF weighted by W_{pn} for $K > 1$. Right: (d) the PSF weighted by T_{pn} for $K < 1$. The central spike of the PSF has been suppressed here to enhance the display of the small values at the periphery.

The weights for those pixels that lie within the central flat zone of the PSF are set to 1, with the net result normalised, as shown in Fig. 4b. Weights W_{pn} and T_{pn} multiply, point to point, the raw PSF image values,

$$
\text{PSF}^+ = \begin{cases} \text{PSF} \odot W_{pn}, & \text{if } K > 1 \\ \text{PSF} \odot T_{pn}, & \text{if } K < 1 \end{cases} \tag{6}
$$

where $\odot$ denotes the element-wise multiplication of 2D vectors.

4   Reconstruction Results for De-convolution of Back-Projected Images Using the Weighted PSF

We applied the weightings W_{pn} and T_{pn} to the PSF (as shown in Fig. 4c,d) and recorded the PSNR for images that were reconstructed, using equations (3) and (6), for varying numbers of projected views. Here all angle sets are comprised of the shortest (p, q) vectors. The results are given in Table 1 for image sizes of 63×63 and 127×127, respectively. At the Katz point, $K = 1$, the PSNR values for reconstruction from projections after adding normally distributed noise have

Table 1. PSNR values for two images of sizes of 63×63 and 127×127, reconstructed using a weighted PSF, as a function of the number of projected views (shortest (p, q) vector angle sets)

Image size	K (Katz' value)	Projections count	Weight W_{pn}	Weight T_{pn}
	0.59	20	18.89	18.67
	0.81	24	19.98	19.93
	1	28	21.63	21.63
63×63	1.22	32	22.92	22.73
	2.52	52	27.61	26.76
	3.67	64	30.08	28.54
	6.46	96	34.34	31.06
	9.89	128	35.74	31.62
	0.5	28	17.77	17.78
	0.61	32	18.90	18.75
	0.73	36	19.30	19.38
	0.84	40	20.30	20.09
127×127	0.98	44	21.35	20.92
	1.11	48	22.54	21.66
	3.2	96	29.70	26.95
	4.91	128	32.74	28.55
	9.11	192	35.01	29.44

also been included. Those results provide further confirmation of the robustness of our direct inversion method.

We observed that weight W_{pn} performs better for $K > 1$, whilst the reconstruction results for weight T_{pn} become slightly better than for weight W_{pn} for $K < 1$, especially in larger images. For example, a 509×509 portion of the Lena image, reconstructed using the 96 shortest angles, where $K = 0.8$, yields a PSNR of 19.80 for T_{pn} and 18.41 for W_{pn}.

The only other tool used to regularise the inverse of the PSF is to apply a threshold test to the Fourier coefficient values before taking their inverse. If the Fourier coefficient of the (weighted) PSF at (u, v) was less than a selected fixed value, that coefficient was set to the mean of all above-threshold 3×3 neighbouring coefficients. The choice of this threshold value turned out to be relatively insensitive; optimising its value made relatively small differences to the final PSNR values. When reconstructing any images, we keep track of the number of times the threshold is reached, to indicate if the threshold needs to be modified. For example, at $K = 0.59$, for 20 angles, PSNR = 18.89 improves to 19.45 after scanning across a range of threshold values to optimise the PSNR.

The size of the zero-padded region can be adjusted to be made (symmetrically) slightly smaller or larger. This can change the reconstructed PSNR (again, only slightly), as it may affect the degree of aliasing by the discretisation of the finite Fourier transform at specific frequencies.

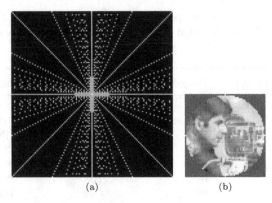

(a)   (b)

Fig. 5. Reconstruction of an image from a clustered set of 52 projection angles $(\pm 1, i), (\pm i, 1); 0 \leq i \leq 13\}$, for which $K = 3.29$. Left: (a) the raw clustered view PSF (125×125). Right: (b) the reconstructed image (63×63, using weight T_{pn}), with PSNR $= 23.33$. After adding normally distributed noise, PSNR $= 23.1 \pm 0.1$.

5   Reconstruction Using Different Distributions of Projected View Angles

5.1   Clustered Projection Angles

We reconstructed images from sets of projected views that are clustered around $0°$ or around $90°$. We do this by selecting discrete angles (p, q) corresponding to $(\pm 1, i)$ and $(\pm i, 1)$, for integers $0 \leq i \leq n$. To maintain four-fold angle symmetry, we usually increment the number of angles in steps of 4. The strongly non-uniform distribution of angles makes the shape of the PSF less uniform and thus more difficult to correct. Here the discrete weight T_{pn} (PSNR $= 23.33$) performs better than weight W_{pn} (PSNR $= 20.08$), even for $K > 1$. An example reconstructed image is shown in Fig. 5.

5.2   Randomly Distributed Projection Angles

We generated a random set of M projection angles selected from a range of (p, q) vectors that was three times larger than for the shortest angle set for M projections. As p and q can now be much larger than for the shortest set, the number of bins in these projections, as given by $n_{\text{bin}} = (|p| + |q|)(N - 1) + 1$, is also larger.

Whilst randomising the projection angles generally decreases the reconstructed image quality because the PSF is less uniform, the use of large p and q values generally improves the discrete reconstructions. The large $|p| + |q|$ discrete projections are less heavily summed, because there are more projection bins, but the projected ray passing into each bin intersects fewer pixels. Forcing the inclusion of the projections $(1, 0)$, $(0, 1)$ and $(\pm 1, 1)$ as part of the "random" set improves the results, as those projections carry significant information about the image.

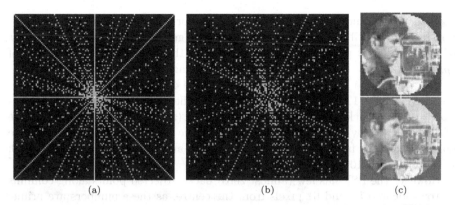

Fig. 6. (a-b): PSF (125×125) for two random sets of 52 projection angles. (a): Including $(1,0)$, $(0,1)$ and $(\pm 1, 1)$. (b): Excluding $(1,0)$, $(0,1)$ and $(\pm 1, 1)$. (c): Corresponding 63×63 reconstructed images using filter W_{pn}. Top image, PSNR = 31.31, bottom image, PSNR = 26.33. After adding normally distributed noise PSNR = 27 ± 1 and 24.5 ± 1 respectively.

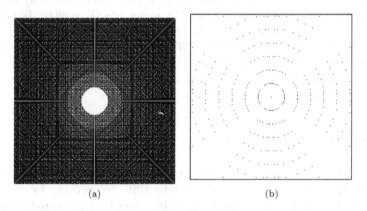

Fig. 7. (a): Raw discrete PSF (129×129) for the 440 shortest (p, q) angles. The highlighted portion of the rows and columns 60 pixels away from the PSF centre are heavily tiled by back-projection. In contrast, pixels along the rows and columns a prime distance from the centre (e.g. 59 and 61) are sparsely tiled. (b): Differences in the discrete PSF for the projection sets comprised of the 440 and 416 shortest angles emphasize the bias towards better reconstruction along the horizontal and vertical directions.

The PSF weight W_{pn} performs slightly better here (mean PSNR = 28.5 ± 1.5) than does weight T_{pn} (mean PSNR = 25.5 ± 1.5), as the view angles are, on average, more uniform and $K > 1$. Here $K = 4.5 \pm 0.3$, see Fig. 6.

5.3 Discrete Image Reconstruction Errors

Our reconstruction errors occur predominantly along the image rows, columns and diagonals (as in Fig. 3). Partly this effect may be due to the method we used to weight and invert the PSF, but another contribution comes from the non-random distribution of (p, q) points as they get back-projected. The number of back-projected points and their inclusion in the PSF depends on the set of (p, q) view angles, but also on the size of the image ROI to be reconstructed.

Fig. 7a shows a PSF back-projected from the 440 discrete shortest (p, q) angles (as a 129×129 pixel image). Entries in the PSF that arise from projection (p, q) are back-projected as points located (np, nq) pixels from the centre of the PSF. Many back-projected points will lie on columns or rows that are 60 pixels distant from the centre, as n can be any of the many factors of 60; $n = \{2, 3, 4, 5, 6, 10, 12, 15, 20, 30\}$.

However the PSF has few (almost zero) back-projected points along columns or rows located 59 and 61 pixels from the centre, as these numbers are prime. A 61×61 portion from the image in Fig. 3c reconstructs (using an unweighted PSF at 440 shortest views) with a PSNR of 41.76, whereas a 59×59 portion of the same data reconstructs with PSNR $= 47.68$, using identical views. Including or excluding row and column 60 from the reconstructed ROI makes a large difference to the PSF that, in turn, affects the reconstructed PSNR.

Strong, local changes at the edges of the PSF may also explain why we obtain reconstructions for 65×65 images using 416 angles (PSNR $= 46.62$, last $(p, q) = (12, 17)$) that are, uncharacteristically, slightly better than those for a much larger angle set of 440 views (PSNR $= 46.41$, last $(p, q) = (4, 21)$). For 63×63 images we obtain PSNR $= 45.52$ for 416 views, PSNR $= 46.69$ at 440 views, because part of the outer circle of back-projected points shown in Fig. 7b (arising from the $(4, 21)$ view), are now excluded.

6   Summary, Conclusions and Future Work

We presented, in this work, an approach to reconstructions that uses a weighted version of the raw back-projected discrete PSF to recover, by direct deconvolution, the original digital image from its direct back-projected reconstruction. As this is a linear system, the same approach could equally be achieved by filtering the Mojette data in 1D before back-projection, as was originally suggested by Andrew Kingston. Correction of 3D back-projected images, via the inverse of the 3D PSF, is a natural extension of this 2D approach.

This filtered back-projection approach also tolerates the presence of significant levels of noise in the projected data. Computation of these filtered back-projected images is fast to compute, especially if the Fourier transform of the discrete PSF and the associated weight functions (T_{pn} or W_{pn}) are known.

The relatively high accuracy of our reconstructions and the rapidity with which they can be obtained would provide high quality initial-image estimates that may enhance the convergence rate for slower, statistical iterative reconstruction methods.

The use of an exact, algebraic approach to pre-filter each projection also seems possible. Nicolas Normand has shown that direct inversion is possible for projection data cast as a Vandermonde matrix and has also shown that

Moore-Penrose pseudo-inverse techniques may be used to invert projection matrices (unpublished work). These new techniques complement the existing "row-solving" techniques [1,2] or methods based on the central–slice theorem [9] that map between Mojette (or FRT) data and image space.

An enhanced ability to invert arbitrary sets of Mojette data will permit detailed studies to be undertaken to understand how the grey-level quantisation of a digital image affects image reconstruction, as seen via the Katz criterion. Our method has already been applied to reconstruct binary and ternary image data from sparse sets of discrete projections (unpublished work).

Acknowledgments. The IVC group hosted a visit by Imants Svalbe to work in their laboratory from November to December, 2013. Polytech Nantes provided financial support to him as a Visiting Invited Professor for that period.

References

1. Chandra, S., Svalbe, I.D.: Exact image representation via a number-theoretic Radon transform. IET Computer Vision (accepted November 2013)
2. Chandra, S., Svalbe, I.D., Guédon, J., Kingston, A.M., Normand, N.: Recovering missing slices of the discrete Fourier transform using ghosts. IEEE Trans. Image Process. 21(10), 4431–4441 (2012)
3. Guédon, J., Normand, N.: Direct Mojette Transform. In: Guédon, J. (ed.) The Mojette Transform: Theory and Applications, pp. 37–60. ISTE-WILEY (2009)
4. Herman, G.T., Kuba, A. (eds.): Advances in Discrete Tomography and its Applications. Birkhäuser, Boston (2007)
5. Hou, W., Zhang, C.: Parallel-Beam CT reconstruction based on Mojette transform and compressed sensing. Int. J. of Computer and Electrical Eng. 5(1), 83–87 (2013)
6. Katz, M.B.: Questions of uniqueness and resolution in reconstruction from projections. Springer, Berlin (1978)
7. Kingston, A.M., Svalbe, I.D.: Generalised finite Radon transform for N×N images. Image and Vision Computing 25(10), 1620–1630 (2007)
8. Lun, D., Hsung, T., Shen, T.: Orthogonal discrete periodic Radon transforms, Parts I & II. Signal Processing 83(5), 941–971 (2003)
9. Matúš, F., Flusser, J.: Image representation via a finite Radon transform. IEEE Trans. Pattern Anal. Mach. Intell. 15(10), 996–1006 (1993)
10. Normand, N., Kingston, A., Évenou, P.: A Geometry Driven Reconstruction Algorithm for the Mojette Transform. In: Kuba, A., Nyúl, L.G., Palágyi, K. (eds.) DGCI 2006. LNCS, vol. 4245, pp. 122–133. Springer, Heidelberg (2006)
11. Servières, M., Idier, J., Normand, N., Guédon, J.: Conjugate gradient Mojette reconstruction. In: SPIE Medical Imaging, pp. 2067–2074. SPIE (2005)
12. Servières, M., Normand, N., Guédon, J.: Interpolation method for the Mojette transform. In: SPIE Medical Imaging, SPIE, San Diego, CA (2006)
13. Svalbe, I.D., Kingston, A.M., Guédon, J., Normand, N., Chandra, S.: Direct Inversion of Mojette Projections. In: Int. Conf. on Image Processing, Melbourne, Australia, pp. 1036–1040 (2013)

Discrete Tomography Reconstruction Algorithms for Images with a Blocking Component

Stefano Bilotta and Stefano Brocchi

Dipartimento di Matematica e Informatica 'U. Dini', Università di Firenze,
Viale Morgagni, 65 - 50134 Firenze, Italy
{stefano.bilotta,stefano.brocchi}@unifi.it

Abstract. We study a problem involving the reconstruction of an image from its horizontal and vertical projections in the case where some parts of these projections are unavailable. The desired goal is to model applications where part of the x-rays used for the analysis of an object are blocked by particularly dense components that do not allow the rays to pass through the material. This is a common issue in many tomographic scans, and while there are several heuristics to handle quite efficiently the problem in applications, the underlying theory has not been extensively developed. In this paper, we study the properties of consistency and uniqueness of this problem, and we propose an efficient reconstruction algorithm. We also show how this task can be reduced to a network flow problem, similarly to the standard reconstruction algorithm, allowing the determination of a solution even in the case where some pixels of the output image must have some prescribed values.

1 Introduction

Discrete tomography is the discipline that studies the reconstruction of discrete sets from the partial information deriving from their projections. Its main motivation arises from applications that tackle the problem of obtaining information about an object by examining the data obtained from the x-rays projected through the material, as in medical scans. Differently from computerized tomography, in discrete tomography we suppose that the pixels forming the original image may have a limited set of discrete values, often only 0 or 1. This assumption is reasonable in many cases where the object has a uniform density, and allows the definition of efficient algorithms even upon availability of a limited number of x-rays, as in [6].

In applications, often heuristic algorithms allow an efficient reconstruction, but on the other hand determining exact algorithms for discrete tomography reconstruction is often a hard task, as the involved problems in many cases are highly undetermined or computationally intractable [12]. To tackle these problems and to include some other information that may model effectively properties of the image to be rebuilt, often discrete tomography problems include

E. Barcucci et al. (Eds.): DGCI 2014, LNCS 8668, pp. 250–261, 2014.

some prior knowledge that gives birth to many variations of the reconstruction problem. Some examples that have been studied in literature include connectivity and convexity [4,2], cell coloring [10,3] or skeletal properties [14,15]. Often, with appropriate assumptions, the arising problems result to be connected to other fields of study as timetabling [18], image compression [1], network flow [5], graph theory [7] and combinatorics [13]. An excellent survey that can be examined for some classical results is [16].

In this paper, we model a situation where, due to causes such as a very dense block of material, some x-rays are blocked and do not provide measurements for these zones of the object. This is a common problem in many tomographical applications that may cause artifacts in the rebuilt images, that in some cases may reduce the quality of the reconstruction up to the point of making the resulting image unusable, for example for diagnostic purposes in the case of a medical scan. Many heuristics exists to reduce such artifacts, based on algebraic approaches [20], statistical analysis [8], linear interpolation [17], partial differential equations [9], and image impainting [11]. The described approaches are often sufficient in applications, however to the best of our knowledge there is no theoretical study that determines the basic properties of consistency and uniqueness of these problems. In this paper, we answer these questions and we propose an efficient reconstruction algorithm that uses the horizontal and vertical projections. Differently from the cited works, this article does not have an immediate practical application, but is aimed to develop the theory underlying these problems. Here we consider as input of the problem projections in only two directions; this assumption is quite typical in discrete tomography, as the reconstruction problem of binary matrices from three of more projections is know to be NP-complete [12].

The paper is organized as follows. In Section 2 we give some preliminaries and describe the adopted notation. In Section 3 we describe some basic properties of the problem, and how they relate to the classical reconstruction problems without the blocking component. In Section 4 we describe an efficient reconstruction algorithm for the problem. In Section 5 we show how the reconstruction is connected with a flow problem; this result is a natural extension of the well known reduction of the standard problem, and allows us to solve the reconstruction even in the cases where some zones of the resulting image must be fixed to some prescribed values. Finally, in Section 6 we discuss and draw some conclusions with an insight on future developments.

2   Notation and Preliminaries

In this section we give a formal description of the studied problem. Classically, a solution of a reconstruction problem is represented by a binary matrix containing the values 0 and 1; we refer to the entries of the matrix as *cells*. In this paper, we suppose that for some zones of an image the horizontal and vertical projections that would account for some zones are unavailable, as if in a tomographical scan the area corresponding to these pixels blocked the x-rays. We refer to this set of cells as a *blocking component*; the image that we aim to rebuild hence also contains a special symbol * in correspondence to the cells in this component.

Without loss of generality, the blocking component can be positioned in the last columns and rows of the matrix, as we could relate to this situation any other configuration by a rearrangement of the columns and of the rows. We also suppose that the blocking component has a rectangular shape, as any other shape would introduce some cells that do not contribute to neither the horizontal nor the vertical projections, and that are not relevant to the formulation of the problem. The problem is hence defined as follows:

Input: two integers k^h and k^v representing the size of the blocking component, and two vectors of projections $H = (h_1, \ldots, h_{n-k^v})$, $V = (v_1, \ldots, v_{m-k^h})$.

Output: an $n \times m$ matrix $A = (a_{i,j})$, such that:
- $\forall\, 1 \le i \le n - k^v$, $\sum_{j=1}^{m} a_{i,j} = h_i$;
- $\forall\, 1 \le j \le m - k^h$, $\sum_{i=1}^{n} a_{i,j} = v_j$;
- $\forall\, a_{i,j} : i > n - k^v, j > m - k^h$ then $a_{i,j} = *$.

An interesting feature of our problem is that, due to the missing projections, the sum of the horizontal and vertical projections may differ, but even in this case, there may be a solution fitting the constraints. We define this difference as $D = \sum_j v_j - \sum_i h_i$, and without loss of generality we consider $D \ge 0$. In Figure 1, we depicted an example of the problem with horizontal and vertical projections, and with a blocking component covering two rows and two columns. The vectors H and V represent the input of the problem, while the content of the cells of the matrix represents one of the possible solutions.

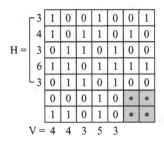

Fig. 1. An instance of the considered problem and a possible solution

3   Properties

From classical results in literature, we dispose of a theorem which guarantees the existence of a solution of the standard reconstruction problem. In this case, we want to rebuild a binary matrix with horizontal and vertical projections equal to H and V. Without loss of generality, hereafter we consider the vectors H and V to be ordered, so $h_1 \geq h_2 \geq h_3 \ldots$ and $v_1 \geq v_2 \geq v_3 \ldots$

Definition 1. *For two vectors A, B of length n, then $A \leq_d B$ if for every j we have $\sum_{i=1}^{j} a_i \leq \sum_{i=1}^{j} b_i$.*

Theorem 1. *(Ryser [19]) The reconstruction problem with projections H, V has a solution if and only if $H \leq_d V^*$, where $V^* = (v_1^*, v_2^*, \ldots, v_n^*)$ is defined as $v_i^* = |\{v_j : v_j \geq i\}|$. The problem admits a unique solution if and only if $H = V^*$.*

In this section we formally define the reconstruction problem involving blocking components. For a simpler notation, we define a partition of the cells of a matrix A in three different submatrices C, X and Y in order to be able to identify immediately the relative position of each cell with a blocking component. These three matrices are dependent from A, however we omit this dependency in the notation. The cells in C are those who give contribution to both the horizontal and vertical projections, while those in X and Y are only counted in one of the two. We have:

- $C = (c_{i,j})$ of size $(n - k^v) \times (m - k^h)$ with $c_{i,j} = a_{i,j}$ for $i \leq n - k^v$ and $j \leq m - k^h$;
- $X = (x_{i,j})$ of size $(n - k^v) \times k^h$ with $x_{i,j} = a_{i,m-k^h+j}$ for $1 \leq j \leq k^h$;
- $Y = (y_{i,j})$ of size $k^v \times (m - k^h)$ with $y_{i,j} = a_{n-k^v+i,j}$ for $1 \leq i \leq k^v$.

We define the operators $H(M)$ and $V(M)$ that extract the horizontal and vertical projections of a generic matrix M. We use the following operators to count the number of ones in each zone; we define $N^C(A)$ as $\sum c_{i,j}$, and similarly we refer to $N^X(A)$ and $N^Y(A)$; also $N(A) = \sum_{a_{i,j} \neq *} a_{i,j}$. When the argument of these operators is unambiguous, we omit the argument (A) of the function, referring simply, for example, to N^C or N^X. Finally, $N^H = \sum_i h_i$ and $N^V = \sum_j v_j$.

One of the additional difficulties of our problem is that the number of cells in H may be different from the ones in V, since some cells contribute only one of the horizontal or vertical projections. In order to define a reconstruction algorithm, one of the first steps is to determine correct values for N^X and N^Y. As some first trivial conditions, it must stand $N^C + N^X + N^Y = N$, and $N^Y \geq D$. The maximal number of cells M^X (symmetrically, M^Y) in the group X (resp. Y) for any solution of the problem is given by the following property.

Property 1. For any matrix A having V as vertical projections, we have $N^Y \leq M^Y$, where M^Y is defined by $M^Y = \sum_j \min(v_j, k^v)$.

Proof. To prove this simple property, it is sufficient to observe that $N^Y > M^Y$ would imply that on some column the number of cells equal to 1 would exceed the height k^v of the matrix Y or a vertical projection, bringing to a contradiction. $\qquad\square$

Note that while the N operators have as argument a matrix, the M operators are in dependence from the input of the problem. From this property and the trivial condition $N^Y = N^X + D$ we obtain the following corollary.

Property 2. For any matrix A solution of a reconstruction problem with blocking component, we have $N^Y \leq \min(M^Y, M^X + D)$.

3.1   Switching Components and Unique Solutions

We recall that the cells inside the blocking component of a matrix A, i.e. the elements $a_{i,j}$ such that $i > n - k^v$ and $j > m - k^h$, are represented with the special symbol $*$. We now define some *switching operations* that, starting from a solution of a problem, enable us to build other matrices satisfying the problem constraints. These operations are similar to the standard switches found in literature, but the presence of a blocking component leads to other types of switches involving the symbol $*$. The possible switches are shown in Figure 2 (left); it is easy to verify that all these operations do not alter the horizontal and vertical projections of a matrix.

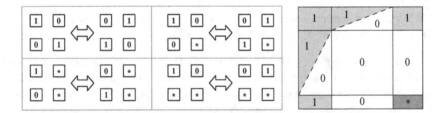

Fig. 2. (Left) Possible switches in our instances. The one in the upper left is the classical switching operation in the standard reconstruction problem, while the other three are introduced by the existence of a blocking component in the problem. (Right) The only possible structure of a matrix with no switching components, assuming rows and columns are ordered by projections.

Considering these four types of switch, we can state a uniqueness result. In that aim, we first show the structure of a matrix A that does not contain any of the switching components. We recall that since w.l.g. $h_1 \geq h_2 \geq \ldots$ and $v_1 \geq v_2 \geq \ldots$, then we have:

Theorem 2. *Let A be a solution of a blocking component reconstruction problem. The matrix A does not contain any switching components if and only if its partitions C, X and Y satisfy the conditions below, for the generic values $i \in [1, n]$ and $j, j' \in [1, m]$.*

1. $\forall i, j, j'$ then $x_{i,j} = x_{i,j'}$;
2. $\forall i, i', j$ then $y_{i,j} = y_{i',j}$;

3. $\forall x_{i,j} = y_{i',j'}$ *then* $c_{i,j'} = x_{i,j} = y_{i',j'}$;
4. $H(C) = V^*(C)$.

Proof. Observe that to avoid switching components involving two equal symbols * every row of X and every column of Y must contain only one of the values (points 1 and 2). To prohibit switches with one *, we impose Condition 3, while rule 4 must be adopted in order to prevent the standard switching operation. □

In Figure 2 (right) we depicted a graphical representation of a matrix A without any switching components. An example of a unique solution on a specific problem instance can be seen in Figure 3.

In the standard reconstruction problem, the absence of switching components is sufficient to prove that a matrix is the unique solution. In this problem, however, we may speculate about this fact, as two different solutions may exist, but that there is no series of switching operations that transform one of them into the other. The following theorem is necessary to show how this conjecture turns out to be false, and furthermore its proof exhibits a procedure allowing the determination of the unique solution if it exists.

Theorem 3. *If a matrix with no switching components exists for a blocking component reconstruction problem, then the given matrix is the only solution of the problem.*

Proof. Consider as input of the problem the dimension of the blocking component described by k^v and k^h and the vectors of projections $H = (h_1, \ldots, h_{n-k^v})$, $V = (v_1, \ldots, v_{m-k^h})$. We shall determine, if possible, a value $s \in [0, v_1 - k^v]$ such that for the k^v elements in V^* from v^*_{s+1} to $v^*_{s+k^v}$, we have $v^*_{s+1} = \ldots = v^*_{s+k^v} = r$ for some value r, and such that $H = (v^*_1 + k^h, \ldots, v^*_s + k^h, v^*_{s+k^v+1}, \ldots, v^*_{v_1})$. The solution for the blocking component reconstruction problem is given by a matrix A whose submatrices are defined uniquely by the following:

- $H(C) = V^*(C) = (v^*_1, \ldots, v^*_s, v^*_{s+k^v+1}, \ldots, v^*_{v_1})$;
- $H(X) = (x_1, \ldots, x_s)$ where $x_i = k^h$, $1 \le i \le s$;
- $V(Y) = (y_1, \ldots, y_r)$ where $y_j = k^v$, $1 \le j \le r$.

Clearly, $H(C) + H(X) = H$ and $V(C) + V(Y) = V$; note that knowing $H(X)$ and $V(Y)$ the matrices X and Y are trivially determined. By Theorem 2 each solution with the previous structure does not admit any switching components. Supposing that such solution exists, then we show that it is unique by proving that there is only one possible value for $s \in [0, v_1 - k^v]$. Let us compute the number of ones in the matrices C and X:

$$\sum_{i=1}^{v_1} v^*_i - \sum_{i=s}^{s+k^v} v^*_i + sk^h = \sum_{i=1}^{n-k^v} h_i.$$

Since V^* is decreasing, then $\sum_{i=s}^{s+k^v} v^*_i$ is decreasing in s, and further sk^h grows with s. From these properties, the left side of the equation is strictly increasing with s, so this variable can have only one value to satisfy the equation. □

In Figure 3 we depicted a graphical representation of the unique solution for the blocking component reconstruction problem having as input $k^v = 3$, $k^h = 2$, $H = (10, 8, 7, 7, 6, 6, 5, 2, 2, 1, 1)$ and $V = (14, 12, 10, 6, 4, 2, 1, 1)$.

4   The Reconstruction Algorithm

In this section we describe the reconstruction algorithm for a problem with a blocking component. The core of the procedure is defined by Algorithm 1 that determines the vector $V(Y)$ given a fixed number $N^Y = \sum_i v_i(Y)$. We remark that the array $V(Y)$ contains the vertical projections of Y. Our goal is to define $V(Y)$ such that the vector $V - V(Y)$ is the minimum that we can obtain with respect to the dominance ordering; this property guarantees that if $V - V(Y)$ and $H - H(X)$ are not consistent with a solution for the standard reconstruction problem for matrix C, then any other configuration of X and Y cannot yield a solution. Without loss of generality, we impose that the elements of both V and $V - V(Y)$ are ordered. The idea of the procedure consists in determining a value p such that for every element $v_i \geq p$ we may set $v_i(Y) \in \{k^v, v_i - p, v_i - p + 1\}$, and for every element $v_i < p$ then $v_i(Y) = 0$. To compute p, the procedure uses a vector T, containing the maximum number of cells that could be contained in $V(Y)$ for every possible choice of p. Doing so, placing k^v elements in the first columns of $V(Y)$ and 0 in the last ones, we maximize the vector $V - V(Y)$ in the dominance ordering. In Figure 4 is shown an example of the execution of Algorithm 1, where $k^v = 3$ and $N^Y = 23$.

It may be possible to compute the vectors T and Z with a closed formula, but this algorithmic formulation simplifies the following proof.

Theorem 4. *Consider a vector V and two fixed integers N^Y and k^v, and call $V - V(Y)$ the output of Algorithm 1 with input V, N^Y, k^v. For any ordered*

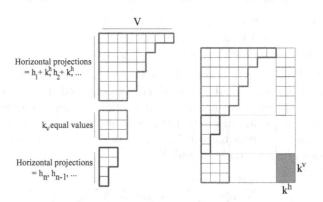

Fig. 3. Computing the only solution without switching components. To the left, we have the graphical representation of V, while to the right we can see the unique solution of the problem.

Algorithm 1. Determination of $V(C)$

---

1. Input: a vector V, two integers N^Y, k^v
2. $Z = [0, \ldots, 0], T = [0, \ldots, 0], p = 0$
3. **for** $i = 1$ **to** $m - k^h$ **do**
4.   **for** $j = v_i$ **downto** $\max(v_i - k^v + 1, 1)$ **do**
5. $z_j = z_j + 1$
6.   **end for**
7. **end for**
8. $t_{v_1} = z_{v_1}$
9. **for** $i = v_1 - 1$ **downto** 1 **do**
10. $t_i = t_{i+1} + z_i$
11. **end for**
12. $p = \max_i(t_i \geq N^Y)$
13. **for** $i = 1$ **to** $m - k^h$ **do**
14. $v_i(y) = v_i - \max(v_i - k^v, p)$
15. **end for**
16. $r = N^Y - \sum_i v_i(Y)$
17. $b = \max_i(v_i \geq p)$
18. **for** $i = b$ **downto** $b - r + 1$ **do**
19. $v_i(Y) = v_i(Y) + 1$
20. **end for**
21. **return** $V - V(Y)$

---

vector $V'(Y)$ such that $\sum_i v_i'(Y) = \sum_i v_i(Y)$ and $\forall i, v_i'(Y) \leq \min(k^v, v_i)$, we have $V - V(Y) \leq_d V - V'(Y)$.

Proof. (Sketch) Call $K = V - V(Y)$ and $L = V - V'(Y)$, and suppose by contradiction that it does not stand that $K \leq_d L$, hence that for some index i, $\sum_{w=1}^i k_w > \sum_{w=1}^i l_w$. Since $\sum_w k_w = \sum_w l_w$ this implies that for some $j > i$, $k_j < l_j$. Consider the value p computed in the procedure, and name $a = \max_w(v_w > p + k_w)$ and $b = \min_w(v_w < p)$ (b is the same value as the one computed in Algorithm 1). For $w \leq a$ then $V(Y) = k^v$, hence $i > a$. For $w \geq b$, then $V(Y) = 0$, and $i < j < b$. By construction of the algorithm, it follows that $[k_{a+1}, \ldots, k_{b-1}] = [p, \ldots, p, p-1, \ldots, p-1]$, hence no indexes i and j can be found in this interval that satisfy the conditions $\sum_{w=1}^i k_w > \sum_{w=1}^i l_w$ and $k_j < l_j$, and also maintain the vector K ordered, hence this brings us to our contradiction. □

At this point we are ready to define formally our reconstruction procedure; since we have proved that given a fixed N^Y, Algorithm 1 executes an optimal choice for the cells in $V(Y)$ (resp. $H(X)$). Thanks to this property, to solve the problem it would suffice to find an appropriate value for N^Y (resp. $N^Y - D$). Unfortunately, we have not found yet a compact formula to determine this value. For this reason, the following procedure, described by Algorithm 2, iterates on all possible values until one yielding a solution is found.

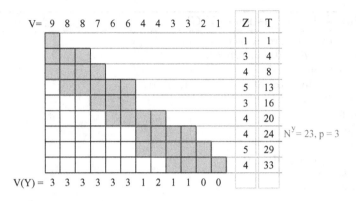

Fig. 4. The optimal choice for vector $V(Y)$; in grey, the maximum number of cells that could be placed in each column given $k^v = 3$. In this example, $p = 3$, hence every grey cell at height > 3 contributes to $V(Y)$, while every cell of height < 3 doesn't.

Algorithm 2. Reconstruction algorithm

---

1. Input: two vectors H and V, two integers k^h and k^v
2. **for** $c = D$ **to** $\min(M^y, M^x + D)$ **do**
3.   Determine $V(Y)$ and $H(X)$ containing c and $c - D$ cells using Algorithm 1
4.   **if** $(V - V(Y))^* \leq_d H - H(X)$ **then**
5.   Rebuild C with Ryser's algorithm, fill X and Y according to $H(X)$ and $V(Y)$, exit for
6.   **end if**
7. **end for**
8. If a solution has not been yet determined, return NO SOLUTION

---

5   A Network Flow Approach to Reconstruction

In this section, we show how the blocking component reconstruction problem may be solved by means of a reduction to a network flow problem. It is well known that the standard reconstruction problem can be solved by a network flow approach; the procedure consists in considering a source for every h_i with capacity equal to the projection, a sink for every v_j again with the appropriate capacity, and a node for every cell (i, j) connected to the related source and sink with arcs of capacity 1.

Beyond showing an interesting connection with an important research field, this reduction also enables us to define a simple algorithm to solve the problem in the case where we have some forbidden positions, i.e. cells that can not have a value of 1. This can be useful in a variety of cases in applications, for example if we already know the configuration of some areas of the image and we want to include this information in the solution; note that positions where a cell must have value 1 can be included in the formulation of the problem with forbidden positions, by simply subtracting 1 from the related projections and setting the cell to a forbidden position. Using the network flow equivalence, we can solve the problem by simply removing the nodes deriving from forbidden positions. Hence by giving a similar reduction in our problem, we also trivially solve the blocking component reconstruction problem with forbidden positions.

The idea of the construction is to compensate the difference in the projections D with an artificial source. Since at least D cells of Y must be equal to 1 in order to obtain a solution, we connect the source with capacity D to all of the cells in Y; further, since for every cell $x_{i,j} = 1$, there must be an additional cell in Y, we also connect the cells in X to every cell in Y.

Formally, the flow problem is composed by the following elements. To simplify the notation, we refer to a node with the name of the related cell; every pair of connected nodes has an edge of capacity 1 linking them. We call the following problem the *associate flow problem* of the original reconstruction one:

- $|H|$ sources of capacity $h_1, \ldots, h_{n-k^v}$;
- 1 source of capacity D;
- $|V|$ sinks of capacity $v_1, \ldots, v_{m-k^h}$;
- $|C|$ nodes, where $c_{i,j}$ is connected with source h_i and sink v_j;
- $|Y|$ nodes, where $y_{i,j}$ is connected to the source D, to the sink v_j and with every node in X;
- $|X|$ nodes, where $x_{i,j}$ is connected to each source h_i and with every node in Y.

The problem configuration is depicted in Figure 5, where the cells are represented by groups, and an edge labelled r represents a series of edges connecting all nodes related by a column or row (as $c_{i,j}$ with h_i) while unlabelled edges represent connections between all possible pairs of the two groups.

Theorem 5. *A reconstruction problem P admits a solution if and only if the associate flow problem F has a solution.*

Proof. (Sketch) From a solution of the flow problem, we can build a matrix that is solution of P by simply setting to 1 every cell in F where the flow enters and

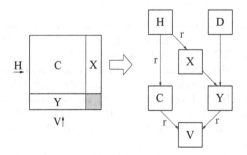

Fig. 5. Reducing an instance of the reconstruction problem to a flow problem

exits. It is easy to verify that the projections match the input vectors H and V, as in F every sink and source receives (emits) a quantity of flow equivalent to the contribution of the cell to the related projection.

Conversely, starting from a solution of P, we can build a solution of F in a similar fashion, but while the edges connecting the groups H, C or C, V or H, X are uniquely determined, we may have many ways to connect the cells in X, Y and D, Y. Any choice yields a correct solution, as long as the number of cells to select in the Y group is equal to D plus the number of nodes to select in the X group, and this property follows immediately from the definition of D. $\square$

6   Conclusions

We have studied how the standard reconstruction problem in discrete tomography can be extended to the case where some vertical and horizontal projections are unavailable, as if a component of the scanned object blocked the x-rays used to study the material. We defined a criteria to determine when a problem has a unique solution, and we furnished two polynomial reconstruction algorithms. One allows us to reduce the problem to the standard reconstruction, allowing the usage of the efficient algorithms known in literature; the other transforms the problem in a flow problem, allowing us to solve it even when we want some prescribed values in the output image.

Since the existence of a blocking component is a recurring problem in some applications, in future works it will be interesting to consider also other assumptions on the image that has to be rebuilt, describing realistic environments. For example, it would be interesting to study how a reconstruction algorithm could work if the image must represent a convex polyomino as in [4], or more generally if we have of some skeletal information as in [15]. Introducing the right assumptions, this line of research could indeed obtain results that could be applied in tomographic applications dealing with reconstruction artifacts caused by blocking components.

References

1. Barcucci, E., Brlek, S., Brocchi, S.: PCIF: an algorithm for lossless true color image compression. In: Wiederhold, P., Barneva, R.P. (eds.) IWCIA 2009. LNCS, vol. 5852, pp. 224–237. Springer, Heidelberg (2009)
2. Barcucci, E., Brocchi, S.: Solving multicolor discrete tomography problems by using prior knowledge. Fundamenta Informaticae 125, 313–328 (2013)
3. Barcucci, E., Brocchi, S., Frosini, A.: Solving the two color problem - An heuristic algorithm. In: Aggarwal, J.K., Barneva, R.P., Brimkov, V.E., Koroutchev, K.N., Korutcheva, E.R. (eds.) IWCIA 2011. LNCS, vol. 6636, pp. 298–310. Springer, Heidelberg (2011)
4. Barcucci, E., Del Lungo, A., Nivat, M., Pinzani, R.: Reconstructing convex polyominoes from horizontal and vertical projections. Theoretical Computer Science 155, 321–347 (1996)

5. Batenburg, K.J.: A network flow algorithm for reconstructing binary images from discrete X-rays. Journal of Mathematical Imaging and Vision 27(2), 175–191 (2013)
6. Batenburg, K.J., Sijbers, J.: DART: a practical reconstruction algorithm for discrete tomography. IEEE Transactions on Image Processing 20(9), 2542–2553 (2011)
7. Costa, M.-C., de Werra, D., Picouleau, C., Schindl, D.: A solvable case of image reconstruction in discrete tomography. Discrete Applied Mathematics 148(3), 240–245 (2005)
8. De Man, B., Nuyts, J., Dupont, P., Marchal, G., Suetens, P.: Reduction of metal streak artifacts in x-ray computed tomography using a transmission maximum a posteriori algorithm. In: Nuclear Science Symposium, pp. 850–854 (1999)
9. Duan, X., Zhang, L., Xiao, Y., Cheng, J., Chen, Z., Xing, Y.: Metal artifact reduction in CT images by sinogram TV inpainting. In: Nuclear Science Symposium Conference, pp. 4175–4177 (2008)
10. Durr, C., Guinez, F., Matamala, M.: Reconstructing 3-Colored Grids from Horizontal and Vertical Projections is NP-Hard, A Solution to the 2-Atom Problem in Discrete Tomography. SIAM J. Discrete Math. 26(1), 330–352 (2012)
11. Faggiano, E., Lorenzi, T., Quarteroni, A.: Metal Artifact Reduction in Computed Tomography Images by Variational Inpainting Methods, MOX-report No. 16/2013 (2013)
12. Gardner, R.J., Gritzmann, P., Pranenberg, D.: On the computational complexity of reconstructing lattice sets from their X-rays. Discrete Mathematics 202(1-3), 45–71 (1999)
13. Guinez, F.: A formulation of the wide partition conjecture using the atom problem in discrete tomography. Discrete Applied Mathematics (2013) (in press)
14. Hantos, N., Balazs, P.: A uniqueness result for reconstructing hv-convex polyominoes from horizontal and vertical projections and morphological skeleton. In: 8th International Symposium on Image and Signal Processing and Analysis (ISPA). IEEE (2013)
15. Hantos, N., Balazs, P.: The Reconstruction of Polyominoes from Horizontal and Vertical Projections and Morphological Skeleton is NP-complete. Fundamenta Informaticae 125(3), 343–359 (2013)
16. Herman, G., Kuba, A.: Advances in discrete tomography and its applications. Birkhauser, Boston (2007)
17. Kalender, W.A., Hebel, R., Ebersberger, J.: Reduction of CT artifacts caused by metallic implants. Radiology 164(2), 576–577 (1987)
18. Picouleau, C., Brunetti, S., Frosini, A.: Reconstructing a binary matrix under timetabling constraints. Electronic Notes in Discrete Mathematics 20, 99–112 (2005)
19. Ryser, H.J.: Combinatorial properties of matrices of zeros and ones. Canadian Journal of Mathematics 9, 371–377 (1957)
20. Wang, G., Snyder, D.L., O'Sullivan, J., Vannier, M.: Iterative deblurring for ct metal artifact reduction. IEEE Transactions on Medical Imaging 15(5), 657–664 (1996)

An Entropic Perturbation Approach to TV-Minimization for Limited-Data Tomography

Andreea Deniţiu[1,2], Stefania Petra[1], Claudius Schnörr[2],
and Christoph Schnörr[1]

[1] Image and Pattern Analysis Group, University of Heidelberg, Germany
{denitiu,petra,schnoerr}@math.uni-heidelberg.de
[2] Hochschule München, Fakultät für Informatik und Mathematik,
München, Germany
denitiu@hm.edu, schnoerr@cs.hm.edu

Abstract. The reconstruction problem of discrete tomography is studied using novel techniques from compressive sensing. Recent theoretical results of the authors enable to predict the number of measurements required for the unique reconstruction of a class of cosparse dense 2D and 3D signals in severely undersampled scenarios by convex programming. These results extend established ℓ_1-related theory based on sparsity of the signal itself to novel scenarios not covered so far, including tomographic projections of 3D solid bodies composed of few different materials. As a consequence, the large-scale optimization task based on total-variation minimization subject to tomographic projection constraints is considerably more complex than basic ℓ_1-programming for sparse regularization. We propose an entropic perturbation of the objective that enables to apply efficient methodologies from unconstrained optimization to the perturbed dual program. Numerical results validate the theory for large-scale recovery problems of integer-valued functions that exceed the capacity of the commercial MOSEK software.

Keywords: discrete tomography, compressed sensing, underdetermined systems of linear equations, cosparsity, phase transitions, total variation, entropic perturbation, convex duality, convex programming.

1 Introduction

This paper addresses the problem of reconstructing compound solid bodies u from few tomographic projections: $Au = b$. Theoretical guarantees of reconstruction performance relate to current research in the field of compressive sensing, concerned with *real* sensor matrices A that do *not* satisfy commonly made assumptions like the restricted isometry property.

By adopting the cosparse signal model [1] that conforms to our tomographic scenario, the authors [2] recently established the existence of "phase transitions" that relate the required limited(!) number of measurements to the cosparsity level of u in order to guarantee unique recovery. This result was extensively validated

E. Barcucci et al. (Eds.): DGCI 2014, LNCS 8668, pp. 262–274, 2014.
© Springer International Publishing Switzerland 2014

numerically using the commercial MOSEK software, to avoid that numerical optimization issues affect the validation.

For our 3D problems, a dedicated numerical optimization algorithm is necessary, however, because MOSEK cannot handle medium and large problem sizes. We present in this paper such an approach by utilizing the fact that adequate perturbations of the optimization problem may lead to a simpler dual problem [3,4]. We work out a corresponding approach to our specific reconstruction problem

$$\min_u \text{TV}(u) \quad \text{subject to} \quad Au = b \, , \tag{1}$$

that minimizes the total variation $\text{TV}(u)$ subject to the projection constraints.

Organization. Section 2 briefly reports the above-mentioned phase transitions, followed by presenting a reformulation in Section 3 together with uniqueness results. A suitable perturbation is worked out in Section 4 that favorably compares to MOSEK for relevant problem sizes (Section 5).

Basic Notation. For $n \in \mathbb{N}$, we denote $[n] = \{1, 2, \ldots, n\}$. The complement of a subset $J \subset [n]$ is denoted by $J^c = [n] \setminus J$. For some matrix A and a vector z, A_J denotes the submatrix of rows indexed by J, and z_J the corresponding subvector. A_i will denote the ith row of A. $\mathcal{N}(A)$ and $\mathcal{R}(A)$ denote the nullspace and the range of A, respectively. Vectors are column vectors and indexed by superscripts. Sometimes, we will write e.g. $(u, v)^\top$ instead of the more correct $(u^\top, v^\top)^\top$. $A^\top$ denotes the transposed of A. $\mathbb{1} = (1, 1, \ldots, 1)^\top$ denotes the one-vector whose dimension will always be clear from the context. The dimension of a vector z is $\dim(z)$. $\langle x, z \rangle$ denotes the standard scalar product in $\mathbb{R}^n$ and we write $x \perp z$ if $\langle x, z \rangle = 0$. The number of nonzeros of a vector x is denoted by $\|x\|_0$. The indicator function of a set C is denoted by $\delta_C(x) := \begin{cases} 0, & \text{if } x \in C \\ +\infty, & \text{if } x \notin C \, . \end{cases}$
$\sigma_C(x) := \sup_{y \in C} \langle y, x \rangle$ denotes the support function of a nonempty set C. $\partial f(x)$ is the subdifferential of f at x and $\text{int}\, C$ and $\text{rint}\, C$ denote the interior and the relative interior of a set C. By f^* we denote the conjugate function of f. We refer to [5] for related properties.

In what follows, we will work with an anisotropic discretization of the TV operator in (1) given by

$$\text{TV}(u) := \|Bu\|_1, \quad B := \begin{pmatrix} \partial_1 \otimes I \otimes I \\ I \otimes \partial_2 \otimes I \\ I \otimes I \otimes \partial_3 \end{pmatrix} \in \mathbb{R}^{p \times n} \, , \tag{2}$$

where $\otimes$ denotes the Kronecker product and ∂_i, $i = 1, 2, 3$, are derivative matrices forming the forward first-order differences of u wrt. the considered coordinates. Implicitly, it is understood that u on the r.h.s. of (2) is a vector representing all voxel values in appropriate order.

2   Weak Phase Transitions for TV-Based Reconstruction

We summarize in this section an essential result from [2] concerning the unique recovery from tomographic projection by solving problem (1), depending on the cosparsity level ℓ of u (Definition 1) and the number m of measurements (projections) of u. This result motivates the mathematical programming approach discussed in Section 3 and the corresponding numerical optimization approach presented in Section 4.

Definition 1 (cosparsity, cosupport). *The* cosparsity *of* $u \in \mathbb{R}^n$ *with respect to* $B \in \mathbb{R}^{p \times n}$ *is*

$$\ell := p - \|Bu\|_0 , \tag{3}$$

and the cosupport *of* u *with respect to* B *is*

$$\Lambda := \{r \in [p] : (Bu)_r = 0\}, \qquad |\Lambda| = \ell . \tag{4}$$

Thus, $B_\Lambda u = 0$. In view of the specific operator B given by (2), ℓ measures the homogeneity of the volume function u. A large value of ℓ can be expected for u corresponding to solid bodies composed of few homogeneous components.

Proposition 1 ([2, Cor. 4.6]). *For given $A \in \mathbb{R}^{m \times n}$ and $B \in \mathbb{R}^{p \times n}$, suppose the rows of $\left(\begin{smallmatrix} A \\ B \end{smallmatrix}\right)$ are linearly independent. Then a ℓ-cosparse solution u to the measurement equations $Au = b$ will be unique if the number of measurements satisfies*

$$m \geq 2n - (\ell + \sqrt{2\ell + 1} - 1) \qquad \text{(in 2D)} , \tag{5a}$$

$$m \geq 2n - \frac{2}{3}\left(\ell + \sqrt[3]{3\ell^2} + 2\sqrt[3]{\frac{\ell}{3}} - 2\right) \qquad \text{(in 3D)} . \tag{5b}$$

The above lower bounds on the number of measurements m required to recover a ℓ-cosparse vector u, imply that recovery can be carried out by solving $\min_u \|Bu\|_0$ subject to $Au = b$. Replacing this objective by the convex relaxation (2) yields an excellent agreement of empirical results with the prediction (5), as shown in [2], although the independency assumption made in Prop. (1) does not strictly hold for the sensor matrices A and the operator B (2) used in practice.

This motivates to focus on efficient and sparse numerical optimization techniques that scale up to large problem sizes.

3   TV-Recovery by Linear Programming

We consider the discretized TV-term (2), an additional nonnegative constraint on image u and express $Bu = z$. Thus, (1) becomes

$$\min_{u,z} \|z\|_1 \quad \text{s.t.} \quad Bu = z, \quad Au = b, \quad u \geq 0 . \tag{6}$$

3.1 Primal Linear Program and Its Dual

By splitting the variable z in its positive $v^1 := \max\{0, z\}$ and negative part $v^2 := -\min\{0, z\}$ we convert problem (6) into a linear program in normal form. With

$$M := \begin{pmatrix} B & -I & I \\ A & 0 & 0 \end{pmatrix}, \quad q := \begin{pmatrix} 0 \\ b \end{pmatrix}, \tag{7}$$

and the polyhedral set

$$\mathcal{P} := \{x \in \mathbb{R}^{n+2p} : Mx = q, \ x \geq 0\}, \quad x := (u, v^1, v^2)^\top, \tag{8}$$

problem (6) becomes the linear program (P)

$$(P) \quad \min_{x \in \mathcal{P}} \langle c, x \rangle = \min_{(u, v^1, v^2) \in \mathcal{P}} \langle \mathbb{1}, v^1 + v^2 \rangle, \quad c = (0, \mathbb{1}, \mathbb{1})^\top. \tag{9}$$

We further assume that $\mathcal{P} \neq \emptyset$, i.e. a feasible solution always exists. Due to $c \geq 0$, the linear objective in (P) is bounded on $\mathcal{P}$. Thus (P) always has a solution under the feasibility assumption. In view of basic linear programing theory, compare [5, 11.43], the dual program also has a solution. The dual program (D) reads

$$(D) \quad \min_y -\langle q, y \rangle, \quad M^\top y \leq c. $$

With

$$y = \begin{pmatrix} y_0 \\ y_b \end{pmatrix}, \quad M^\top = \begin{pmatrix} B^\top & A^\top \\ -I & 0 \\ I & 0 \end{pmatrix}, \quad M^\top y = \begin{pmatrix} B^\top y_0 + A^\top y_b \\ -y_0 \\ y_0 \end{pmatrix}, \tag{10}$$

this reads

$$\min_{y_0, y_b} -\langle b, y_b \rangle \quad \text{s.t.} \quad B^\top y_0 + A^\top y_b \leq 0, \quad -1 \leq y_0 \leq 1. \tag{11}$$

Moreover, both primal and dual solutions $(\bar{x}, \bar{y})$ will satisfy the following optimality conditions

$$0 \leq c - M^\top y \perp x \geq 0, \tag{12}$$

$$Mx = q. \tag{13}$$

3.2 Uniqueness of Primal LP

A classical argument for replacing $\|\cdot\|_0$ by $\|\cdot\|_1$ and solving for (6) is uniqueness of the minimal ℓ_1 (thus LP) solution. Let $\bar{x} = (\bar{u}, \bar{v}) = (\bar{u}, \bar{v}^1, \bar{v}^2)$ be ℓ-cosparse and solve (9). We assume throughout

$$\bar{u}_i > 0, \ i \in [n]. \tag{14}$$

Based on $\bar{x}$, we define the corresponding support set

$$J := \{i \in [\dim(x)] : \bar{x}_i \neq 0\} = \mathrm{supp}(\bar{x}), \quad \bar{J} := J^c = [\dim(x)] \setminus J. \tag{15}$$

Denoting $k := p - \ell$, the cardinality of the index sets J and $\bar{J}$ is

$$|\bar{J}| = 2\ell + k = p + \ell, \qquad |J| = n + 2p - |\bar{J}| = n + k \ , \qquad (16)$$

compare [2, Lem. 5.2]. This shows that $x \in \mathbb{R}^{n+2p}$ is a $(n+k)$-sparse vector.

Theorem 1 ([6, Thm. 2(iii)]). *Let $\bar{x}$ be a solution of the linear program (9). The following statements are equivalent:*

(i) $\bar{x}$ is unique.
(ii) There exists no x satisfying

$$Mx = 0, \quad x_{\bar{J}} \geq 0, \quad \langle c, x \rangle \leq 0, \quad x \neq 0 \ . \qquad (17)$$

Theorem (1) can be turned into a *nullspace condition* w.r.t. the sensor matrix A, for the unique solvability of problems (9) and (6).

Proposition 2 ([2, Cor. 5.3]). *Let $\bar{x} = (\bar{u}, \bar{v}^1, \bar{v}^2)$ be a solution of the linear program (9) with component $\bar{u}$ that has cosupport Λ with respect to B. Then $\bar{x}$, resp. $\bar{u}$, are unique if and only if*

$$\forall x = \begin{pmatrix} u \\ v \end{pmatrix}, \ v = \begin{pmatrix} v^1 \\ v^2 \end{pmatrix} \quad s.t. \quad u \in \mathcal{N}(A) \setminus \{0\} \quad and \quad Bu = v^1 - v^2 \qquad (18)$$

the condition

$$\|(Bu)_\Lambda\|_1 > \langle (Bu)_{\Lambda^c}, \operatorname{sign}(B\bar{u})_{\Lambda^c} \rangle \qquad (19)$$

holds. Furthermore, any unknown ℓ-cosparse vector u^*, with $Au^* = b$, can be uniquely recovered as solution $\bar{u} = u^*$ to (6) if and only if, for all vectors u conforming to (18), the condition

$$\|(Bu)_\Lambda\|_1 > \sup_{\Lambda \subset [p]:\ |\Lambda|=\ell} \ \sup_{\bar{u} \in \mathcal{N}(B_\Lambda)} \langle (Bu)_{\Lambda^c}, \operatorname{sign}(B\bar{u})_{\Lambda^c} \rangle \qquad (20)$$

holds.

Remark 1. Conditions (19) and (20) clearly indicate the direct influence of cosparsity on the recovery performance: if $\ell = |\Lambda|$ increases, then these conditions will more likely hold. On the other hand, these results are mainly theoretical since numerically checking (20) is infeasible. However, we will assume that uniqueness of (6) is given, provided that the cosparsity ℓ of the unique solution $\bar{u}$ satisfies the conditions in (5a) and (5b). This assumption is motivated by the comprehensive experimental assessment of recovery properties reported in [2].

Remark 2. We note that, besides the condition for uniqueness from Thm. (1), a LP solution is unique if there is a unique feasible point. For high cosparsity levels ℓ, this seems to be often the case.

Let $\bar{x}$ be a (possibly unique) primal solution of (P) and $\bar{y}$ a dual solution. In view of (15) and (12) we have

$$(c - M^\top \bar{y})_i = 0, \quad \forall i \in J \ . \qquad (21)$$

We note that non-degeneracy of the primal-dual pair $(\bar{x}, \bar{y})$ implies uniqueness of the dual variable $\bar{y}$.

4   Recovery by Perturbed Linear Programming

Preliminaries: Fenchel Duality Scheme. We will use the following result.

Theorem 2 ([5]). *Let* $f : \mathbb{R}^n \to \mathbb{R} \cup \{+\infty\}$, $g : \mathbb{R}^m \to \mathbb{R} \cup \{+\infty\}$ *and* $A \in \mathbb{R}^{m \times n}$. *Consider the two problems*

$$\inf_{x \in \mathbb{R}^n} \varphi(x), \qquad \varphi(x) = \langle c, x \rangle + f(x) + g(b - Ax), \tag{22a}$$

$$\sup_{y \in \mathbb{R}^m} \psi(y), \qquad \psi(y) = \langle b, y \rangle - g^*(y) - f^*(A^\top y - c) . \tag{22b}$$

where the functions f *and* g *are proper, lower-semicontinuous (lsc) and convex.*
Suppose that

$$b \in \text{int}(A \, \text{dom} \, f + \text{dom} \, g), \tag{23a}$$

$$c \in \text{int}(A^\top \, \text{dom} \, g^* - \text{dom} \, f^*) . \tag{23b}$$

Then the optimal solutions $\bar{x}, \bar{y}$ *are determined by*

$$0 \in c + \partial f(\bar{x}) - A^\top \partial g(b - A\bar{x}), \qquad 0 \in b - \partial g^*(\bar{y}) - A \partial f^*(A^\top \bar{y} - c) \tag{24a}$$

and connected through

$$\bar{y} \in \partial g(b - A\bar{x}), \qquad \bar{x} \in \partial f^*(A^\top \bar{y} - c), \tag{25a}$$

$$A^\top \bar{y} - c \in \partial f(\bar{x}), \qquad b - A\bar{x} \in \partial g^*(\bar{y}) . \tag{25b}$$

Furthermore, the duality gap vanishes: $\varphi(\bar{x}) = \psi(\bar{y})$.

Entropic Perturbation and Exponential Penalty. In various approaches
to solving large-scale linear programs, one regularizes the problem by adding to
the linear cost function a separable nonlinear function multiplied by a small positive parameter. Popular choices of this nonlinear function include the quadratic
function, the logarithm function, and the $\langle x, \log(x) \rangle$-entropy function. Our main
motivation in following this trend is that by adding a strictly convex and separable perturbation function, the dual problem will become unconstrained and
differentiable. Consider

$$(P_\varepsilon) \qquad \min \langle c, x \rangle + \varepsilon \langle x, \log x - \mathbb{1} \rangle \quad \text{s.t.} \quad Mx = q, x \geq 0 . \tag{26}$$

The perturbation approach by the entropy function was studied by Fang et
al. [4,7] and, from a dual exponential penalty view, by Cominetti et al. [8].

The Unconstrained Dual. We write (P_ε) (26) in the form (22a)

$$\min \varphi(x), \quad \varphi(x) := \langle c, x \rangle + \underbrace{\varepsilon \langle x, \log x - \mathbb{1} \rangle + \delta_{\mathbb{R}^n_+}(x)}_{:=f(x)} + \delta_0(q - Mx) . \tag{27}$$

With $g := \delta_0$, we get $g^* \equiv 0$, since $\delta_C^* \equiv \sigma_C$ and thus

$$g^*(y) = \delta_0^*(y) = \sigma_0(y) = \sup_{z=0}\langle y, z \rangle = 0, \quad \forall y \in \mathbb{R}^n$$

holds. On the other hand, we have $f^*(y) = \varepsilon \langle \mathbb{1}, e^{\frac{y}{\varepsilon}} \rangle$. Now (22b) gives immediately the dual problem

$$\sup \psi(y), \quad \psi(y) := \langle q, y \rangle - \varepsilon \langle \mathbb{1}, e^{\frac{M^\top y - c}{\varepsilon}} \rangle \ . \tag{28}$$

We note that ψ is unconstrained and twice differentiable with

$$\nabla \psi(y) = q - M e^{\frac{M^\top y - c}{\varepsilon}} \quad \text{and} \tag{29a}$$

$$\nabla^2 \psi(y) = -\frac{1}{\varepsilon} M \operatorname{diag} e^{\frac{M^\top y - c}{\varepsilon}} M^\top \ . \tag{29b}$$

Moreover, $-\nabla^2 \psi \succ 0$ for all y, with $e^{\frac{M^\top y - c}{\varepsilon}} \in \mathcal{R}(M) = \mathcal{N}(M)^\perp$, in view of (29b). Note that if ψ has a solution, then it is unique and the strictly feasible set must be nonempty, see (29a), thus $\operatorname{rint} \mathcal{P} = \{x \colon Mx = q, x > 0\} \neq \emptyset \Leftrightarrow q \in M(\mathbb{R}_{++}^n)$. Further, we can rewrite (28) in a more detailed form in view of (10)

$$(D_\varepsilon) \quad \min_{y_0, y_b} -\langle b, y_b \rangle + \varepsilon \langle \mathbb{1}_n, e^{\frac{B^\top y_0 + A^\top y_b}{\varepsilon}} \rangle + \varepsilon \langle \mathbb{1}_p, e^{\frac{-y_0 - \mathbb{1}_p}{\varepsilon}} \rangle + \varepsilon \langle \mathbb{1}_p, e^{\frac{y_0 - \mathbb{1}_p}{\varepsilon}} \rangle \ . \tag{30}$$

Connecting Primal and Dual Variables. With $\operatorname{dom} g = 0$, $\operatorname{dom} g^* = \mathbb{R}^n$, $\operatorname{dom} f^* = \mathbb{R}^n$ and $\operatorname{dom} f = \mathbb{R}_+^n$, the assumptions (23) become $q \in \operatorname{int} M(\mathbb{R}_+^n) = M(\operatorname{int} \mathbb{R}_+^n) = M(\mathbb{R}_{++}^n)$, compare [5, Prop. 2.44], and $c \in \operatorname{int} \mathbb{R}^n = \mathbb{R}^n$. Thus, under the assumption of a strictly feasible set, we have no duality gap. Moreover both problems (27) and (28) have a solution.

Theorem 3. *Denote by x_ε and y_ε a solution of (P_ε) and (D_ε) respectively. Then the following statements are equivalent:*

(a) $q \in M(\mathbb{R}_{++}^n)$, thus the strictly feasible set is nonempty.
(b) The duality gap is zero $\psi(y_\varepsilon) = \varphi(x_\varepsilon)$.
(c) Solutions x_ε and y_ε of (P_ε) and (D_ε) exist and are connected through

$$x_\varepsilon = e^{\frac{M^\top y_\varepsilon - c}{\varepsilon}} \ . \tag{31}$$

Proof. (a) $\Rightarrow$ (b): holds due to Thm. 2. On the other hand, (b) implies solvability of ψ and thus (a), as noted after Eq. (29b). (a) $\Rightarrow$ (c): The assumptions of Thm. 2 hold. Now $\partial f^*(y) = \{\nabla f^*(y)\} = \{e^{\frac{y}{\varepsilon}}\}$ and the r.h.s. of (25a) gives (c). Now, (c) implies $M x_\varepsilon = q$ and thus (a). $\qquad \square$

The following result shows that for $\varepsilon \to 0$ and under the nonempty strictly feasible set assumption, x_ε given by (31) approaches the least-entropy solution of (P), if y_ε is a solution of (D_ε). The proof follows along the lines of [9, Prop. 1].

Theorem 4. *Denote the solution set of* (9) *by* $\mathcal{S}$. *Assume* $\mathcal{S} \neq \emptyset$ *and* rint $\mathcal{P} \neq \emptyset$. *Then, for any sequence of positive scalars* (ε_k) *tending to zero and any sequence of vectors* (x_{ε_k}), *converging to some* x^*, *we have* $x^* \in \text{argmin}_{x \in S}\langle x, \log x - 1\rangle$. *If* $\mathcal{S}$ *is a singleton, denoted by* $\bar{x}$, *then* $x_{\varepsilon_k} \to \bar{x}$.

Partial Perturbation. In the case of a unique and sparse feasible point $\bar{x}$ the assumption $q \in M(\mathbb{R}^n_{++})$ does not hold. With $J = \text{supp}(\bar{x})$ the primal reads

$$\min\langle c, x\rangle + \varepsilon\langle x_J, \log x_J - \mathbb{1}_J\rangle \quad \text{s.t.} \quad Mx = q, x_{J^c} = 0, x \geq 0,$$

and the dual becomes

$$\max_y \langle q, y\rangle - \varepsilon\langle \mathbb{1}, e^{\frac{(M^\top)_J y - c_J}{\varepsilon}}\rangle \; .$$

However, the solution support J is unknown. Using (21), one can show that an approximative solution y_ε of (D_ε), i.e. $\|\nabla\psi(y_\varepsilon)\| \leq \tau_\varepsilon$, with $\tau_\varepsilon > 0$ small, can be used to construct x_ε according to (31), such that $x_\varepsilon \to \bar{x}$.

Exponential Penalty Method. We discussed above how problem (P_ε) tends to (P) as $\varepsilon \to 0$. Likewise, (D_ε) tends to (D). This was shown by Cominetti et al. [8, Prop. 3.1]. The authors noticed that the problem (D_ε) is a exponential penalty formulation of (D), compare (10) and (30).

They also investigated the asymptotic behavior of the trajectory y_ε and its relation with the solution set of (D). They proved the trajectory y_ε is approximatively a straight line directed towards the center of the optimal face of (D), namely $y_\varepsilon = y^* + \varepsilon d^* + \eta(\varepsilon)$, where y^* is a particular solution of (D). Moreover, the error $\eta(\varepsilon)$ goes to zero exponentially fast, i.e. at the speed of $e^{\frac{-\mu}{\varepsilon}}$ for some $\mu > 0$. See the proof of [8, Prop. 3.2].

5   Numerical Experiments

In this section, we illustrate the performance of our perturbation approach compared to the LP solver MOSEK, in noisy and non-noisy environments, for 2D and 3D cases. Besides the proposed entropic approach, we implemented a quadratic perturbation approach for comparison purposes. We solved the perturbed dual formulations by a conventional unconstrained optimization approach, the *Limited Memory BFGS* algorithm [10], which scales to large problem sizes. In all experiments, the perturbation parameters were kept fixed to $\varepsilon = 1/50$ and $\alpha = 1$, see Fig. 3 for a justification. In the following $\frac{1}{2\alpha}$ denotes the perturbation parameter of a quadratic term applied to (6). We allowed a maximum number of 1500 iterations and stopped when the norm of the gradient of the perturbed dual function satisfies $\|\nabla\psi(y^k)\| \leq 10^{-4}$.

The first performance test was done on 2D $d \times d$ images of randomly located ellipsoids with random radii along the coordinate axes. See Fig. 1 (bottom row)

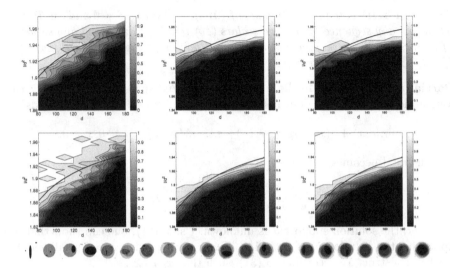

Fig. 1. Phase transitions for the 2D case, $d = 80 : 10 : 180$, 4 cameras (top plots) and 6 cameras (bottom plots), computed for the noiseless case with MOSEK (plots on left), our approach (plots in the middle) and our approach for the noisy case (plots on the right). The black solid line corresponds to the theoretical curve (5a). The last row displays some of the random $d \times d$ images used in the experiments in order of decreasing cosparsity ℓ. MOSEK performs better on smaller images and worse on larger ones.

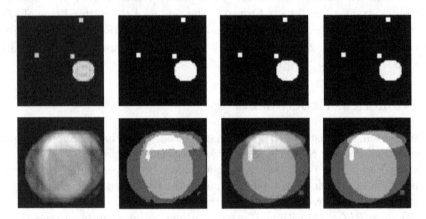

Fig. 2. Comparison between the quadratic perturbation approach (left two columns) and entropic perturbation approach (right two columns) for two relative cosparsity levels. Two 80×80 images, are projected along 6 directions. For both $\rho = \ell/d^2 = 1.7786$ (top row) and $\rho = \ell/d^2 = 1.8586$ (bottom row) reconstruction should in theory be *exact*. Result (left column) and rounded result (second left column) of the quadratic perturbation approach for $\alpha = 1$. Results for the entropic perturbation approach (right two columns) with $\varepsilon = 1/50$. Here the rounded result *exactly equals* the original image (right column). Hence, the approximate solution by the entropic approach is closer to the original solution.

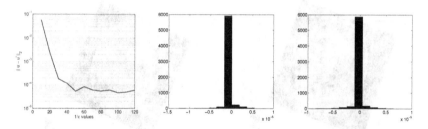

Fig. 3. Experimental finite perturbation property of the entropic approach. Here $\varepsilon = 1/50$ is a reasonable value in 2D since the reconstruction error varies insignificantly (left). The histograms of $(u - u^*)$ for $\varepsilon = 1/50$ (middle) and $\varepsilon = 1/120$ (right) are highly similar.

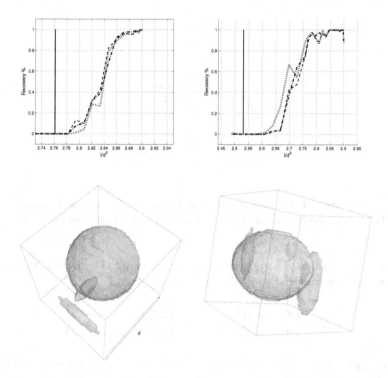

Fig. 4. Phase transitions for the 3D case, 3 cameras (top left) and 4 cameras (top right) and random example of perfectly reconstructed images $d = 31$ (bottom). The average performance of MOSEK (dotted line) for the noiseless case, and the entropic approach in the noiseless (dashed line) and noisy (dash-dot line) case for $\varepsilon = 1/50$ as a variation of relative cosparsity. The black solid line corresponds to the theoretical curve (5b). Measurements were corrupted by Poisson noise of SNR = 50db.

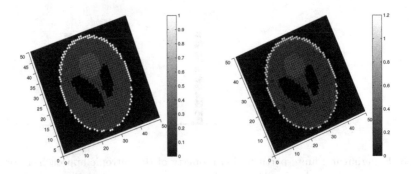

Fig. 5. Slices through the 3D volume ($d = 51$) of an original Shepp-Logan image (left) and the reconstructed image from 7 *noisy* projecting directions via the entropic perturbation approach, satisfying $\|u - u^*\|_\infty < 0.5$ (right). This shows that the approach is also stable for low noise levels as opposed to MOSEK. Measurements were corrupted by Poisson noise of SNR = 50db.

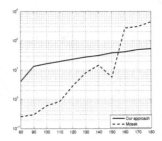

Fig. 6. Comparison between computation times of the proposed approach and MOSEK. We average results for 90 2D test images and vary $d = 80 : 10 : 180$.

for two sample images. The relative cosparsity is denoted by $\rho := \frac{\ell}{n}$. Parameters p and n vary for two- and three-dimensional images as

$$n = \begin{cases} d^2 & \text{in 2D} \\ d^3 & \text{in 3D} \end{cases} , \quad p = \begin{cases} 2d(d-1) & \text{in 2D} \\ 3d^2(d-1) & \text{in 3D} \end{cases} . \tag{32}$$

Our parametrization relates to the design of the projection matrices $A \in \mathbb{R}^{m \times n}$, see [2] for details.

The phase transitions in Fig. 1 display the empirical probability of exact recovery over the space of parameters that characterize the problem. Here we performed 90 tests for each (ρ, d) parameter combination.

We analyzed the influence of the image cosparsity, also for 3D images, see Fig. 4. In 3D, for each problem instance defined by a (ρ, d)-point, we generated 60 random images. In both 2D and 3D, we declared a random test as successful if $\|u - u^*\|_\infty < 0.5$, which leads to perfect reconstruction after rounding. Both Fig. 1 and Fig. 4 display a phase transition and exhibit regions where exact image reconstruction has probability equal or close to one and closely match the solid green line in the plots, which stands for the theoretical curve (5a). In the noisy case, projection data was corrupted by Poisson noise of $SNR = 50$db. The perturbation parameter has been set as in the noiseless case, i.e. $\varepsilon = 1/50$ and $\alpha = 1$. MOSEK however was unable to solve the given problem, stating that either the primal or the dual might be infeasible. Thus our perturbation approach is also stable to low noise levels as opposed to MOSEK. Moreover the proposed algorithm scales much better with the problem size and is significantly more efficient for large problem sizes that are relevant to applications. In particular, problem sizes can be handled where MOSEK stalls, see Fig. 6. Finally, we underline that the entropic perturbation approach performs significantly better than quadratic perturbation as shown in Fig. 2.

6 Conclusion

We presented a mathematical programming approach based on perturbation that copes with large tomographic reconstruction problems of the form (1). While the perturbation enables to apply efficient sparse numerics, it does not compromise reconstruction accuracy. This is a significant step in view of the big data volumes of industrial scenarios.

Our further work will examine the relation between the geometry induced by perturbations on the u-space and the geometry of Newton-like minimizing paths, and the potential for parallel implementations.

Acknowledgement. SP gratefully acknowledges financial support from the Ministry of Science, Research and Arts, Baden-Württemberg, within the Margarete von Wrangell postdoctoral lecture qualification program. AD and the remaining authors appreciate financial support of this project by the Bavarian State Ministry of Education, Science and Arts.

References

1. Nam, S., Davies, M., Elad, M., Gribonval, R.: The Cosparse Analysis Model and Algorithms. Applied and Computational Harmonic Analysis 34(1), 30–56 (2013)
2. Deniţiu, A., Petra, S., Schnörr, C., Schnörr, C.: Phase Transitions and Cosparse Tomographic Recovery of Compound Solid Bodies from Few Projections. ArXiv e-prints (November 2013)
3. Ferris, M.C., Mangasarian, O.L.: Finite perturbation of convex programs. Appl. Math. Optim. 23, 263–273 (1991)
4. Fang, S.C., Tsao, H.S.J.: Linear programming with entropic perturbation. Math. Meth. of OR 37(2), 171–186 (1993)
5. Rockafellar, R., Wets, R.J.B.: Variational Analysis, 2nd edn. Springer (2009)
6. Mangasarian, O.L.: Uniqueness of Solution in Linear Programming. Linear Algebra and its Applications 25, 151–162 (1979)

7. Fang, S.C., Tsao, H.S.J.: On the entropic perturbation and exponential penalty methods for linear programming. J. Optim. Theory Appl. 89, 461–466 (1996)
8. Cominetti, R., San Martin, J.: Asymptotic analysis of the exponential penalty trajectory in linear programming. Math. Progr. 67, 169–187 (1994)
9. Tseng, P.: Convergence and Error Bound for Perturbation of Linear Programs. Computational Optimization and Applications 13(1-3), 221–230 (1999)
10. Bonnans, J.F., Gilbert, Lemaréchal, C., Sagastizábal, C.: Numerical Optimization – Theoretical and Practical Aspects. Springer, Berlin (2006)

Fourier Inversion of the Mojette Transform

Andrew Kingston[1], Heyang Li[1], Nicolas Normand[2], and Imants Svalbe[3]

[1] Dept. Applied Maths, RSPE, The Australian National University,
Canberra ACT 2600, Australia
`andrew.kingston@anu.edu.au`
[2] IRCCyN, École Polytechnique de l'Université de Nantes,
La Chantrerie, Nantes 44306, France
[3] School of Physics, Monash University, Clayton VIC 3800, Australia

Abstract. The Mojette transform is a form of discrete Radon transform that maps a 2D image ($P \times Q$ pixels) to a set of I 1D projections. Several fast inversion methods exist that require $O(PQI)$ operations but those methods are ill-conditioned. Several robust (or well-conditioned) inversion methods exist, but they are slow, requiring $O(P^2 Q^2 I)$ operations. Ideally we require an inversion scheme that is both fast and robust to deal with noisy projections. Noisy projection data can arise from data that is corrupted in storage or by errors in data transmission, quantisation errors in image compression, or through noisy acquisition of physical projections, such as in X-ray computed tomography. This paper presents a robust reconstruction method, performed in the Fourier domain, that requires $O(P^2 Q \log P)$ operations.

Keywords: Radon transform, Mojette transform, Fourier inversion, tomography.

1 Introduction

The Mojette transform is a discrete form of the Radon transform that maps a 2D image ($P \times Q$ pixels) to a set of I 1D projections. It was first developed by Guedon *et al* in 1996 [5] in the context of psychovisual image coding. Due to its distribution, redundancy, and invertibility properies, it has since found applications in distributed data storage on disks and networks, network packet data transmission, tomographic reconstruction, image watermarking, image compression, and image encryption. A summary of the early development of the Mojette transform was the subject of an invited paper to DGCI in 2005 [6] and a book on the Mojette transform was published in 2009 [4].

The Radon transform has many inversion algorithms based on the form of the input data and the *a priori* information available. Similarly, the Mojette transform has many inverse algorithms. There are fast methods that use local back-projection (LBP) [10] and geometrically driven LBP (GLBP) [11]. These recover the original data in $O(PQI)$ operations, however, they are ill-conditioned being essentially forms of Gaussian elimination; any errors in the determination of a variable value are propagated and compounded throughout the remaining

E. Barcucci et al. (Eds.): DGCI 2014, LNCS 8668, pp. 275–284, 2014.

variables. There are also well-conditioned, or robust, methods that have been explored in the context of tomographic reconstruction. These include a conjugate gradient back-projection (CGBP) [12] that requires $O(P^2Q^2I)$ operations, and an exact back-projection (EBP) method that requires an inordinate number of projections ($I \approx 12PQ/\pi^2$ projections) and therefore $O(P^2Q^2)$ operations.

To date there has been no Mojette inversion approach that is fast, exact given noise-free projections, and well-conditioned given noisy projections. This paper presents an inversion algorithm based on the classic Fourier inversion (FI) technique for the Radon transform. It solves rows of Fourier space using a conjugate gradient algorithm that minimises the L2 norm of the error (arising from noise). Assuming I is $O(P)$, Mojette FI is slower than LBP and GLBP requiring $O(P^2Q \log P)$ operations but more robust and is faster than CGBP and EBP but less robust.

The remainder of the paper is outlined as follows: Section 2 defines the Mojette transform and the requirements for a unique reconstruction (in the noise-free case). The Fourier properties of the Mojette projection data are outlined in Section 3. The Fourier inversion based on these properties is developed in Section 4. Some results are presented in Section 5 followed by some concluding remarks in Section 6.

2   The Mojette Transform

The Mojette transform of a 2D function, $f(k,l)$, is comprised of a set of I projections. Each projection has an associated projection direction vector, (p_i, q_i), and is comprised of a set of parallel discrete sums over f along lines defined by this vector, i.e., $b = kq_i - lp_i$. The value of the Mojette transform in each bin, b, of a projection is then defined as follows:

$$\text{proj}_{p_i,q_i}(b) = \sum_{k=-P/2}^{P/2} \sum_{l=-Q/2}^{Q/2} f(k,l)\delta(b - kq_i + lp_i), \tag{1}$$

where $\delta(x) = 1$ if $x = 0$ and is 0 otherwise. Note that $p_i \in \mathbb{Z}$, and $q_i \in \mathbb{N}$ such that $\gcd(p_i, q_i) = 1$. The distance between adjacent parallel lines (or sampling rate) of a projection varies with the direction vector, (p_i, q_i), as $1/\sqrt{p_i^2 + q_i^2}$. The number of bins per projection also varies with (p_i, q_i), as follows:

$$B_i = |p_i|Q + q_iP \tag{2}$$

The set of I projections is arbitrary both in cardinality and direction vectors. Therefore, a criterion is required to determine when a sufficient number of projections have been acquired in order to ensure a unique solution for a $P \times Q$ dataset. Katz determined the following criterion [7]:

$$\sum_{i=0}^{I-1} |p_i| \geq P \quad \text{or} \quad \sum_{i=0}^{I-1} q_i \geq Q. \tag{3}$$

For the remainder of this paper we will assume that $\sum |p_i| \geq P$. Below Katz criterion there is no unique solution, therfore the condition number is infinite; when Katz criterion is just satisfied, $i.e.$, $\sum |p_i| = P$, the condition number is very large so any noise is amplified and propagated into the reconstruction; as additional projections are added, the condition number decreases rapidly at first but then with diminishing returns.

3   Fourier Properties of Mojette Projections

Before developing a Fourier inversion technique it is beneficial to establish how Mojette projection data maps into discrete Fourier space. First we define the 1D discrete Fourier transform (DFT) of $g(k)$ as follows:

$$\mathcal{F}^1\{g\} = \hat{g}(u) = \sum_{k=-P/2}^{P/2} g(k)\exp(-i2\pi ku/P), \tag{4}$$

and define the 2D DFT of $g(k,l)$ as follows:

$$\mathcal{F}^2\{g\} = \hat{g}(u,v) = \sum_{k=-P/2}^{P/2}\sum_{l=-Q/2}^{Q/2} g(k,l)\exp(-i2\pi ku/P)\exp(-i2\pi lv/Q). \tag{5}$$

A discrete form of the Fourier slice theorem for classical Radon projections was first presented in [3] using the Z-transform. It was first applied to the nD Mojette transform by Verbert and Guédon in [13] also using the Z-transform. Here we demonstrate it for the 2D Mojette transform using the DFT.

Theorem 1. *The 1D DFT of a Mojette projection of f maps to a discrete line (or "slice") through the origin of the 2D DFT of f as follows:*

$$\mathcal{F}^1\{\text{proj}_{p_i,q_i}\}(w) = \hat{f}(\tfrac{q_i P}{B_i}w, \tfrac{-p_i Q}{B_i}w) \tag{6}$$

Proof.

$$\mathcal{F}^1\{\text{proj}_{p_i,q_i}\} = \sum_{b=-B_i/2}^{B_i/2}\sum_{k=-P/2}^{P/2}\sum_{l=-Q/2}^{Q/2} f(k,l)\delta(b-kq_i+lp_i)\exp(-i2\pi bw/B_i)$$

$$= \sum_{k=-P/2}^{P/2}\sum_{l=-Q/2}^{Q/2} f(k,l)\exp(-i2\pi(kq_i-lp_i)w/B_i)$$

$$= \sum_{k=-P/2}^{P/2}\sum_{l=-Q/2}^{Q/2} f(k,l)\exp(-i2\pi\tfrac{k}{P}(q_i Pw)/B_i)$$

$$\exp(-i2\pi\tfrac{l}{Q}(-p_i Qw)/B_i)$$

$$= \hat{f}(\tfrac{q_i P}{B_i}w, \tfrac{-p_i Q}{B_i}w)$$

$\square$

Figure 1 provides a depiction of the theorem. The sampling rate of the DFT of projection data in frequency space is $\sqrt{q_i^2 P^2 + p_i^2 Q^2}/B$. An important property of the slices that arises is the following:

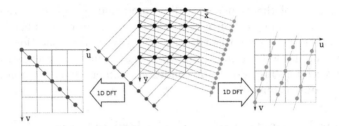

Fig. 1. A depiction of Theorem 1 for $proj_{-1,1}$ and $proj_{3,1}$ on a 4×4 image

Lemma 2. *Given the slice in Fourier space formed from a projection with direction vector (p_i, q_i), any horizontal line, $v = $ const, intersects the slice exactly $|p_i|$ times and any vertical line, $u = $ const, intersects the slice exactly q_i times.*

Proof. We have $w \in \mathbb{Z}_{B_i}$ and from Theorem 1 the sampling step between each successive point in frequency space is $(\frac{q_i P}{B_i}, \frac{-p_i Q}{B_i})$. Therefore, the total sampling displacement in frequencey space is $(\frac{q_i P}{B_i}, \frac{-p_i Q}{B_i})B_i = (q_i P, -p_i Q) \equiv (0, 0)$ mod (P, Q). There is no overlap in the sampling since p_i and q_i are coprime, hence implying the theorem.   □

4   Mojette Fourier Inversion

The frequency data from the I slices gives us a nonuniform sampling of discrete Fourier space. Before continuing, we shall introduce the nonuniform Fourier transform (NFT). We define the 1D FT of irregularly sampled data to J frequencies as:

$$\widehat{g}(w_j) = \sum_{k=-N/2}^{N/2} g(k) \exp(-i2\pi k w_j) \tag{7}$$

for $w_j \in [-0.5, 0.5)$ and $j \in \mathbb{Z}_J$ The NFT can be performed in $O(N \log N)$ operations, see [8]. Iterative inversion of the NFT can be achieved through the conjugate gradient method which converges within N iterations. Therefore, the inverse NFT (INFT) can be obtained in $O(N^2 \log N)$ operations. In practice this can be much faster if the right pre-conditioner is used, see [9].

Robust Mojette Fourier inversion can of course be achieved by taking the frequency data from the I slices and performing a 2D INFT. However, this will be performed in $O(P^2 Q^2 \log PQ)$ operations. This has been done for projections with 5% noise for comparison with the proposed method in sect. 5. In what follows we describe a method to achieve a less robust inversion in only $O(P^2 Q \log Q)$ operations by a deliberate sampling of the slice data to lie on horizontal lines enabling the 1D INFT to be applied to each line followed by an inverse fast Fourier transform (IFFT) applied to the columns.

4.1   Exact Resampling of Projection Slices

The slice data can be exactly resampled to lie on common horizontal lines in 2D discrete Fourier space, similar to [3,13] and Theorem 1, but using a specific Chirp z-transform of projections. First let us define the 1D Chirp z-transform as follows:

$$\mathcal{Z}^1\{g\} = \tilde{g}(w) = \sum_{k=-N/2}^{N/2} g(k)z^{wk}, \qquad w \in \mathbb{Z}_M. \tag{8}$$

Here $z \in \mathbb{C}$ and typically $|z| = 1$. As was done by Bailey and Swarztrauber in [1] we define $z = \exp(-i2\pi\alpha)$ for $w \in \mathbb{Z}$ and $\alpha \in \mathbb{R}$ to give:

$$\mathcal{Z}_\alpha^1\{g\} = \hat{g}^\alpha(w) = \sum_{k=-N/2}^{N/2-1} g(k)\exp(-i2\pi kw\alpha) = \hat{g}(N\alpha w). \tag{9}$$

So $\alpha = 1/N$ gives the frequency data of the 1D DFT. This can be done in $O(N \log N)$ if α is rational, (see [2,1]). Here we use $\alpha = 1/(|p_i|Q)$ so that the frequency sampling rate is 1.0 in the v-direction of $\hat{f}(u,v)$ as follows:

Corollary 3.

$$\mathcal{Z}^1_{\frac{1}{|p_i|Q}}\{\text{proj}_{p_i,q_i}\}(w) = \hat{f}(\tfrac{q_iP}{|p_i|Q}w, -\text{sgn}(p_i)w) \tag{10}$$

Proof.

$$\mathcal{Z}^1_{\frac{1}{|p_i|Q}}\{\text{proj}_{p_i,q_i}\}(w) = \sum_{b=-B_i/2}^{B_i/2} \sum_{k=-P/2}^{P/2-1} \sum_{l=-Q/2}^{Q/2-1} f(k,l)\delta(b - kp_i + lp_i)$$
$$\exp(-i2\pi bw/|p_i|Q)$$
$$= \sum_{k=-P/2}^{P/2-1} \sum_{l=-Q/2}^{Q/2-1} f(k,l)\exp(-i2\pi(kq_i - lp_i)w/|p_i|Q)$$
$$= \sum_{k=-P/2}^{P/2-1} \sum_{l=-Q/2}^{Q/2-1} f(k,l)\exp(-i2\pi\tfrac{k}{P}(q_iPw)/|p_i|Q)$$
$$\exp(-i2\pi\tfrac{l}{Q} - \text{sgn}(p_i)w)$$
$$= \hat{f}(\tfrac{q_iP}{|p_i|Q}w, -\text{sgn}(p_i)w)$$
$$\qquad\qquad\qquad\qquad\qquad\qquad\qquad\qquad\qquad\qquad \square$$

From Lemma 2 there are $|p_i|$ intersections with any horizontal line, thus each slice samples $|p_i|Q$ frequencies in $\hat{f}$. This causes all frequency data from the slices to lie on lines $v = c$ for $c \in \mathbb{Z}_Q$ and all projection slice data combined samples $\sum|p_i|Q \geq PQ$ frequencies provided each sample point is unique. The problem is that at $v = 0$ all lines intersect giving redundant data, hence insufficient unique data for exact inversion. This can be overcome by offsetting the sampling in the v-direction by a fraction $\phi \in (0,1)$. An offset of frequency data by ϕ can of course be done as a phase shift on the projections prior to applying the 1D Chirp z-transform. An example of the resulting data is depicted in Fig. 2.

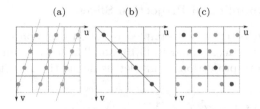

Fig. 2. A depiction of the chirp z-transforms of (a) $proj_{-1,1}$ and (b) $proj_{3,1}$ on a 4×4 image with $\phi = 0.5$. α is selected to give data on $v \in \{-1.5, -0.5, 0.5, 1.5\}$ frequencies. (c) Results from (a) and (b) overlayed in the Fourier domain; note there are $\sum_I |p_i|$ samples in each row and Q rows of samples. According to Katz criterion (3) this is a reconstructible sample set.

Ideally we would set $\phi = 0.5$ for our slice frequency data points to be as far from the origin as possible. However slices may intersect at points other than the origin. We need to check if the frequency data points sampled by the slices intersect at $v = c + \phi$ for $c \in \mathbb{Z}_Q$. We must identify some sufficient conditions to ensure no overlapping data points in Fourier space. Using the result from Corollary 3, the coordinate in fourier space for the projection with a direction vector (p_i, q_i) on a $P \times Q$ image with phaseshift of ϕ is:

$$(c + \phi)(q_i P/p_i Q, 1) \bmod (P, Q), \quad \text{for } c \in [0, q_i P - 1]. \tag{11}$$

In order for two Fourier data points to meet, they must come from two different projections. Let these two projections have direction vectors (p_1, q_1) and (p_2, q_2) with $c = c_1$ and c_2 respectively. Placing this into (11) gives the following:

$$(c_1 + \phi)(q_1 P/p_1 Q, 1) = (c_2 + \phi)(q_2 P/p_2 Q, 1) \bmod (P, Q), \tag{12}$$

for $c_1 \in \mathbb{Z}_{p_1 Q}$ and $c_2 \in \mathbb{Z}_{p_2 Q}$. This equation implies $c_1 = c_2 \bmod (Q)$ and $(c_1 + \phi)(q_1 P/p_1 Q) = (c_2 + \phi)(q_2 P/p_2 Q) \bmod (P)$. Therefore we have

$$(c_1 + \phi)\left(\frac{q_1}{p_1 Q}\right) - (c_2 + \phi)\left(\frac{q_2}{p_2 Q}\right) \quad \text{is an integer.} \tag{13}$$

This becomes:

$$x/p_1 p_2 Q + \phi \frac{q_1 p_2 - q_2 p_1}{p_1 p_2 Q} \quad \text{for} \quad x \in \mathbb{Z}. \tag{14}$$

We can now present a lemma that guarantees no point of intersection:

Proposition 4. *Setting $\phi = \frac{k}{2 \cdot k + 1}$ with $k = \max_{i,j} |p_i q_j|$ would have no point of intersection between two different projection set in Fourier space.*

Proof. Setting $\phi = \frac{k}{2 \cdot k + 1}$ with $k \geq \max_{i,j} |p_i q_j|$ forces $q_1 p_2 - q_2 p_1 = 0$ as otherwise (13) is false. Since the $\gcd(p_i, q_i) = \gcd(p_j, q_j) = 1$, we have $p_i = p_j$ and $q_i = q_j$ implies there is no point of intersection in Fourier space. With this shift, we can also guarantee a minimum seperation in Fourier space of $\min_{i,j} \frac{P}{Q p_i p_j (2k+1)}$ with $i \neq j$. $\square$

Placing the offset, resampled, slice frequency data from each projection into 2D Fourier space now gives us data as depicted in Fig. 2c.

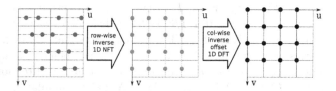

Fig. 3. A depiction of the process to reconstruct a 4×4 image from the chirp z-transforms of $proj_{-1,1}$ and $proj_{3,1}$

4.2   Reconstruction Process

This section outlines the Fourier inversion process, as depicted in Fig. 3, to recover f from the offset, resampled, slice frequency data, *e.g.*, Fig. 2c. First we propose the following theorem:

Theorem 5. *For a set of Mojette projections that satisfies Katz Criterion (3), there is sufficient slice data to be inverted in Fourier space after the chirp Z-transform with $\alpha = 1/(|p_i|Q)$ and no intersecting data points.*

Proof. Here we are assuming Katz Criterion is satisfied by $\sum |p_i| \geq P$. By Lemma 2 and Corollary 3 the Chirp z-transform of all projections gives Q rows of data points with $\sum |p_i| \geq P$ elements in each row. Since none of these points intersect, each row is invertible using the 1D INFT. There are Q data points in each column that are invertible using the 1D IDFT.   □

Proposition 4 guarantees no intersecting data points in Fourier space with particular choice of phase shift, ϕ. Therefore we can satisfy the requirement of no intersecting data points in Theorem 5.

The entire reconstruction process, given I Mojette projections, can be summarised in the following 6 steps:

1. Calculate $\phi = \frac{k}{2 \cdot k + 1}$ with $k = \max_{i,j} |p_i q_j|$;
2. Apply the phase shift, ϕ, to all projection data;
3. Fast chirp z-transform each projection with $\alpha = 1/(|p_i|Q)$;
4. Remap frequency data to 2D Fourier domain;
5. Invert nonuniform frequency data in rows. Q rows $\times$ maximum P iterations of INFT.
6. Inverse Fast Fourier transform of P columns (with a phase shift of $-\phi$);

5   Results

In this section we demonstrate the performance of the proposed algorithm with various noise conditions and various number and distribution of projection direction vectors. We investigate 0%, 1% and 5% Gaussian noise added to the projection data, *i.e.*, noise following a normal distribution with a standard deviation equal to $\eta\%$ of the average value. We also investigate three types of angle distributions, F_N^{180}, F_N^{90}, M_N. Here, F_N denotes the direction vectors formed by the Farey series, *e.g.*, $F_3 = 0/1, 1/3, 1/2, 2/3, 1/1$ corresponds to direction vectors $(1,0), (3,1), (2,1), (3,2), (1,1)$. The 90 and 180 signify 90 degree or 180 degree symmetries, *e.g.*, for direction vector (a,b), 90 includes $(-a,b)$ while 180 also includes (b,a) and $(-b,a)$. M_N denotes the set of direction vectors $(1,0), (1,1), (-1,1), (2,1), (-2,1), \ldots, (N,1), (-N,1)$. All simulations are for a 256×256 image of Lena shown in Fig. 4a.

(a)   (b) 5% noise

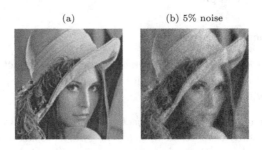

Fig. 4. (a) original image used for simulations. (b) Reconstruction using full 2D-MFI for Mojette data with angle set F_8^{180} and 5% Gaussian noise added. RMSE 17.1.

Firstly, using the F_8^{180} angle set we investigated the effect of 0%, 1%, and 5% noise added to the projections. The results are presented in Fig. 5a. It can be seen that the proposed method can deal with a moderate amount of noise. It is not as robust as CGM (see Fig. 4b); the RMSE data shows that noise is amplified in the reconstruction. Even the noise-free input data has a non-zero RMSE. This seems to be due to round-off errors, paricularly a problem for the INFT of rows $v = \phi$ and $v = \phi - 1$ of Fourier space. Smaller arrays with $P, Q \leq 32$ do not suffer from this.

Secondly, by adding a fixed level of 1% noise to the projection data, we establish the effect of the number of projections, and more specifically the value of $\sum_i p_i$ as it increases above Katz limit (3). Figure 5b gives the reconstructions using F_7^{180}, F_8^{180}, and F_9^{180} with $\sum_i p_i = 275$, 371, and 533 respectively. As expected, the results improve as $\sum_i p_i$ increases and the condition number of the system decreases.

Lastly, by again adding 1% noise to the projection data, we explore the effect of angle distribution. We use three angle sets that each have a similar $\sum_i p_i$, F_{9a}^{90}, F_7^{180}, and M_{16} with $\sum_i p_i = 283$, 275, and 273 respectively. (Note 9a indicates we are using F_8^{90} and adding only direction vectors $(\pm 9, 1)$ and $(\pm 9, 2)$ from F_9). With $\sum_i p_i$ very close to Katz limit (3) this is very unstable (as described in sect. 2). Results are presented in Fig. 5c and show that reconstruction with a more symmetric angle set with greater redundancy performs better than limited angle datasets.

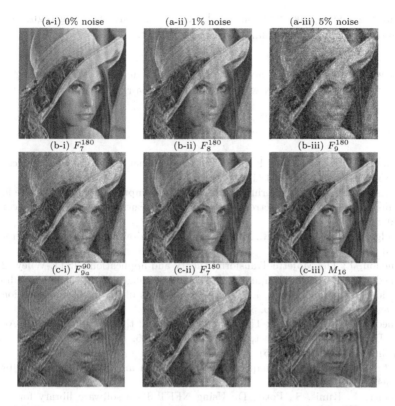

(a-i) 0% noise   (a-ii) 1% noise   (a-iii) 5% noise

(b-i) F_7^{180} (b-ii) F_8^{180} (b-iii) F_9^{180}

(c-i) F_{9a}^{90} (c-ii) F_7^{180} (c-iii) M_{16}

Fig. 5. (a) Reconstruction using MFI for Mojette data with angle set F_8^{180}. Amount of Gaussian noise added as indicated. RMSE (i) 6.0 (ii) 16.2 (iii) 26.7. (b) Reconstruction using MFI for Mojette data with 1% Gaussian noise added using angle set as indicated. RMSE (i) 19.5 (ii) 16.2 (iii) 13.1. (c) Reconstruction using MFI for Mojette data with 1% Gaussian noise added using angle set as indicated. RMSE (i) 28.3 (ii) 19.5 (iii) 29.1.

6   Conclusion

We have presented a reconstruction technique for Mojette projection data that is based on the Fourier inversion technique for the classical Radon transform. We use exact frequency data resampling to make the slice data lie on a set of parallel lines in Fourier space. This enables inversion to be broken into a set of 1D INFT and a set of 1D IFFT and speeds up the process. Given the Mojette transform data of a $P \times Q$ image with I projections, our Fourier inversion algorithm requires $O(P^2 Q \log P)$ operations. This technique is a compromise between fast, $O(PQI)$, but ill-conditioned local back-projection algorithms and slow, $O(P^2 Q^2 I)$ but well-conditioned techniques. Results show that this method can tolerate a moderate amount of noise and that redundant and symmetrical projection data is preferable for increased performance. Future work includes determining a stopping criterion for the INFT and investigating regularisation or weighting techniques to make MFI more robust.

Acknowledgements. AK acknowledges l'Univ. de Nantes for funding a one month visit to IVC-IRCCyN where much of this work was undertaken.

References

1. Bailey, D., Swarztrauber, P.: The fractional fourier transform and applications. SIAM Review 33, 389–404 (1991)
2. Bluestein, L.: A linear filtering approach to the computation of the discrete fourier transform. Northeast Electronics Research and Engineering Meeting Record 10 (1968)
3. Dudgeon, D., Mersereau, R.: Multidimensional Digital Signal Processing. Prentice-Hall (1983)
4. Guédon, J.: The Mojette Transform: theory and applications. ISTE-Wiley (2009)
5. Guédon, J., Barba, D., Burger, N.: Psychovisual image coding via an exact discrete Radon transform. In: Wu, L.T. (ed.) Proceedings of Visual Communication and Image Processing 1995, pp. 562–572 (May 1995)
6. Guédon, J., Normand, N.: The Mojette transform: the first ten years. In: Andrès, É., Damiand, G., Lienhardt, P. (eds.) DGCI 2005. LNCS, vol. 3429, pp. 79–91. Springer, Heidelberg (2005)
7. Katz, M.: Questions of uniqueness and resolution in reconstruction from projections. Springer (1977)
8. Keiner, J., Kunis, S., Potts, D.: Using NFFT 3 - a software library for various nonequispaced fast Fourier transforms. ACM Trans. Math. Software 36, 1–30 (Article 19) (2009)
9. Kunis, S., Potts, D.: Stability results for scattered data interpolation by trigonometric polynomials. SIAM J. Sci. Comput. 29, 1403–1419 (2007)
10. Normand, N., Guédon, J., Philippe, O., Barba, D.: Controlled redundancy for image coding and high-speed transmission. In: Ansari, R., Smith, M. (eds.) Proceedings of SPIE Visual Communications and Image Processing 1996, vol. 2727, pp. 1070–1081. SPIE (February 1996)
11. Normand, N., Kingston, A., Évenou, P.: A geometry driven reconstruction algorithm for the Mojette transform. In: Kuba, A., Nyúl, L.G., Palágyi, K. (eds.) DGCI 2006. LNCS, vol. 4245, pp. 122–133. Springer, Heidelberg (2006)
12. Servières, M., Idier, J., Normand, N., Guédon, J.: Conjugate gradient Mojette reconstruction. In: Fitzpatrick, J., Reinhardt, J. (eds.) Proceedings of SPIE Medical Imaging 2005: Image Processing, vol. 5747, pp. 2067–2074 (April 2005)
13. Verbert, P., Guédon, J.: N-dimensional Mojette transfrom. Application to multiple description. IEEE Discrete Signal Processing 2, 1211–1214 (2002)

Uniqueness Regions under Sets
of Generic Projections in Discrete Tomography

Paolo Dulio[1], Andrea Frosini[2], and Silvia M.C. Pagani[1,*]

[1] Dipartimento di Matematica "F. Brioschi", Politecnico di Milano,
Piazza Leonardo da Vinci 32, I-20133 Milano, Italy
{paolo.dulio,silviamaria.pagani}@polimi.it
[2] Dipartimento di Matematica e Informatica "U. Dini", Università di Firenze,
viale Morgagni 67/A, 50134 Firenze, Italy
andrea.frosini@unifi.it

Abstract. In Discrete Tomography, objects are reconstructed by means of their projections along certain directions. It is known that, for any given lattice grid, special sets of four valid projections exist that ensure uniqueness of reconstruction in the whole grid. However, in real applications, some physical or mechanical constraints could prevent the use of such theoretical uniqueness results, and one must employ projections fitting some further constraints. It turns out that global uniqueness cannot be guaranteed, even if, in some special areas included in the grid, uniqueness might be still preserved.

In this paper we address such a question of local uniqueness. In particular, we wish to focus on the problem of characterizing, in a sufficiently large lattice rectangular grid, the sub-region which is uniquely determined under a set S of generic projections. It turns out that the regions of local uniqueness consist of some curious twisting of rectangular areas. This deserves a special interest even from the pure combinatorial point of view, and can be explained by means of numerical relations among the entries of the employed directions.

Keywords: Discrete Tomography, lattice grid, projection, uniqueness region.

1 Introduction

Image reconstruction is one of the main topics in Discrete Tomography, i.e., the discipline that studies how to infer geometric properties of a discrete unknown object, regarded as a finite set of points in the $2D$ or $3D$ lattice, from quantitative data about the number of its primary constituents along a set of fixed discrete lines, say projections. The considered image is subject to a discretization operation and therefore is seen as a matrix, whose entries, say *pixels*, are integer numbers and correspond to the different colors, or grey levels, of the

---

* Corresponding author.

E. Barcucci et al. (Eds.): DGCI 2014, LNCS 8668, pp. 285–296, 2014.
© Springer International Publishing Switzerland 2014

original image. The dimensions of the image are those of the minimal bounding rectangle, i.e., the sizes of the obtained matrix.

So, the vector of projections along a discrete direction u sums the values of the pixels that lie on each line parallel to u.

Motivated by practical applications, one of the most challenging problems in Discrete Tomography is the *faithful* reconstruction of unknown images from projections along a set S of fixed lattice directions, say *Reconstruction Problem* (for an overview of the topic see [9]). In general, this problem is not quickly solvable (here quickly means in polynomial time with respect to the dimensions of the grid) even for three directions [7]. However, this result, as many similar ones in the field, relies on the possibility of choosing unknown images of arbitrarily large dimensions. On the other hand, for suitably selected sets of directions, each pixel can be uniquely determined if we confine the problem in special bounded lattice grids (see for instance [4]).

Another challenging aspect of the reconstruction process is that the images consistent with a given set of projections are, usually, a huge class (see for instance [1, 12]), whose members can share few (possibly no) points, so the faithful reconstruction of an unknown image is, in general, a hopeless task. This statement becomes clear if one considers the $n \times n$ images having homogeneous 1 projections, i.e., projections with constant entries 1, along the two coordinate directions; an easy check shows that their number is $n!$, i.e., as many as the number of permutations of n, and some of them have no common elements. To gain uniqueness, different strategies have been considered, usually relying on additional geometrical or combinatorial information about the unknown image.

Using a different perspective, in [4] it has been shown how far an accurate choice of four directions of projections can push the dimensions of the unknown image preserving the uniqueness of the reconstruction process.

In this paper, starting from the results in [4], we move towards the solution of the question of how much one can enlarge the size of the unknown image preserving both uniqueness and polynomial time reconstruction from a given set of projections. In particular, here we consider the related problem of characterizing, in a sufficiently large image, the shape and dimensions of the area uniquely determined by the projections along two given directions: we will show that such a shape varies according to the slopes of the two directions. This result also allows to infer a polynomial strategy to detect the elements inside the area. It is worth noting that this paper belongs to a more general and theoretical perspective, since the uniqueness region is determined regardless of the features of the image; in particular, it is not forced to be binary or with few grey levels. Our approach can be considered as a counterpart of the mathematical morphology tools described, for instance, in [5, 11, 13]; however, our aim is to investigate the problem without using the Mojette transform. A full comparison between the two methods seems to be an interesting issue, and could be treated in a separate paper.

The paper is organized as follows. In Section 2, we provide the basic notions and definitions of discrete tomography, together with the theoretical results

about uniqueness on a grid. Section 3 concerns the uniqueness problem restricted
to two directions of projections: the shape of the sub-picture uniquely determined
by the projections is characterized according to the mutual relation of the slopes
of the directions. Section 4 gives some perspectives for future work and concludes
the paper.

2   Definition and Known Results

A digital image E naturally represents a finite set of points in the $2D$ lattice. To
each point an integer number, which corresponds to a color or a grey level of the
original image, is attached. Each set will be considered up to translation. The
size of the image E is that of its minimal bounding rectangle (or *lattice grid*)
$\mathcal{A} = [m] \times [n]$, where, for $p \in \mathbb{N}$, $[p] = \{0, 1, ..., p - 1\}$. We can define a function
$f : \mathcal{A} \longrightarrow \mathbb{Z}$, which maps each pixel of $\mathcal{A}$ onto its corresponding value.

Let $u = (a, b)$ be a discrete direction, i.e., a couple of integers such that
$\gcd\{a, b\} = 1$, with the further assumption that if $a = 0$, then $b = 1$ and
viceversa. We say that a set $S = \{(a_k, b_k)\}_{k=1}^{d}$ of d lattice directions is *valid* for
a finite grid $\mathcal{A} = [m] \times [n]$, if

$$\sum_{k=1}^{d} |a_k| < m, \qquad \sum_{k=1}^{d} |b_k| < n.$$

We recall the standard notion of *projection* of a finite set of points E along
the direction u as the vector $P_u = (p_1, \ldots, p_k)$, with $k \geq 1$, such that

$$p_t = \sum_{(i,j) \in \mathcal{A}, aj = bi + t} f(i, j),$$

being $aj = bi + t$ the t-th lattice line intersecting the grid. Since we assume that
E is finite, then k is. An example is shown in Figure 1, where the black pixels
of the binary image correspond to the value 1, and white pixels to the value 0.
The projection of the set of black pixels is considered in the direction $u = (1, 1)$;
there $P_u = (0, 0, 0, 0, 0, 4, 5, 4, 4, 5, 7, 5, 3, 2, 3, 3, 2, 1, 0, 0, 0)$.

As a matter of fact, two sets E and F may have the same projections along a
set S of fixed directions, say they are *(tomographically) equivalent*. On the other
hand, if E does not share its projections with any other different set, then it is
said to be *S-unique*.

In general the class of images having the same projections along a set S of
directions is really huge. Consider for instance, as already observed, the $[n] \times [n]$
images having projections with all entries equal to 1 along the two axes directions
$(1, 0)$ and $(0, 1)$: there are $n!$ possible outputs.

So, the problem of the faithful reconstruction of an unknown (finite) set of
points, that is of primary relevance in the field of Discrete Tomography, is not
well posed, and it can not be considered without some prior knowledge about
the object itself.

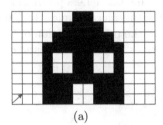 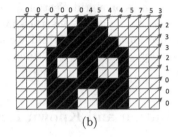

(a)   (b)

Fig. 1. (a) A binary image. (b) Projection of the set of black pixels in the direction $u = (1, 1)$.

Studying the characterization of the projections (directions and entries) that guarantee the uniqueness of the reconstruction process, a theory about the configurations of points that prevent such a property was developed.

In a first seminal study, H. J. Ryser [12] focused on the axes directions, and called these configurations *interchanges*. Later, himself and other authors refined the notion till reaching that of *switching component* or *bad configuration*. It represents a simple switching along a generic set of discrete directions S, i.e., a rearrangement of some elements of a set that preserves the projections along S (see [9] for a survey). An algebraic treatment of switching components was also developed in [8]. In [6] it was shown that a set E is S-unique if and only if it contains no bad configuration along the directions of S. On the other hand, since [10], we know that there are sets of points that are non-unique with respect to any given set S of directions. Because of this result, we wish to focus on the related problem of characterizing, in a sufficiently large lattice rectangular grid, the sub-region formed by pixels which are uniquely determined under a set S of generic projections. As a first step, the case $|S| = 2$ is considered.

3   Characterization of the Uniqueness Regions by Two Directions

From the previous considerations it follows that S-uniqueness in a lattice grid $\mathcal{A} = [m] \times [n]$ is guaranteed whenever no switching component exists in $\mathcal{A}$ with respect to the set S of projections. The following result, which slightly restates [3, Theorem 6], proves that this can be achieved by means of suitable choices of four lattice directions.

Theorem 1. *Let $S = \{u_1, u_2, u_3, u_4 = u_1 + u_2 \pm u_3\}$ be a valid set of four directions for the grid $\mathcal{A} = [m] \times [n]$, where u_1, u_2, u_3 are chosen arbitrarily, while u_4 is obtained as a combination of the other three. Let moreover*

$$\sum_{r=1}^{4} |a_r| = h \quad and \quad \sum_{r=1}^{4} |b_r| = k,$$

being $(a_r, b_r) = u_r$ where $r = 1, ..., 4$. Let $D = \pm S \cup \widehat{S}$, where $\pm S = \{\pm u : u \in S\}$, and $\widehat{S} = \{\pm (u_i - u_4) : u_i \in \{u_1, u_2, u_4 - u_1 - u_2\}\}$. Moreover, let $A = \{(a, b) \in D : |a| > |b|\}$, and $B = \{(a, b) \in D : |b| > |a|\}$. Then, $\mathcal{A}$ contains no switching component with respect to S if and only if

$$\min_{|a|} A \geq \min\{m - h, n - k\}, \quad \min_{|b|} B \geq \min\{m - h, n - k\}, \tag{1}$$

and

$$m - h < n - k \Rightarrow \forall (a, b) \in B \ (|a| \geq m - h \text{ or } |b| \geq n - k),$$
$$n - k < m - h \Rightarrow \forall (a, b) \in A \ (|a| \geq m - h \text{ or } |b| \geq n - k),$$

where, if one of the sets A, B is empty, the corresponding condition in (1) drops.

For examples, see [3, Example 8] and [4, Example 1].

However, in real applications, some physical or mechanical constraints could prevent the use of these suitable sets of four lattice directions. For instance, due to the employed tomographic procedure, only projections confined in a limited angle could be considered, or these should be selected trying to minimize the local uncertainty. There is a wide literature concerning this problem, see for instance [2, 14–16]. It turns out that global uniqueness in the whole assigned grid, in general, cannot be guaranteed. Nevertheless, in some special regions included in the grid, uniqueness might be still preserved.

Definition 1. *Let S be a set of valid directions for a grid $\mathcal{A}$. The region of uniqueness (ROU) of $\mathcal{A}$ is the set of pixels of $\mathcal{A}$ which are uniquely determined by the projections along the elements of S.*

Therefore, it is worth trying to characterize the shape of the ROU, determined, inside a given rectangular grid $\mathcal{A} = [m] \times [n]$, by any set of projections. As a first step towards this challenging problem we wish to investigate the shape of the ROU determined by a pair of valid directions.

Let us denote by (a, b) and (c, d) the employed lattice directions, where we set $a, c < 0$ and $b, d > 0$. Basing on these choices, we are led to construct the ROU by filling the grid $\mathcal{A} = [m] \times [n]$ from its bottom-left corner, and by symmetry, from its upper-right corner. Due to symmetry, it suffices to argue only on one of these two regions, say the bottom-left one. Also, we can always assume $-a > b$.

Remark 1. Note that our approach is w.l.o.g., since, for different choices of the signs of a, b, c, d, the arguments are quite similar, just the ROU fills different corners of $\mathcal{A}$.

We denote by $P = (p_1, p_2, ..., p_{s-1}, p_s)$ a SE to NW zig-zag path, with alternating horizontal and vertical steps of lengths $p_1, p_2, ..., p_{s-1}, p_s$, being the first one (of length p_1) a vertical step, and the last one (of length p_s) a horizontal step (see Figure 2).

Also, we denote by $R(x, y)$ a rectangle having horizontal and vertical sides of lengths x and y respectively.

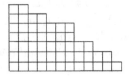

Fig. 2. The lattice region delimited by the zig-zag path $P = (1, 1, 1, 2, 1, 2, 2, 3, 1, 2, 1, 2)$

The following result shows that, in the case when the directions (a, b), (c, d) satisfy $|a| > |c|$ and $b < d$, then the ROU is simply an L-shaped area.

Theorem 2. *Let (a, b), (c, d) be two lattice directions such that $a, c < 0$, $b, d > 0$, $-c < -a < -2c$ and $2b < d$. Then, the associated ROU is delimited by the zig-zag path $P = (b, a, d, c)$.*

Proof. Consider a rectangle $R(-c, d)$ whose bottom-left corner is placed in the bottom-left corner of the grid. Then, all lattice points contained in $R(-c, d)$ are uniquely determined along the direction (c, d). Consider now a rectangle $R(-a + c, b)$ whose bottom-left corner is adjacent to the bottom-right corner of $R(-c, d)$. Any point belonging to $R(-a + c, b)$ is switched along (a, b) outside $R(-c, d)$ from its left vertical side, so that it is uniquely determined in the grid. Let $R(-c, b)$ be such that its bottom-left corner is adjacent to the bottom-right corn er of $R(-a+c, b)$. Its switching by (a, b) is completely included in $R(-c, d)$, and any further switching moves it outside the grid. Now, the rectangle $R(-c, b)$ placed above $R(-c, d)$, can be uniquely determined by means of direction (c, d), since it is mapped inside the previously determined uniqueness region.   □

Corollary 1. *Let (a, b), (c, d) be two lattice directions such that $a, c < 0$, $b, d > 0$, $|a| > |c|$ and $b < d$. Then, the associated ROU is delimited by the zig-zag path $P = (b, a, d, c)$.*

Proof. We can argue as in Theorem 2 by means of iterated switching. The only difference relies on the fact that in the remaining cases $-a \geq -2c$ and $2b < d$, or $-c < -a < -2c$ and $2b \geq d$, or $-a \geq -2c$ and $2b \geq d$, more iterations are needed to get the L-shaped region delimited by the zig-zag path $P = (b, a, d, c)$.   □

Example 1. If we assume $(a, b) = (-13, 3)$, $(c, d) = (-7, 11)$, the corresponding ROU is depicted in Figure 3.

The case when $|a| > |c|$ and $b \geq d$ seems to be much more intriguing. In fact, several different shapes appear, depending on the different interplay of the numerical relations among the entries a, b, c, d. The following theorem points out the case when the ROU consists of a big rectangle having two small rectangles adjacent to its right and upper side, respectively.

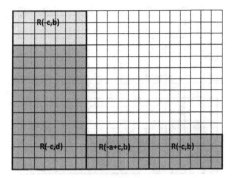

Fig. 3. The ROU associated to the pair $(-13,3), (-7,11)$, delimited by the zig-zag path $P = (3, 13, 11, 7)$

Theorem 3. *Let (a, b), (c, d) be two lattice directions such that $|a| > |c|$ and $b > d$. Let λ, μ be the quotients of the divisions between $-a, -c$ and b, d, respectively. If $\lambda \neq \mu$, then the associated ROU is the region delimited by the zig-zag path $P = (d, c, b - d, -a + c, d, c)$.*

Proof. Assume for instance $\lambda < \mu$. The rectangle $R_0 = R(-a, b)$ is the bottom-left zone of the ROU by means of the direction (a, b). Consider its bottom-right adjacent rectangle $R_1 = R(-c, d)$. Since $\lambda < \mu$, we can translate R_1 along (c, d) for λ times, still remaining inside R_0; a further switching along (c, d) moves (a part of) the translated rectangle outside the grid, on the left of R_0, before reaching its upper side. This implies that the whole R_1 is added to the ROU. Now, by using the direction (a, b), R_1 is mapped to a same sized rectangle adjacent to the upper-left corner of R_0, which is added to the ROU. This ends the construction of the ROU, which consequently is delimited by the zig-zag path $P = (d, c, b - d, a - c, d, c)$.

For the case $\lambda > \mu$, repeat the argument by starting from the $R(-c, d)$ rectangle above R_0 and moving SE along (c, d). □

Example 2. If we assume $(a, b) = (-20, 17)$, $(c, d) = (-11, 5)$, we have $20 = 11 \times 1 + 9$ and $17 = 5 \times 3 + 2$, so that $\lambda = 1 \neq 3 = \mu$. Therefore the assumptions of Theorem 3 are fulfilled, and the corresponding ROU is depicted in Figure 4.

When $|a| > |c|$, $b > d$ and $\lambda = \mu$, differently from Theorem 3, also the remainders of the divisions have to be taken into account. Here the situation becomes much more intricate. As a first contribution to clarify the matter, we present a result where the shape of the ROU consists of a rectangular *erosion* of the previously obtained configurations.

Theorem 4. *Let (a, b), (c, d) be two lattice directions such that $|a| > |c|$ and $b > d$. Assume $-a = \lambda(-c) + r$ and $b = \mu d + S$, where $0 < r < -c$ and $0 < s < d$. If $\lambda = \mu$, $r > -c/2$ and $s < d/2$, then the associated ROU is the region delimited by the zig-zag path $P = (d - s, -c - r, s, r, b - d, -a + c, d - s, -c - r, s, r)$.*

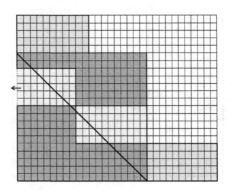

Fig. 4. The ROU associated to the pair $(-20, 17), (-11, 5)$, delimited by the zig-zag path $P = (5, 11, 12, 9, 5, 11)$

Proof. The rectangle $R_0 = R(-a, b)$ is the bottom-left zone of the ROU by means of the direction (a, b). Since $\lambda = \mu$, we cannot repeat the same argument as in Theorem 3. Consider then the right adjacent rectangle of R_0 given by $R_1 = R(-c-r, s)$. Since $\lambda = \mu$, by switching R_1 for μ times along (c, d), we reach the upper side of R_0 still remaining inside R_0, and a further switching moves the rectangle outside the grid. Therefore R_1 is added to the ROU. A similar argument applies to the rectangle $R_2 = R(r, d-s)$ placed adjacent to the upper-left corner of R_0, so the ROU increases of R_2. Then we apply alternatively the switching (a, b) and (c, d) on the resulting new parts of the ROU. Since $r > -c/2$ and $s < d/2$, we add pixels to the ROU as follows.

- Starting from R_2, this rectangle is mapped along (a, b) to the region R_3 containing R_1 in its bottom-left corner, and $R_3 \setminus R_1$ is added to ROU. Note that $R_3 \setminus R_1$ can be decomposed as the sum of three rectangles, namely $R_3 \setminus R_1 = R(2r + c, s) \cup R(-c - r, d - 2s) \cup R(2r + c, d - 2s)$.
- The sub-rectangle $R(2r + c, s)$ of $R_3 \setminus R_1$ is mapped along (c, d), outside $R(-a, b)$, to a congruent rectangle placed upper-left above R_2, which is added to the ROU. Analogously, the sub-rectangle $R(-c - r, d - 2s)$ of $R_3 \setminus R_1$ is mapped along (c, d), outside $R(-a, b)$, to a congruent rectangle placed bottom-right adjacent to R_2, which is added to the ROU. Note that the sub-rectangle $R(2r + c, d - 2s)$ does not contribute to the ROU since it is mapped twice in the complement of the ROU contained in the grid.
- We repeat the previous construction on the new parts of the ROU so obtaining further new parts of the same size $R(2r+c, s)$ and $R(-c-r, d-2s)$.
- This can be applied until the ROU becomes the region delimited by the zig-zag path $P = (d - s, -c - r, s, r, b - d, -a + c, d - s, -c - r, s, r)$.

□

Example 3. Let us consider the pair of directions $(a, b) = (-13, 7)$ and $(c, d) = (-8, 5)$. Since $13 = 8 \times 1 + 5$, and $7 = 5 \times 1 + 2$, we have $\lambda = \mu = 1$, $r = 5$ and $s = 2$, so that all the assumptions of Theorem 4 are fulfilled. The construction sketched in the proof can be followed in Figure 5. Note that the ROU is obtained by means of alternating switchings along the directions (a, b) and (c, d). For a better reading, we preserved the colors of a region when it is translated along (a, b), while the color has been changed under the translations along (c, d).

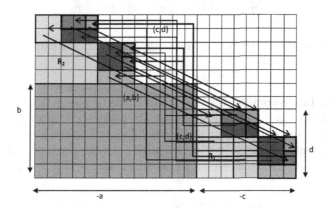

Fig. 5. The ROU associated to the pair $(-13, 7), (-8, 5)$, delimited by the zig-zag path $P = (3, 3, 2, 5, 2, 5, 3, 3, 2, 5)$

Note that the zig-zag path provided by Theorem 4 can also be written as follows:

$$P = (\max\{s, d - s\}, \min\{-c - r, r\}, \min\{s, d - s\}, \max\{r, -c - r\}, b - d, \quad (2)$$
$$-a + c, \max\{s, d - s\}, \min\{-c - r, r\}, \min\{s, d - s\}, \max\{r, -c - r\}).$$

This implies that reversing both the inequalities involving r, s leads to a similar result.

Theorem 5. *Let* (a, b), (c, d) *be two lattice directions such that* $|a| > |c|$ *and* $b > d$. *Assume* $-a = \lambda(-c) + r$ *and* $b = \mu d + s$, *where* $0 < r < -c$ *and* $0 < s < d$. *If* $\lambda = \mu$, $r < -c/2$ *and* $s > d/2$, *then the associated ROU is the region delimited by the zig-zag path* $P = (s, r, d - s, -c - r, b - d, -a + c, s, r, d - s, -c - r)$.

Proof. Replace s with $d - s$, and r with $-c - r$ in the proof of Theorem 4. □

Example 4. Let us consider the pair of directions $(a, b) = (-13, 7)$ and $(c, d) = (-11, 4)$. Since $13 = 11 \times 1 + 2$, and $7 = 4 \times 1 + 3$, we have $\lambda = \mu = 1$, $r = 2$ and $s = 3$, so that $r < -c/2$, $s > d/2$ and all the assumptions of Theorem 5 are fulfilled. In Figure 6 the corresponding ROU is represented.

The assumptions $r > -c/2$ and $s < d/2$ in Theorem 4, or $r < -c/2$ and $s > d/2$ in Theorem 5, are essential in order that the ROU is delimited by the zig-zag path as in (2). Differently, the shape of the ROU changes. Below we provide examples in a few cases.

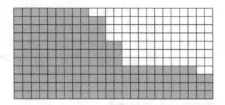

Fig. 6. The ROU associated to the pair $(-13,7),(-11,4)$, delimited by the zig-zag path $P = (3,2,1,9,3,2,3,2,1,9)$

Example 5. Let us consider the pair of directions $(a,b) = (-13,7)$ and $(c,d) = (-9,5)$. Being $13 = 9 \times 1 + 4$, and $7 = 5 \times 1 + 2$, we have $r = 4$, $s = 2$, so that the assumption $r > -c/2$ of Theorem 4 is not fulfilled. In Figure 7 the corresponding ROU is represented. Note that it is delimited by a zig-zag path different from $P = (d-s, -c-r, s, r, b-d, -a+c, d-s, -c-r, s, r) = (3,5,2,4,2,4,3,5,2,4)$.

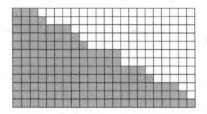

Fig. 7. The ROU associated to the pair $(-13,7),(-9,5)$, delimited by the zig-zag path $P = (1,1,1,1,1,3,1,1,1,3,2,4,1,1,1,1,1,3,1,1,1,3)$

Example 6. Let us consider the pair of directions $(a,b) = (-13,7)$ and $(c,d) = (-7,4)$. Being $13 = 7 \times 1 + 6$, and $7 = 4 \times 1 + 3$, we have $r = 6$, $s = 3$, so that the assumption $s < d/2$ of Theorem 4 is not fulfilled. In Figure 8 the corresponding ROU is represented.

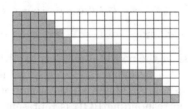

Fig. 8. The ROU associated to the pair $(-13,7),(-7,4)$, delimited by the zig-zag path $P = (1,1,1,1,1,1,1,4,3,6,1,1,1,1,1,1,1,4)$

The shape of the ROU varies when changing the considered directions; however, there is a common feature. We define a region of a grid to be *horizontally convex* (respectively, *vertically convex*) if the intersection of each row (respectively, column) of the grid with the region is a connected set. The proof of the following result can be easily obtained by induction, after noticing that the base step concerns the ROU determined by a single direction, i.e., a rectangle.

Theorem 6. *The ROU is horizontally and vertically convex.*

4   Conclusions and New Directions of Research

We addressed the problem of reconstructing the shape of the region of uniqueness (ROU) determined in a preassigned lattice rectangular grid $\mathcal{A} = [m] \times [n]$ by a generic choice of two valid directions. We characterized the ROU in various cases, showing that it consists of some curious displacement of rectangular areas, delimited by a SE-NW zig-zag path, whose edges have lengths depending on numerical relations among the entries of the employed directions. Several improvements could be considered. First of all, the case when (a, b) and (c, d) are selected so that $|a| > |c|$ and $b > d$ must be investigated when r and s do not satisfy the inequalities as in Theorem 4 and Theorem 5. This will provide a complete characterization of the ROU in the case of two directions. Also, a unifying picture of all the treated cases would be desirable. We feel that such a general approach should exist, probably based on some intertwining between switching operations and integer division.

A further step is the extension of such a characterization when data come from more than two directions. Experiments carried out with three projections show that the path delimiting the ROU presents a much more fragmented profile. Just as an example, Figure 9 shows what happens with the choice of projections $(-13, 7)$, $(-9, 5)$ and $(-8, 3)$.

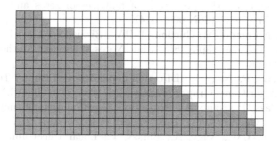

Fig. 9. The ROU associated to the triple $(-13, 7), (-9, 5), (-8, 3)$

Finally, in view of real applications, an explicit reconstruction algorithm of the ROU can be investigated, and exploited to get the reconstruction of a lattice set inside a grid of uniqueness along four directions. We have successfully developed some preliminary programs, running for special sets of four random directions, and we are confident to be able to unify them in a general strategy.

References

1. Barcucci, E., Del Lungo, A., Nivat, M., Pinzani, R.: Reconstructing convex poly-ominoes from horizontal and vertical projections. Theoret. Comput. Sci. 155(2), 321–347 (1996) ISSN 0304-3975
2. Batenburg, K.J., Palenstijn, W.J., Balázs, P., Sijbers, J.: Dynamic Angle Selection in Binary Tomography. Comput. Vis. Image Underst. 117(4), 306–318 (2013) ISSN 1077-3142
3. Brunetti, S., Dulio, P., Peri, C.: Discrete Tomography determination of bounded lattice sets from four X-rays. Discrete Applied Mathematics 161(15), 2281–2292 (2013)
4. Brunetti, S., Dulio, P., Peri, C.: On the Non-additive Sets of Uniqueness in a Finite Grid. In: Gonzalez-Diaz, R., Jimenez, M.-J., Medrano, B. (eds.) DGCI 2013. LNCS, vol. 7749, pp. 288–299. Springer, Heidelberg (2013)
5. Chandra, S., Svalbe, I.D., Guédon, J.-P.: An Exact, Non-iterative Mojette Inversion Technique Utilising Ghosts. In: Coeurjolly, D., Sivignon, I., Tougne, L., Dupont, F. (eds.) DGCI 2008. LNCS, vol. 4992, pp. 401–412. Springer, Heidelberg (2008)
6. Fishburn, P.C., Shepp, L.A.: Sets of uniqueness and additivity in integer lattices. In: Discrete Tomography. Appl. Numer. Harmon. Anal., pp. 35–58. Birkhäuser, Boston (1999)
7. Gardner, R.J., Gritzmann, P., Prangenberg, D.: On the computational complexity of reconstructing lattice sets from their X-rays. Discrete Math. 202(1-3), 45–71 (1999) ISSN 0012-365X
8. Hajdu, L., Tijdeman, R.: Algebraic aspects of discrete tomography. J. Reine Angew. Math. 534, 119–128 (2001) ISSN 0075-4102
9. Kuba, A., Herman, G.T.: Discrete tomography: a historical overview. In: Discrete Tomography. Appl. Numer. Harmon. Anal., pp. 3–34. Birkhäuser, Boston (1999)
10. Lorentz, G.G.: A problem of plane measure. Amer. J. Math. 71, 417–426 (1949) ISSN 0002-9327
11. Normand, N., Kingston, A., Évenou, P.: A Geometry Driven Reconstruction Algorithm for the Mojette Transform. In: Kuba, A., Nyúl, L.G., Palágyi, K. (eds.) DGCI 2006. LNCS, vol. 4245, pp. 122–133. Springer, Heidelberg (2006)
12. Ryser, H.J.: Combinatorial properties of matrices of zeros and ones. Canad. J. Math. 9, 371–377 (1957) ISSN 0008-414X
13. Servieres, M., Normand, N., Guédon, J.-P.V., Bizais, Y.: The Mojette transform: Discrete angles for tomography. Electronic Notes in Discrete Mathematics 20, 587–606 (2005)
14. Varga, L., Balázs, P., Nagy, A.: Direction-dependency of Binary Tomographic Reconstruction Algorithms. Graph. Models 73(6), 365–375 (2011) ISSN 1524-0703
15. Varga, L., Balázs, P., Nagy, A.: Projection Selection Dependency in Binary Tomography. Acta Cybern. 20(1), 167–187 (2011) ISSN 0324-721X
16. Varga, L., Nyúl, L.G., Nagy, A., Balázs, P.: Local Uncertainty in Binary Tomographic Reconstruction. In: Proceedings of the IASTED International Conference on Signal Processing, Pattern Recognition and Applications, IASTED. ACTA Press, Calgary (2013) ISBN 978-0-88986-954-7

Adaptive Grid Refinement
for Discrete Tomography

Tristan van Leeuwen[1] and K. Joost Batenburg[1,2,3]

[1] Centrum Wiskunde & Informatica, Amsterdam, The Netherlands
[2] Vision-Lab., University of Antwerp, Antwerp, Belgium
[3] Dept. of Mathematics, University of Leiden, Leiden, The Netherlands
{Tristan.van.Leeuwen,kbatenbu}@cwi.nl

Abstract. Discrete tomography has proven itself as a powerful approach to image reconstruction from limited data. In recent years, algebraic reconstruction methods have been applied successfully to a range of experimental data sets. However, the computational cost of such reconstruction techniques currently prevents routine application to large data-sets. In this paper we investigate the use of adaptive refinement on QuadTree grids to reduce the number of pixels (or voxels) needed to represent an image. Such locally refined grids match well with the domain of discrete tomography as they are optimally suited for representing images containing large homogeneous regions. Reducing the number of pixels ultimately promises a reduction in both the computation time of discrete algebraic reconstruction techniques as well as reduced memory requirements. At the same time, a reduction of the number of unknowns can reduce the influence of noise on the reconstruction. The resulting refined grid can be used directly for further post-processing (such as segmentation, feature extraction or metrology). The proposed approach can also be used in a non-adaptive manner for region-of-interest tomography. We present a computational approach for automatic determination of the locations where the grid must be defined. We demonstrate how algebraic discrete tomography algorithms can be constructed based on the QuadTree data structure, resulting in reconstruction methods that are fast, accurate and memory efficient.

Keywords: Tomography, adaptive refinement, QuadTree grids, algebraic reconstruction techniques.

1 Introduction

We consider a linear tomography problem

$$\mathbf{p} = W\mathbf{x} + \mathbf{n}, \tag{1}$$

where $W \in \mathbb{R}^{M \times N}$ is the projection matrix, $\mathbf{x} \in \mathbb{R}^N$ is the image, $\mathbf{p} \in \mathbb{R}^M$ are the projections and $\mathbf{n}$ is additive noise. The goal is to retrieve $\mathbf{x}$ from the noisy projections $\mathbf{p}$, which is typically done by solving a least-squares problem:

$$\min_{\mathbf{x}} ||W\mathbf{x} - \mathbf{p}||_2^2. \tag{2}$$

E. Barcucci et al. (Eds.): DGCI 2014, LNCS 8668, pp. 297–308, 2014.
© Springer International Publishing Switzerland 2014

The system of equations is often underdetermined (i.e., $M < N$) due to the limited number of measurements (small M) and the demand for high-resolution images (large N). The resulting non-uniqueness can be partially mitigated by adding prior knowledge, either in the form of a regularization penalty or by employing a tailored reconstruction algorithm that enforces the prior.

In discrete tomography, the prior is particularly strong – the object consists of only a few different materials – and the reconstruction problem can be formulated as a discrete optimization problem [1,2]. Solving such problems exactly is not feasible for large-scale problems due to their combinatorial nature and often not desirable due to noise. Many heuristic reconstruction have been developed over the years, which fall into two basic classes: methods that aim directly at solving the discrete optimization problem [3,4] and methods that solve (a series) of continuous optimization problems [5,6]. Even state-of-the-art iterative algorithms such as DART [6,7] are computationally very expensive as they are based on iterative reconstruction algorithms. Not only the costs of forward and backward projection and the memory usage scale linearly with N, the number of iterations required is also expected to scale linearly with N. To reduce the computational costs and the required number of iterations of iterative reconstruction methods, many authors have considered multi-scale or multi-grid methods for general, continuous tomography problems. [8] proposes a multi-level strategy that coarsens the projection images by averaging or subsampling the detector pixels. The use of classical multi-grid algorithms is discussed by [9,10,11]. Algorithms of a more heuristic nature are discussed by [12], who propose coarsening in both the image and data space, and [13], who develop a two-level approach. The closest in spirit to the current work is [14] who present an adaptive refinement strategy using QuadTree grids but use a much simpler refinement criterion.

Multi-scale reconstruction approaches aimed specifically at binary tomography have also been proposed. In [3] the authors use a simulated annealing approach in conjunction with uniform refinement. The use of QT grids in a similar context is explored by [4], who proposes refinement of the edges of the object.

In this paper, we investigate the use of QuadTree (QT) grids for iterative image reconstruction in discrete tomography. If the original object consists of large homogeneous regions, each consisting of a single material, QuadTree grids can strongly reduce the number of pixels needed to represent the image. The use of QT grids serves a double purpose in the case of discrete tomography; it can help to regularize the problem and to reduce the computational cost. For the interior of the object, coarse grid pixels can be used, thereby implicitly enforcing the discrete tomography constraint of constant gray levels for these interior regions. As a consequence, even when algorithms for continuous tomography (i.e. allowing all Gray values) are applied to the QT representation, the resulting reconstructions will contain large homogeneous regions and therefore this choice of image representation allows standard iterative methods from continuous tomography to be applied to DT problems successfully.

To illustrate the main ideas we consider the following toy example. A binary phantom and its corresponding (optimal) QT grid are shown in figure 1 (a-b). In this case the phantom consists of $128^2 = 16384$ pixels, while the QT grid allows us to represent the same image with only 25 pixels. To illustrate the potential benefits, we assume for the moment that we know the optimal QT grid and use it for reconstruction. We consider three scenarios: *i)* A benchmark reconstruction with 32 angles between 0 and 180 degrees and no noise; *ii)* a reconstruction with 32 angles and 20% Gaussian noise and finally, *iii)* a reconstruction with only 5 angles and no noise. The results are shown in figure 2. These examples clearly illustrate the potential benefits of reducing the number of unknowns; the results are more stable with respect to noise and it allows us to recover from severely limited data *using a conventional algebraic reconstruction algorithm.* Moreover, since the computational cost of the forward projection and the required memory is proportional to the number of pixels, using QT grids may also lead to significant computational savings.

Of course, we do not know the optimal QT grid a-priori in practice. Therefore, we propose adaptive refinement strategy that allows us to construct a QT grid as part of the iterative reconstruction. By starting from a coarse grid and refining only in areas of high variability, we never introduce more unknowns than needed and are able to construct an efficient representation of the reconstruction directly from the projection data. We apply the proposed method on 3 (binary) phantoms.

The outline of the paper is as follows. First, we discuss multi-level reconstruction and adaptive refinement in section 3. Numerical experiments where we apply the standard SIRT algorithm for continuous tomography to discrete image reconstruction on a QT grid are presented in section 4. Finally, we present conclusions and discuss possible future extensions as well as open questions in section 6.

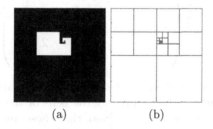

(a)   (b)

Fig. 1. (a) Spiral phantom and (b) corresponding QuadTree grid

2   Algorithm

We represent the image as a piece-wise constant function on a QuadTree grid. An example of a QT grid is shown in figure 3. A QT grid is represented by a collection of triples (i, j, s) which store the location of the upper-left corner of each cell as well as its size, both w.r.t. an underlying fine grid. An image on this

32 angles, no noise – 32 angles, 20% Gaussian noise – 5 angles, no noise

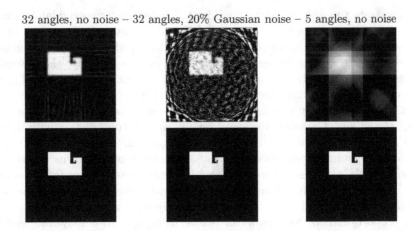

Fig. 2. Reconstructions of the phantom depicted in figure 1 (a) for different scenarios. The top row shows reconstructions on a fine grid with 16384 pixels while the bottom row the results for the reconstructions on the QuadTree grid with 25 pixels (cf. figure 1 (b)).

grid is represented with a single number for each cell. Although it is in principle possible to work with the image directly in this representation, is often more convenient to work with images that are represented on a uniform fine grid. For this purpose, we introduce the *mapping* matrix V that maps from a given QT grid to the underlying fine grid. For the QT grid depicted in figure 3 this matrix is given by

$$V^T = \frac{1}{2} \begin{pmatrix} 1\,1\,\cdot\,\cdot\,1\,1\,\cdot\,\cdot\,\cdot\,\cdot\,\cdot\,\cdot\,\cdot\,\cdot\,\cdot\,\cdot \\ \cdot\,\cdot\,1\,1\,\cdot\,\cdot\,1\,1\,\cdot\,\cdot\,\cdot\,\cdot\,\cdot\,\cdot\,\cdot\,\cdot \\ \cdot\,\cdot\,\cdot\,\cdot\,\cdot\,\cdot\,\cdot\,\cdot\,2\,\cdot\,\cdot\,\cdot\,\cdot\,\cdot\,\cdot\,\cdot \\ \cdot\,\cdot\,\cdot\,\cdot\,\cdot\,\cdot\,\cdot\,\cdot\,\cdot\,2\,\cdot\,\cdot\,\cdot\,\cdot\,\cdot\,\cdot \\ \cdot\,\cdot\,\cdot\,\cdot\,\cdot\,\cdot\,\cdot\,\cdot\,\cdot\,\cdot\,1\,1\,\cdot\,\cdot\,1\,1 \\ \cdot\,\cdot\,\cdot\,\cdot\,\cdot\,\cdot\,\cdot\,\cdot\,\cdot\,\cdot\,\cdot\,\cdot\,2\,\cdot\,\cdot\,\cdot \\ \cdot\,\cdot\,\cdot\,\cdot\,\cdot\,\cdot\,\cdot\,\cdot\,\cdot\,\cdot\,\cdot\,\cdot\,\cdot\,2\,\cdot\,\cdot \end{pmatrix}. \tag{3}$$

Here, the underlying fine grid has 4×4 pixels and dictates the finest level of the QT grid. The columns of V represent the cells of the QT grid and couple the corresponding cells in the fine grid. Note that these matrices are normalized such that for any given image $\mathbf{x}$ on the fine grid $V^T \mathbf{x}$ is the best approximation in the euclidean norm of this image on the QT grid.

For a given QT grid, we can pose the reconstruction problem as

$$\min_{\mathbf{x}} \|WV\mathbf{x} - \mathbf{p}\|_2^2,$$

where $\mathbf{x}$ represents the image on the QT grid defined by V. The resulting reconstruction problem can be solved with any conventional reconstruction algorithm, such as SIRT or ART. If the QT grid has fewer cells than measurements, this reconstruction problem is overdetermined and much better posed than the original reconstruction problem.

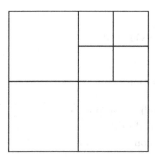

Fig. 3. Example of a QT grid

2.1 Adaptive Refinement

To construct a QT representation of a given image with as few cells as possible we propose the following refinement procedure, starting from an initial coarse grid:

1. refine all cells on the finest level;
2. compute the error between the current reconstruction and the reconstruction on the refined grid;
3. for each refined cell, keep the refinement if the local error is bigger than some threshold, coarsen otherwise;

This approach is different from traditional adaptive refinement strategies which typically refine after the fact based on local image gradients [14]. A problem with such approaches is that the image gradient needs to be estimated on the current (coarse) grid, making it difficult to detect features that are not properly resolved on this grid. Our procedure circumvents this problem by refining before measuring the error and reverting back to the coarse grid if the difference is small.

A more detailed description this procedure is shown in Algorithm 1. Here, refine(V_k) refines all cells on the finest level by splitting them into 4. refine($V, \mathcal{I}$) refines only the cells in the index set $\mathcal{I}$. By refining only the cells on the finest level, we avoid having to refine the same cells over and over again. The algorithm automatically terminates when we have reached the finest level corresponding to the underlying fine grid. So, for an underlying fine grid with $N = n^2$ pixels, we have a total of $K = \log_2(n)$ levels.

We compute the difference between the reconstructions on the coarse and refined grids V_c and V_f as

$$\Delta \mathbf{x}_f = \mathbf{x}_f - V_f^T V_c \mathbf{x}_c.$$

Algorithm 1. Adaptive refinement algorithm
___
Require:
  $\mathbf{x}$ - input image
  V_0 - basis for initial subspace
  δ - tolerance for refinement
Ensure:
  $\mathbf{x}_K$ - final representation on QT grid
  V_K - corresponding matrix

  $\mathbf{x}_0 = V_0^T \mathbf{x}$ {initial reconstruction}
  for $k = 0$ **to** $K-1$ **do**
    $\widetilde{V} = \mathsf{refine}(V_k, \mathcal{I}_k)$ {refinement proposal}
    $\widetilde{\mathbf{x}} = \widetilde{V}^T \mathbf{x}$ {map onto refined grid}
    $\widetilde{\mathbf{e}} = \mathsf{error}(\widetilde{\mathbf{x}} - \widetilde{V}^T V_k \mathbf{x}_k)$ {compute error}
    $\widetilde{\mathcal{I}} = \{ i \mid \widetilde{e}_i > \delta \}$
    $V_{k+1} = \mathsf{refine}(V_k, \widetilde{\mathcal{I}})$ {update}
    $\mathbf{x}_{k+1} = V_{k+1}^T \widetilde{V} \widetilde{\mathbf{x}}$
  end for

We then define a quantity $\mathbf{e}_c = \mathsf{error}(\Delta \mathbf{x}_f)$ on the coarse grid that contains the accumulated contributions for each grid cell such that

$$\mathbf{e}_c^T \mathbf{1} = \| \Delta \mathbf{x}_f \|_2^2. \tag{4}$$

All cells i on the coarse grid for which $e_{c,i} > \delta$ are subsequently refined. Thus, when the algorithm terminates (i.e., when $e_{c,i} \leq \delta \, \forall i$) we have $\| \Delta \mathbf{x}_f \|_2^2 \leq \delta N_c$ where N_c is the number of cells in the coarse grid.

An example of a series of adaptively refined grids for the Shepp-Logan phantom is shown in figure 4. On the finest level, we perfectly reconstruct the original image with only 1948 cells (compared to 16384 for the original image).

2.2 Reconstruction

We can adapt the above described refinement algorithm for reconstruction by replacing the mapping of the true image onto the refined QT grids in Algorithm 1 by a mapping of the projection data onto the refined QT grid. This can be achieved by

$$\widetilde{\mathbf{x}} = \left(W \widetilde{V} \right)^\dagger \mathbf{p},$$

where † denotes the pseudo-inverse. In practice, we never compute the pseudo inverse explicitly, but instead perform a tomographic reconstruction on the refined grid using the previous iterate $\mathbf{x}_k$ as initial guess. The resulting algorithm is stated in Algorithm 2. Here, $\mathbf{x}_1 = \mathsf{reconstruction}(W, \mathbf{p}, \mathbf{x}_0, L, \epsilon)$ performs up to L iterations of an iterative reconstruction technique starting from initial guess $\mathbf{x}_0$ with stopping criterion $\| W \mathbf{x}_1 - \mathbf{p} \|_2 \leq \epsilon \| \mathbf{p} \|_2$.

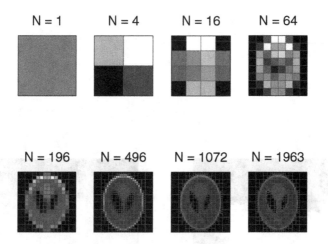

Fig. 4. Example of adaptive refinement

Algorithm 2. Adaptive multi-scale reconstruction algorithm

Require:
   W - projection operator
   $\mathbf{p}$ - projection data
   V_0 - basis for initial subspace
   L - iteration count for iterative reconstruction
   ϵ - tolerance for iterative reconstruction
   δ - tolerance for refinement
Ensure:
   $\mathbf{x}_K$ - final reconstruction

   $\mathbf{x}_0 = \text{reconstruction}(WV_0, \mathbf{p}, 0, L, \epsilon)$ {initial reconstruction}
   for $k = 0$ **to** $K - 1$ **do**
      $\widetilde{V} = \text{refine}(V_k, \mathcal{I}_k)$ {refinement proposal}
      $\widetilde{\mathbf{x}} = \text{reconstruction}(W\widetilde{V}, \mathbf{p}, \widetilde{V}^T V_k \mathbf{x}_k, L, \epsilon)$ {new reconstruction}
      $\widetilde{\mathbf{e}} = \text{error}(\widetilde{\mathbf{x}} - \widetilde{V}^T V_k \mathbf{x}_k)$ {compute error}
      $\widetilde{\mathcal{I}} = \{i \mid \widetilde{e}_i > \delta\}$
      $V_{k+1} = \text{refine}(V_k, \widetilde{\mathcal{I}})$ {update}
      $\mathbf{x}_{k+1} = V_{k+1}^T \widetilde{V} \widetilde{\mathbf{x}}$
   end for

---

3   Numerical Results

We conduct numerical experiments on three phantoms. For the phantoms, the
projection data is generated on a 256×256 grid with 128 detectors and 64
projections. The reconstruction is done on an underlying fine grid of 128×128
in order to avoid the *inverse crime*. We use the ASTRA toolbox to compute the
forward and backward projections [15]. For the adaptive method the mapping
matrices as discussed above are used to map to and from the QT grids to the

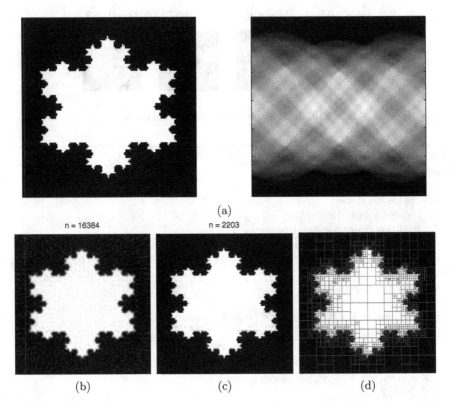

(a)

(b)   (c)   (d)

Fig. 5. (a) Ground truth and corresponding projection data, (b) SIRT reconstruction, (c) multi-scale reconstruction and (d) corresponding QT grid overlaying the ground truth

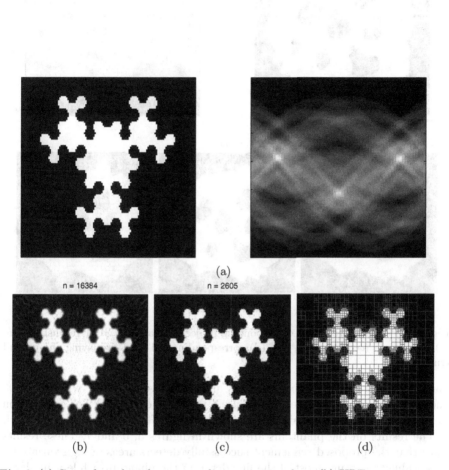

Fig. 6. (a) Ground truth and corresponding projection data, (b) SIRT reconstruction, (c) multi-scale reconstruction and (d) corresponding QT grid overlaying the ground truth

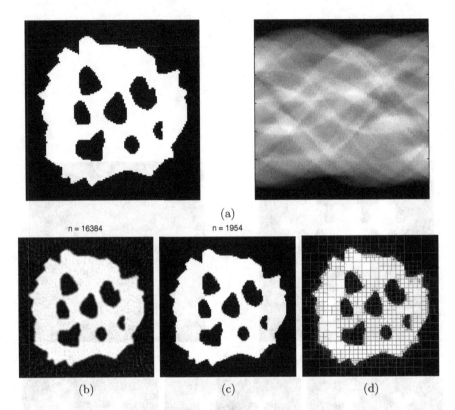

Fig. 7. (a) Ground truth and corresponding projection data, (b) SIRT reconstruction, (c) multi-scale reconstruction and (d) corresponding QT grid overlaying the ground truth

underlying fine grid. As reconstruction algorithm, we use SIRT with $L = 200$ and $\epsilon = 10^{-3}$. For the refinement we use a tolerance of $\delta = 0.2$. For comparison we also show the result obtained when applying SIRT directly on the finest grid.

The results on the phantoms are shown in figures 5, 6 and 7. These results show that the proposed refinement successfully detects areas of high variability and is able to capture most of the fine detail in the phantoms while using large cells in homogeneous areas. The resulting reconstructions are (almost) binary, showing the regularizing properties of the QT grid.

4   Conclusions and Discussion

We have presented an adaptive refinement strategy for tomographic reconstruction on QuadTree grids. The algorithm starts from a coarse grid and adaptively refines those cells where the reconstruction error is above some threshold. If we combine the QT grid approach with the standard SIRT algorithm for continuous tomography, and apply it to discrete images, the resulting grid represents the reconstructed image with only a fraction of the number of pixels otherwise required. We expect this approach to be useful in a wide range of applications where high resolution is required and where the images are characterized by large homogeneous regions, which is typically the case in discrete tomography. We also envision that QuadTree grids will be useful for region-of-interest tomography. In this case, the QT grids can be reconstructed a-priori based on a simple FBP reconstruction to identify regions of interest.

To optimally benefit from the reduction of the number of pixels when using QuadTree grids, the projection operator will have to compute the projections directly based on the QuadTree representation of the image, without mapping to an underlying fine grid first (as we did in this paper). Since the cost of forward projection is proportional to the number of pixels, this would directly reduce the computations by an order of magnitude. For reconstruction, we expect that having less unknowns will lead to less iterations, promising another reduction of the computational cost. An extension of the proposed algorithm to 3D reconstruction using OcTree grids is straightforward. Future research is aimed at including regularization to explicitly enforce desirable properties (such as discreteness) on the reconstruction and application to continuous tomography. For the latter, we expect that moving away from the piece-wise constant representation (e.g., by using linear basis functions on the QT grid) will be beneficial.

Acknowledgements. This work was supported by the Netherlands Organisation for Scientific Research (NWO), programme 639.072.005. Networking support was provided by the EXTREMA COST Action MP1207.

References

1. Herman, G.T., Kuba, A.: Discrete Tomography: Foundations, Algorithms, and Applications. Birkhäuser (1999)
2. Herman, G.T., Kuba, A.: Advances in discrete tomography and its applications. Birkhäuser (2007)
3. Ruskó, L., Kuba, A.: Multi-resolution method for binary tomography. Electronic Notes in Discrete Mathematics 20, 299–311 (2005)
4. Gerard, Y.: Elementary algorithms for multiresolution geometric tomography with strip model of projections. In: 2013 8th International Symposium on Image and Signal Processing and Analysis (ISPA), pp. 600–605. University of Trieste and University of Zagreb (2013)
5. Schule, T., Schnorr, C., Weber, S., Hornegger, J.: Discrete tomography by convex–concave regularization and D.C. programming. Discrete Applied Mathematics 151(1-3), 229–243 (2005)
6. Batenburg, K.J., Sijbers, J.: DART: a practical reconstruction algorithm for discrete tomography. IEEE Transactions on Image Processing 20(9), 2542–2553 (2011)
7. Batenburg, K., Sijbers, J.: Dart: A Fast Heuristic Algebraic Reconstruction Algorithm for Discrete Tomography. In: IEEE International Conference on Image Processing, pp. IV-133–IV-136. IEEE (2007)
8. Herman, G.T., Levkowitz, H., Tuy, H.: Multilevel Image Reconstruction. In: Rosenfeld, A. (ed.) Multiresolution Image Processing and Analysis. Springer Series in Information Sciences, vol. 12, pp. 121–135. Springer, Heidelberg (1984)

9. Henson, V.E., Limber, M.A., McCormick, S.F., Robinson, B.T.: Multilevel Image Reconstruction with Natural Pixels. SIAM Journal on Scientific Computing 17(1), 193–216 (1996)
10. Kostler, H., Popa, C., Ummer, M., Rude, U.: Towards an algebraic multigrid method for tomographic image reconstruction-improving convergence of ART. In: European Conference on Computational Fluid Dynamics, pp. 1–12 (2006)
11. Cools, S., Ghysels, P., van Aarle, W., Vanroose, W.: A multilevel preconditioned Krylov method for algebraic tomographic reconstruction. arXiv, 26 (2013)
12. Bouman, C., Webb, K.: Multigrid tomographic inversion with variable resolution data and image spaces. IEEE Transactions on Image Processing 15(9), 2805–2819 (2006)
13. De Witte, Y., Vlassenbroeck, J., Van Hoorebeke, L.: A multiresolution approach to iterative reconstruction algorithms in X-ray computed tomography. IEEE Transactions on Image Processing: A Publication of the IEEE Signal Processing Society 19(9), 2419–2427 (2010)
14. Schumacher, H., Heldmann, S., Haber, E., Fischer, B.: Iterative Reconstruction of SPECT Images Using Adaptive Multi-level Refinement. In: Tolxdorff, T., Braun, J., Deserno, T.M., Horsch, A., Handels, H., Meinzer, H.P. (eds.) Bildverarbeitung für die Medizin 2008, pp. 318–322. Springer, Heidelberg (2008)
15. Palenstijn, W.J., Batenburg, K.J., Sijbers, J.: Performance improvements for iterative electron tomography reconstruction using graphics processing units (GPUs). Journal of Structural Biology 176(2), 250–253 (2011)

Exact Evaluation of Stochastic Watersheds: From Trees to General Graphs

Filip Malmberg, Bettina Selig, and Cris L. Luengo Hendriks

Centre for Image Analysis,
Uppsala University and Swedish University of Agricultural Sciences, Sweden
{filip,bettina,cris}@cb.uu.se

Abstract. The *stochastic watershed* is a method for identifying salient contours in an image, with applications to image segmentation. The method computes a probability density function (PDF), assigning to each piece of contour in the image the probability to appear as a segmentation boundary in seeded watershed segmentation with randomly selected seedpoints. Contours that appear with high probability are assumed to be more important. This paper concerns an efficient method for computing the stochastic watershed PDF exactly, without performing any actual seeded watershed computations. A method for exact evaluation of stochastic watersheds was proposed by Meyer and Stawiaski (2010). Their method does not operate directly on the image, but on a compact tree representation where each edge in the tree corresponds to a watershed partition of the image elements. The output of the exact evaluation algorithm is thus a PDF defined over the edges of the tree. While the compact tree representation is useful in its own right, it is in many cases desirable to convert the results from this abstract representation back to the image, e.g, for further processing. Here, we present an efficient linear time algorithm for performing this conversion.

Keywords: Stochastic watershed, Watershed cut, Minimum spanning tree.

1 Introduction

The *stochastic watershed*, proposed by Angulo and Jeulin [1], is a method for identifying salient contours in an image, with applications to image segmentation. This method is based on the seeded watershed [14], which partitions the image into regions according to a set of *seedpoints*, so that every region contains precisely one seed and the boundaries between regions are optimally aligned with strong gradients in the image [6]. The stochastic watershed estimates the strength of edges in the image by repeatedly performing seeded watershed segmentation with randomly selected seedpoints. Each repetition will find a different subset of edges, but more important edges will be found more frequently. The output of the method is a probability density function (PDF), assigning to each piece of contour in the image the probability to appear as a segmentation boundary in seeded watershed segmentation with N randomly selected seedpoints.

E. Barcucci et al. (Eds.): DGCI 2014, LNCS 8668, pp. 309–319, 2014.

Here, we study watersheds on edge weighted graphs. In this context, a digital image is commonly represented by its *pixel adjacency graph*, i.e., a graph where each image element corresponds to a vertex in the graph, and adjacent image elements are connected by graph edges. Consequently, the seedpoints used for watershed segmentation consist of a set of vertices in the graph, and the watershed itself is defined as a *watershed cut* [6,7]. Informally, a cut is a set of edges that, if removed from the graph, separates it into two or more connected components. The stochastic watershed PDF is then defined as a mapping over the edges of the graph, i.e., every edge is considered to be a "piece of contour".

In the original paper by Angulo and Jeulin [1], the PDF was estimated by Monte Carlo simulation, i.e., repeatedly selecting N random seedpoints and performing seeded watershed segmentation. The drawback of this approach is that a large number of watershed segmentations must be performed to obtain a good estimate of the PDF. Meyer and Stawiaski [15] showed that the PDF can be calculated *exactly*, without performing any Monte Carlo simulations. Their work was later extended by Malmberg and Luengo [13], who proposed an efficient (pseudo-linear) algorithm for computing the exact PDF.

The exact evaluation method does not operate directly on the pixel adjacency graph G, but on a minimum spanning tree of G. The minimum spanning tree provides a compact representation of all watershed cuts on G in the sense that, under certain conditions, there is a one-to-one correspondence between the watershed cuts on G and the watershed cuts on its minimum spanning tree. The output of the exact evaluation algorithm is thus a map from the edges of the minimum spanning tree to $[0, 1]$ such that the value of the map for a given edge equals the probability of that edge being included in the watershed cut on the tree for a set of randomly selected seedpoints.

The compact tree representation is useful in its own right. For example, it is straightforward to obtain a segmentation directly from the tree by removing a set of edges with high probability values, and performing connected component labeling on the resulting forest. In general, the tree representation can be treated as any other segmentation hierarchy for morphological segmentation [8]. Nevertheless, there are cases where it might be useful to extend the stochastic watershed PDF to all edges in the graph. Malmberg and Luengo [13] suggested that such an extension could be performed using the *saliency map* approach of Najman and Schmitt [17,9]. As will be shown, however, this does not lead to the correct PDF for stochastic watersheds. Here, we propose an efficient algorithm for performing the correct extension. We show that for sparse graphs, common in image analysis applications, the time complexity of the proposed algorithm is $\mathcal{O}(|V|)$, where $|V|$ is the number of vertices in the graph.

2   Preliminaries

2.1   Edge Weighted Graphs

We will formulate our results in the framework of edge weighted graphs. In this context, a digital image is represented by its *pixel adjacency graph*, where each image element corresponds to a vertex in the graph, and adjacent image elements are connected by graph edges. In this section, we introduce some basic definitions to handle edge weighted graphs.

We define a *graph* as a triple $G = (V, E, \lambda)$ where

- V is a finite set.
- E is a set of unordered pairs of distinct elements in V, i.e., $E \subseteq \{\{v, w\} \subseteq V \mid v \neq w\}$.
- λ is a map $\lambda : E \to \mathbb{R}$.

The elements of V are called *vertices* of G, and the elements of E are called *edges* of G. When necessary, V, E, and λ will be denoted $V(G)$, $E(G)$, and $\lambda(G)$ to explicitly indicate which graph they belong to. An edge spanning two vertices v and w is denoted $e_{v,w}$. If $e_{v,w}$ is an edge in E, the vertices v and w are *adjacent*. The *neighborhood* of a vertex v it the set of all vertices adjacent to v, and is denoted by $\mathcal{N}(v)$.

For any edge $e \in E$, $\lambda(e)$ is the *weight* or *altitude* of e. Throughout the paper, we will assume that the value of $\lambda(e)$ represents the dissimilarity between the vertices spanned by e. Thus, we assume that the salient contours are located on the highest edges of the graph. In the context of image processing, we may define the edge weights as, e.g.,

$$\lambda(e_{v,w}) = |I(v) - I(w)|, \tag{1}$$

where $I(v)$ and $I(w)$ are the intensities of the image elements corresponding to the vertices v and w, respectively.

The seeded watershed method, and hence also the method presented here, depends on an increasing order of the edge weights in a graph, but not on their exact value [6,7]. To ensure the uniqueness of the seeded watershed segmentation, we will only consider graphs where each edge has a unique weight, thereby ensuring a unique increasing order. Graphs that do not fulfill this property can be easily be converted to the correct format as follows:

1. Fix an increasing ordering of the graph edges, i.e., find a map $O : E \to \mathbb{Z}$ such that $e_i \neq e_j \Rightarrow O(e_i) \neq O(e_j)$ and $O(e_i) < O(e_j) \Rightarrow \lambda(e_i) \leq \lambda(e_j)$ for all $e_i, e_j \in E$.
2. For all $e \in E$, set $\lambda(e) \leftarrow O(e)$.

Let G be a graph. A *path* in G is an ordered sequence of vertices $\pi = \langle v_i \rangle_{i=1}^k = \langle v_1, v_2, \ldots, v_k \rangle$ such that $e_{v_i, v_{i+1}} \in E$ for all $i \in [1, k-1]$. We denote the origin v_1 and the destination v_k of π by $\text{org}(\pi)$ and $\text{dst}(\pi)$, respectively. The set of vertices $\{v_1, v_2, \ldots, v_k\}$ along π is denoted $V(\pi)$, and the set of edges $\{e_{v_i, v_{i+1}} \mid i \in [1, k-1]\}$ along π is denoted $E(\pi)$. A path that has no repeated vertices is said to be *simple*. Two vertices v and w are *linked* in G if there exists a path π in G such that $\text{org}(\pi) = v$ and $\text{dst}(\pi) = w$. The notation $v \underset{G}{\sim} w$ will here be used to indicate that v and w are linked on G. If all pairs of vertices in G are linked, then G is *connected*, otherwise it is *disconnected*.

Let G and H be two graphs. If $V(H) \subseteq V(G)$ and $E(H) \subseteq E(G)$, then H is a *subgraph* of G. If H is a connected subgraph of G and $v \underset{G}{\not\sim} w$ for all vertices $v \in H$ and $w \notin H$, then H is a *connected component* of G.

Let G be a graph and let π be a path in G. If $\text{dst}(\pi) = \text{org}(\pi)$, then π is a *cycle*. A *cycle* is *simple* if it has no repeated vertices other than the endpoints. If G has no simple cycles, then G is a *forest*. A connected forest is called a *tree*.

Let G be a graph, and let T be a subgraph of G such that T is a tree and $V(G) = V(T)$. Then T is a *spanning tree* for G. The *weight* of a tree is the sum of all edge weights in the tree. A *minimum spanning tree* of G is a spanning tree with weight less than or equal to the weight of every other spanning tree of G.

2.2   Exact Stochastic Watersheds

In this section, we briefly review the method of Meyer and Stawiaski [15] for exact evaluation of stochastic watersheds.

Let G be a graph. In the method of Meyer and Stawiaski, the hierarchy of watershed segmentations is represented by a *flooding tree*, T, which in our context is equivalent to a minimum spanning tree T of the graph G. Since T has no simple cycles, every set of edges $S \subseteq E(T)$ forms a cut on T. Moreover, every set of edges $S \subseteq E(T)$ corresponds to a cut S' in G, given by

$$S' = \{e_{v,w} \in E(G) \mid \underset{(V(T), E(T) \setminus S)}{v \not\sim w}\}. \tag{2}$$

We say that S' is the cut on G *induced* by S. If S is a watershed cut on T with respect to some set of seedpoints, then the cut induced by S is a watershed cut on G with respect to the same seedpoints [6,7]. In fact, there is a one-to-one correspondence between the set of all watershed cuts on T and the set of all watershed cuts on G. Meyer and Stawiaski [15] used this correspondence to compactly represent the stochastic watershed PDF as a mapping $P : E(T) \to [0,1]$, where $P(e)$ is the probability of e being included in the watershed cut on T for N randomly selected seedpoints. This mapping can be computed in $\mathcal{O}(|E|\alpha(|V|))$ time using the algorithm proposed by Malmberg and Luengo [13], where α is the extremely slow-growing inverse of the *Ackermann* function [5].

3   Method

Let G be a graph, and let T be a minimum spanning tree for G. For all $e \in E(T)$, let $P(e)$ be the probability of e being included in the watershed cut on T for N randomly selected seedpoints. We will now show how the mapping P defined over the edges of T can be used to compute a mapping $P' : E(G) \to [0,1]$ such that $P'(e)$ is equal to the probability of e being included in the watershed cut on G for N randomly selected seedpoints.

Malmberg and Luengo [13] suggested that the mapping P could be extended to all edges in $E(G)$ using the saliency map approach of Najman and Schmitt [17] For every edge $e_{v,w} \in E(G)$, there is a unique path, denoted $\pi_{v,w}$, on T connecting the vertices spanned by the edge. According to the saliency map approach, the mapping P' would be defined as

$$P'(e_{v,w}) = \max_{e \in E(\pi_{v,w})} (P(e)) . \tag{3}$$

By this definition, however, $P'(e)$ does *not* equal the probability of e being included in the watershed cut on G for N randomly selected seedpoints, as shown below.

Theorem 1. *Let S be a cut on T, and let S' be the cut on G induced by S. For every edge $e_{v,w} \in E(G)$, it holds that $e_{v,w} \in S'$ if and only if $S \cap E(\pi_{v,w}) \neq \emptyset$.*

Proof. If $E(\pi_{v,w}) \cap S = \emptyset$, then $\pi_{v,w}$ is a path between v and w on $(V, E(T) \setminus S)$, i.e., $\underset{((V,E(T)\setminus S))}{v \sim w}$, and so $e_{v,w} \notin S'$. Conversely, if $E(\pi_{v,w}) \cap S \neq \emptyset$, then $\underset{((V,E(T)\setminus S))}{v \nsim w}$, and so $e_{v,w} \in S'$.

The probability of $P'(e_{v,w})$ of $e_{v,w}$ being included in the watershed cut on G is therefore equal to the probability that the watershed cut on T contains at least one edge along the path $\pi_{v,w}$ connecting v and w on T. This probability is given by

$$P'(e_{v,w}) = 1 - \prod_{e \in E(\pi_{v,w})} (1 - P(e)) . \tag{4}$$

The derivation of Equation 4 is straightforward. Since the probability of an edge along $\pi_{v,w}$ being part of the watershed cut on T is independent of that for other edges along $\pi_{v,w}$, the probability that no edge in $E(\pi_{v,w})$ belongs to the watershed cut can be computed by multiplication of the individual probabilities. Note that for all edges $e_{v,w} \in E(T)$, Equation 4 reduces to $P'(e_{v,w}) = P(e_{v,w})$ as expected.

Based on the above results, we can formulate a naive approach for calculating $P'(e_{v,w})$ as follows:

Algorithm 1. Calculate $\Phi(v, r)$ for all vertices v in a tree with root r

---

Input: A tree $T = (V, E)$, rooted at a vertex r. A map $P : E \to [0, 1]$
Auxiliary: Two sets of vertices C, D
Output: A map $F : V \to \mathbb{R}$, such that $F(v) = \Phi(v, r)$ for all $v \in V$.

1 Set $C \to r$, $D \to \emptyset$, and $F(r) \leftarrow 1$;
2 **while** $C \neq \emptyset$ **do**
3 | Select any vertex $v \in C$;
4 | Set $C \leftarrow C \setminus \{v\}$ and $D \leftarrow D \cup \{v\}$;
5 | **foreach** $w \in \mathcal{N}(v) \setminus D$ **do**
6 | | Set $F(w) \leftarrow F(v) \cdot (1 - P(e_{v,w}))$;
7 | | Set $C \leftarrow C \cup \{w\}$

---

1. Find the unique path $\pi_{v,w}$ connecting v and w on T using, e.g., breadth-first search.
2. Use Equation 4 to calculate $P'(e_{v,w})$.

Performing these calculations for all edges in a graph, however, is prohibitively slow. We will now present an efficient algorithm for computing $P'(e)$ for all $e \in E(G)$. First, we define the function $\Phi(a, b)$ as

$$\Phi(a, b) = \prod_{e \in E(\pi_{a,b})} (1 - P(e)), \qquad (5)$$

Next, we designate an arbitrary vertex $r \in V$ to be the *root* of T. The choice of r does not affect the output of the algorithm. For a pair of vertices $v, v' \in V$, we say that v is an *ancestor* of v' if v lies along the path from v' to r on T. The *lowest common ancestor* $\mathrm{LCA}(v, w)$ of two vertices v and w is the vertex located farthest from the root that is an ancestor of both v and w. With these definitions in place, we can rewrite $\Phi(v, w)$ as

$$\Phi(v, w) = \frac{\Phi(v, r)\Phi(w, r)}{\Phi(\mathrm{LCA}(v, w), r)^2} . \qquad (6)$$

For a fixed root r, the value of $\Phi(v, r)$ for all vertices $v \in V$ can be computed using Algorithm 1. Computationally, this algorithm is equivalent to breadth-first search, and so has the same $\mathcal{O}(|V| + |E(T)|) = \mathcal{O}(|V|)$ time complexity.

Once the value of $\Phi(v, r)$ has been calculated for all $v \in V$, Algorithm 2 can be used to calculate the desired probability $P'(e)$ for all $e \in E(G)$. Algorithm 2 iterates over all edges of G, and computes the LCA of the vertices spanned by the edge. The time complexity of Algorithm 2 is therefore $\mathcal{O}(|E| \cdot X)$, where X is the cost of finding the LCA between a pair of vertices. As shown by Harel and Tarjan [12], the LCA of a pair of vertices can be found in $\mathcal{O}(1)$ time, after performing an $\mathcal{O}(|V|)$ preprocessing step. An algorithm satisfying these bounds, while also being suitable for practical implementation, was proposed by Bender and Farach-Colton [3].

Algorithm 2. Calculate $P'(e)$ for all edges e in a graph

Input: An edge-weighted graph $G = (V, E, \lambda)$. A minimum spanning tree T of G, rooted at a vertex r. A map $F : V \to \mathbb{R}$, such that $F(v) = \Phi(v, r)$ for all $v \in V$.

Output: A map $\Pi : E \to [0, 1]$ such that $\Pi(e) = P'(e)$ for all $e \in E$.

1 **foreach** $e_{v,w} \in E$ **do**
2   Find $c = LCA(v, w)$;
3   Set $\Pi(e_{v,w}) \leftarrow 1 - \frac{F(v)F(w)}{F(c)^2}$;

In total, the time complexity of the proposed method for calculating P' given P is therefore $\mathcal{O}(|V| + |E(G)|)$. For the sparse graphs typically encountered in image processing it holds that $\mathcal{O}(|E(G)|) = \mathcal{O}(|V|)$, and for such graphs the total time complexity of the proposed method is therefore $\mathcal{O}(|V|)$.

4   Visualizing the Probability Density Function

In the common case where G is a pixel adjacency graph, i.e., the vertices of G correspond to image elements, it may be of interest to visualize the stochastic watershed PDF on the vertices of the graph. To this end, we introduce the notion of a *boundary operator*. A boundary operator δ is a mapping $\delta : V \to \mathcal{P}(V)$, such that $\delta(v) \subseteq \mathcal{N}(v)$ for all $v \in V$. Given a cut S on G and a boundary operator δ, we say that a vertex v is a *boundary vertex* for S and δ if any of the vertices in $\delta(v)$ are separated from v by the cut S. The probability $P''(v)$ of v being a boundary vertex is then given by

$$P''(v) = 1 - \prod_{w \in \delta(v)} (1 - P'(e_{v,w})) \,. \tag{7}$$

The derivation of Equation 7 is analogous to the derivation of Equation 4.

5   Experiments

As shown in Section 3, the asymptotic time required for computing the stochastic watershed PDF on the edges of the MST of a graph is the same as that required for computing the PDF on all edges of the graph. In this section, we investigate how much the the computation increases in practice when the PDF is computed for all edges, rather than on the edges of the MST only.

We calculate stochastic watershed PDFs on a set of two-dimensional images of varying sizes, generated by scaling up an original low resolution image using bi-cubic interpolation. Pixel adjacency graphs were constructed from the images using a standard 4-connected adjacency relation with edge weights defined according to Equation 1. For every pixel $\mathbf{x}$ in an image, the boundary operator

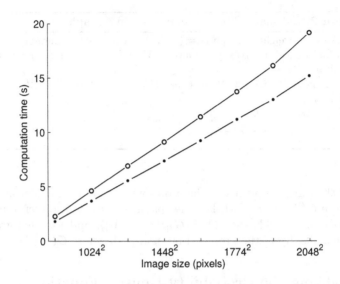

Fig. 1. Computation time for calculating the exact stochastic watershed PDF on pixel adjacency graphs for 2D images of varying sizes. The bottom curve shows the computation time for calculating the exact stochastic watershed PDF on the MST of the pixel adjacency graph using the method proposed by Malmberg and Luengo [13]. The top curve shows the total computation time for calculating the stochastic watershed PDF for all edges in the pixel adjacency graph, using the proposed method to extend the PDF from the edges of the MST to all edges in the graph.

$\delta(\mathbf{x})$ was defined as $\delta(\mathbf{x}) = \{\mathbf{x}+(1,0), \mathbf{x}+(0,1)\}$. For all pixel adjacency graphs, the exact stochastic watershed was first calculated on the MST of the pixel adjacency graph using the method of Malmberg and Luengo [13], and the PDF was then extended to all edges in the graph using the proposed method. For all experiments, N = 20 random seedpoints were used. Figure 1 shows the total computation time for both these steps, as well as the computation time for the first step only. As the figure shows, the overhead for extending the PDF to all edges is small – the average increase in computation time is about 25%. Figure 2 shows the stochastic watershed PDF computed on the original low resolution 2D image. For reference, the same PDF approximated by Monte Carlo simulation with 1000 repetitions is also shown. At each repetition, we calculate the watershed cut corresponding to a random set of seedpoints and identify the boundary pixels using the boundary operator defined above. The final PDF is obtained by counting the number of times each pixel appears as a boundary pixel, and dividing this number by the total number of repetitions.

Fig. 2. The stochastic watershed PDF computed on a 2D image. The contrast of the images has been adjusted for display purposes. (Top) Original image. (Bottom left) PDF obtained by the proposed exact method. (Bottom right) Reference PDF obtained by Monte Carlo simulation with 1000 repetitions.

6   Conclusions

The stochastic watershed method has found some applications, such as segmentation of multi-spectral satellite images [18,2], characterization of the grain structure of nuclear fuel pellets [4], study of angiogenesis [19], segmentation of granular materials in 3D microtomography [10,11], and detection of the optic disc in fundus images [16]. However, the computational cost of the Monte-Carlo simulation that estimates the PDF is a barrier to more wide-spread use. In this view, the prospect of efficiently computing the exact stochastic watershed PDF, without resorting to Monte Carlo simulation, is appealing. An important step in this direction was taken by Malmberg and Luengo [13], who proposed a pseudo-linear algorithm for computing the exact stochastic watershed PDF for all edges of a tree.

Here, we have extended the work of Malmberg and Luengo by presenting a method for calculating the exact stochastic watershed PDF for all edges in a graph, given that the PDF is known for all edges included in the minimum spanning tree of the graph. Additionally we have presented a method that, via the concept of a boundary operator, transfers the PDF from the edges to the vertices of a graph. This allows exact stochastic watersheds to be computed on the pixels of an image, rather than on the abstract tree representation used in previous exact evaluation methods [15,13]. We believe this makes the exact evaluation approach more attractive for use in practical applications. For sparse graphs, typical in image processing applications, the proposed method terminates in $\mathcal{O}(|V|)$ time, which is asymptotically smaller than the $\mathcal{O}(|E|\alpha(|V|))$ time required for calculating the PDF over the edges of the minimum spanning tree of the graph. Our experiments demonstrate that in practice, the computational cost of extending the PDF to all edges in a graph is small compared to the cost of computing the PDF on the edges of the MST of the graph.

In the original paper by Angulo and Jeulin [1], the PDF obtained by Monte-Carlo simulation was convolved with a Gaussian function to obtain a smooth estimate of the true PDF. An interesting direction for future research is to incorporate this kind of relaxation in the proposed method by allowing larger boundary operators, i.e. not constraining the boundary operator to be a subset of the neighborhood of a vertex, and assigning appropriate weights to the elements of the boundary operator.

References

1. Angulo, J., Jeulin, D.: Stochastic watershed segmentation. In: Intern. Symp. on Mathematical Morphology, vol. 8, pp. 265–276 (2007)
2. Angulo, J., Velasco-Forero, S.: Semi-supervised hyperspectral image segmentation using regionalized stochastic watershed. In: Proceedings of SPIE Symposium on Defense, Security, and Sensing: Algorithms and Technologies for Multispectral, Hyperspectral, and Ultraspectral Imagery XVI. Proceedings of SPIE, vol. 7695, p. 76951F. SPIE, Bellingham (2010)

3. Bender, M.A., Farach-Colton, M.: The LCA problem revisited. In: Gonnet, G.H., Viola, A. (eds.) LATIN 2000. LNCS, vol. 1776, pp. 88–94. Springer, Heidelberg (2000)
4. Cativa Tolosa, S., Blacher, S., Denis, A., Marajofsky, A., Pirard, J.P., Gommes, C.J.: Two methods of random seed generation to avoid over-segmentation with stochastic watershed: application to nuclear fuel micrographs. Journal of Microscopy 236(1), 79–86 (2009)
5. Cormen, T.H., Leiserson, C.E., Rivest, R.L., Stein, C.: Introduction to algorithms. MIT Press (2001)
6. Cousty, J., Bertrand, G., Najman, L., Couprie, M.: Watershed cuts: Minimum spanning forests and the drop of water principle. IEEE Transactions on Pattern Analysis and Machine Intelligence 31(8), 1362–1374 (2009)
7. Cousty, J., Bertrand, G., Najman, L., Couprie, M.: Watershed cuts: Thinnings, shortest path forests, and topological watersheds. IEEE Transactions on Pattern Analysis and Machine Intelligence 32(5), 925–939 (2010)
8. Cousty, J., Najman, L., Perret, B.: Constructive links between some morphological hierarchies on edge-weighted graphs. In: Hendriks, C.L.L., Borgefors, G., Strand, R. (eds.) ISMM 2013. LNCS, vol. 7883, pp. 86–97. Springer, Heidelberg (2013)
9. Cousty, J., Najman, L.: Incremental algorithm for hierarchical minimum spanning forests and saliency of watershed cuts. In: Soille, P., Pesaresi, M., Ouzounis, G.K. (eds.) ISMM 2011. LNCS, vol. 6671, pp. 272–283. Springer, Heidelberg (2011)
10. Faessel, M., Jeulin, D.: Segmentation of 3D microtomographic images of granular materials with the stochastic watershed. Journal of Microscopy 239(1), 17–31 (2010)
11. Gillibert, L., Jeulin, D.: Stochastic multiscale segmentation constrained by image content. In: Soille, P., Pesaresi, M., Ouzounis, G.K. (eds.) ISMM 2011. LNCS, vol. 6671, pp. 132–142. Springer, Heidelberg (2011)
12. Harel, D., Tarjan, R.E.: Fast algorithms for finding nearest common ancestors. SIAM Journal on Computing 13(2), 338–355 (1984)
13. Malmberg, F., Hendriks, C.L.L.: An efficient algorithm for exact evaluation of stochastic watersheds. Pattern Recognition Letters (2014)
14. Meyer, F., Beucher, S.: Morphological segmentation. Journal of Visual Communication and Image Representation 1(1), 21–46 (1990)
15. Meyer, F., Stawiaski, J.: A stochastic evaluation of the contour strength. In: Goesele, M., Roth, S., Kuijper, A., Schiele, B., Schindler, K. (eds.) DAGM 2010. LNCS, vol. 6376, pp. 513–522. Springer, Heidelberg (2010)
16. Morales, S., Naranjo, V., Angulo, J., Alcaniz, M.: Automatic detection of optic disc based on PCA and mathematical morphology. IEEE Transactions on Medical Imaging 32(4), 786–796 (2013)
17. Najman, L., Schmitt, M.: Geodesic saliency of watershed contours and hierarchical segmentation. IEEE Transactions on Pattern Analysis and Machine Intelligence 18(12), 1163–1173 (1996)
18. Noyel, G., Angulo, J., Jeulin, D.: Classification-driven stochastic watershed: application to multispectral segmentation. In: IS&T's Fourth European Conference on Color in Graphics Imaging, and Vision (CGIV 2008), pp. 471–476 (2008)
19. Noyel, G., Angulo, J., Jeulin, D.: Regionalized random germs by a classification for probabilistic watershed application: Angiogenesis imaging segmentation. In: Fitt, A.D., Norbury, J., Ockendon, H., Wilson, E. (eds.) Progress in Industrial Mathematics at ECMI 2008. Mathematics in Industry, pp. 211–216. Springer (2010)

On Making nD Images Well-Composed
by a Self-dual Local Interpolation

Nicolas Boutry[1,2], Thierry Géraud[1], and Laurent Najman[2]

[1] EPITA Research and Development Laboratory (LRDE), France
[2] Université Paris-Est, LIGM, Équipe A3SI, ESIEE, France
firstname.lastname@lrde.epita.fr, l.najman@esiee.fr

Abstract. Natural and synthetic discrete images are generally not well-composed, leading to many topological issues: connectivities in binary images are not equivalent, the Jordan Separation theorem is not true anymore, and so on. Conversely, making images well-composed solves those problems and then gives access to many powerful tools already known in mathematical morphology as the Tree of Shapes which is of our principal interest. In this paper, we present two main results: a characterization of 3D well-composed gray-valued images; and a counter-example showing that no local self-dual interpolation satisfying a classical set of properties makes well-composed images with one subdivision in 3D, as soon as we choose the mean operator to interpolate in 1D. Then, we briefly discuss various constraints that could be interesting to change to make the problem solvable in nD.

Keywords: Digital topology, gray-level images, well-composed sets, well-composed images.

1 Introduction

Natural and synthetic images are usually not well-composed. This fact raises many topological issues. As an example, the Jordan Separation theorem, stating that a simple closed curve in $\mathbb{R}^2$ separates the space in only two components is not true anymore for binary 2D discrete images [5]. To solve this problem, we have to juggle with two complementary connectivities: 4 for the background and 8 for the foreground, or the inverse. 2D well-composed binary images have the fundamental property that 4- and 8-connectivities are equivalent. Hence, such topological issues vanish. In the same manner, well-composed nD images, with $n > 2$, have $2n$- and $(3^n - 1)$-connectivities equivalent [11]. Other advantages of well-composed images are for example the preservation of topological properties by a rigid transform [10], simplification of thinning algorithms [8] and simplification of graph structures resulting from skeleton algorithms [5]. Also, and it is our most important goal, one can compute the Tree of Shapes [9,2] of a well-composed image with a quasi-linear algorithm [3]. An introduction to the Tree of Shapes in the continuous case can be found in [1].

E. Barcucci et al. (Eds.): DGCI 2014, LNCS 8668, pp. 320–331, 2014.

Section 2 recalls the definitions of 2D and 3D well-composed sets and gray-valued images, and introduces a characterization of 3D gray-valued well-composed images. Because we do not want to deteriorate the initial signal, we use an interpolation that produces a well-composed image. Furthermore, in order to treat in the same manner bright components on dark background and dark components over bright background, this interpolation process will be self-dual. We present in Section 3 a general scheme that recursively defines local interpolations satisfying a classical set of properties with one subdivision. We show that such interpolations, with the added property of being self-dual, fail in 3D (and then further) to make well-composed images. We conclude in Section 4 with some perspectives that could work in nD even if $n > 2$ (in local and non-local ways).

2   A Characterization of 3D Well-Composed Gray-Valued Images

2.1   2D WC Sets and Gray-Valued Images

Let us begin by the definitions of a block of $\mathbb{Z}^n$. We will then be able to recall the definition and the characterization of 2D well-composed sets and images.

A block in nD associated to $z \in \mathbb{Z}^n$ is the set S_z defined such that $S_z = \{z' \in \mathbb{Z}^n \mid \|z - z'\|_\infty \leq 1 \text{ and } \forall i \in [1, n], z_i' \geq z_i\}$ (where z_i represents the ith coordinate of z). Moreover, we call blocks of $\mathcal{D} \subseteq \mathbb{Z}^n$ any element of the set $\{S_z \mid \exists z \in \mathcal{D}, S_z \subseteq \mathcal{D}\}$.

Definition 1 (2D WC Sets [5]). *A set X is weakly well-composed if any 8-component of X is a 4-component. X is well-composed if both X and its complement $X^c = \mathbb{Z}^2 \setminus X$ are weakly well-composed.*

Proposition 1 (Local Connectivity and No Critical Configurations [5]). *A set $X \subseteq \mathbb{Z}^2$ is well-composed iff it is locally 4-connected. Also, a set X is well-composed if none of the critical configurations $\begin{pmatrix} 1 & 0 \\ 0 & 1 \end{pmatrix}$ or $\begin{pmatrix} 0 & 1 \\ 1 & 0 \end{pmatrix}$ appears in X.*

Definition 2 (Cuts in nD). *For any $\lambda \in \mathbb{R}$ and any gray-valued map $u : \mathcal{D} \subseteq \mathbb{Z}^n \mapsto \mathbb{R}$, we denote by $[u > \lambda]$ and $[u < \lambda]$ the sets $[u > \lambda] = \{M \in \mathcal{D} \mid u(M) > \lambda\}$ and $[u < \lambda] = \{M \in \mathcal{D} \mid u(M) < \lambda\}$. We call them respectively upper strict cuts and lower strict cuts [3].*

We remark that an image $u : \mathcal{D} \subseteq \mathbb{Z}^2 \mapsto \mathbb{R}$ with a finite domain $\mathcal{D}$ can only be well-composed if $\mathcal{D}$ is itself well-composed (since $[u < \max(u) + 1] = \mathcal{D}$).

Definition 3 (Gray-valued WC 2D Maps [5]). *A gray-level map $u : \mathbb{Z}^2 \mapsto \mathbb{R}$ is well-composed iff for every $\lambda \in \mathbb{R}$, the strict cuts $[u > \lambda]$ and $[u < \lambda]$ result in well-composed sets.*

We recall that the *interval value* of the couple $(x, y) \in \mathbb{R}^2$ is defined as $\text{intvl}(x, y) = [\min(x, y), \max(x, y)]$.

Proposition 2 (Characterization of 2D WC maps [5]). *A gray-level map* $u : \mathbb{Z}^2 \mapsto \mathbb{R}$ *is well-composed iff for every 2D block* S *such that* $u|_S = \begin{pmatrix} a & b \\ c & d \end{pmatrix}$, *the interval values satisfy* $\mathrm{intvl}(a, d) \cap \mathrm{intvl}(b, c) \neq \emptyset$.

2.2 3D WC Sets and Gray-Valued Maps

As we will see, for $n = 3$, the equivalence between local connectivity and well-composedness is no longer true. This led Latecki [4] to introduce the continuous analog.

Fig. 1. Illustration of the bdCA of a set containing a critical configurations of type 1 (left), and of type 2 (right)

Fig. 2. A set locally 6-connected but not well-composed

Definition 4 (CA and bdCA [4]). *The* continuous analog $\mathrm{CA}(z)$ *of a point* $z \in \mathbb{Z}^3$ *is the closed unit cube centered at this point with faces parallel to the coordinate planes, and the continuous analog of a set* $X \subseteq \mathbb{Z}^3$ *is defined as* $\mathrm{CA}(X) = \bigcup \{\mathrm{CA}(x) | x \in X\}$. *The (face) boundary of the continuous analog* $\mathrm{CA}(X)$ *of a set* $X \subseteq \mathbb{Z}^3$ *is noted* $\mathrm{bdCA}(X)$ *and is defined as the union of the set of closed faces each of which is the common face of a cube in* $\mathrm{CA}(X)$ *and a cube not in* $\mathrm{CA}(X)$.

Definition 5 (Well-composedness in 3D [4]). *A 3D set* $X \subseteq \mathbb{Z}^3$ *is well-composed iff* $\mathrm{bdCA}(X)$ *is a 2D manifold, i.e., a topological space which is locally Euclidian.*

Proposition 3 (No Critical Configurations [4]). *A set* $X \subseteq \mathbb{Z}^3$ *is well-composed iff the following critical configurations of cubes of type 1 or of type 2 (modulo reflections and rotations) do not occur in* $\mathrm{CA}(X)$ *or in* $CA(X^c)$ *(see Figure 1).*

We remark that if a set $X \subseteq \mathbb{Z}^3$ is well-composed, then X is locally 6-connected. The converse is false (see Figure 2).

Definition 6 (WC Gray-valued Maps). *We say that a 3D real-valued map* $u : \mathcal{D} \subseteq \mathbb{Z}^3 \mapsto \mathbb{R}$ *is well-composed if its strict cuts* $[u > \lambda]$ *and* $[u < \lambda]$, $\forall \lambda \in \mathbb{R}$, *are well-composed.*

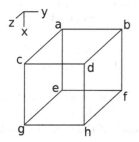

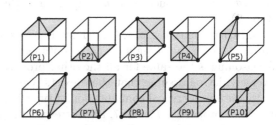

Fig. 3. The restriction $u\big|_S$ of u to a 3D block S

Fig. 4. The ten characteristical relations of well-composedness of a gray-valued image u restricted to a 3D block S

To characterize 3D gray-level well-composed images, we first give two lemmas concerning the detection of the critical configurations of respectively type 1 and type 2.

Lemma 1. *The strict cuts $[u > \lambda]$ and $[u < \lambda]$, $\lambda \in \mathbb{R}$, of a gray-valued image u defined on a block S, such as depicted in Figure 3, do not contain any critical configurations of type 1 iff the six following properties hold:*
intvl$(a,d) \bigcap$ intvl$(b,c) \neq \emptyset$ $(P1)$,   intvl$(e,h) \bigcap$ intvl$(g,f) \neq \emptyset$ $(P2)$
intvl$(a,f) \bigcap$ intvl$(b,e) \neq \emptyset$ $(P3)$,   intvl$(c,h) \bigcap$ intvl$(g,d) \neq \emptyset$ $(P4)$
intvl$(a,g) \bigcap$ intvl$(e,c) \neq \emptyset$ $(P5)$,   intvl$(b,h) \bigcap$ intvl$(f,d) \neq \emptyset$ $(P6)$

Proof : Let us assume that one of these properties $(P_i), i \in [1,6]$, is false. Let us say it is the case of $(P1)$. Then two cases are possible: either $\max(a,d) < \min(b,c)$, and that means that there exists $\lambda = (\max(a,d) + \min(b,c))/2$ such that $[u < \lambda]$ contains the critical configuration $\{a,d\}$ (of type 1), or $\min(a,d) > \max(b,c)$, and there exists $\lambda = (\min(a,d) + \max(b,c))/2$ such that one more time $[u > \lambda]$ contains the critical configuration $\{a,d\}$. The reasoning is the same for all the other properties. Conversely, let us assume that there exists $\lambda \in \mathbb{R}$ such that either $[u > \lambda]$ or $[u < \lambda]$ contains a critical configuration of type 1. That means immediately that one of the 6 properties $P_i, i \in [1,6]$, corresponding to each of the six faces of the block S, is false (see Figure 4 for the faces and the corners concerned by the properties).   □

Recall that the span of a set of values $E \subseteq \mathbb{R}$ is span$(E) = [\inf(E), \sup(E)]$.

Lemma 2. *The strict cuts $[u > \lambda]$ and $[u < \lambda]$, $\lambda \in \mathbb{R}$, of a gray-valued image u defined on a block S such as depicted in Figure 3, do not contain any critical configuration of type 2 iff the four following properties are true:*
intvl$(a,h) \bigcap$ span$\{b,c,d,e,f,g\} \neq \emptyset$ $(P7)$
intvl$(b,g) \bigcap$ span$\{a,c,d,e,f,h\} \neq \emptyset$ $(P8)$
intvl$(c,f) \bigcap$ span$\{a,b,d,e,g,h\} \neq \emptyset$ $(P9)$
intvl$(d,e) \bigcap$ span$\{a,b,c,f,g,h\} \neq \emptyset$ $(P10)$

Proof : Let us assume that one of these properties $(P_i), i \in [7,10]$, is false. Let us say it is the case of $(P7)$. Then two cases are possible:

- either $\max(a, h) < \min(b, c, d, e, f, g)$. Then there exists $\lambda = (\max(a, h) + \min(b, c, d, e, f, g))/2$ such that $[u < \lambda]$ contains the critical configuration $\{a, h\}$ (of type 2),
- or $\min(a, h) > \max(b, c, d, e, f, g)$. Then there exists $\lambda = (\max(b, c, d, e, f, g) + \min(a, h))/2$ such that (again) $[u > \lambda]$ contains the critical configuration $\{a, h\}$. The reasoning is the same for all the other properties.

Conversely, let us assume that there exists $\lambda \in \mathbb{R}$ such that either $[u > \lambda]$ or $[u < \lambda]$ contains a critical configuration of type 2. That means immediately that one of the 4 properties $P_i, i \in [7, 10]$, corresponding to each of the four diagonals of the block S, is false (see Figure 4). □

We are now ready to state the main theorem of this section, characterizing the well-composedness on a 3D gray-valued image.

Theorem 1 (Characterization of well-composedness in 3D). *Let us suppose that $\mathcal{D}$ is a hyperrectangle in $\mathbb{Z}^3$. A gray-valued 3D image $u : \mathcal{D} \mapsto \mathbb{R}$ is well-composed on $\mathcal{D}$ iff on any block $S \subseteq \mathcal{D}$, $u\big|_S$ satisfies the properties $(P_i), i \in [1, 10]$.*

3   Local Interpolations

Using interpolations with one subdivision does not deteriorate the initial signal. The size of the original image is multiplied by a factor of 2^n, where n is the dimension of the space of the image. Figure 5 illustrates this subdivision process.

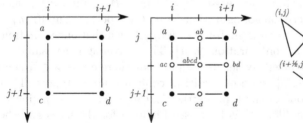 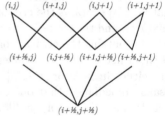

Fig. 5. Illustration of the subdivision process on a block S

Fig. 6. $s(S) \subseteq \left(\frac{\mathbb{Z}}{2}\right)^n$ as a poset

3.1   Subdivision of a Domain and $\left(\frac{\mathbb{Z}}{2}\right)^n$ as a Poset

The subdivision of a block allows us to provide an order to the elements. Using this order, the subdivided space is a poset.

Let z be a point in $\mathbb{Z}^n$, and S_z its associated block. We define the subdivision of S_z as $s(S_z) = \{z' \in \left(\frac{\mathbb{Z}}{2}\right)^n \mid \|z - z'\|_\infty \leq 1 \text{ and } \forall i \in [1, n], z'_i \geq z_i\}$.

The subdivision of a domain $\mathcal{D} \subseteq \mathbb{Z}^n$ is the union of the subdivisions of the blocks contained in $\mathcal{D}$, i.e., $s(\mathcal{D}) = \cup_{S \subseteq \mathcal{D}} s(S)$.

Definition 7 (Order of a point of $\left(\frac{\mathbb{Z}}{2}\right)^n$). *Assume e_i is a fixed basis of $\mathbb{Z}^n$. We note $\frac{1}{2}(z) = \{i \in [1, n] | z_i \in \frac{\mathbb{Z}}{2} \setminus \mathbb{Z}\}$. The sets $\mathbb{E}_k$, for $k \in [0, n]$, are defined such that $\mathbb{E}_k = \{z \in \left(\frac{\mathbb{Z}}{2}\right)^n \mid |\frac{1}{2}(z)| = k\}$ (where $|E|$ denotes the cardinal of the set E), and represent a partition of $\left(\frac{\mathbb{Z}}{2}\right)^n$. We call order of a point z the value k such that $z \in \mathbb{E}_k$ and we note it $\mathbb{o}(z)$.*

Definition 8 (Parents in $\left(\frac{\mathbb{Z}}{2}\right)^n$). *Let z be a point of $\left(\frac{\mathbb{Z}}{2}\right)^n$. The set of the parents of $z \in \left(\frac{\mathbb{Z}}{2}\right)^n$, noted $\mathbb{P}(z)$, is defined by $\mathbb{P}(z) = \cup_{i \in \frac{1}{2}(z)} \{z - \frac{e_i}{2}, z + \frac{e_i}{2}\}$. The parents of $z \in \left(\frac{\mathbb{Z}}{2}\right)^n$ of order 0 are $\mathbb{P}^0(z) = \{z\}$ and of order $k > 0$ are defined recursively by $\mathbb{P}^k(z) = \cup_{p \in \mathbb{P}(z)} \mathbb{P}^{k-1}(p)$.*

Definition 9 ($\mathcal{G}(z)$ and $\mathbb{A}(z)$). *Let z be a point of $\left(\frac{\mathbb{Z}}{2}\right)^n$. The ancestors of $z \in \left(\frac{\mathbb{Z}}{2}\right)^n$ are $\mathbb{A}(z) = \mathbb{P}^{\mathbb{o}(z)}(z)$. We set $\mathcal{G}(z) = \cup_{k \in [0, \mathbb{o}(z)]} \mathbb{P}^k(z)$.*

Notice that $\mathbb{A}(z) \subseteq \mathbb{Z}^n$ and that any point $z \in \mathbb{E}_k$, $k \in [1, n]$, has its parents in $\mathbb{E}_{k-1}$. Hence $\{\mathbb{E}_k\}_{k \in [0, n]}$ is a (hierarchical) partition of $\left(\frac{\mathbb{Z}}{2}\right)^n$, and $(\left(\frac{\mathbb{Z}}{2}\right)^n, \mathbb{P})$ is a poset (see Figure 6).

Definition 10 (Opposites). *Let z be a point of $\left(\frac{\mathbb{Z}}{2}\right)^n$. The (set of) opposites of $z \in \left(\frac{\mathbb{Z}}{2}\right)^n$ is the family of pairs of points $\text{opp}(z) = \cup_{i \in \frac{1}{2}(z)} \{\{z - \frac{e_i}{2}, z + \frac{e_i}{2}\}\}$.*

3.2 Interpolations with One Subdivision

Let us recall that the convex hull $\text{convhull}(Z)$ of a set of m points $Z = \{z^1, \ldots, z^m\} \subseteq \mathbb{Z}^n$ is:

$$\text{convhull}(Z) = \left\{ \sum_{i=1}^{m} \alpha_i z^i \mid \sum_{i=1}^{m} \alpha_i = 1 \text{ and } \forall i \in [1, m], \alpha_i \geq 0 \right\}$$

Definition 11 (Subdivision of edges, faces, and cubes). *Let $\mathcal{E} = \{z_1, z_2\}$ be an edge in $\mathbb{Z}^n$. The subdivision of $\mathcal{E}$ is $s(\mathcal{E}) = \{z \in \left(\frac{\mathbb{Z}}{2}\right)^n | z \in \text{convhull}(\mathcal{E})\}$. The subdivision of a face $\mathcal{F} = \{z_1, z_2, z_3, z_4\}$ is $s(\mathcal{F}) = \{z \in \left(\frac{\mathbb{Z}}{2}\right)^n | z \in \text{convhull}(\mathcal{F})\}$. The subdivision of a cube $\mathcal{C} = \{z_1, \ldots, z_8\}$ is $s(\mathcal{C}) = \{z \in \left(\frac{\mathbb{Z}}{2}\right)^n | z \in \text{convhull}(\mathcal{C})\}$.*

3.3 A Set of Properties That an Interpolation Has to Satisfy

An *interpolation* of a map $u : \mathcal{D} \subseteq \mathbb{Z}^n \mapsto \mathbb{R}$ to a map $\mathfrak{I}(u) : s(\mathcal{D}) \subseteq \left(\frac{\mathbb{Z}}{2}\right)^n \mapsto \mathbb{R}$ is a transformation such that $\mathfrak{I}(u)|_S = u|_S$ for any block $S \subseteq \mathcal{D}$.

Let $u : \mathcal{D} \subseteq \mathbb{Z}^3 \mapsto \mathbb{R}$ be any 3D gray-valued image. We say that an interpolation $\mathfrak{I} : u \mapsto \mathfrak{I}(u)$ is *self-dual* iff $\mathfrak{I}(-u) = -\mathfrak{I}(u)$. A self-dual interpolation does not overemphasize bright components at the expense of the dark ones, or conversely.

An interpolation $\mathfrak{I} : u \mapsto \mathfrak{I}(u)$ in 3D is said *ordered* if the new values are inserted firstly at the centers of the subdivided edges, secondly at the centers of the subdivided faces, and finally at the centers of the subdivided cubes.

An ordered interpolation is said *in between* iff it puts the values at a point z in between the values of its parents $\mathbb{P}(z)$.

Finally, we say that an interpolation is *well-composed* iff the image $\mathfrak{I}(u)$ resulting from the interpolation of u is well-composed for any given image u.

We are interested in interpolations $\mathfrak{I}$ with the following properties.

$$(\mathcal{P}) \Leftrightarrow \begin{cases} \mathfrak{I} \text{ is invariant by translations, } \frac{\pi}{2}\text{'s rotations and axial symmetries} \\ \mathfrak{I} \text{ is ordered} \\ \mathfrak{I} \text{ is in-between} \\ \mathfrak{I} \text{ is self-dual} \\ \mathfrak{I} \text{ is well-composed} \end{cases}$$

3.4   The Scheme of Local Interpolations Verifying $\mathcal{P}$

A *local interpolation* $\mathfrak{I}$ is an interpolation such as for any block $S \subseteq \mathcal{D}$, $\mathfrak{I}(u)$ on $s(S)$ is computed only from its nearest neighbours belonging to $\mathbb{E}_0$ (we see an image as a graph). For convenience, we will write u' instead of $\mathfrak{I}(u)$ for local interpolations in the sequel.

Fig. 7. From left to right: an image, and its interpolations with the *median*, the *mean/median*, the *min* and the *max*

Lemma 3 (Local interpolation scheme). *Any local interpolation $\mathfrak{I}$ on $\left(\frac{\mathbb{Z}}{2}\right)^n$ verifying $\mathcal{P}$ can be characterized by a set of functions $\{f_k\}_{k \in [1,n]}$ such that:*

$$\forall z \in \left(\frac{\mathbb{Z}}{2}\right)^n, \ u'(z) = \begin{cases} u(z) & \text{if } z \in \mathbb{E}_0 \\ f_k(u|_{\mathbb{A}(z)}) & \text{if } z \in \mathbb{E}_k, \ k \in [1,n] \end{cases}$$

We denote such an interpolation $\mathfrak{I}_{f_1,\dots,f_n}$.

Proof : The interpolation process on the subdivided edges depends only on the values of u at the vertices of the original edges due to the locality of the method. Furthermore it has to be invariant by axial symmetries and rotations. Hence, there is a unique function f_1 characterizing the interpolation on the subdivided edges. The reasoning is the same on the faces and the cubes respectively for f_2 and f_3.   □

Notice that it is an implication and not an equivalence: an interpolation verifying this scheme does not verify all the properties in $\mathcal{P}$.

3.5 I_0, I_{WC}, and I_{sol} for Local Interpolations

Let us introduce some useful sets to express recursively the local interpolations satisfying the properties $\mathcal{P}$.

Definition 12 (I_0 and definition of a local in-between interpolation).
Let $u : \mathcal{D} \mapsto \mathbb{R}$ be a gray-valued map, let z be a point of $s(\mathcal{D}) \setminus \mathbb{E}_0$, and let $\mathfrak{I}$ be a given local interpolation. We define the set $I_0(u, z)$ associated to $\mathfrak{I}$ by:

$$I_0(u, z) \stackrel{(def)}{=} \bigcap_{\{z^-, z^+\} \in \mathrm{opp}(z)} \mathrm{intvl}(u'(z^-), u'(z^+))$$

Then, an ordered local interpolation $\mathfrak{I}$ is said in-between iff $u'(z) \in I_0(u, z)$ for any image $u : \mathcal{D} \mapsto \mathbb{R}$ and $z \in s(\mathcal{D}) \setminus \mathbb{E}_0$.

Definition 13 (I_{WC} and I_{sol}). *Let $u : \mathcal{D} \mapsto \mathbb{R}$ be a gray-valued image, z be a point of $s(\mathcal{D}) \setminus \mathbb{E}_0$, and $\mathfrak{I}$ be a given local interpolation. We define the set $I_{WC}(u, z)$ associated to $\mathfrak{I}$ such as for any $z \in \mathbb{E}_1$, $I_{WC}(u, z) = \mathbb{R}$ and for any $z \in \mathbb{E}_k$ with $k \geq 2$:*

$$I_{WC}(u, z) = \{ v \in \mathbb{R} \mid u'(z) = v \Rightarrow u'|_{\mathcal{G}(z)} \text{ is well-composed} \}$$

Last, let us denote $I_{sol}(u, z) = I_0(u, z) \cap I_{WC}(u, z)$.

The following scheme is necessary to satisfy $\mathcal{P}$ (but not sufficient).

Theorem 2. *Any local interpolation $\mathfrak{I}$ satisfying $\mathcal{P}$ is such that:*
$$\forall z \in \left(\tfrac{\mathbb{Z}}{2}\right)^n, u'(z) = \begin{cases} u(z) & \text{if } z \in \mathbb{E}_0 \\ f_k(u|_{\mathbb{A}(z)}) \in I_{sol}(u, z) & \text{if } z \in \mathbb{E}_k, k \in [1, n] \end{cases}$$

Notice that such a local interpolation $\mathfrak{I}$ is ordered, in-between, well-composed, but not necessarily self-dual.

3.6 Determining f_1 for Self-dual Local Interpolations

Let us begin with the study of f_1, *i.e.*, the function setting the values at the centers of the subdivided edges. This function has to be self-dual, symmetrical, and in-between. We choose one of the most common function satisfying these constraints: the mean operator $f_1 : \mathbb{R}^2 \mapsto \mathbb{R} : (v_1, v_2) \mapsto f_1(v_1, v_2) = (v_1 + v_2)/2$.

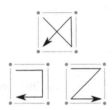

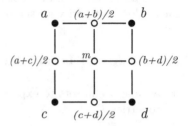

Fig. 8. The 3 possible configurations in 2D (modulo reflections and rotations)

Fig. 9. $u'|_{\mathcal{G}(z)}$ for $z \in \mathbb{E}_2$ for any self-dual local interpolation after the application of f_1 (with m any value $\in \mathbb{R}$)

3.7   Equations of f_2 for Self-dual Local Interpolations

Concerning f_2, *i.e.*, the function which sets the values of u' at the centers of the subdivided faces, let us compute $I_0(u, z)$ and $I_{WC}(u, z)$ for any given $z \in \mathbb{E}_2$ to deduce $I_{sol}(u, z)$. Their values depend on the *configurations* of $u|_{A(z)}$.

Let us assume that $u|_{A(z)} = \begin{pmatrix} a & b \\ c & d \end{pmatrix}$. Then a total of $4! = 24$ increasing orders are possible for these 4 values. Modulo reflections and axial symmetries, we obtain a total of 3 possible configurations: the α-*configurations* correspond to the relation $a \leq d < b \leq c$, the U-*configurations* to $a \leq b \leq d \leq c$, and the Z-*configurations* to $a \leq b \leq c \leq d$ (see Figure 8).

Lemma 4. *Let z be a point in $\mathbb{E}_2$. Modulo reflections and symmetries, an α-configuration implies that $u|_{A(z)}$ is not well-composed, whereas a U- or Z-configuration implies that $u|_{A(z)}$ is well-composed.*

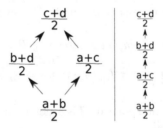

Fig. 10. The Hasse diagrams for the α- and the U-configurations (left) and for the Z-configuration (right)

Let us begin with the computation of $I_0(u, z)$ for $z \in \mathbb{E}_2$. From the values already set in u' on $\mathbb{P}(z) \subseteq \mathbb{E}_1$ by f_1 during the recursive process (see Figure 9), we can compute $I_0(u, z)$ using the Hasse diagram[1] for each configuration (see Figure 10). We obtain finally that $I_0(u, z) = \text{intvl}(\frac{a+c}{2}, \frac{b+d}{2})$ for the

---

[1] Recall that a Hasse diagram is used to represent finite partially ordered sets with the biggest elements at the top.

three configurations, with one important property: the median value of $u|_{A(z)}$ always belongs to $I_0(u, z)$.

Let us follow with the computation of $I_{WC}(u, z)$, where $u'|_{G(z)}$ (see Figure 9) satisfies four conditions:

$$\text{intvl}(a, m) \cap \text{intvl}((a + b)/2, (a + c)/2) \neq \emptyset, \tag{1}$$
$$\text{intvl}((a + b)/2, (b + d)/2) \cap \text{intvl}(m, b) \neq \emptyset, \tag{2}$$
$$\text{intvl}((a + c)/2, (c + d)/2) \cap \text{intvl}(m, c) \neq \emptyset, \tag{3}$$
$$\text{intvl}(m, d) \cap \text{intvl}((c + d)/2, (b + d)/2) \neq \emptyset. \tag{4}$$

In the case of the α-configuration, $(2) \Rightarrow m \leq \frac{b+d}{2}$ and $(4) \Rightarrow m \geq \frac{b+d}{2}$. That implies that $m = \frac{b+d}{2}$, which also satisfies (1) and (3). Consequently, $I_{WC}(u, z) = \{\text{med}\{u|_{A(z)}\}\}$, and because $I_{WC}(u, z) \subseteq I_0(u, z)$, $I_{sol}(u, z) = \{\text{med}\{u|_{A(z)}\}\}$ in the not well-composed case.

In the cases of the U- and the Z-configurations, we obtain that $I_{WC}(u, z) = [\frac{a+b}{2}, \frac{c+d}{2}] \supseteq I_0(u, z)$, so we conclude that $I_{sol}(u, z) = I_0(u, z)$.

Theorem 3. *Given an image $u : \mathcal{D} \mapsto \mathbb{R}$, any local interpolation $\mathfrak{I}_{f_1, f_2, f_3}$ satisfying $\mathcal{P}$ is such that $\forall z \in s(\mathcal{D}) \cap \mathbb{E}_2$:*

$$f_2(u|_{A(z)}) = \text{med}\{u|_{A(z)}\} \quad \text{if } u|_{A(z)} \text{ is not W.C.,}$$
$$f_2(u|_{A(z)}) \in I_0(u, z) \quad \text{otherwise.}$$

Let z be a point in $s(\mathcal{D}) \cap \mathbb{E}_2$. Amongst the applications f_2 satisfying $\mathcal{P}$, there exists (at least) the *median method* (see Figure 7), consisting in setting the value of $u'(z)$ at $\text{med}\{u|_{A(z)}\}$ (in this case f_2 is an operator and not only a function), the *mean/median method of Latecki* [6] consisting in setting the value $u'(z)$ at $\text{mean}\{u|_{A(z)}\}$ in the well-composed case and to $\text{med}\{u|_{A(z)}\}$ otherwise, and also the *min/max method*, consisting in setting the value $u'(z)$ at $\frac{1}{2}(\min\{u|_{A(z)}\} + \max\{u|_{A(z)}\})$ in the well-composed case and to $\text{med}\{u|_{A(z)}\}$ otherwise.

3.8   Equations of f_3 for Local Self-dual Interpolations

Theorem 4. *No local interpolation satisfies $\mathcal{P}$ for $n \geq 3$ with one subdivision as soon as we chose the mean operator to interpolate in 1D.*

Proof: Let z be the center of a subdivided cube. We have $u'|_{A(z)}$ as in the Figure 11 (on the left). We apply the first interpolating function f_1, i.e., we set the values of u' at the centers of the subdivided edges at the mean of the values on the vertices. Then we apply the second interpolating function f_2, which fixes the values of u' at the centers of the subdivided faces at the median of the values of u' at the four corresponding corners (because u is well-composed on none of the faces of the cube). Finally, referring to the properties that a function u' has to satisfy to be well-composed (see theorem 1), f_3 must also satisfy the constraints $c \geq 3$ and $c \leq 1$ (both are constraints of type 2) that are incompatible. So, no local interpolation of this sort can satisfy the set of constraints $\mathcal{P}$ as soon as $n \geq 3$. □

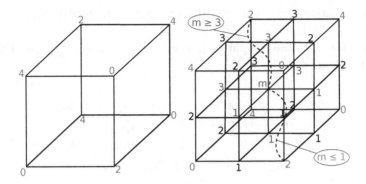

Fig. 11. A counter-example proving that a local interpolation satisfying $\mathcal{P}$ with one subdivision can not ensure well-composedness (the values of u' on $\mathbb{E}_0$ are in green, the ones on $\mathbb{E}_1$ are in blue, the ones on $\mathbb{E}_2$ are in red, and the ones on $\mathbb{E}_3$ are in purple)

4   Conclusion

In this paper, we have presented a characterization of well-composedness for 3D gray-valued images. We proved that no local interpolation satisfying $\mathcal{P}$ with one subdivision is able to make 3D well-composed images as soon as we choose the mean operator as interpolation in 1D.

Although our formalism is developped in the continuous domain (the interpolations take their values in $\mathbb{R}$), it is in fact a discrete setting. Indeed, the image u' can easily be computed in $\mathbb{Z}$ as soon as the space image of u is also $\mathbb{Z}$. We just have to multiply the values of the original image u by a factor $k \in \mathbb{Z}$ where k depends on the interpolation we use (e.g., $k = 2$ for the median method and $k = 4$ for the mean/median method in 2D). Another way to deal with images having values in $\mathbb{Z}/k$ is to use a generic image processing library [7].

Future research should tackle the two following directions. The first direction is to use an alternative to f_1 such as $(a, b) \mapsto \text{med}(a, b, c)$ (where c is the center of the space of the image u). The second direction is to use a non-local approach, e.g., a front propagation algorithm. In that case, we do not have to use any systematic operator f_1 anymore, nor to use an ordered interpolation. First results in this second direction are promising.

References

1. Caselles, V., Monasse, P.: Geometric Description of Images as Topographic Maps. Lecture Notes in Mathematics, vol. 1984. Springer (2009)
2. Géraud, T.: Self-duality and discrete topology: Links between the morphological tree of shapes and well-composed gray-level images. Journée du Groupe de Travail de Géométrie Discrète (June 2013), http://jgeodis2013.sciencesconf.org/conference/jgeodis2013/program/JGTGeoDis2013Geraud.pdf
3. Géraud, T., Carlinet, E., Crozet, S., Najman, L.: A quasi-linear algorithm to compute the tree of shapes of nD images. In: Hendriks, C.L.L., Borgefors, G., Strand, R. (eds.) ISMM 2013. LNCS, vol. 7883, pp. 98–110. Springer, Heidelberg (2013)

4. Latecki, L.J.: 3D well-composed pictures. Graphical Models and Image Processing 59(3), 164–172 (1997)
5. Latecki, L.J., Eckhardt, U., Rosenfeld, A.: Well-composed sets. Computer Vision and Image Understanding 61(1), 70–83 (1995)
6. Latecki, L.J.: Well-composed sets. In: Advances in Imaging and Electron Physics, vol. 112, pp. 95–163. Academic Press (2000)
7. Levillain, R., Géraud, T., Najman, L.: Writing reusable digital topology algorithms in a generic image processing framework. In: Köthe, U., Montanvert, A., Soille, P. (eds.) WADGMM 2010. LNCS, vol. 7346, pp. 140–153. Springer, Heidelberg (2012)
8. Marchadier, J., Arquès, D., Michelin, S.: Thinning grayscale well-composed images. Pattern Recognition Letters 25, 581–590 (2004)
9. Najman, L., Géraud, T.: Discrete set-valued continuity and interpolation. In: Hendriks, C.L.L., Borgefors, G., Strand, R. (eds.) ISMM 2013. LNCS, vol. 7883, pp. 37–48. Springer, Heidelberg (2013)
10. Ngo, P., Passat, N., Kenmochi, Y., Talbot, H.: Topology-preserving rigid transformation of 2D digital images. IEEE Transactions on Image Processing 23(2), 885–897 (2014)
11. Rosenfeld, A.: Connectivity in digital pictures. Journal of the ACM 17(1), 146–160 (1970)

Implicit Digital Surfaces
in Arbitrary Dimensions

Jean-Luc Toutant[1], Eric Andres[2], Gaelle Largeteau-Skapin[2], and Rita Zrour[2]

[1] Clermont Université, Université d'Auvergne, ISIT, UMR CNRS 6284, BP 10448,
F-63000 Clermont-Ferrand, France
`j-luc.toutant@udamail.fr`
[2] Université de Poitiers, Laboratoire XLIM, SIC, UMR CNRS 7252, BP 30179,
F-86962 Futuroscope Chasseneuil, France
`{eric.andres,gaelle.largeteau.skapin,rita.zrour}@univ-poitiers.fr`

Abstract. In this paper we introduce a notion of digital implicit surface in arbitrary dimensions. The digital implicit surface is the result of a morphology inspired digitization of an implicit surface $\{\mathbf{x} \in \mathbb{R}^n : f(\mathbf{x}) = 0\}$ which is the boundary of a given closed subset of $\mathbb{R}^n$, $\{\mathbf{x} \in \mathbb{R}^n : f(\mathbf{x}) \leq 0\}$. Under some constraints, the digital implicit surface has some interesting properties, such as k-tunnel freeness. Furthermore, for a large class of the digital implicit surfaces, there exists a very simple analytical characterization.

Keywords: Implicit Curve, Implicit Surface, Digital Object, Flake Digitization.

1 Introduction

In computer graphics, implicit surfaces $\{\mathbf{x} \in \mathbb{R}^n : f(\mathbf{x}) = 0\}$ play a fundamental role because of their powerful expressiveness for modeling and their ability to describe general closed manifolds [1, 2]. It is a very convenient way to define surfaces or more generally isosurfaces. The question regarding implicit surfaces in the discrete space is a long standing problem that has been studied mainly because it allows the visualization of often (topologically) complicated surfaces [3, 4]. Different rasterization algorithms for implicit curves and surfaces have been proposed [4–9]. Many of the rasterization methods dealing with implicit curves and surfaces are associated with some subdivision scheme in order to deal with all the singularities and topological inconsistencies that may appear at a given scale. None of the methods however, to the authors best knowledge, have defined a digital implicit surface in arbitrary dimension in a simple mathematical way.

In this paper we address the problem of defining a digital implicit equivalent to an implicit surface in arbitrary dimensions. The rasterization process itself is not addressed although a naive method, consisting in testing all the voxels in a given box, is trivial to implement with our proposed analytical characterization.

E. Barcucci et al. (Eds.): DGCI 2014, LNCS 8668, pp. 332–343, 2014.

In the paper we investigate the topological properties of the so defined digital implicit surface and show that we may achieve properties such as k-tunnel freeness. Our analytical characterization is however not completely general. We show under which conditions, the analytical characterization is not accurate. This happens mainly when the curvature is large and/or when the surface circumvolution details are small compared to the size of a voxel. Precise criteria on the r-regularity of the surface are provided. One of the forthcoming works will consist in subdividing the grid at such places in order to increase the precision and remove the topological errors in the digitization.

The digitization method is a morphology inspired digitization scheme with structuring elements called *adjacency flakes*. They have been introduced in [10] in order to analytically characterize minimal (with respect to set inclusion) and k-separating digital hyperspheres. Using adjacency flakes as structuring elements in the digitization scheme provides the offset region defining the digital object with quite simple analytical characterization, while preserving important topological properties. This allows us to analytically characterize k-tunnel free implicit digital surfaces.

In section 2, we will present the digitization models and present a somewhat simplified flake family than the one proposed in [10]. In section 3, we will discuss the conditions under which topological properties are preserved by the digitization process as we propose it. In section 4, we show that, under these conditions, the digital implicit surface can be, correctly and simply, analytically characterized. We conclude this paper, in section 5, with a short discussion and some perspectives.

Now, let us end this introduction with some recalls and notations.

1.1   Recalls and Notations

Let $\{\mathbf{e_1}, \ldots, \mathbf{e_n}\}$ be the canonical basis of the n-dimensional Euclidean vector space. We denote by x_i the i-th coordinate of a point, or a vector, $\mathbf{x}$, that is its coordinate associated to $\mathbf{e_i}$. A *digital object* is a set of integer points. A *digital inequality* is an inequality with coefficients in $\mathbb{R}$ from which we retain only the integer coordinate solutions. A *digital analytical object* is a digital object defined by a finite set of digital inequalities.

To each integer point $\mathbf{v}$, a region is associated denoted by $\mathcal{V}(\mathbf{v})$ and called a *voxel*. It corresponds to the Voronoï cell of $\mathbf{v}$ in the Voronoï partition of the Euclidean space $\mathbb{R}^n$, with $\mathbb{Z}^n$ as seeds. Geometrically, a voxel is the unit hypercube (ball of radius $1/2$ based on the ℓ^∞-norm) centered on $\mathbf{v}$.

For all $k \in \{0, \ldots, n-1\}$, two integer points $\mathbf{v}$ and $\mathbf{w}$ are said to be *k-adjacent* or *k-neighbors*, if for all $i \in \{1, \ldots, n\}$, $|v_i - w_i| \leq 1$ and $\sum_{j=1}^n |v_j - w_j| \leq n - k$. In the 2-dimensional plane, the 0- and 1-neighborhood notations correspond respectively to the classical 8- and 4-neighborhood notations. In the 3-dimensional space, the 0-, 1- and 2-neighborhood notations correspond respectively to the classical 26-, 18- and 6-neighborhood notations.

A k-*path* is a sequence of integer points such that every two consecutive points in the sequence are k-adjacent. A digital object E is k-connected if there exists a k-path in E between any two points of E. A maximum k-connected subset of E is called a k-*connected component*. Let us suppose that the complement of a digital object E, $\mathbb{Z}^n \setminus E$, admits exactly two k-connected components C_1 and C_2, or, in other words, that there exists no k-path joining integer points of C_1 and C_2. Then, E is said to be k-*separating* in $\mathbb{Z}^n$.

Let $\oplus$ be the dilation, known as Minkowski addition, such that $\mathcal{A} \oplus \mathcal{B} = \cup_{\mathbf{b} \in \mathcal{B}} \{\mathbf{a} + \mathbf{b} : \mathbf{a} \in \mathcal{A}\}$. Let $\ominus$ be the erosion, such that $\mathcal{A} \ominus \mathcal{B} = \cap_{\mathbf{b} \in \mathcal{B}} \{\mathbf{a} - \mathbf{b} : \mathbf{a} \in \mathcal{A}\}$.

The Gauss digitization, denoted by $\mathbb{G}(\mathcal{E})$, and the Supercover digitization, denoted by $\mathbb{S}(\mathcal{E})$, of a set $\mathcal{E} \subseteq \mathbb{R}^n$ are defined as follows:

$$\mathbb{G}(\mathcal{E}) = \{\mathbf{v} \in \mathbb{Z}^n : \mathbf{v} \in \mathcal{E}\},$$
$$\mathbb{S}(\mathcal{E}) = \{\mathbf{v} \in \mathbb{Z}^n : \mathcal{V}(\mathbf{v}) \cap \mathcal{E} \neq \emptyset\}.$$

The Gauss digitization is the set of integer points lying in the initial set whereas the Supercover is the set of integer points for which the associated voxel shares at least one point with the initial set.

2   Digitization Model

Let us consider a closed subset $\mathcal{E}$ in $\mathbb{R}^n$ ($n \geq 2$). we denote $\partial \mathcal{E}$ the boundary of $\mathcal{E}$. Right now it does not really matter but the aim of course is to suppose that the boundary can be implicitly described by $\{\mathbf{x} \in \mathbb{R}^n : f(\mathbf{x}) = 0\}$ and the closed set $\mathcal{E}$ by $\{\mathbf{x} \in \mathbb{R}^n : f(\mathbf{x}) \leq 0\}$.

In the sequel of the paper, such a boundary is called a *surface* $\mathcal{S} = \partial \mathcal{E}$. It induces a partition of $\mathbb{R}^n$ in three subsets, the interior of $\mathcal{E}$, $\mathcal{E}^o = \mathcal{E} \setminus \partial \mathcal{E}$, the complement (or exterior) of $\mathcal{E}$, $\mathcal{E}^c = \mathbb{R}^n \setminus \mathcal{E}$ and of course $\partial \mathcal{E}$ itself.

2.1   The Closed Centered Digitization Model

The Gauss digitization of a surface $\mathcal{S}$ has not, in general, enough integer points to ensure good topological properties such as separation of the space. The discrete points belonging to a continuous straight line, for instance, have no reason to form a connected discrete object: $\{(x_1, x_2) \in \mathbb{Z}^n : ax_1 + bx_2 + c = 0\}$. In order to obtain a digital surface, one first dilates $\mathcal{S}$ with a structuring element to define a region $\mathcal{O}(\mathcal{S})$, located around $\mathcal{S}$ and called *offset region*. The digitization of $\mathcal{S}$ is then the Gauss digitization of the offset region, i.e. the set of integer coordinate points lying in $\mathcal{O}(\mathcal{S})$.

Used in conjonction with a closed connected structuring element $\mathcal{A}$ symmetric about the origin, we call this digitization scheme the *closed centered model* and denote it by $D_{\mathcal{A}}(\mathcal{S})$:

$$D_{\mathcal{A}}(\mathcal{S}) = \mathbb{G}(\mathcal{O}_{\mathcal{A}}(\mathcal{S})) = \mathbb{G}\left(\mathcal{S} \oplus \mathcal{A}\right).$$

An equivalent definition of $\mathcal{O}_{\mathcal{A}}(\mathcal{S})$, useful in the sequel of the paper, is:

$$\mathcal{O}_{\mathcal{A}}(\mathcal{S}) = (\mathcal{E} \oplus \mathcal{A}) \setminus (\mathcal{E} \ominus \mathcal{A}).$$

Alternative models have been introduced to overcome some limitations of the closed centered model [10] (open or semi-open models, exterior or interior Gaussian models, etc.). For the sake of clarity, we here only focus on the closed centered model. Many of the properties described in this paper are also verified for those other models.

2.2   Structuring Elements

The structuring elements we will consider are called *adjacency flakes* and can be described as the union of a finite number of straight segments centered on the origin.

Definition 1 (Adjacency flakes). *Let n be the dimension of the space and $0 \le k < n$. The minimal k-adjacency flake, $F_k(\rho)$ with radius $\rho \in \mathbb{R}^+$ is defined by:*

$$F_k(\rho) = \left\{ \lambda \mathbf{u} : \lambda \in [0, \rho], \mathbf{u} \in \{-1, 0, 1\}^n, \sum_{i=1}^{n} |u_i| = (n - k) \right\}.$$

Fig. 1 shows the different adjacency flakes in 2- and 3-dimensional spaces.

An important property is that two integer points $\mathbf{v}$ and $\mathbf{w}$ are k-adjacent if $(\mathbf{v} \oplus F_k(1/2)) \cap (\mathbf{w} \oplus F_k(1/2)) \ne \emptyset$.

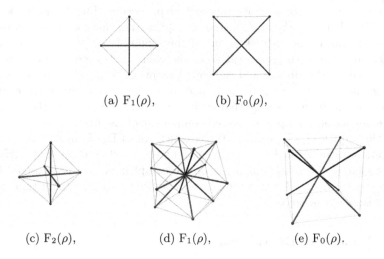

(a) $F_1(\rho)$,   (b) $F_0(\rho)$,

(c) $F_2(\rho)$,   (d) $F_1(\rho)$,   (e) $F_0(\rho)$.

Fig. 1. Adjacency flakes $F_1(\rho)$, $F_0(\rho)$ in the 2-dimensional space and $F_2(\rho)$, $F_1(\rho)$, $F_0(\rho)$ in the 3-dimensional space

Definition 2. *The F_k-digitization of a surface S is the closed centered digitization with structuring element $F_k(\frac{1}{2})$ of S. We denote it by $D_k(S)$ and its offset region by $O_k(S)$.*

In the sequel of the paper, we only consider the F_k-digitizations of surfaces.

2.3 Digital Implicit Surface

Now, lets us introduce the definition of digital implicit surface in arbitrary dimensions. We suppose that we deal with an implicit surface $S = \{ \mathbf{x} \in \mathbb{R}^n : f(\mathbf{x}) = 0 \}$ which is the boundary of a closed set $\mathcal{E}$ such that, for all $\mathbf{x} \in \mathcal{E}^o$, we have $f(\mathbf{x}) > 0$ and for all $\mathbf{x} \in \mathcal{E}^c$, we have $f(\mathbf{x}) < 0$.

Definition 3. *A F_k-digital implicit surface is the F_k-digitization of an implicit surface S.*

It is easy to see that that the F_0-digital implicit hyperplane corresponds to the supercover of a hyperplane [11] and that the F_0-digital implicit hyperspheres are particular cases of the digital hypersperes described in [10]. See Figure 4 for some examples of digital implicit surfaces.

In the next section we are going to examine the topological properties of such surface digitization and in Section 4, we are going to propose an analytical characterization for a large class of digital implicit surfaces.

3 Preserving Topology

The purpose of this section is to give conditions on S to ensure that the digitization preserves some of its topological properties. Ideally, we look for an equivalence between the surface and its digitization, for something close to an homeomorphism. Such a task is out of the scope of the present paper and we restrict our goal to the preservation of the connected components between the complement of the surface and the complement of its digitization. By preservation, we mean that there is a one-to-one correspondence between the connected components of both sets such that each connected component of the second is a proper subset of a unique connected component of the first.

In the sequel, $D_k(S)^c$ denotes the complement of $D_k(S)$ in $\mathbb{Z}^n$.

First, we study the k-tunnel freeness of the F_k-digital implicit surfaces to ensure that connected components of the complement are not merged by the digitization process. Then, we focus on conditions to guarantee that none of them disappear or split.

3.1 Tunnel-Free Digitization

The notion of k-separating set is too restrictive when dealing with surface digitization. In our case, the underlying object $\mathcal{E}$ can be composed of more than one connected component and thus the digital analog of its boundary may separate the digital space in more than only two k-connected components. The notion of *tunnel-free* digitization allows to overcome this limitation [12].

Definition 4 (Tunnel-free digitization [12]). *A digitization* $D(\mathcal{S})$ *of a surface* $\mathcal{S}$ *is said to be* k-*tunnel-free if for all* $\mathbf{v}$, $\mathbf{w} \in D(\mathcal{S})^c$ *such that* $\mathbf{v}$ *and* $\mathbf{w}$ *are* k-*adjacent, the straight segment* $[\mathbf{vw}]$ *does not cross* $\mathcal{S}$. *If such a couple of voxel exists, it is called a* k-*tunnel of* $D(\mathcal{S})$.

To prove that our digitizations satisfy this property, we first need to introduce the notion of regular set.

Definition 5 (Regular set [13]). *Let* $\mathcal{E} \subseteq \mathbb{R}^n$ *be a closed set such that for all* $\mathbf{x} \in \partial \mathcal{E}$ *it is possible to find two osculating open balls [13] of radius* r, *one lying entirely in* $\mathcal{E}^o$ *and the other lying entirely in* $\mathcal{E}^c$. *Then* $\mathcal{E}$ *is a* r-regular *set*.

Proposition 1. *Let* n *be the dimension of the space. The* F_k-*digital implicit surface,* $D_k(\mathcal{S})$, *is a* k-*tunnel-free digitization of* $\mathcal{S} = \partial \mathcal{E}$ *if* $\mathcal{E}$ *is a* r-regular *set with* $r > \sqrt{n - k}/4$.

Proof. Let us consider two integer points $\mathbf{v}$, $\mathbf{w}$ k-adjacent and such that the straight segment $[\mathbf{vw}]$ intersects $\mathcal{S}$ in a point $\mathbf{s}$. if $\mathbf{s} = \mathbf{v}$, then directly $\mathbf{v} \in D_k(\mathcal{S})$. The same occurs for $\mathbf{w}$ if $\mathbf{s} = \mathbf{w}$. In other cases, any open ball of radius $r > \sqrt{n - k}/4$ through $\mathbf{s}$ (i.e. the center of the ball is at a distance r of $\mathbf{s}$) contains at least one point of the union of k-adjacency flakes centered at $\mathbf{v}$ and $\mathbf{w}$. Moreover, by definition of $\mathcal{E}$ as a r-regular set with $r > \sqrt{n - k}/4$ (an osculating open ball of radius r entirely in $\mathcal{E}^o$ and another one entirely in $\mathcal{E}^c$ for each boundary point), it exists a path in $(\mathbf{v} \oplus F_k(1/2)) \cup (\mathbf{w} \oplus F_k(1/2))$ with one end-point in $\mathcal{E}^c$ and the other one in $\mathcal{E}^o$. This path necessarily intersects $\mathcal{S} = \partial \mathcal{E}$ in at least one point $\mathbf{s}'$. By symmetry of the adjacency flake, either $\mathbf{v}$ or $\mathbf{w}$ belongs to $\mathbf{s}' \oplus F_k(1/2)$ and thus belongs to $D_k(\mathcal{S})$.

It does not exist a couple of k-adjacent integer points $(\mathbf{v}, \mathbf{w})$ outside $D_k(\mathcal{S})$ such that the straight segment $[\mathbf{vw}]$ intersects $\mathcal{S}$.   □

This result means that, under non very restrictive conditions, whatever the supporting surface $\mathcal{S}$, a k-connected component of the digital complement of the F_k-digital implicit surface $D_k(\mathcal{S})$ only contains points belonging to a unique connected component of $\mathcal{S}^c$: two connected components of the complement of the initial surface cannot be merged by the digitization. Nevertheless, some connected components of $\mathcal{S}^c$ may have no representative in $D_k(\mathcal{S})^c$: they can be deleted by the digitization. Or, on the contrary, some may have representatives in several k-connected components of $D_k(\mathcal{S})^c$: they can be split by the digitization.

The following part discusses conditions to obtain a one to one correspondence between the connected components of $\mathcal{S}^c$ and the k-connected components of $D_k(\mathcal{S})^c$, i.e. no collapses and no splits occur.

3.2   Preserving Connected Components by Digitization

We work in two steps. We first introduce a condition to ensure that the connected components of $\mathcal{S}^c$ are preserved by the dilation. Then, we study the condition ensuring that the Gauss digitization also preserves them.

An immediate result concerns the first step:

Proposition 2. *The connected components are preserved between the complement of $\mathcal{S} = \partial\mathcal{E}$ and the complement of $\mathcal{O}_k(\mathcal{S})$, if $\mathcal{E}$ is a r-regular set with $r > \sqrt{n-k}/2$.*

Proof. By definition of $\mathcal{E}$ as regular set, any connected component $\mathcal{C}$ of $\mathcal{S}^c$ contains at least a closed Euclidean ball of radius r and center $\mathbf{c}$. Since, $F_k(1/2) \subset \mathcal{B}(\sqrt{n-k}/2)$, $\mathbf{c}$ is not in $\mathcal{O}_k(\mathcal{S})$, and $\mathcal{C} \setminus \mathcal{O}_k(\mathcal{S})$ is not empty. $\mathcal{C}$ is itself a connected r-regular set with $r > \sqrt{n-k}/2$. The set resulting of its erosion by $F_k(1/2)$ is connected. □

The next step is to ensure that the Gauss digitization of each connected components of $\mathcal{O}_k(\mathcal{S})^c$ is not empty.

First, Lemma 1 gives a sufficient condition on a connected set to ensure that it contains at least one digital point, i.e. it does not collapse. Then, Lemma 2 and Proposition 3 state that the Gauss digitization of such a set is always a $(n-1)$-connected set, i.e. it is not split.

Lemma 1. *Let $r > r' > \sqrt{n}/2$. Let $\mathcal{A}$ be a connected r-regular set. Let $\mathcal{A}'$ be the open interior of the erosion of $\mathcal{A}$ by a closed Euclidean ball of radius r'. Then, one has $\emptyset \subset \mathbb{S}(\mathcal{A}') \subset \mathcal{A}$.*

Proof. $\mathcal{A}$ is a r-regular set, so it contains at least one ball of radius r. The center of this ball lies necessarily in $\mathcal{A}'$ since $r' < r$. It ensures that $\mathcal{A}'$ is not reduced to the empty set. Since the supercover of a non empty set is not empty, $\mathbb{S}(\mathcal{A}')$ contains at least one integer point, or, in other words, $\emptyset \subset \mathbb{S}(\mathcal{A}')$.

Let us now suppose that $\mathbf{x} \in \mathcal{A}'$ and $\mathbf{y} \in \mathbb{R}^n \setminus \mathcal{A}$. One has $d(\mathbf{x},\mathbf{y}) > \sqrt{n}/2$. $\mathbf{x} \in \mathcal{V}(\mathbf{y})$ would imply that $d(\mathbf{x},\mathbf{y}) \leq \sqrt{n}/2$. Thus $\mathbf{x}$ belongs necessarily only to voxels with center in $\mathcal{A}$ and one has $\mathbb{S}(\mathcal{A}') \subset \mathcal{A}$. □

Lemma 2. *Let $\mathcal{A}$ and $\mathcal{A}'$ be defined as in Lemma 1. Then, $\mathbb{S}(\mathcal{A}')$ is $(n-1)$-connected.*

Proof. $\mathcal{A}$ is a connected r-regular set. Thus $\mathcal{A}'$ is an open, connected set [13]. The supercover of a connected set is a $(n-1)$-connected set [14]. □

Proposition 3. *Let $\mathcal{A}$ and $\mathcal{A}'$ be defined as in Lemma 1. Then, the Gauss digitization of $\mathcal{A}$, $\mathbb{G}(\mathcal{A})$, is a $(n-1)$-connected set.*

Proof. Let $\mathbf{v}$ be any integer point in $\mathcal{A}$. Let $\mathcal{B}(r)$ be the open ball of radius r based on the Euclidean norm. Then there exists a point $\mathbf{c}$ such that $\mathcal{B}(r) \oplus \{\mathbf{c}\} \subseteq \mathcal{A}$ and $\mathbf{v} \in \mathcal{B}(r) \oplus \{\mathbf{c}\}$. $\mathbf{c}$ lies in $\mathcal{A}'$ and belongs to $\mathcal{V}(\mathbf{w})$ (possibly $\mathbf{v} = \mathbf{w}$). Consider the supercover of the segment $[\mathbf{vc}]$. Every integer point in it is in the ball $\mathcal{B}(r) \oplus \{\mathbf{c}\}$. So there exists a $(n-1)$-connected path linking $\mathbf{v}$ and $\mathbf{w}$, thus $(n-1)$-connecting $\mathbf{v}$ to $\mathbb{S}(\mathcal{A}')$. □

By combining results of Proposition 1, Proposition 2 and Proposition 3, we state the main theorem of this section:

Theorem 1. *if $\mathcal{E}$ is a r-regular set with $r > (\sqrt{n-k} + \sqrt{n})/2$, then the connected components are preserved between $\mathcal{S}^c$ and $D_k(\mathcal{S})^c$, according to the k-adjacency relationship.*

4 Analytical Characterization of a Digital Implicit Surface

Let us denote the set of end-points of the segments composing the adjacency flake $F_k(\rho)$ by $V_k(\rho) = \{\mathbf{x} : \mathbf{x} \in \{-\rho, 0, \rho\}^n, \sum_{i=1}^{n} |x_i| = \rho(n-k)\}$. The following technical lemma shows that, under the condition of Theorem 1, we only need to consider the end-points of the flake line segments to characterize the offset zone.

Lemma 3. *Let $\mathcal{S}$ be a surface satisfying the condition of Theorem 1. Let also $\mathbf{x} \in \mathcal{O}_k(\mathcal{S})$. Then, there exists $\mathbf{y}, \mathbf{y}' \in (\mathbf{x} \oplus V_k(1/2))$ such that $\mathbf{y} \in \mathcal{E}$ and $\mathbf{y}' \in cl(\mathcal{E}^c)$, where $cl(\mathcal{E}^c)$ is the closure of $\mathcal{E}^c$.*

Proof. By definition of $\mathcal{O}_k(\mathcal{S})$, $(\mathbf{x} \oplus F_k(1/2)) \cap \mathcal{S} \neq \emptyset$. Due to the condition of Theorem 1, the number of intersections between a segment of $\mathbf{x} \oplus F_k(1/2)$ and $\mathcal{S}$ is lower or equal to 2. Let us consider a segment of $\mathbf{x} \oplus F_k(1/2)$ intersecting $\mathcal{S}$. Either one of its end-points is in $\mathcal{E}$ and the other in $cl(\mathcal{E}^c)$, or both are in $\mathcal{E}^o$ or in $\mathcal{E}^c$. In the first case, the result is immediate. In the second case, there is necessarily another segment of $\mathbf{x} \oplus F_k(1/2)$ which satisfies the first case. $\square$

Figure 2 illustrates the lemma. It shows, on the left, a case where considering only the end-points of the flake is equivalent to considering the whole flake and, on the right, a case where it is not equivalent. This leads immediately to the following theorem which allows a very simple analytical characterization for a large class of implicit digital surfaces:

Theorem 2. *Let $\mathcal{S} = \{\mathbf{x} \in \mathbb{R}^n : f(\mathbf{x}) = 0\}$ be an implicit surface (boundary of a closed set $\mathcal{E}$ such that, for all $\mathbf{x} \in \mathcal{E}^o$, we have $f(\mathbf{x}) > 0$ and for all $\mathbf{x} \in \mathcal{E}^c$, we have $f(\mathbf{x}) < 0$) satisfying the condition of Theorem 1 ($\mathcal{E}$ is a r-regular set with $r > (\sqrt{n-k}+\sqrt{n})/2$). Then, the F_k-digital implicit surface $D_k(\mathcal{S})$ is analytically characterized as follows:*

$$D_k(\mathcal{S}) = \left\{\mathbf{v} \in \mathbb{Z}^n : \begin{array}{c} \min\{f(\mathbf{x}) : \mathbf{x} \in (\mathbf{v} \oplus V_k(1/2))\} \leq 0 \\ \text{and} \ \max\{f(\mathbf{x}) : \mathbf{x} \in (\mathbf{v} \oplus V_k(1/2))\} \geq 0 \end{array}\right\}.$$

Proof. According to Lemma 3, for any $\mathbf{v} \in D_k(\mathcal{S})$, it exists $\mathbf{x}, \mathbf{x}' \in (\mathbf{v} \oplus V_k(1/2))$ such that $\mathbf{x} \in \mathcal{E}$ and $\mathbf{x}' \in cl(\mathcal{E}^c)$. Since for all $\mathbf{x} \in \mathcal{E}^o$, we have $f(\mathbf{x}) > 0$ and for all $\mathbf{x} \in \mathcal{E}^c$, we have $f(\mathbf{x}) < 0$, the analytic formulation is immediate. $\square$

A F_k-digital implicit surface is thus entirely defined by the knowledge of the function f and of the value k.

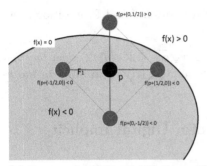

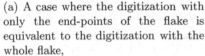

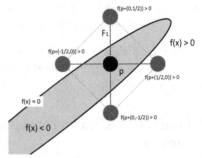

(a) A case where the digitization with only the end-points of the flake is equivalent to the digitization with the whole flake,

(b) A case where the two digitizations lead to different results.

Fig. 2. Illustration of the limits of the analytical characterization

5   Discussion, Conclusion and Perspectives

In the present paper, we have introduced a simple analytical definition of digital implicit surfaces. They are built as the digitizations of an implicit surface $\mathcal{S} = \{\mathbf{x} \in \mathbb{R}^n : f(\mathbf{x}) = 0\}$ satisfying some specific conditions. Namely, $\mathcal{S}$ should be the boundary of a closed set $\mathcal{E} = \{\mathbf{x} \in \mathbb{R}^n : f(\mathbf{x}) \leq 0\}$ which is r-regular with $r > (\sqrt{n-k} + \sqrt{n})/2$.

In addition to the analytical characterization, these conditions ensures the k-tunnel freeness of the digital implicit surfaces. They also preserve the connected components between the complement of the implicit surface and the complement

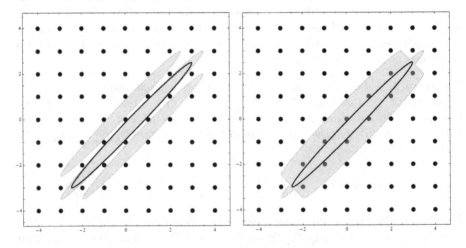

Fig. 3. In yellow, on the left, the offset region obtained by only considering $V_k(1/2)$ as structuring element, on the right, the real offset region of the two dimensional implicit conic $201x^2 - 398xy + 201y^2 - 200x + 200y + 20 = 0$

of its digitization with regard to the k-adjacency relationship. These conditions are of course sufficient but not necessary.

Figure 3 presents an extreme case where the analytical characterization fails to correctly represent the F_0-digital implicit surface. On the left of the figure, we see that the offset region that we obtain with the analytical characterization is composed by three distinct connected components and contains, in this case, no integer points. According to the analytical characterization, the digitization of the ellipse is an empty set. The Condition of Theorem 1, and more specifically the condition of Proposition 1 (to ensure a tunnel-free digitization), is not met. On the right of the figure, we have the correct result corresponding to the F_0-digital implicit ellipse. This is the limit of the analytical characterization proposed in this paper : if the condition of Theorem 1 (or a quite less restrictive, but very similar one) is not met, the proposed analytical characterization is just an approximation of the digital implicit surface, since not only the end-points of the flake line segments contribute to the boundary of the offset zone. It is necessary to compute the intersection between said line segments and the implicit surface. This can be accomplished in several ways. It can be done with help of the derivatives or by direct intersection computation ; however exact computations can only be done for a limited class of surfaces and the result will not be as simple as the one proposed in this paper. Another obvious limitation of the method is that it is limited to $(n-1)$-dimensional surfaces in dimension n. It does not work for 3D curves for instance; but intersecting a surface with a flake line segment or intersecting a curve with the faces of an adjacency norm ball is a problem of somewhat similar nature and gives a good, although not simple, way of proposing three-dimensional curves of specific connectivity. We plan to propose such descriptions in a forthcoming paper.

Another problem occurs even in the corrected version of Figure 3: the interior of the ellipse disappeared during the digitization process. The condition of Theorem 1, and more specifically the condition of Proposition 2, is not met. This is inherently a problem of grid size. One way around the problem is to locally refine the grid size. This is not new and is actually the way the digitization of implicit curves and surfaces have been done most of the time [1, 2, 4–9]. Our aim is to explore such subdivision methods for surfaces in dimension n.

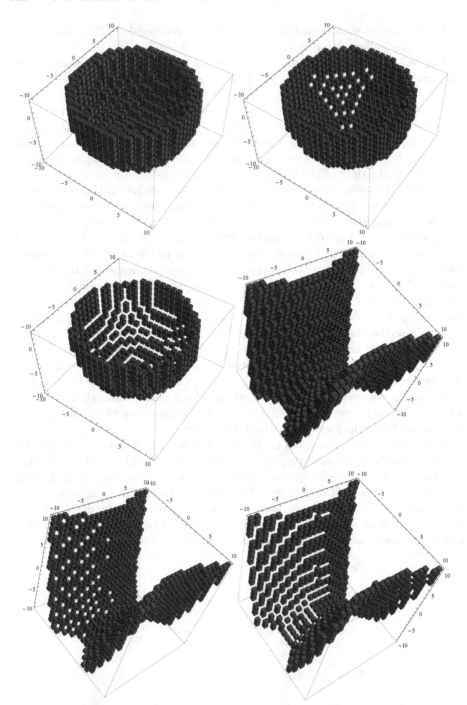

Fig. 4. Examples of digital implicit surface: F_0-, F_1- and F_2-digital implicit spheres of radius 9 (cut in order to see the tunnels) and digital implicit quadric $9x^2 - 4y^2 - 36z - 180 = 0$

References

1. Bloomenthal, J., Wyvill, B. (eds.): An Introduction to Implicit Surfaces. Morgan Kaufmann Publishers Inc., San Francisco (1997)
2. Velho, L., Gomes, J., de Figueiredo, L.H. (eds.): Implicit Objects in Computer Graphics. Springer (2002)
3. Stolte, N.: Arbitrary 3D resolution discrete ray tracing of implicit surfaces. In: Andrès, É., Damiand, G., Lienhardt, P. (eds.) DGCI 2005. LNCS, vol. 3429, pp. 414–426. Springer, Heidelberg (2005)
4. Emeliyanenko, P., Berberich, E., Sagraloff, M.: Visualizing arcs of implicit algebraic curves, exactly and fast. In: Bebis, G., et al. (eds.) ISVC 2009, Part I. LNCS, vol. 5875, pp. 608–619. Springer, Heidelberg (2009)
5. Stolte, N., Caubet, R.: Comparison between Different Rasterization Methods for Implicit Surfaces. In: Visualization and Modeling, ch. 10. Academic Press (1997)
6. Stolte, N., Kaufman, A.: Novel techniques for robust voxelization and visualization of implicit surfaces. Graphical Models 63(6), 387–412 (2001)
7. Sigg, C.: Representation and Rendering of Implicit Surface. Phd thesis, diss. eth no. 16664, ETH Zurich, Switzerland (2006)
8. Taubin, G.: An accurate algorithm for rasterizing algebraic curves. In: Proceedings of the 2nd ACM Solid Modeling and Applications, pp. 221–230 (1993)
9. Taubin, G.: Rasterizing algebraic curves and surfaces. IEEE Computer Graphics and Applications 14(2), 14–23 (1994)
10. Toutant, J.-L., Andres, E., Roussillon, T.: Digital circles, spheres and hyperspheres: From morphological models to analytical characterizations and topological properties. Discrete Applied Mathematics 161(16-17), 2662–2677 (2013)
11. Andres, E.: The supercover of an m-flat is a discrete analytical object. Theoretical Computer Science 406(1-2), 8–14 (2008)
12. Cohen-Or, D., Kaufman, A.E.: Fundamentals of surface voxelization. CVGIP: Graphical Model and Image Processing 57(6), 453–461 (1995)
13. Stelldinger, P., Köthe, U.: Towards a general sampling theory for shape preservation. Image and Vision Computing 23(2), 237–248 (2005)
14. Brimkov, V.E., Andres, E., Barneva, R.P.: Object discretizations in higher dimensions. Pattern Recognition Letters 23(6), 623–636 (2002)

Algorithms for Fast Digital Straight Segments Union

Isabelle Sivignon[*]

Gipsa-lab, CNRS, UMR 5216, F-38420, France
isabelle.sivignon@gipsa-lab.grenoble-inp.fr

Abstract. Given two Digital Straight Segments (DSS for short) of known minimal characteristics, we investigate the union of these DSSs: is it still a DSS ? If yes, what are its minimal characteristics ? We show that the problem is actually easy and can be solved in, at worst, logarithmic time using a state-of-the-art algorithm. We moreover propose a new algorithm of logarithmic worst-case complexity based on arithmetical properties. But when the two DSSs are connected, the time complexity of this algorithm is lowered to $\mathcal{O}(1)$ and experiments show that it outperforms the state-of-the art one in any case.

Keywords: Digital geometry, Union, Digital straight segment.

1 Introduction

Digital Straight Lines (DSL) and Digital Straight Segments (DSS) have been used for many years in many pattern recognition applications involving digital curves. Whether it be for polygonal approximation or to design efficient and precise geometric estimators, a basic task is the so-called DSS recognition problem: given a set of pixels, decide whether this set is a DSS and compute its characteristics. Many linear-in-time algorithms have been proposed to solve this problem through the years. Furthermore, Constructive Solid Geometry-like operations have been considered for these objects: the intersection of two DSL has been studied in [15,16,8,5], algorithms for the fast computation of subsegments were described in [9,17]. Surprisingly enough, the union of DSSs has not yet been studied. The problem was raised in [2] in the context of parallel recognition of DSSs along digital contours. The recognition step was followed by a merging step where the problem of DSSs union appeared. In this work, we show how to solve this problem, both using state-of-the-art algorithm, and proposing a new and faster algorithm.

---

[*] This work was partially founded by the French *Agence Nationale de la Recherche* (Grant agreement ANR-11-BS02-009).

E. Barcucci et al. (Eds.): DGCI 2014, LNCS 8668, pp. 344–357, 2014.

2   General Considerations

2.1   Preliminary Definitions

A *Digital Straight Line* (DSL for short) of integer characteristics (a, b, μ) is the infinite set of digital points $(x, y) \in \mathbb{Z}^2$ such that $0 \leq ax - by + \mu < max(|a|, |b|)$ [4]. These DSL are 8-connected and often called *naive*. The fraction $\frac{a}{b}$ is the slope of the DSL, and $\frac{\mu}{b}$ is the shift at the origin. In the following, without loss of generality, we assume that $0 \leq a \leq b$. The remainder of a DSL of characteristics (a, b, μ) for a given digital point (x, y) is the value $ax - by + \mu$. The *upper (resp. lower) leaning line* of a DSL is the straight line $ax - by + \mu = 0$ (resp. $ax - by + \mu = b - 1$). Upper (resp. lower) leaning points are the digital points of the DSL lying on the upper (resp. lower) leaning lines. A *Digital Straight Segment* (DSS) is a finite 8-connected part of a DSL.

If we consider the digitisation process related to this DSL definition, the points of the DSL **L** of parameters (a, b, μ) are simply the grid points (x, y) lying below or on the straight line $l : ax - by + \mu = 0$ (Object Boundary Quantization), and such that the points $(x, y + 1)$ lie above l. Otherwise said, line l separates the points X of the DSL from the points $X + (0, 1)$ [14], and is called *separating line*. More generally, for an arbitrary set of digital points X, the separating lines are the lines that separate the points X from the points $X + (0, 1)$. In other words, the separating lines separate the upper convex hull of X from the lower convex hull of $X + (0, 1)$. Computing the set of separating lines of two polygons is a very classical problem of computational geometry. It is well known that specific lines called critical support lines can be defined: there are the separating lines passing through a point of each polygon boundary. Critical support points are the points of the polygons belonging to critical support lines [12].

All the separating lines of a DSL have the same slope, but this is not true for arbitrary sets of digital points. The *minimal characteristics* of a set of digital points X are the characteristics of the separating line of minimal b and minimal μ. The set of separating lines of a DSS is well known, and the critical support points are exactly defined by the DSS leaning points: they define the minimal characteristics of the DSS.

The set of separating lines of a set of points X can also conveniently be defined in a dual space, also called parameter space. In this space a straight line $l : \alpha x - y + \beta = 0$ is represented by the 2D point (α, β). Given a set of digital points X, a line $l : \alpha x - y + \beta = 0$ is a separating line if and only if for all $(x, y) \in X, 0 \leq \alpha x - y + \beta < 1$. This definition is strictly equivalent to the one given previously. The preimage of X is the representation of its separating lines in the dual space and is defined as $\mathcal{P}(X) = \{(\alpha, \beta), 0 \leq \alpha \leq 1, 0 \leq \beta \leq 1 | \forall (x, y) \in X, 0 \leq \alpha x - y + \beta < 1\}$. The set of separating lines of a set of pixels is an open set in the digital space, but it is a convex polygon in the dual space. In this work, this dual space will not be used explicitly in the algorithms, but we will see that this representation is convenient in some proofs. Moreover, the arrangement of all the constraints for any pixel (x, y) with $y \leq x \leq n$ is called Farey Fan [10] of order n: each cell of this arrangement is the preimage of a DSS of length n. Figure 1 is an illustration of the separating lines of a DSS, both in the digital space and in the dual space: they separate the points X of the DSS in black, from the points $X + (0, 1)$ in white. Note that the edges of the preimage of a DSS are exactly supported by the dual representation of its leaning points, or equivalently its critical support points.

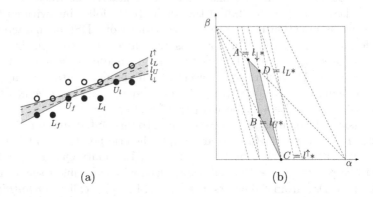

(a)   (b)

Fig. 1. (a) DSS of minimal characteristics $(1, 3, 1)$ with its leaning points U_f, U_l, L_f, L_l. (b) Each vertex of the preimage maps to a straight line in the digital space. The vertex $B(\frac{1}{3}, \frac{1}{3})$ maps to the upper leaning line, the characteristics of which are the minimal characteristics of the DSS.

2.2   Setting the Problem and Useful Properties

Consider now the following problem :

Problem 1. Given two DSSs $S_1 = [P_1Q_1]$ and $S_2 = [P_2Q_2]$ of known minimal characteristics, decide if there exists a DSL containing both S_1 and S_2. If yes, compute the minimal characteristics of $S_1 \cup S_2$.

If S_1 and S_2 do not belong to the same octant, then it is easy to conclude that there is no DSL containing both S_1 and S_2. Thus, in the following S_1 and S_2 belong to the first octant, i.e. we have $S_1(a_1, b_1, \mu_1)$ and $S_2(a_2, b_2, \mu_2)$ with $0 \leq a_1 \leq b_1$ and $0 \leq a_2 \leq b_2$. We denote by $r_i(x, y) = a_i x - b_i y + \mu_i, i \in \{1, 2\}$ the remainder function of each DSS.

In what follows, we moreover suppose that the leaning points of S_1 and S_2 are known as input data. This is not a strong requirement since the most efficient recognition algorithms actually compute this data on the fly.

By convention, we also suppose that the abscissa of Q_2 is greater than the abscissa of Q_1. Note that we make no assumption on the connectivity of S_1 and S_2: the abscissa of P_2 can be lower than, equal to or greater than the abscissa of Q_1. If the abscissa of P_2 is lower than the abscissa of P_1, then the problem is trivial since S_1 is either a subsegment of S_2 or the union is impossible.

Consequently, we also assume that the abscissa of P_2 is greater than the abscissa of P_1.

The first part of Problem 1 consists in deciding if there exists a separating line for the set $S_1 \cup S_2$: we will say that the union is *possible* in this case. If so, then among all the separating lines, the final goal is to find the one with minimal characteristics.

Property 1. The preimage of $S_1 \cup S_2$ is equal to the intersection between the preimages of S_1 and S_2.

Proof. The proof is straightforward since the lines that are separating for $S_1 \cup S_2$ are the ones that are separating for S_1 and S_2.

Corollary 1. *The critical support points of the set of separating lines of $S_1 \cup S_2$ are either upper leaning points or lower leaning points translated by $(0, 1)$ of S_1 and S_2. Thus, to compute the set of separating lines of $S_1 \cup S_2$, it is enough to update the set of separating lines of S_1 with the leaning points of S_2 (or conversely).*

Proof. The critical support points are, in the dual space, lines supporting the edges of the preimage. From Property 1, the lines supportting the edges of $\mathcal{P}(S_1 \cup S_2)$ are lines supporting the edges of $\mathcal{P}(S_1)$ or $\mathcal{P}(S_2)$. However, since S_1 and S_2 are DSSs, the edges of their preimages are supported by the dual representation of either upper leaning points or lower leaning points translated by $(0, 1)$.

3   Fast Union of DSSs: An Arithmetical Algorithm

3.1   Fast Computation of the Set of Separating Lines

A first straightforward solution to compute the set of separating lines of $S_1 \cup S_2$ is to use the state-of-the-art algorithm of O'Rourke [11], re-interpreted in the digital space by Roussillon [14]. Whether it be in the dual space or in the digital space, these algorithms update the critical support points iteratively for each point added. Since at most four points have to be considered in our case, the algorithm is already quite efficient compared to the classical arithmetical recognition algorithm for instance. However, we propose an algorithm that is both faster and simpler to implement, in the spirit of the arithmetical recognition algorithm.

The idea is the following: if we know that the slopes of the separating lines of $S_1 \cup S_2$ are greater/lower than the slopes (given by the minimal characteristics) of S_1 and S_2 respectively, then we can conclude that some leaning points of S_1 or S_2 cannot be critical support points for $S_1 \cup S_2$.

Property 2. Let S_1 be a DSS of minimal characteristics (a_1, b_1, μ_1). Let L_{1f}, L_{1l}, U_{1f}, U_{1l} be its first and last, lower and upper leaning points. If all the separating lines of $S_1 \cup S_2$ have a slope greater (res. lower) than $\frac{a_1}{b_1}$, then U_{1l} (resp. U_{1f}) and L_{1f} (resp. L_{1l}) are not critical support points for $S_1 \cup S_2$.

Proof. If all the separating lines of $S_1 \cup S_2$ have a slope lower than $\frac{a_1}{b_1}$, then, in the dual space and from Property 1, $\mathcal{P}(S_1 \cup S_2)$ is a subpart of the triangle defined by the vertices ABD (see Figure 1(b)). In particular, the edges $[BC]$ and $[DC]$ of $\mathcal{P}(S_1)$ supported by U_{1f}^* and L_{1l}^* respectively cannot be edges of $\mathcal{P}(S_1 \cup S_2)$. Therefore, the leaning points U_{1f} and L_{1l} are not critical support points for $S_1 \cup S_2$. The proof is the same if we suppose that the separating lines all have a slope greater than $\frac{a_1}{b_1}$.

Note that if S_1 has three leaning points only, let's say for instance only one lower leaning point L_1, then setting L_{1l} and L_{1f} to L_1 (L_1 is "duplicated"), the property is also valid. A similar result holds when the leaning points of S_2 are considered. However, guessing the slope of the union can be tricky, and taking into account only the DSS slopes is not enough. For example, it is easy to exhibit cases where the slope of S_2 is greater than the slope of S_1, and the slope of $S_1 \cup S_2$ is nevertheless lower than both the slope of S_1 and the slope of S_2 (see Figure 2(a)). We establish hereafter some properties linking the remainder of the leaning points of S_2 and the slope of the separating lines for $S_1 \cup S_2$ if they exist.

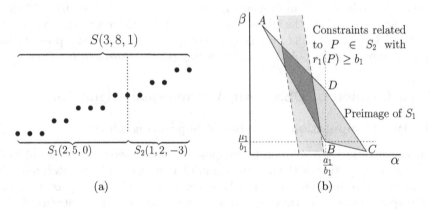

Fig. 2. (a) The slope of S_1 is equal to $\frac{2}{5}$ and lower than the slope of S_2, which is equal to $\frac{1}{2}$. However, the slope of $S_1 \cup S_2$ is equal to $\frac{3}{8}$ and smaller than both. (b) Illustration of Property 4.

Let's start with a very simple consideration.

Property 3. If for all the leaning points P of S_2 (resp. S_1), we have $0 \le r_1(P) < b_1$ (resp. $0 \le r_2(P) < b_2$), then there exists a DSL containing $S_1 \cup S_2$ and its minimal characteristics are the one of S_1 (resp. S_2).

Proof. In the dual space, the points B and D of the preimage of S_1 satisfy all the constraints related to the points P defined as above, which ends the proof.

With the following property, we investigate the other cases.

Property 4. Let P be a leaning point of S_2.

- if $\mathbf{r_1(P) \geq b_1}$, the slope of all the separating lines for $S_1 \cup S_2$, if any, is **lower** than the slope of S_1
- if $\mathbf{r_1(P) < 0}$, the slope of all the separating lines for $S_1 \cup S_2$, if any, is **greater** than the slope of S_1

Let P be a leaning point of S_1.

- if $\mathbf{r_2(P) \geq b_2}$, the slope of all the separating lines for $S_1 \cup S_2$, if any, is **greater** than the slope of S_2
- if $\mathbf{r_2(P) < 0}$, the slope of all the separating lines for $S_1 \cup S_2$, if any, is **lower** than the slope of S_2

Proof. We prove the first item, for a leaning point of S_2 with a remainder greater than or equal to b_1. Proving the other cases is similar. Consider a point $P \in S_2$ such that $r_1(P) \geq b_1$. Let us consider the stripe defined by the constraints related to this point in the dual space. It is very simple to see that the point $(\frac{a_1}{b_1}, \frac{\mu_1}{b_1})$ is above this stripe (see Figure 2(b)). Since P does not belong to S_1, and with the assumptions made in Section 2.2 on the relative position of S_1 and S_2, its abscissa is greater than the abscissas of all the leaning points of S_1. This means that the intersection, if not empty, between the stripe and $\mathcal{P}(S_1)$ lies in the subspace $\alpha < \frac{a_1}{b_1}$.

Table 1 summarises the computation of the four possible critical support points combining Properties 2 and 4. Figure 6 in Appendix illustrates the first line of this table.

Table 1. Possible critical support points according to remainder values

remainder	value	< 0	≥ b
$P \in S_2$	$r_1(P)$	$(U_f, L_f) = (U_{1f}, L_{1l})$	$(U_f, L_f) = (U_{1l}, L_{1f})$
$P \in S_1$	$r_2(P)$	$(U_l, L_l) = (U_{2l}, L_{2f})$	$(U_l, L_l) = (U_{2f}, L_{2l})$

At this point, we have identified four points denoted by U_f, U_l, L_f and L_l, that may be critical support points for $S_1 \cup S_2$. However, they may not be all critical support points. Since the preimage of $S_1 \cup S_2$ is a convex polygon, it has at least three edges and thus, at least three out of the four possible points are indeed critical support points. Property 5 gives a way to decide whether the four points are critical support points or not.

Property 5. U_f and U_l (resp. L_f and L_l) are both critical support points if and only if L_f and L_l (resp. U_f and U_l) belong to the DSL of directional vector $U_l - U_f$ (resp. $L_l - L_f$) and upper (resp. lower) leaning points U_f and U_l (resp. L_f and L_l).

Proof. U_f and U_l are both critical support points is equivalent to say that the straight line $(U_f U_l)$ is separating for $S_1 \cup S_2$. This is also equivalent to the fact that L_f and L_l belong to the DSL as defined in the property statement.

If the four points are not all critical support points, the three critical support points are identified using Property 6.

Property 6. Let U_f, U_l, L_f and L_l be the four possible critical support points for $S_1 \cup S_2$. If they are not all critical support points, then:

- if U_f and U_l are both critical support points then:
  - if the slope of $(U_f U_l)$ is lower than the slope of $(L_f L_l)$, L_f is the third critical support point ;
  - otherwise, L_l is the third critical support point ;
- if L_f and L_l are both critical support points then:
  - if the slope of $(L_f L_l)$ is greater than the slope of $(U_f U_l)$, U_f is the third critical support point ;
  - otherwise, U_l is the third critical support point ;

Proof. We write the proof for the case where U_f and U_l are both critical support points. The other case is similar.

Consider the dual representation of the points U_f and U_l, denoted by U_f^* and U_l^*. By hypothesis, these two lines support edges of $\mathcal{P}(S_1 \cup S_2)$. Consider now the dual representation of the points $L_f + (0, 1)$ and $L_l + (0, 1)$, denoted by L_{f+}^* and L_{l+}^*. The third edge of $\mathcal{P}(S_1 \cup S_2)$ is a segment of either L_{f+}^* or L_{l+}^*. We suppose now that the slope of $(U_f U_l)$ is lower than the slope of $(L_f L_l)$ and illustrate the rest of the proof with Figure 3. Then, the abscissa of point $D = L_{f+}^* \cap L_{l+}^*$ is greater than the abscissa of point $B = U_f^* \cap U_l^*$ (it lies in the light-gray half-space on Figure 3). It is now easy to see that if D is above the line U_f^*, then both L_{f+}^* and L_{l+}^* support edges of $\mathcal{P}(S_1 \cup S_2)$, which is not possible by hypothesis. Then, D is below the line U_f^*, which implies that the third edge of the preimage is a segment of L_{f+}^*, and equivalently L_f is the third critical support point.

3.2 Pulling Out the Minimal Characteristics

In the previous section, we showed how to efficiently compute the three or four critical support points of $S_1 \cup S_2$. These points also define the preimage of $S_1 \cup S_2$. Until now, the results were valid whether the two DSSs were connected or not. In order to compute the minimal characteristics, we have to consider several cases.

Input DSSs Are Connected. We consider here the case where the first point of S_2 is either a point of S_1 or 8-connected to the last point of S_1. In this case, if there exists a DSL containing $S1 \cup S_2$, then $S_1 \cup S_2$ is a DSS of length n, the difference of abscissa between the first point of S_1 and the last point of S_2. As a consequence, $\mathcal{P}(S_1 \cup S_2)$ is a cell of the Farey Fan of order n, with very well-known properties. In particular, the critical support points computed in Section 3.1 are exactly the leaning points of the DSS.

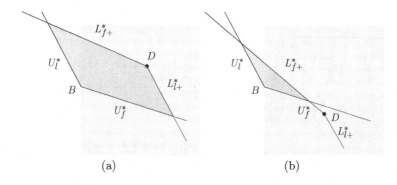

Fig. 3. Illustration of the proof of Property 6

S_1 Last Point and S_2 First Point Have the Same Ordinate. We show that this case is actually as easy as the previous one.

Property 7. If the last point of S_1 and the first point of S_2 have the same ordinate, then $\mathcal{P}(S_1 \cup S_2)$ is a unique cell of the Farey Fan of order n, the difference of abscissa between the last point of S_2 and the first point of $S1$.

Proof. In a DSS, the edges of the preimage are defined by the leaning points only. Actually, the preimage of a DSS is equal to the preimage of its leaning points, all the other points make no contribution. If the last point of S_1 and the first point of S_2 have the same ordinate, then all the missing points between these two points also have this same ordinate. Consequently, they cannot be neither lower nor upper leaning points for any DSL containing $S_1 \cup S_2$. This proves that $\mathcal{P}(S_1 \cup S_2)$ is the same as the preimage of the set of pixels composed of S_1, S_2 and all the missing points between the two. Then, $\mathcal{P}(S_1 \cup S_2)$ is the preimage of a DSS of length n, the difference of abscissa between the last point of S_2 and the first point of $S1$, which is similar to the previous case.

Disconnected DSSs. This case is trickier since $\mathcal{P}(S_1 \cup S_2)$ may not be a unique cell but can be a union of adjacent cells of a Farey Fan (see Figure 4 for an example). The characteristics given by the critical points may not be the minimal ones. However, from the critical support points we can easily infer the range of slopes of the separating lines. If we denote s_{low} and s_{up} the minimum and maximum slopes of the separating lines, the slope of the line of minimal characteristics is given by the fraction of smallest denominator between s_{low} and s_{up}. It is finally easy to decide which one of either U_f or U_l is an upper leaning point of the line of minimal characteristics (see Algorithm 1 for more details).

3.3   General Algorithm

All the properties presented above are put together to design the fast union algorithm described in Algorithm 1. The algorithm returns the minimal characteristics of $S_1 \cup S_2$ if the union is possible. The result is given as a directional

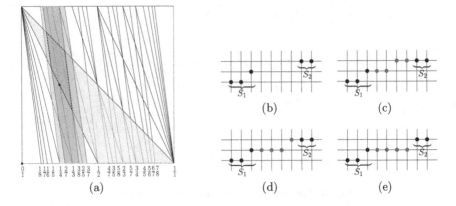

Fig. 4. When S_1 and S_2 are not connected, like the ones depicted in (b), $\mathcal{P}(S_1 \cup S_2)$ may be the union of several cells: in (a), $\mathcal{P}(S_1)$ is depicted in light gray, and the two constraints related to the leaning points of S_2 are depicted in red and blue. The intersection is bordered by a dotted black line: it is composed of three cells, each one being the preimage of a DSS containing $S_1 \cup S_2$, depicted in (c), (d), and (e).

vector (b, a) and an upper leaning point. The algorithm can be decomposed into three main parts. Between line 1 and 2, the four possible critical support points are computed. The function `initCriticalSupportPoint` is the implementation of Table 1 and is detailed in Algorithm 2 presented in Appendix. The string parameter given in input is there to discriminate between the two lines of Table 1. At the same time, easy cases where $S_1 \cup S_2$ has the same minimal characteristics as S_1 or S_2 are treated: in such cases, the variable `inDSL` is set to true by `initCriticalSupportPoint` and we can conclude directly. Then, between lines 3 and 4, the exact critical support points are computed. The function `isSolution?` implements Property 5 and is detailed in Algorithm 3 in Appendix: here, the string parameter tells if the upper leaning points are tested w.r.t the lower leaning points or conversely. The function `thirdPoint` implements Property 6: if the variable `solU` is true, then the first item of the property is concerned, otherwise `solL` is true, and the second item is concerned. The last part, between lines 5 and 7 returns the minimal characteristics of $S_1 \cup S_2$. On line 6, if S_1 and S_2 are connected or if the last point of S_1 and the first point of S_2 have the same ordinate, the result is straightforward from the critical support points. Otherwise, as explained in Section 3.2 the line of minimal characteristics is computed among all the separating lines. Function `minimalSlope` is then called. It can be implemented in several way, using for instance the decomposition into continued fractions, or, like in [16]-Algorithm 3 (see also [7,6]) using the Stern-Brocot tree.

3.4   Complexity Analysis

Lemma 1. *The complexity of Algorithm 1 is $\mathcal{O}(1)$ when S_1 and S_2 are connected or when the last point of S_1 and the first point of S_2 have the same ordinate. Its*

Algorithm 1. `FastArithmeticalDSSUnion(DSS` S_1`, DSS` S_2`)`

U_f, U_l, L_f, L_l critical support points of $S_1 \cup S_2$;
boolean inDSL ← false; boolean solU, solL;
connected ← true if S_1 and S_2 are connected or the last point of S_1 and the first point of S_2 have the same ordinate, false otherwise;

1 (U_l, L_l, inDSL) ← `initCriticalSupportPoints(`S_1, S_2, "after")
   if inDSL = *true* **then** **return** DSS(a_1,b_1,U_{1f})
   else

      (U_f, L_f, inDSL) ← `initCriticalSupportPoints(`S_2, S_1, "before")
      if inDSL = *true* **then** **return** DSS(a_2,b_2,U_{2l})
      else
         `// Four possible critical support points are known`
3   (solL,a_L,b_L) ← isSolution? (L_f,L_l,U_f,U_l,"lower")
         (solU,a_U,b_U) ← isSolution? (U_f,U_l,L_f,L_l,"upper")
         if solU = *false* and solL = *false* **then**
            return DSS(0,0,Point(0,0)) `// Union of` S_1 `and` S_2 `is not possible`
         else
            if solU = *false* or solL = *false* **then**
               `// Three points only are critical`
               if solU = *true* **then** (a,b) ← (a_U,b_U) **else** (a,b) ← (a_L,b_L)
               (U_f,U_l,L_f,L_l) = thirdPoint $(U_f,U_l,L_f,L_l$,solU,solL$)$
4   **else** (a,b) ← (a_U,b_U) `// The four points are critical`
            `// At this point, the exact critical support points are known`
5   **if** connected = *true* **then**
            return DSS(a,b,U_f)
         else
6   (a,b) ← minimalSlope (U_f,U_l,L_f,L_l)
            if $U_l = U_f$ **then** U ← U_f
            else
               if $slope((U_f \ U_l)) > \frac{a}{b}$ **then** U ← U_f **else** U ← U_l
7   **return** DSS(a,b,U)

---

complexity is $\mathcal{O}(log(n))$ *otherwise, where* n *is the difference of abscissa between the last point of* S_2 *and the first point of* S_1.

Proof. If we assume a computing model where the standard arithmetic operations are done in constant time, then all the operations from line 1 to line 5 are also done in constant time. Whichever the algorithm chosen, the function `minimalSlope` on line 6 always requires, in a more or less direct way, the computation of the continued fractions of two fractions $\frac{p}{q}$ with $p \leq q \leq n$, and n is the difference of abscissa between the last point of S_2 and the first point of S_1. This is done in $\mathcal{O}(log(n))$ time (see [6] for instance).

4   Experimental Results

Algorithm 1 was implemented in C++ using the open-source library DGtal [1]. We refer to it as the `FastArithmetical` algorithm in the following. We compare our algorithm with two other ones. The first one is the well-known arithmetical recognition algorithm [4], implemented in DGtal (called `Arithmetical` algorithm in what follows). As stated at the beginning of Section 3.1, the algorithm of O'Rourke [11] can be used to compute the set of separating lines. It was implemented in DGtal by T. Roussillon as the `StabbingLine` algorithm. The `Arithmetical` algorithm works only when the two DSSs are connected and is used as follows : the minimal characteristics are initialised with the ones of S_1, and updated as the points of S_2 are added one by one. Concerning the `StabbingLine` algorithm, the preimage is initialised with the one of S_1 and updated as the leaning points of S_2 are added (Corollary 1). The result is the set of critical support points of $S_1 \cup S_2$.

The experimental setup is the following:

- a DSL of characteristics (a, b, μ) is picked up at random ;
- the abscissas x_1 and x_2 of the first and last points of S_1 are randomly selected ;
- the abscissa x_3 of the first point of S_2 is either equal to x_2+1 in the connected case, or randomly selected and greater than x_2 in the disconnected one ;
- the abscissa of the last point is set at a fixed distance from x_3 ;

Two parameters govern this setup : `maxb` is the maximal value of b ; `valX` is the length of S_2. b is randomly picked in the interval $[1, \text{maxb}]$, a is drawn in the interval $[1, b]$ and such that a and b are relatively prime, and μ in the interval $[0, 2\text{maxb}]$. The value of x_1 is drawn in the interval $[0, \text{maxb}]$. The length of S_1 (*i.e.* $x_2 - x_1$) is randomly selected in the interval $[\text{valX}, 2\text{valX}]$, so that S_1 is always longer than S_2. In our test, `maxb` is set to 1000 and `valX` varies from 10 to 2maxb. For each value of `valX`, 2000 pairs of values (a, b) are drawn. For each of them, 5 different values of μ are picked up, and then 10 different values of x_1 are tested, for a total of 10^5 tests.

When the two DSSs are connected, the first test we perform consists in verifying that the three algorithms actually compute the same minimal characteristics. Then, the performances in terms of computation time are compared. Figure 4 shows the results (logarithmic scale for both axis): the x-axis represents the value of `valX`, the y-axis is the mean CPU computation time for a pair of DSSs, and for each algorithm.

First we can observe that the experimental behaviour of `FastArithmetical` algorithm confirms the constant-time complexity. Unsurprisingly, the `Arithmetical` algorithm has a linear-time complexity. Concerning the `StabbingLine` algorithm, its performances are slightly worse than the `FastArithmetical` algorithm, and a slight increase of the mean computation time is observed for larger DSS lengths: this is due to the fact that a post-treatment has to be done on the result returned by this algorithm in order to compute the minimal characteristics. This post-treatment involves a *gcd* computation, which explains the plot.

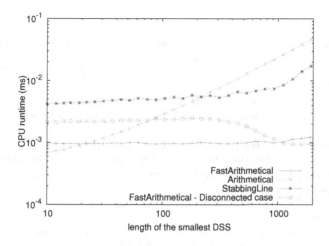

Fig. 5. Experimental results

However, the main information is that the `FastArithmetical` algorithm gets faster than the classical `Arithmetical` one when the length of the smallest DSS is greater than 20. In comparison, the `StabbingLine` algorithm becomes faster for lengths greater than 200 only. This means that what could appear as a small gain on a constant term in comparison to the `StabbingLine` algorithm makes the `FastArithmetical` relevant in practice compared to the `Arithmetical` algorithm. Last, the `FastArithmetical` remains faster than the `StabbingLine` algorithm even when the two DSSs are not connected. The slight decrease of the mean computation time for long DSSs is related to the fact that the longer the DSSs, the more easy cases appear.

5   Conclusion

In this work, we have shown that the union of two DSSs can be very efficiently computed since it is enough to "update" the minimal characteristics of the first segment with the leaning points of the second one. To do so, we have demonstrated that a state-of-the-art algorithm - the stabbing line algorithm - can be used to compute the union in logarithmic time. Moreover, we have exhibited a number of simple arithmetical properties to design an even faster algorithm. This algorithm runs in $\mathcal{O}(log(n))$ worst-time complexity, and $\mathcal{O}(1)$ for easy cases and the experiments have shown that the implementation concretises this complexity.

Now, an interesting question remains: what if the union is not possible ? Can we measure the "distance" between the two DSSs ? A solution would be to consider "thicker" DSSs and to compute the thickness necessary for the union to be possible. This problem seems actually to be very close the the blurred DSS recognition algorithms [3,13], and this trail seems worthy to be followed.

356.Sivignon

References

1. DGtal: Digital Geometry Tools and Algorithms Library, http://libdgtal.org
2. Damiand, G., Coeurjolly, D.: A generic and parallel algorithm for 2D digital curve polygonal approximation. Journal of Real-Time Image Processing (JRTIP) 6(3), 145–157 (2011)
3. Debled-Rennesson, I., Feschet, F., Rouyer-Degli, J.: Optimal blurred segments decomposition of noisy shapes in linear time. Computers & Graphics 30(1), 30–36 (2006)
4. Debled-Rennesson, I., Reveillès, J.P.: A linear algorithm for segmentation of digital curves. Inter. Jour. of Pattern Recog. and Art. Intell. 9(6), 635–662 (1995)
5. Debled-Rennesson, I., Reveillès, J.P.: A new approach to digital planes. In: SPIE - Vision Geometry III (1994)
6. Graham, R.L., Knuth, D.E., Patashnik, O.: Concrete Mathematics. Addisson-Wesley (1994)
7. Harel, D., Tarjan, R.E.: Fast algorithms for finding nearest common ancestor. SIAM Journal on Computing 13(2), 338–355 (1984)
8. Jacob, M.A.: Applications quasi-affines. Ph.D. thesis, Université Louis Pasteur, Strasbourg, France (1993)
9. Lachaud, J.O., Said, M.: Two efficient algorithms for computing the characteristics of a subsegment of a digital straight line. Discrete Applied Mathematics 161(15), 2293–2315 (2013)
10. McIlroy, M.D.: A note on discrete representation of lines. AT&T Technical Journal 64(2), 481–490 (1985)
11. O'Rourke, J.: An on-line algorithm for fitting straight lines between data ranges. Commun. ACM 24(9), 574–578 (1981)
12. O'Rourke, J.: Computational Geometry in C. Cambridge University Press (1998)
13. Roussillon, T., Tougne, L., Sivignon, I.: Computation of binary objects sides number using discrete geometry, application to automatic pebbles shape analysis. In: Int. Conf. on Image Analysis and Processing, pp. 763–768 (2007)
14. Roussillon, T.: Algorithmes d'extraction de modèles géométriques discrets pour la représentation robuste des formes. Ph.D. thesis, Université Lumière Lyon 2 (2009)
15. Said, M., Lachaud, J.O., Feschet, F.: Multiscale analysis of digital segments by intersection of 2D digital lines. In: ICPR 2010, pp. 4097–4100 (2010)
16. Sivignon, I., Dupont, F., Chassery, J.M.: Digital intersections: minimal carrier, connectiviy and periodicity properties. Graphical Models 66(4), 226–244 (2004)
17. Sivignon, I.: Walking in the farey fan to compute the characteristics of a discrete straight line subsegment. In: Gonzalez-Diaz, R., Jimenez, M.-J., Medrano, B. (eds.) DGCI 2013. LNCS, vol. 7749, pp. 23–34. Springer, Heidelberg (2013)

Appendix

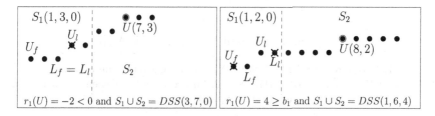

Fig. 6. Illustration of the first line of Table 1: the leaning points of S_1 marked with a cross cannot be critical support points if the point U of S_2 is added

---

Algorithm 2. InitCriticalSupportPoints(DSS S, DSS S', string *position*)

U_f, U_l, L_f, L_l the leaning points of S, r remainder function of S
inDSL a boolean; inDSL ← true
foreach *leaning point P of S' and if* inDSL = *true* **do**
  if $r(P) < 0$ **then**
    if position = *"after"* **then** $U \leftarrow U_f$, $L \leftarrow L_l$ **else** $U \leftarrow U_l$, $U \leftarrow U_f$
    inDSL ← false
  else
    if $r(P) \geq b$ **then**
      if position = *"after"* **then** $U \leftarrow U_l$, $L \leftarrow L_f$ **else** $U \leftarrow U_f$, $L \leftarrow L_l$
      inDSL ← false
    else
      if $r(P) = 0$ **then** $U \leftarrow P$
      if $r(P) = b - 1$ **then** $L \leftarrow P$
end
return (U,L,inDSL)

Algorithm 3. isSolution?(P_f,P_l,Q_f,Q_l, string *type*)

// compute the characteristics defined by the points P_f and P_l
$a \leftarrow P_l.y - P_f.y$, $b \leftarrow P_l.x - P_f.x$
if type = *"lower"* **then** $\mu \leftarrow b - 1 - aP_f.x + bP_f.y$
else $\mu \leftarrow -aP_f.x + bP_f.y$
// check the position of Q_f and Q_l w.r.t these characteristics
Let $r(Q) = aQ.x - bQ.y + \mu$
if $0 \leq r(Q_f) < b$ *and* $0 \leq r(Q_l) < b$ **then return** (true,a, b, μ)
else
  return (false,a, b, μ)

Digital Geometry from a Geometric Algebra Perspective

Lilian Aveneau, Laurent Fuchs, and Eric Andres

Laboratoire XLIM-SIC UMR CNRS 7252, Université de Poitiers,
Bld Marie et Pierre Curie, BP 30179, 86962, Futuroscope Chasseneuil Cedex, France
{lilian.aveneau,laurent.fuchs,eric.andres}@univ-poitiers.fr

Abstract. To model Euclidean spaces in computerized geometric calculations, the Geometric Algebra framework is becoming popular in computer vision, image analysis, etc. Focusing on the Conformal Geometric Algebra, the claim of the paper is that this framework is useful in digital geometry too. To illustrate this, this paper shows how the Conformal Geometric Algebra allow to simplify the description of digital objects, such as k-dimensional circles in any n-dimensional discrete space. Moreover, the notion of duality is an inherent part of the Geometric Algebra. This is particularly useful since many algorithms are based on this notion in digital geometry. We illustrate this important aspect with the definition of k-dimensional spheres.

Keywords: Digital Geometry, Geometric Algebra, Conformal Model, Digital Object.

1 Introduction

The purpose of this paper is to introduce the computational and mathematical framework of Geometric Algebra (GA) in digital geometry. GA form a powerful mathematical language for expressing and representing geometric objects, transformations or even for working in dual spaces [10,17,6]. GA are becoming popular in various computer imagery sub-fields such as computer vision or image analysis, and even more largely in fields like physics and engineering [15,5,7]. The reason for such a popularity is that the mathematical framework of GA is well adapted for handling geometric data of any dimension in a very intuitive way.

GA represent a natural extension of complex numbers and quaternions in arbitrary dimension. Each instance of GA is an associative algebra (known as Clifford algebra) of a real vector space equipped with a given quadratic form. In this paper, we consider the Conformal Geometric Algebra (CGA), defined over the Minkowski space $\mathbb{R}^{n+1,1}$. It offers a very natural representation for circles and spheres (points, lines and planes are simply particular cases of the former) and extension of these geometric primitives in any dimension. CGA also provides a way to represent transformations such as translations, reflections or rotations.

E. Barcucci et al. (Eds.): DGCI 2014, LNCS 8668, pp. 358–369, 2014.

In Section 2, we present the conformal geometric algebra and some general results on geometric algebra. This is intended as an introduction to CGA for the readers that are not familiar with GA. In Section 3, we apply CGA to define discrete primitives in any dimension. First, based on a previous work [4], discrete hyperspheres and hyperplanes are presented as they often play a special role in algorithms and definitions. The more general case of discrete k-sphere (discrete spheres of dimension k in a n-dimensional space) is also introduced. Discrete lines, planes and more generally discrete k-flats are particular cases of discrete k-spheres. Finally, Section 4 proposes a conclusion and some perspectives: One of the hopes in the future, is that generation and recognition algorithms of circles and lines can be somewhat unified in such a general framework.

2 An Overview of Geometric Algebra

The basic idea of GA is to use vector subspaces which can be geometrically interpreted as Euclidean geometric primitives, and manipulated with some transformations. Thus, in any dimension, geometric primitives (lines, circles, planes, spheres) and their transformations are represented by vector subspaces.

In this paper we use the conformal model of GA, usually denoted as Conformal Geometric Algebra. The following presentation is mainly based on the book of L. Dorst $et\ al.$ [6] which provides an accessible and deep description of GA. For a shorter introduction, the reader can consult [12].

2.1 Building the Conformal Geometric Algebra

The Vector Space Structure. The starting point to build the CGA is a Euclidean space $\mathbb{R}^n$ of dimension n with an orthonormal basis $\{e_1, e_2, \ldots, e_n\}$. This Euclidean space is naturally equipped with a scalar product such that $e_i^2 = 1$ for $i \in \{1, 2, \ldots, n\}$.

This Euclidean space is extended with two extra basis vectors e_+ and e_- such that $e_+^2 = 1$ and $e_-^2 = -1$, and such that $\{e_1, e_2, \ldots, e_n, e_+, e_-\}$ is an orthogonal basis. This gives the conformal space $\mathbb{R}^{n+1,1}$. Due to its particular scalar product, this space has a particular metric, which is a key point to obtain an interpretation of its vector subspaces as geometric primitives and geometric transformations.

In the rest of this paper, for simplicity, we use the basis $\{n_o, e_1, e_2, \ldots, e_n, n_\infty\}$ where $n_o = \frac{1}{2}(e_+ + e_-)$ and $n_\infty = e_- - e_+$.

Introducing the Outer Product. Starting from this conformal space, the CGA is built using the GRASSMANN or exterior or outer product, denoted by "$\wedge$". Among others properties, the outer product is anticommutative, meaning that for two vector a and b, we have $a \wedge b = -b \wedge a$ and $a \wedge a = 0$. This product generates new elements from the vectors of $\mathbb{R}^{n+1,1}$. For example, the outer product $a \wedge b$ of two independent vectors a and b is a new element, called a 2-vector, which lies in a new vector space. The outer product $a \wedge b \wedge c$ of three independent vectors a, b and c generates again a vector in a new vector space, and so on until the $(n + 2)$-vector space.

Considering the linear combinations of such elements[1], we obtain the algebra of vector subspaces of $\mathbb{R}^{n+1,1}$. This algebra $\bigwedge \left(\mathbb{R}^{n+1,1} \right)$ is a graded algebra:

$$\bigwedge \left(\mathbb{R}^{n+1,1} \right) = \bigwedge{}^{0} \left(\mathbb{R}^{n+1,1} \right) \oplus \bigwedge{}^{1} \left(\mathbb{R}^{n+1,1} \right) \oplus \cdots \oplus \bigwedge{}^{n+2} \left(\mathbb{R}^{n+1,1} \right)$$

where $\bigwedge^{k} \left(\mathbb{R}^{n+1,1} \right)$ is the vector space of k-vectors. The dimension of each of such subspaces is $\binom{n+2}{k}$, and then the dimension of the algebra is 2^{n+2}. The space $\bigwedge^{0} \left(\mathbb{R}^{n+1,1} \right)$ is of dimension 1 and corresponds to the space of the scalars. The space $\bigwedge^{n+2} \left(\mathbb{R}^{n+1,1} \right)$ is the space of pseudo-scalars; it is also a space of dimension 1 spanned by the pseudo-scalar $I_{n+1,1} = n_o \wedge \mathbf{I}_n \wedge n_\infty$ where $\mathbf{I}_n = e_1 \wedge \cdots \wedge e_n$ is called the Euclidean pseudo-scalar. Hence, by duality, $\bigwedge^{n+2} \left(\mathbb{R}^{n+1,1} \right)$ is isomorphic to the scalar space. This principle is extended to each k-vector space, which is by duality isomorphic to the $(n + 2 - k)$-vector space.

Elements of the k-vector space $\bigwedge^{k} \left(\mathbb{R}^{n+1,1} \right)$ are called *multivectors*. For a multivector A, the part in $\bigwedge^{k} \left(\mathbb{R}^{n+1,1} \right)$ is denoted by $\langle A \rangle_k$ and is called the part of *grade* k of A.

Elements that can be written as a product $a_1 \wedge a_2 \wedge \cdots \wedge a_k$ of 1-vectors are called k-*blades*[2] and are of special interest because they can be interpreted as geometric primitives. These elements represent the vector subspaces of the conformal space $\mathbb{R}^{n+1,1}$, since the outer product of k independent vectors spans a k-dimensional vector subspace. Hence, for a particular vector subspace $\mathcal{A}$ of dimension k, generated by a k-blade $A = a_1 \wedge \cdots \wedge a_k$, we can determine a k-blade I such that $A = \lambda I$. The real coefficient λ is the (relative) *weight* of A to the chosen k-blade I. The *attitude* of A is the equivalence class λA for any $\lambda \in \mathbb{R}$ and the (relative) *orientation* of A is the sign of λ. These three quantities are well known for a vector line; for any vector line defined by a vector a we can define a (unit) vector i such that $a = \lambda i$. In that case the attitude corresponds to the direction of the line.

Introducing the Geometric Product. The *geometric product* is first defined for two vectors a and b, $ab = a \cdot b + a \wedge b$ where "$\cdot$" is the scalar product of $\mathbb{R}^{n+1,1}$. The geometric product has no symbol to denote it. This product is then linearly extended to any algebra element using the following properties for all scalars α and multivectors A, B and C:

$$1A = A1 = A, \qquad A(B + C) = AB + AC, \qquad (B + C)A = BA + CA,$$
$$(AB)C = A(BC), \qquad (\alpha A)B = A(\alpha B) = \alpha(AB).$$

---

[1] Recall that an algebra is a vector subspace equipped with a product. Hence, addition of vector subspaces and scalar multiplication of vector subspaces are available operations.

[2] In some textbooks, they are also called *decomposable k-vectors*.

This product is the fundamental product of the Geometric Algebra, the other products can be defined from it. The definitions of the previous outer product and the left contraction used below from the geometric product are:

outer product : $A_k \wedge B_l = \langle A_k B_l \rangle_{k+l}$ **left contraction** : $A_k \rfloor B_l = \langle A_k B_l \rangle_{l-k}$ where the indexes denote the grade of the multivectors A and B.

The most important notion introduced by the geometric product is the possibility to compute an inverse of a k-blade that has nonzero norm. For example, the inverse of $I_{n+1,1}$ is given by $I_{n+1,1}^{-1} = n_o \wedge \mathbf{I}_n^{-1} \wedge n_\infty$ where $\mathbf{I}_n^{-1} = (-1)^{n(n-1)/2}\mathbf{I}_n$. This lets us define the *dual* of a multivector A as $A^* = AI_{n+1,1}^{-1}$, or equivalently $A^* = A\rfloor I_{n+1,1}^{-1}$. If A_k is a k-blade then A_k^* is an $(n+2-k)$-blade which represents the orthogonal complement[3] of the k-blade A_k.

Euclidean Point Representation in CGA. Any Euclidean point $\mathbf{p}$ of $\mathbb{R}^n$ is represented by a vector p of the conformal space $\mathbb{R}^{n+1,1}$ by:

$$p = F(\mathbf{p}) = n_o + \mathbf{p} + \frac{1}{2}\mathbf{p}^2 n_\infty.$$

This vector p is normalized as the coefficient of n_o is 1. Else, it can be normalized using $\frac{p}{-n_\infty \cdot p}$. In a general setting, the coefficient of n_o is the weight of the vector p and, using $n_o \cdot n_\infty = -1$, is equal to $-n_\infty \cdot p$.

The dot product of two vectors p and q representing two Euclidean points $\mathbf{p}$ and $\mathbf{q}$ is directly linked to their Euclidean distance $d_2(\mathbf{p}, \mathbf{q})$:

$$p \cdot q = -\frac{1}{2}(\mathbf{p} - \mathbf{q})^2. \tag{1}$$

Hence, for a vector $p \in \mathbb{R}^{n+1,1}$ representing a Euclidean point $\mathbf{p}$, it follows $p \cdot p = 0$ and $-n_\infty \cdot p \neq 0$; so, Euclidean points are represented by *null vectors* (*i.e.* vectors that square to zero). Moreover, the normalization condition and the equality $n_\infty \cdot n_\infty = 0$ tell us that n_∞ is a point and can be geometrically interpreted as *the point at infinity*. The vector n_o is also a null vector, so it represents a point and can be geometrically interpreted as *the point at the origin* in the chosen representation[4] of Euclidean points in the conformal space.

2.2   Representing Geometric Elements

Hyperspheres and Hyperplanes. The equation (1) immediately gives us the equation of a hypersphere with the Euclidean point $\mathbf{c}$ as center and radius ρ. For a Euclidean point $\mathbf{x}$ of the hypersphere we have:

$$x \cdot c = -\frac{1}{2}\rho^2$$

---

[3] This orthogonality notion refers to the particular metric we have defined on $\mathbb{R}^{n+1,1}$.
[4] Actually any other finite point p can be chosen as point at origin. Since the normalization condition imposes $-n_\infty \cdot p = 1$, it has the same relation with n_∞ as n_o.

where x and c are the vectors representing the points $\mathbf{c}$ and $\mathbf{x}$. This last equation is equivalent to:

$$x \cdot (c - \frac{1}{2}\rho^2 n_\infty) = 0 \qquad (2)$$

by using the normalizing condition $-n_\infty \cdot x = 1$. The vector $\sigma = c - \frac{1}{2}\rho^2 n_\infty$ represents the *hypersphere with center* $\mathbf{c}$ *and radius* ρ.

As the defining equation of a hypersphere is $x \cdot \sigma = 0$, we say that the vector σ dually represents the hypersphere. From this, a direct representation of a hypersphere can be deduced (see [6]). In brief, in dual representation a hypersphere is represented by a vector which is a 1-dimension vector subspace. Then, taking the dual in $\mathbb{R}^{n+1,1}$ of this vector in leads to a vector subspace Σ of dimension $n+1$. This vector subspace is spanned by $n+1$ vectors that can be chosen as representing vectors of $n+1$ Euclidean points. So we can write $\Sigma = p_1 \wedge p_2 \wedge \cdots \wedge p_{n+1}$ and $\Sigma^* = \sigma$. This means that a Euclidean hypersphere in $\mathbb{R}^n$ is defined by $n+1$ Euclidean points.

Considering equation (2) for a Euclidean point $\mathbf{x}$ not on the hypersphere represented by σ gives

$$x \cdot \sigma = x \cdot (c - \frac{1}{2}\rho^2 n_\infty) = -\frac{1}{2}((\mathbf{x} - \mathbf{c})^2 - \rho^2) \qquad (3)$$

then $x \cdot \sigma > 0$ if the point $\mathbf{x}$ is inside the hypersphere and $x \cdot \sigma < 0$ if the point $\mathbf{x}$ is outside the hypersphere. This gives us a way to determine the relative positions of a point and a hypersphere in any dimension.

In these settings a hyperplane Π is a hypersphere with a point at infinity, hence $\Pi = p_1 \wedge \cdots \wedge p_n \wedge n_\infty$. A Euclidean hyperplane is thus defined with n Euclidean points. It is dually represented by the vector $\Pi^* = \pi = \mathbf{n} + \delta n_\infty$ where $\mathbf{n}$ is the normal to the hyperplane and δ the distance to the origin along $\mathbf{n}$. Hence for a point $\mathbf{x}$ of the hyperplane represented by π we have

$$x \cdot \pi = \mathbf{x} \cdot \mathbf{n} - \delta = 0$$

which is the usual (i.e. in linear algebra setting) equation for an hyperplane of normal vector $\mathbf{n}$ and distance δ to the origin along $\mathbf{n}$. Thus, for a point $\mathbf{x}$ not on the hyperplane, we can determine the relative positions of $\mathbf{x}$ and π using the sign of $x \cdot \pi$.

Flats. Flats (k-flats) are offsets of k-dimensional subspaces of $\mathbb{R}^n$ (i.e. lines, planes, etc). In the CGA framework they are represented by algebra elements of the direct form: $p \wedge \mathbf{A}_k \wedge n_\infty$, where $\mathbf{A}_k$ is a Euclidean blade (i.e. a blade with no vector n_o or n_∞ as factor) and the vector p represents the Euclidean point $\mathbf{p}$ the flat is passing through. Dualizing this expression leads to the dual form of a k-flat: $-p\rfloor(\mathbf{A}_k^* n_\infty)$, where $\mathbf{A}_k^* = \mathbf{A}_k \mathbf{I}_n^{-1}$ is the Euclidean dual of $\mathbf{A}_k$.

Rounds. Rounds are geometric algebra elements representing k-spheres. They can be defined using the outer product of $k+2$ independent vectors representing Euclidean points: $\Sigma = p_1 \wedge \cdots \wedge p_{k+2}$.

Hence a circle (i.e a 1-sphere) is defined by three points and a 2-sphere is defined by four points. The minimum number of points to obtain a round is 2, this corresponds to a 0-sphere which is a point pair. This is easily explained if we use an alternate definition of a k-sphere as the intersection of a hypersphere with a $(k + 1)$-flat. So the intersection of a hypersphere with a line gives a 0-sphere which is the two points of intersection.

In the CGA framework the intersection of an hypersphere $\Sigma^* = \sigma$ and a flat $\Pi^* = \pi$ in dual form is easily computed by:

$$\sigma \wedge \pi = \Sigma^* \wedge \Pi^*$$

using the formula[5] $(\Sigma \cap \Pi)^* = \Sigma^* \wedge \Pi^*$ (see [6]). This leads to the dual form of a round with center c and radius ρ:

$$\Sigma^* = \sigma = (c - \frac{1}{2}\rho^2 n_\infty) \wedge (-c \rfloor ((-1)^{(n-k)} \mathbf{A}_k^\star \, n_\infty))$$

where the change of sign is to maintain coherent orientation. Dualizing this expression leads to another direct form of a round:

$$\Sigma = (c + \frac{1}{2}\rho^2 n_\infty) \wedge (-c \rfloor ((-1)^k \mathbf{A}_k \, n_\infty))$$

with center c and radius ρ.

Imaginary Hyperspheres and Imaginary Rounds. In this last expression the algebra element $(c + \frac{1}{2}\rho^2 n_\infty)$ has a particular meaning. For a Euclidean point $\mathbf{x}$, considering the equation $x.(c + \frac{1}{2}\rho^2 n_\infty) = 0$ leads to $(\mathbf{x} - \mathbf{c})^2 = -\rho^2$. This means that the squared distance of all Euclidean points satisfying $x.(c + \frac{1}{2}\rho^2 n_\infty) = 0$ must be negative. By analogy we say that the vector $(c + \frac{1}{2}\rho^2 n_\infty)$ represents an *imaginary hypersphere*.

If such an imaginary hypersphere is used in the definition of the dual form of a round given above, we obtain the dual form of an *imaginary round*. As only squared distances enter in algebra computations, complex numbers are not needed.

Those elements occur naturally as results of intersections when a real solution does not exist (see figure 1).

3 Discrete Geometric Primitives

Basic discrete primitives such as discrete straight lines, discrete hyperplanes and discrete hyperspheres [2,1,14] have been defined as all the discrete points

---

[5] This formula corresponding to the dual of an intersection, called the *plunge*, is valid because the union of Σ and Π is the whole space. Otherwise, the same formula can be used but the dual must be taken wrt. the union of Σ and Π (more details can be found in [6] chap. 14).

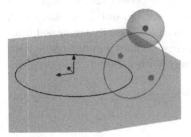
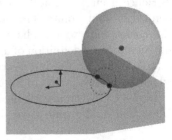

(a) Intersection: imaginary case.   (b) Intersection: real case.

Fig. 1. The intersection of hypersphere $S_{\mathbf{p},\rho}$ (the sphere) and a 1-sphere S_1 (the circle). (1a) When there is no intersection the expression $S^*_{\mathbf{p},\rho} \rfloor S_1$ is an imaginary round (dashed point pair) and the expression $S^*_{\mathbf{p},\rho} \wedge S^*_1$ is a real round (red circle). (1b) When there is an intersection, real (blue points) and imaginary (dashed circle) are interchanged.

verifying a set of inequalities in the classical linear algebra framework. There is however no direct way to define discrete rounds or flats in such a way. A more recent approach proposed a morphological based digitization scheme [16] defined as the intersection of the discrete space and the Minkowski sum of a structuring element and the object points. For a structuring element corresponding to a ball for a given distance, it is equivalent to considering all the discrete points that are at the ball radius distance of the Euclidean primitive.

For instance in nD space, considering the ball $B_2\left(c, \frac{1}{2}\right) = \{x \in \mathbb{R}^n | d_2(x, c) \leq \frac{1}{2}\}$ as structuring element, then the discretisation $D(F)$ of a Euclidean object F is defined by:

$$D(F) = \left(B_2\left(x \in F, \frac{1}{2}\right) \oplus F\right) \cap \mathbb{Z}^n$$

where $\oplus$ is the Minkowski sum operator. This can also be interpreted as

$$D(F) = \left\{X \in \mathbb{Z}^n \mid d_2(X, F) \leq \frac{1}{2}\right\}.$$

The problem is to test efficiently the inequality $d_2(X, F) \leq 1/2$. For instance, let us consider a round R of dimension k defined by

$$R = \left\{v \in \mathbb{R}^n \,\middle|\, \|c - v\|^2 = \rho^2\right\} \cap \left\{v \in \mathbb{R}^n \,\middle|\, v = \sum \lambda_i u_i\right\}$$

where $\{u_i\}$ are k linearly independent vectors. There is no simple immediate expression for the distance $d_2(x, R)$ between a point x and the k-dimensional round R in dimension n.

In the following subsections, we are going to examine how discrete hyperspheres, hyperplanes, k-spheres and k-flats can be described within the CGA framework (see also [4]). The interest of those expressions is that they can directly be computed.

3.1   Discrete Hyperspheres and Discrete Hyperplanes in CGA

Using the Euclidean distance, a discrete hypersphere centered at the point **c** with radius ρ is defined as:

$$\left\{ \mathbf{p} \in \mathbb{Z}^n \,|\, (\rho - d)^2 \leq |\mathbf{c} - \mathbf{p}|^2 < (\rho + d)^2 \right\}$$

where the width d is a positive real number smaller than ρ. This is the set of discrete points close to the Euclidean hypersphere.

Hence, a point lies in the discrete hypersphere if it is inside the Euclidean hypersphere of radius $\rho + d$, and outside the hypersphere of radius $\rho - d$.

In the CGA framework these two hyperspheres are defined in dual form as:

$$\sigma_{\mathbf{c},\rho+d} = c - \tfrac{1}{2}(\rho+d)^2 \, n_\infty \qquad \text{and} \qquad \sigma_{\mathbf{c},\rho-d} = c - \tfrac{1}{2}(\rho-d)^2 \, n_\infty.$$

Now, using equation (3) and a discrete point $\mathbf{p} \in \mathbb{Z}^n$, distances to the hyperspheres $\sigma_{\mathbf{c},\rho+d}$ and $\sigma_{\mathbf{c},\rho-d}$ are checked with the following expressions:

$$p \cdot \sigma_{\mathbf{c},\rho+d} = \frac{1}{2}\left((\rho+d)^2 - (\mathbf{p}-\mathbf{c})^2\right) \tag{4}$$

$$p \cdot \sigma_{\mathbf{c},\rho-d} = \frac{1}{2}\left((\rho-d)^2 - (\mathbf{p}-\mathbf{c})^2\right) \tag{5}$$

expression (4) must be positive and expression (5) must be negative.

Hence, in any dimension, a discrete hypersphere centered at **c** with radius ρ is defined as:

$$\{\mathbf{p} \in \mathbb{Z}^n \,|\, p \cdot \sigma_{\mathbf{c},\rho-d} < 0 \text{ and } p \cdot \sigma_{\mathbf{c},\rho+d} \geq 0\}.$$

The figure 2 shows an example of a discrete hypersphere drawn with this definition.

Using the same development, a discrete hyperplane can be defined as the set of discrete points close to the Euclidean hyperplane $\pi_{\mathbf{n},\delta} = \mathbf{n} + \delta n_\infty$ and we must find the discrete points enclosed between two hyperplanes. To do this, we define two hyperplanes $\pi_{\mathbf{n},\delta-d} = \mathbf{n} + (\delta-d)n_\infty$ and $\pi_{\mathbf{n},\delta+d} = \mathbf{n} + (\delta+d)n_\infty$ translated for

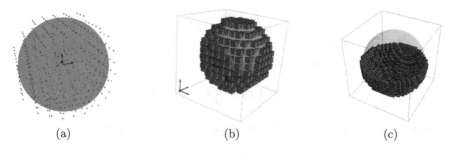

(a)   (b)   (c)

Fig. 2. Example of a discrete hypersphere. (2a) Drawn with the "centers" of the voxels. (2b) Drawn with voxels. (2c) Partial view.

a width d along the normal of the first one. Hence, in any dimension, a discrete hyperplane with normal $\mathbf{n}$ at distance δ to the origin along $\mathbf{n}$ is defined as:

$$\{\mathbf{p} \in \mathbb{Z}^n \mid p \cdot \pi_{\mathbf{n},\delta-d} < 0 \text{ and } p \cdot \pi_{\mathbf{n},\delta+d} \geq 0\}$$

which is basically the same definition as for the discrete hypersphere.

To conclude this section, in the CGA framework discrete hyperspheres and discrete hyperplanes are the same objects, and their definition works in any dimension. It simply consists in checking the signs of two vector dot product expressions.

3.2   Discrete Rounds and Discrete Flats in CGA

In this section we define discrete rounds (i.e. discrete k-spheres) using the structuring element approach. In our case, we use a hypersphere as structuring element. So, we have to check if a given point $\mathbf{p}$ lies into the discrete round by verifying that the hypersphere centered on $\mathbf{p}$ intersects the k-sphere.

Two cases must be distinguished depending on the form of the given round to digitize. If it is in direct form, we use an expression involving the left contraction product. Otherwise, if the round is in dual form, an expression with the outer product is used. In both cases, it is easier to consider the hypersphere in its dual form. This situation is usual in the CGA framework as duality is fully integrated. Once we have the expressions, either mode (direct or dual) is easy to work with, it just depends on the way the data have been given.

Rounds in Direct From. Let S_k be a k-sphere and $S_{\mathbf{p},\rho}$ a hypersphere with center $\mathbf{p}$ and radius ρ. Let Σ_k be a round in direct form representing S_k. The intersection of S_k and $S_{\mathbf{p},\rho}$ is given by the formula $S^*_{\mathbf{p},\rho} \rfloor \Sigma_k$ where $S^*_{\mathbf{p},\rho}$ is the algebra element representing $S_{\mathbf{p},\rho}$ in dual form (see [6]).

Hence, as $S^*_{\mathbf{p},\rho}$ is a blade of grade 1 and Σ_k is a blade of grade k, the intersection must be a blade of grade $k-1$. This is coherent with the usual result when intersecting a k-sphere with a hypersphere we obtain a $(k-1)$-sphere.

Moreover, if no intersection exists, the obtained $(k-1)$-sphere is an imaginary round (see figure 1).

For a round Σ_k in direct form, its squared radius is given by the formula

$$\rho^2 = (-1)^k \frac{\Sigma_k^2}{(n_\infty \rfloor \Sigma_k)^2}.$$

So, to test if a discrete point $\mathbf{p} \in \mathbb{Z}^n$ is in a discrete round we only have to test the sign of the expression

$$(-1)^{k-1} \left((p - \frac{1}{2}\rho^2 n_\infty) \rfloor \Sigma_k \right)^2.$$

This gives us the definition of a discrete round in direct form

$$\left\{ \mathbf{p} \in \mathbb{Z}^n \mid (-1)^{k-1} \left((p - \frac{1}{2}\rho^2 n_\infty) \rfloor \Sigma_k \right)^2 \geq 0 \right\}.$$

Rounds in Dual Form. Let S_k be a k-sphere and $S_{\mathbf{p},\rho}$ an hypersphere with center $\mathbf{p}$ and radius ρ. Let σ_k be a round in dual form representing S_k. The dual of the intersection of σ_k and $S_{\mathbf{p},\rho}$ is given by the formula $(S_k \cap S_{\mathbf{p},\rho})^* = S_{\mathbf{p},\rho}^* \wedge \sigma_k$ where $S_{\mathbf{p},\rho}^*$ is the algebra element representing $S_{\mathbf{p},\rho}$ in dual form (see [6]).

As before, $S_{\mathbf{p},\rho}^*$ is a blade of grade 1 and σ_k is a blade of grade $n + 2 - k$ thus the dual of the intersection is a blade of grade $n + 3 - k$ so its dual is a blade of grade $(k - 1)$. Expression $S_{\mathbf{p},\rho}^* \wedge \sigma_k$ corresponds to a $(k - 1)$-sphere represented by a round in dual form.

The squared radius of a round σ_k in dual form is given by the formula

$$\rho^2 = (-1)^{n+1-k} \frac{\sigma_k^2}{(n_\infty \rfloor \sigma_k)^2}.$$

So, to test if a discrete point $\mathbf{p} \in \mathbb{Z}^n$ is in a discrete round we only have to test the sign of the expression

$$(-1)^{n-k} \left((p - \frac{1}{2}\rho^2 n_\infty) \wedge \sigma_k \right)^2.$$

This gives us the definition of a discrete round in dual form

$$\left\{ \mathbf{p} \in \mathbb{Z}^n \mid (-1)^{n-k} \left((p - \frac{1}{2}\rho^2 n_\infty) \wedge \sigma_k \right)^2 \geq 0 \right\}$$

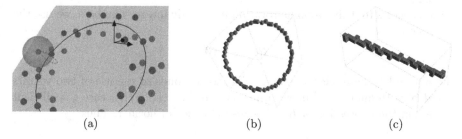

(a)   (b)   (c)

Fig. 3. Examples of discrete k-spheres. (3a) Using a hypersphere as structuring element. The points are the "centers" of the voxels defining a discrete circle. (3b) A discrete circle. (3c) A discrete line generated in the same way as the discrete circle.

To conclude this section about discrete rounds, we have seen that discrete rounds can be defined from rounds either in direct form or dual form. The structure of the definitions is the same, the only difference is in the involved product.

Moreover, as k-flats can be seen as particular rounds passing through infinity, from those definitions, discrete circles, lines and so on can easily be defined in any dimension.

4 Conclusion and Future Works

In this paper, definitions of discrete hyperspheres, hyperplanes and rounds (i.e. circles, lines, spheres, flats in any dimension) have been proposed in the Conformal Geometric Algebra formalism. These definitions are valid in any dimension. The expressions are simple and can be directly used in computations contrary to equivalent definitions in a classical framework. They also propose a unified approach for such discrete objects as k-flats are a special case of k-spheres. One of the hopes, beyond the definitions in dimension n, is that generation and recognition algorithms of k-spheres and flats can be somewhat unified in a more general framework.

However, efficient implementation of the Conformal Geometric Algebra is not an easy task [6,11,13,9] and future work is needed to have a specialized implementation for discrete geometry. For this article, we have used GAviewer [8] and the Mathematica package [3].

First experimentation has been conducted about discrete rotations. In Conformal Geometric Algebra, plane rotations are defined in a simple way as:

$$R = \mathbf{ba} = \mathbf{b} \cdot \mathbf{a} + \mathbf{b} \wedge \mathbf{a} = cos(\phi/2) - sin(\phi/2)\mathbf{I}$$

where $\mathbf{a}$ and $\mathbf{b}$ are two purely Euclidean vectors (i.e. without n_o and n_∞) ϕ is the rotation angle (i.e. $\phi/2$ is the angle from $\mathbf{a}$ to $\mathbf{b}$) and $\mathbf{I}$ is the unit bivector for the plane $\mathbf{a} \wedge \mathbf{b}$ (i.e. $\mathbf{I}$ is such that $\mathbf{a} \wedge \mathbf{b} = \beta\mathbf{I}$ with $\beta > 0$). Then RpR^{-1} is the rotation of a point represented by the vector p and $R^{-1} = cos(\phi/2) + sin(\phi/2)\mathbf{I}$.

If one wants to define a plane rotation using only integer numbers, we use the expression

$$R = \alpha - \beta\mathbf{I}$$

where α and β are integer numbers. This corresponds to the use of two vectors $\mathbf{a}'$ and $\mathbf{b}'$ with angle $\phi/2$ but with integer coordinates. In that case a coefficient appears in the expression of R^{-1} because R is not of norm 1. Thus

$$R^{-1} = \frac{1}{\alpha^2 + \beta^2}(\alpha + \beta\mathbf{I})$$

Now, if the vector p represents a point $\mathbf{p}$ with integer coordinates, its rotation is computed by $(\alpha^2 + \beta^2)RpR^{-1}$. In dimension 2 this means that we have to consider the rotation as a function from $\mathbb{Z}^2$ to $\mathbb{Z}^2$ and define a grid $(\alpha^2 + \beta^2)$ times smaller for the image space. Such phenomenon provides a good explanation of why a discrete rotation is not a one-to-one application. Further investigations need to be conducted taking into account not only rotations but also translations in order to be able to handle rigid transforms for example.

References

1. Andres, E., Acharya, R., Sibata, C.: Discrete analytical hyperplanes. Graphical Models and Image Processing 59(5), 302–309 (1997)
2. Andres, E., Jacob, M.-A.: The discrete analytical hyperspheres. IEEE Transactions on Visualization and Computer Graphics 3(1), 75–86 (1997)
3. Aragon-Camarasa, G., Aragon-Gonzalez, G., Aragon, J.L., Rodriguez-Andrade, M.A.: Clifford Algebra with Mathematica. ArXiv e-prints (October 2008)
4. Aveneau, L., Andres, E., Mora, F.: Expressing discrete geometry using the conformal model. Presented at AGACSE 2012, La Rochelle, France (2012), http://hal.archives-ouvertes.fr/hal-00865103
5. Bayro-Corrochano, E., Scheuermann, G. (eds.): Geometric Algebra Computing in Engineering and Computer Science. Springer (2010)
6. Dorst, L., Fontijne, D., Mann, S.: Geometric Algebra for Computer Science: An Object Oriented Approach to Geometry. Morgan Kauffmann Publishers (2007)
7. Dorst, L., Lasenby, J. (eds.): Guide to Geometric Algebra in Practice (Proceedings of AGACSE 2010). Springer (2011)
8. Fontijne, D., Dorst, L., Bouma, T., Mann, S.: GAviewer, interactive visualization software for geometric algebra (2010), Downloadable at http://www.geometricalgebra.net/downloads.html
9. Fuchs, L., Théry, L.: Implementing geometric algebra products with binary trees. Advances in Applied Clifford Algebra (published online first, 2014), http://dx.doi.org/10.1007/s00006-014-0447-3
10. Hestenes, D.: New Foundations for Classical Mechanics. D. Reidel Publ. Co., Dordrecht (1986)
11. Hildenbrand, D.: Foundations of Geometric Algebra Computing. Geometry and Computing, vol. 8. Springer (2013)
12. McDonald, A.: A survey of geometric algebra and geometric calculus (2013), http://faculty.luther.edu/~macdonal/index.html#GA&GC (last checked February 2014)
13. Perwass, C.: Geometric Algebra with Applications in Engineering. Springer (2009)
14. Reveillès, J.-P.: Géométrie Discrète, calculs en nombres entiers et algorithmique. Thèse d'état, Université Louis Pasteur, Strasbourg, France (1991)
15. Sommer, G.: Geometric computing with Clifford algebras: theoretical foundations and applications in computer vision and robotics. Springer (2001)
16. Tajine, M., Ronse, C.: Hausdorff discretizations of algebraic sets and diophantine sets. In: Nyström, I., Sanniti di Baja, G., Borgefors, G. (eds.) DGCI 2000. LNCS, vol. 1953, pp. 99–110. Springer, Heidelberg (2000)
17. Wareham, R., Cameron, J., Lasenby, J.: Applications of conformal geometric algebra in computer vision and graphics. In: Li, H., J. Olver, P., Sommer, G. (eds.) IWMM-GIAE 2004. LNCS, vol. 3519, pp. 329–349. Springer, Heidelberg (2005)

Segmentation of 3D Articulated Components by Slice-Based Vertex-Weighted Reeb Graph

Nilanjana Karmakar[1], Partha Bhowmick[2], and Arindam Biswas[1]

[1] Department of Information Technology,
Indian Institute of Engineering Science and Technology, Shibpur, India
{nilanjana.nk,barindam}@gmail.com
[2] Department of Computer Science and Engineering,
Indian Institute of Technology, Kharagpur, India
bhowmick@gmail.com

Abstract. A fast and efficient algorithm for segmentation of the articulated components of 3D objects is proposed. The algorithm is marked by several novel features, such as DCEL-based fast *orthogonal slicing*, *weighted Reeb graph* with slice areas as vertex weights, and graph cut by *exponential averaging*. Each of the three sets of orthogonal slices obtained from the object is represented by a vertex-weighted Reeb graph of low complexity, as the slicing is done with an appropriate grid resolution. Each linear subgraph in a Reeb graph is traversed from its leaf node up to an articulation node or up to a node whose weight exceeds a dynamically-set threshold, based on exponential averaging of the predecessor weights in the concerned subgraph. The nodes visited in each linear subgraph are marked by a unique component number, thereby helping the inverse mapping for marking the articulated regions during final segmentation. Theoretical analysis shows that the algorithm runs faster for objects with smaller surface area and for larger grid resolutions. The algorithm is stable, invariant to rotation, and leads to natural segmentation, as evidenced by experimentation with a varied dataset.

Keywords: 3D segmentation, DCEL, orthogonal slicing, orthogonal polyhedron, Reeb graph.

1 Introduction

Segmentation of 3D triangulated meshes has an authoritative impact on shape-analytic applications. Hence, 3D segmentation has been attempted over and again in a wide range of ways. For example, segmentation by multi-dimensional scaling and feature points is proposed in [13]. The techniques in [6,10] are based on shape diameter function, skeletons, and randomized cuts. Diffusion distance metric and a variety of medial structures are used in [9]. The works in [13,10] provide segmentation along the natural seams of an object, whereas the work in [6] concentrates on object volume, object posture, and topology. The notion of topological maps and region adjacency graphs has been proposed in [7]. Other

E. Barcucci et al. (Eds.): DGCI 2014, LNCS 8668, pp. 370–383, 2014.

Table 1. Some of the existing algorithms versus the proposed algorithm

Algorithm	Features
Nurre, 1997 [15]	Input: 3D point cloud, organized into slices for segmentation, works only on the object human body
Ju et al., 2000 [11]	Input: 3D point cloud, curvature analysis of profiles, works only on the object human body
Y. Xiao et al., 2003 [17]	Input: 3D point cloud, Reeb graph, works only on the object human body
Katz et al., 2005 [13]	Hierarchical, pose-invariant, feature-based, core extraction by spherical mirroring, natural seam segmentation
Golovinskiy-Funkhouser, 2008 [10]	Hierarchical, pose-invariant, randomized cuts, partition function, natural seam segmentation
Shapira et al., 2008 [6]	Hierarchical, pose-invariant, skeleton extraction by shape diameter function, focuses on invariance of object volume
Goes et al., 2008 [9]	Hierarchical, pose-invariant, diffusion distance metric, medial structure
Proposed algorithm	Orthogonal slicing, rotation-invariant, based on Reeb graph analysis, speedy execution, natural segmentation.

related techniques can be seen in [2,5,8,19]. Table 1 provides a brief comparison of the existing algorithms.

Irrespective of the adopted technique, there always lies a chance of over-segmentation [4,14], which, at times, affects the accuracy of segmentation, and sometimes may also assist in deriving important information about the object [18]. With an objective of obtaining natural segmentation, which is free of skeletonization, and hence fast, robust, and rotationally invariant, we propose here a novel algorithm, which is based on the idea of orthogonal slicing [12], followed by Reeb graph analysis of the slice sets. The only parameter of orthogonal slicing is the grid resolution, g, which is shown to have insignificant impact on the output quality. The proposed segmentation algorithm may be useful for analysis of geometric structures and peripheral topology of objects, which, in turn, may be quite effective for shape matching, collision detection, texture mapping, etc.

Graph-theoretic analysis of 3D objects leads to a stable and dependable performance for 3D shape analysis. Reeb graph, in particular, provides a natural representation that is suitable for object surface representation due to its one-dimensional graph structure and invariance to both global and local transformations [16,1,3]. Given an unorganized cloud of 3D points, identification of human body parts has been attempted in several works. In [15], the data set is organized into a stack of slices so that a specific set of data points belong to a specific part of the object according to its topology. This approach is further improved by curvature analysis [11]. A further modification using Reeb graph is proposed in [17]. However, there is a difference between the work in [17] and the one proposed by us. In [17], a 3D point cloud is used as input, which has the disadvantage of high space complexity; our algorithm, on the contrary, works with triangulated mesh as input. Secondly, the algorithm in [17] is effective only for objects like human body, whereas our algorithm is applicable to a wide range of objects including human body.

2   Preprocessing and Orthogonal Slicing

In our work, a discrete 3D object A is considered as a triangulated mesh in which the triangles truly capture the peripheral topology of the object, i.e., on each edge of the triangulated surface, exactly two triangles are incident. As a preprocessing step, a 3D grid $\mathbb{G}$ of size (resolution) g is imposed on A, and three sets of orthogonal slices (henceforth mentioned as *slices* or *slice polygons*) are obtained by slicing the object along the three coordinate planes. These three slice sets are computed using a combinatorial technique similar to the one proposed in [12]. Each set of slices is processed to identify the subset of slices belonging to each articulated portion and central region of the object. By articulated portion, we mean the meaningful components of the object [1,9,3]. Finally, for each articulated portion, the three subsets of slices are combined to identify the object segment.

The grid $\mathbb{G}$ is represented as a set of *unit grid cubes* (UGCs), each of size g. If a triangle $T_{abc}(v_a, v_b, v_c)$ intersects or lies inside a UGC U_k, then U_k is considered as *object-occupied*. U_k is intersected by T_{abc} if at least one UGC-face $f_k \in U_k$ is intersected by T_{abc}. The projections of f_k and T_{abc} are considered along yz-, zx-, and xy-planes to find the intersection based on the nature of projection of f_k (line segment or square) and that of T_{abc} (line segment or triangle).

Once the object-occupied UGCs are identified, the set of orthogonal slices parallel to each of the yz-, zx-, and xy-planes are determined. The condition of *object occupancy*, as stated in [12], has been modified for this. In [12], a UGC-face is considered as object-occupied if one or both of its adjacent UGCs contains object *voxels*. But in the current work, a UGC-face is considered as object-occupied if exactly one of its adjacent UGCs is intersected by a triangle. The slice polygons are stored in an adjacency list containing n lists, one for each slicing plane, n being the number of slicing planes. Each list contains a sequence of vertices for all the slice polygons on that slicing plane. Apart from this, each slice polygon is stored in two lexicographically sorted lists, L_{xy} and L_{yx} (or L_{yz} and L_{zy}, or L_{zx} and L_{xz}), sorted according to x- and y-(or y- and z-, or z- and x-)coordinates, respectively.

3   Reeb Graph Construction

W.l.o.g., consider the set of slice polygons S_y formed by a set of slicing planes $\Pi = \{\Pi_1, \Pi_2, ..., \Pi_n\}$, parallel to the zx-plane. Note that one or more slice polygons may lie on each slicing plane Π_i. If two slice polygons P' and P'' lie on two consecutive planes, Π_i and Π_{i+1}, so that they have a non-empty intersection when one is projected on the other, then P' and P'' are called *consecutive slices*. If P' lies on Π_i such that exactly one consecutive slice polygon exists either on Π_{i+1} or on Π_{i-1}, then P' is said to be a *leaf slice*.

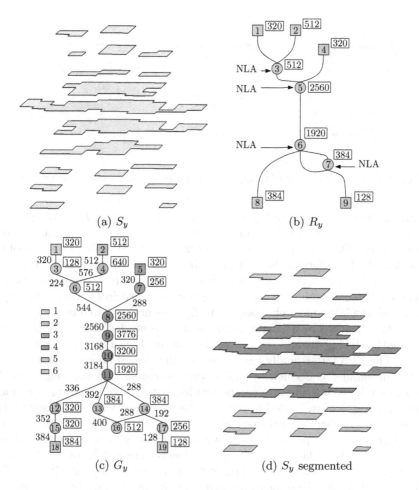

(a) S_y

(b) R_y

(c) G_y

(d) S_y segmented

Fig. 1. Segmentation of slices in S_y. In R_y and G_y, node weights are shown. G_y is segmented with difference in area threshold—shown beside each edge—computed at every slice as exponential average of the previous slice areas.

The set S_y is represented by a (weighted) *slice graph* G_y. Each node of G_y corresponds to a slice polygon of S_y; two nodes have an edge if their corresponding slices are consecutive. The area of a slice is assigned as the weight of its corresponding node. This graph results to *weighted Reeb graph*. As a Reeb graph provides a topological signature of an object, we use it for identification of object articulations based on the peripheral topology captured through three orthogonal sets of slices. A Reeb graph, R_y, corresponding to S_y, is shown in Fig. 1(b). It consists of only leaf nodes and *non-linear articulation nodes* (NLA, of degree ≥ 2). Observe that a critical point of the object, at which the surface topology changes, corresponds to an NLA. The slice graph G_y, on the contrary, contains linear substructures in addition to leaf nodes and NLAs, as illustrated in Fig. 1(c).

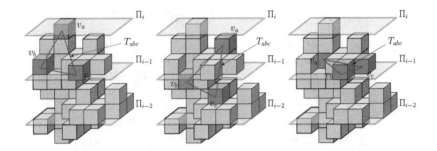

Fig. 2. Different positions of triangles on the object surface w.r.t. slicing planes

During the construction of G_y from S_y, a list of leaf nodes, L_y, is maintained. A leaf node ν_0 is enqueued in a queue Q and BFS starts from ν_0 in G_y. While dequeuing ν_0, the difference of its weight with a dynamically-set threshold τ is checked. If the difference lies within λ times the weight of the last-dequeued node, where λ varies between 0 and 1, then ν_0 is marked by a *component id*, and all the unvisited neighboring nodes of ν_0 are enqueued in Q. Otherwise, ν_0 is considered as a leaf node and appended to L_y. The nodes visited by a single BFS traversal are marked by a unique component id. Next, BFS traversal is started from another leaf node, and the process is continued until all the leaf nodes are visited. The BFS forest finally results to identification of all components.

The threshold τ is decided by *exponential averaging* with the weights of the previous nodes having the same component id. As the traversal starts from a leaf node ν_0, the area of ν_0 is used to initialize the threshold value τ_0. Based on the area of ν_i, at each subsequent node ν_{i+1}, its area is compared with τ_{i+1}, which is computed as $\tau_{i+1} = \rho w_i + (1 - \rho)\tau_i$, where, τ_i denotes the exponential average of the weights of the last i nodes, and w_i denotes the weight of ν_i. The value of ρ can range between 0 and 1; however, we conventionally take it as $\frac{1}{2}$. The node ν_{i+1} belongs to the same component as ν_i if $\mid w_{i+1} - \tau_{i+1} \mid < \lambda.w_i$, where λ ranges between 0 and 1. The use of exponential averaging for dynamically setting the threshold value ensures natural articulation for widely varying component dimensions. Figure 1 shows a demonstration. The threshold values are computed in the direction the algorithm proceeds; the leaf slices are considered in the order $1, 2, 5, 18, 19$, and finally node 8 for the central region.

4   Segmentation by Weighted Reeb Graph

The set of triangles is represented by a doubly connected edge list (DCEL) [12]. For each vertex of a triangle, the neighboring vertices are obtained in constant time, using DCEL. Depending on the positions of its three vertices, a triangle T_{abc} may be intercepted by or may lie on Π_{i-1} (Fig. 2). Hence, a vertex v_a lies on Π_i or between Π_i and Π_{i+1}. Since the set of slices parallel to a coordinate plane are stored as lexicographically sorted lists L_{xy} and L_{yx} (Sec. 2), the coordinate values for the vertex v_a are looked up in L_{xy} and L_{yx} to find the slice polygon

Algorithm 3DSEGMENT$(A, \mathbb{G})$	**Procedure** VERTEX-SEGMENT(V)
01. **for** each coordinate plane t	01. count $\leftarrow 1$
02. $\quad G_t \leftarrow$ graph representing S_t	02. **for** each vertex $v \in V$
03. $\quad L_t \leftarrow$ leaf nodes	03. $\quad$ **if** $visited[v] = 0$
04. $\quad$ GRAPH-SEGMENT(G_t, L_t)	04. $\quad\quad segid[v] \leftarrow -1$
05. $\quad V_t \leftarrow$ triangle vertices	05. $\quad\quad$ **for** each neighbor u of v
$\quad\quad\quad$ with component ids	06. $\quad\quad\quad$ **if** $visited[u] = 1$
06. $\quad$ VERTEX-SEGMENT(V_t)	07. $\quad\quad\quad\quad$ **if**(COMPARE(u, v))
	08. $\quad\quad\quad\quad\quad segid[v] \leftarrow segid[u]$
	09. $\quad\quad$ **if** $segid[v] = -1$
	10. $\quad\quad\quad segid[v] \leftarrow$ count
	11. $\quad\quad\quad$ count $\leftarrow$ count $+ 1$
	12. $\quad\quad visited[v] \leftarrow 1$

Procedure
GRAPH-SEGMENT(G_t, L_t)

01. count $\leftarrow 1$
02. **for** each leaf node $\nu \in L_t$
03. $\quad$ **if** $visited[\nu] = 0$
04. $\quad\quad$ ENQUEUE(Q, ν)
05. $\quad\quad \nu_1 \leftarrow \nu$
06. $\quad\quad$ **while** Q is not empty
07. $\quad\quad\quad \nu \leftarrow$ DEQUEUE(Q)
08. $\quad\quad\quad$ **if** $| \, w[\nu] - \tau_w \, | < \lambda.w[\nu_1]$
09. $\quad\quad\quad\quad compid[\nu] \leftarrow$ count
10. $\quad\quad\quad\quad$ **for** each neighbor ν' of ν
11. $\quad\quad\quad\quad\quad$ **if** $visited[\nu] = 0$
12. $\quad\quad\quad\quad\quad\quad$ ENQUEUE(Q, ν')
13. $\quad\quad\quad\quad\quad\quad visited[\nu] \leftarrow 1$
14. $\quad\quad\quad\quad\quad\quad \nu_1 \leftarrow \nu$
15. $\quad\quad\quad$ **else**
16. $\quad\quad\quad\quad L_t \leftarrow L_t \cup \{\nu'\}$
17. $\quad\quad\quad$ count $\leftarrow$ count $+ 1$

Procedure COMPARE(u, v)

01. **if** $(((c_x^v = c_x^u) \wedge (c_y^v = c_y^u) \wedge (c_z^v = c_z^u))$
$\quad\quad \vee \, (((c_x^v = c_x^u) \wedge (c_y^v = c_y^u) \wedge (c_z^v \neq c_z^u))$
$\quad\quad \wedge \, ((c_x^v \neq -1) \vee (c_y^v \neq -1)))$
$\quad\quad \vee \, (((c_x^v = c_x^u) \wedge (c_y^v \neq c_y^u) \wedge (c_z^v \neq c_z^u))$
$\quad\quad \wedge \, ((c_x^v \neq -1) \wedge (c_y^v = c_z^v = -1))))$
02. $\quad$ **return** 1
03. **else if** $(((c_x^v \neq c_x^u) \wedge (c_y^v \neq c_y^u) \wedge (c_z^v \neq c_z^u))$
$\quad\quad \vee \, (((c_x^v = c_x^u) \wedge (c_y^v = c_y^u) \wedge (c_z^v \neq c_z^u))$
$\quad\quad \wedge \, (c_x^v = c_y^v = -1))$
$\quad\quad \vee \, (((c_x^v = c_x^u) \wedge (c_y^v \neq c_y^u) \wedge (c_z^v \neq c_z^u))$
$\quad\quad \wedge \, ((c_x^v = -1) \wedge (!(c_y^v = c_z^v = -1)))))$
04. $\quad$ **return** 0
05. **return** -1

Fig. 3. The algorithm for segmentation and its related procedures

S to which v_a belongs. Consequently, v_a is assigned the same component id as S. This process is repeated for the vertices of all the triangles, and thus, each such vertex obtains a 3-tuple of component ids corresponding to S_x, S_y, S_z.

Let v be a vertex belonging to triangle T_{abc}. 3-tuples of component ids are assigned to v and all its adjacent vertices in the process outlined above. In order to assign segment id to v, its 3-tuple of component ids, namely (c_x^v, c_y^v, c_z^v), is compared with those of all its neighboring vertices, using DCEL. Then, based on certain combinatorial rules (R1–R4), segment ids are assigned. A segment id signifies the combined id obtained from the 3-tuple of component ids. The procedure COMPARE in Fig. 3 enumerates the combinatorial possibilities. Note that, when the algorithm 3DSEGMENT terminates, we have as many segments as the number of distinct segment ids. It accepts the 3D discrete object A in the form of a triangulated mesh as input and uses the following procedures:

GRAPH-SEGMENT segments the Reeb graph, thereby identifying the slices that belong to each articulated portion or central region along each coordinate plane.

VERTEX-SEGMENT combines the slices belonging to different components along the three coordinate planes, in order to find the final segments of the object.

PROCEDURE COMPARE uses the rules for comparing the component ids of two vertices to decide whether they belong to the same segment or not.

Let (c_x^v, c_y^v, c_z^v) and (c_x^u, c_y^u, c_z^u) denote the 3-tuples of component ids for vertices v and u, and their final segment ids be s_u and s_v, respectively.

R1 If $c_x^v = c_x^u$, $c_y^v = c_y^u$, and $c_z^v = c_z^u$, then s_u is assigned to s_v.

R2A If $c_x^v = c_x^u$, $c_y^v = c_y^u$, and $c_z^v \neq c_z^u$ where $\exists\, c \in \{c_x^v, c_y^v\}$ such that $c \neq -1$, then s_u is assigned to s_v.

R2B If $c_x^v = c_x^u$, $c_y^v = c_y^u$, and $c_z^v \neq c_z^u$ where $\forall\, c \in \{c_x^v, c_y^v\}$ such that $c = -1$, then a new value is assigned to s_v.

R3A If $c_x^v = c_x^u$, $c_y^v \neq c_y^u$, and $c_z^v \neq c_z^u$ such that $c_x^v \neq -1$ and $c_y^v = c_z^v = -1$, then s_u is assigned to s_v.

R3B If $c_x^v = c_x^u$, $c_y^v \neq c_y^u$, and $c_z^v \neq c_z^u$ such that $c_x^v = -1$ and not both of c_y^v and c_z^v are equal to -1, then a new value is assigned to s_v.

R4 If $c_x^v \neq c_x^u$, $c_y^v \neq c_y^u$, and $c_z^v \neq c_z^u$, then a new value is assigned to s_v.

Note that c_x^v, c_y^v, and c_z^v may be used interchangeably in the rules R2A, R2B, R3A, and R3B, where their comparisons with c_x^u, c_y^u, and c_z^u are considered. Vertices belonging to same (or different) components along all the three coordinate planes, belong to the same (or different) segments (R1 and R4). Along each coordinate plane, the central region (indicated by -1) is adjacent to all other components but neither of the components are adjacent to each other. Hence, the rules R2 and R3 are primarily based on whether v or u lie in the central region along any coordinate plane.

4.1   Time Complexity

Let n be the number of UGCs intersected by the surface of the object A, α be the surface area of A, β its volume, and g be the grid resolution. Then, number of UGC-faces on the object surface is $n_f = O(n)$. If A has a sufficiently large area-to-volume ratio, then n UGCs would cover its volume, i.e, $\beta < ng^3$. For a sufficiently small area-to-volume ratio, $\alpha > ng^2$. So, in general, $\beta/g^3 < n < \alpha/g^2$, or, $n_f = O(\alpha/g^2)$.

The number of slices in S_z is $|S_z| = O(n_f)$. Construction of R_z requires identification of consecutive slices from the sorted lists L_{xy} and L_{yx}. Each slice polygon P is traversed exactly once, which needs $O(\log |S_z|) = O(\log n_f)$ time for searching its first vertex in L_{xy} and L_{yx}. Traversal time of P is linear in its perimeter, as DCEL is used. So, Reeb graph construction time is $O(n_f \log n_f)$, since total traversal time is upper-bounded by $O(n_f)$.

The procedure GRAPH-SEGMENT is called thrice, hence requiring $O(n_f)$ time. The procedure VERTEX-SEGMENT executes on the UGCs covering the object surface, hence requiring $O(n)$ time, as Procedure COMPARE executes in constant time. So, the total time complexity is $O(n_f \log n_f) = O(\alpha/g^2 \log(\alpha/g^2))$.

Fig. 4. Results on Dog ($g = 2$). From top-left to bottom-right: Slices parallel to yz-, zx-, xy-planes (with segments marked), and final segmentation

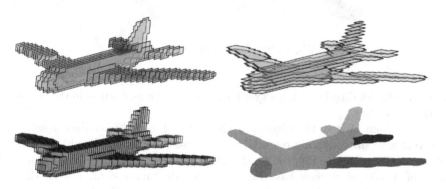

Fig. 5. Results on Airplane ($g = 2$)

5   Results and Conclusion

The proposed algorithm has been implemented in C in Linux Fedora Release 7, Dual Intel Xeon Processor 2.8 GHz, 800 MHz FSB. As found from experimental results (Figs. 4–8), the algorithm can successfully separate out the limb-like articulated portions of the objects from its central region up to a high degree of accuracy. It may be noticed from these results that the final segmentation is a correctly-synthesized output from the three sets of orthogonal slices. The

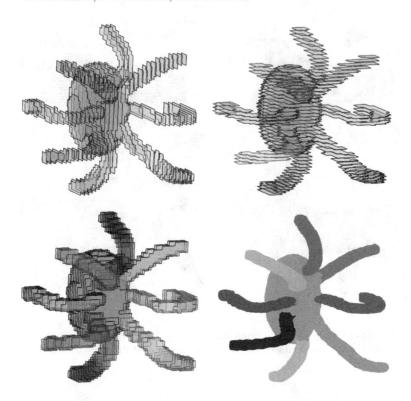

Fig. 6. Results on Octopus ($g = 2$)

dynamically set thresholding for graph cut adds to the robustness in the behavior of the algorithm.

For a given object, the algorithm gives almost equally accurate results for different grid sizes within a reasonable range, and even when the object is tilted at an arbitrary angle (Fig. 8), which demonstrates its stability. An articulated part of a 3D object may have an arbitrary orientation w.r.t. the coordinate system, and hence such a part is identified by correlating its three sets of slices taken along the three coordinate planes. In particular, when the concerned part goes on arbitrarily changing in orientation, no less than three sets of slices can fully capture its shape. This is confirmed by the segmentation results presented in Figs. 4 to 6, and particularly through the results in Figs. 7 and 8. Some more results and statistical data related to our experimentation are given in Appendix. The standard deviation of the segment areas for 'Chair' rotated from 10° through 90° demonstrates the robustness of the algorithm w.r.t. rotation. The accuracy of segmentation with varying threshold values is also shown there for 'Dog' and 'Airplane'.

The CPU time of segmentation increases with increase in the surface area of the object and also with the total number of slices along the three coordinate planes; and with grid size increase, the CPU time falls more than quadratically,

Fig. 7. Results on Chair (top two rows: $g = 2$, bottom two rows: $g = 4$)

Table 2. Statistical results and CPU times for segmentation of some digital objects

Object	Object size		# Slices			CPU time (in secs.)		
	# Vertices	# Faces	yz-plane	zx-plane	xy-plane	Slicing	Segmentation	Total
$g = 2$								
Dog	14862	301435	30	105	82	0.036	0.126	0.162
Octopus	16253	346467	95	98	156	0.062	0.405	0.467
Human	7093	170547	75	93	18	0.020	0.100	0.120
Chair	12787	228913	59	135	46	0.045	0.132	0.177
$g = 4$								
Dog	14862	301435	16	53	39	0.007	0.026	0.033
Octopus	16253	346467	48	46	70	0.011	0.055	0.066
Human	7093	170547	37	45	8	0.005	0.019	0.024
Chair	12787	228913	29	66	22	0.008	0.024	0.032
$g = 6$								
Dog	14862	301435	10	34	26	0.004	0.014	0.018
Octopus	16253	346467	28	27	45	0.005	0.018	0.023
Human	7093	170547	25	30	5	0.003	0.011	0.014
Chair	12787	228913	21	44	15	0.004	0.010	0.014

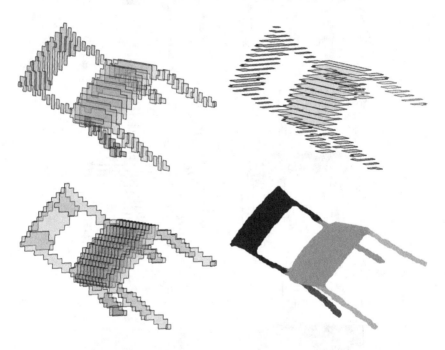

Fig. 8. Results on Chair, tilted ($g = 4$), coloring as in Fig. 4

as evident from Table 2. This conforms to our theoretical analysis on time complexity (Sec. 4.1). For instance, the surface areas for 'Human', 'Dog', and 'Chair' are 193, 337, and 390 units. The area ratios are, therefore, Human:Dog = 0.6, Human:Chair = 0.5, Chair:Dog = 1.18; the corresponding CPU time ratios for $g = 2$ are 0.75, 0.67, 1.09.

Acknowledgement. A part of this research is funded by CSIR, Govt. of India.

References

1. Agathos, A., et al.: 3D articulated object retrieval using a graph-based representation. Vis. Comput. 26(10), 1301–1319 (2010)
2. Attene, M., et al.: Mesh segmentation—A comparative study. In: Proc. SMI 2006, pp. 7–18 (2006)
3. Berretti, S., et al.: 3D Mesh Decomposition using Reeb Graphs. Image Vision Comput. 27(10), 1540–1554 (2009)
4. Chazelle, B., et al.: Strategies for Polyhedral Surface Decomposition: An Experimental Study. Comput. Geom. Theory Appl. 7, 327–342 (1997)
5. Chen, X., et al.: A Benchmark for 3D Mesh Segmentation. ACM Trans. Graph. 28(3), 73:1–73:12 (2009)
6. Cohen-Or, D., et al.: Consistent Mesh Partitioning and Skeletonization using the Shape Diameter Function. The Visual Computer 24, 249–259 (2008)
7. Dupas, A., Damiand, G.: First Results for 3D Image Segmentation with Topological Map. In: Coeurjolly, D., Sivignon, I., Tougne, L., Dupont, F. (eds.) DGCI 2008. LNCS, vol. 4992, pp. 507–518. Springer, Heidelberg (2008)
8. Dupas, A., Damiand, G., Lachaud, J.-O.: Multi-Label Simple Points Definition for 3D Images Digital Deformable Model. In: Brlek, S., Reutenauer, C., Provençal, X. (eds.) DGCI 2009. LNCS, vol. 5810, pp. 156–167. Springer, Heidelberg (2009)
9. de Goes, F., et al.: A Hierarchical Segmentation of Articulated Bodies. In: Proc. SGP 2008, pp. 1349–1356 (2008)
10. Golovinskiy, A., et al.: Randomized cuts for 3D mesh analysis. ACM Transactions on Graphics (Proc. SIGGRAPH ASIA) 27(145) (2008)
11. Ju, X., et al.: Automatic Segmentation of 3D Human Body Scans. In: Proc. Int. Conf. Comp. Graphics & Img., pp. 239–244 (2000)
12. Karmakar, N., Biswas, A., Bhowmick, P.: Fast slicing of orthogonal covers using DCEL. In: Barneva, R.P., Brimkov, V.E., Aggarwal, J.K. (eds.) IWCIA 2012. LNCS, vol. 7655, pp. 16–30. Springer, Heidelberg (2012)
13. Katz, S., et al.: Mesh segmentation using feature point and core extraction. The Visual Computer 21, 649–658 (2005)
14. Mangan, A., et al.: Partitioning 3D surface meshes using watershed segmentation. IEEE TVCG 5, 308–321 (1999)
15. Nurre, J.H.: Locating Landmarks on Human Body Scan Data. In: Proc. 3DIM, pp. 289–295 (1997)
16. Werghi, N.: A robust approach for constructing a graph representation of articulated and tubular-like objects from 3D scattered data. PRL 27(6), 643–651 (2006)
17. Xiao, Y., et al.: A Discrete Reeb Graph approach for the Segmentation of Human Body Scans. In: Proc. 3DIM, pp. 378–385 (2003)
18. Zhang, X., et al.: Salient Object Detection through Over-Segmentation. In: Proc. ICME 2012, pp. 1033–1038 (2012)
19. Zhu, N., et al.: Graph-Based Optimization with Tubularity Markov Tree for 3D Vessel Segmentation. In: Proc. CVPR 2013, pp. 2219–2226 (2013)

Appendix

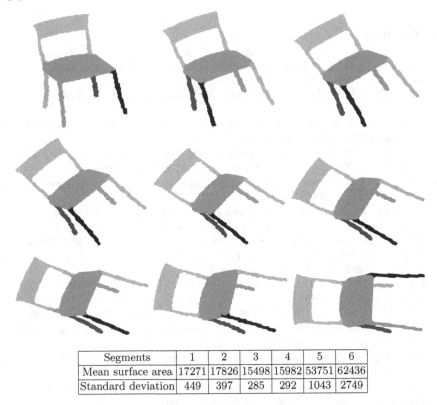

Segments	1	2	3	4	5	6
Mean surface area	17271	17826	15498	15982	53751	62436
Standard deviation	449	397	285	292	1043	2749

Fig. 9. Segmentation results on Chair rotated from $10°$, $20°$, ..., $90°$ demonstrating rotational invariance. Segments are colored in their order of detection. Segment 1: Left fore leg, 2: Right fore leg, 3: Left hind leg, 4: Right hind leg, 5: Back rest, and 6: Seat.

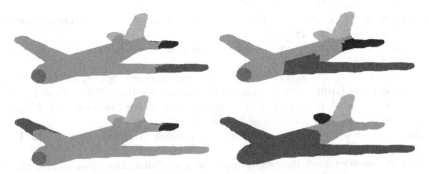

Fig. 10. Segmentation results on Airplane ($g = 2$) for threshold varying with $\rho = 0.4$, 0.5, 0.6, 0.7 (from top-left to bottom-right)

Fig. 11. Segmented articulated portions and central region of Bird, Dolphin, Teddy, Dinosaur, Ant, and Human, at $g = 2$

Fig. 12. Segmentation results on Dog ($g = 2$) for threshold varying with $\rho = 0.3, 0.4, 0.5, 0.6$ (from top-left to bottom-right)

Taylor Optimal Kernel for Derivative Etimation

Henri-Alex Esbelin and Remy Malgouyres

Clermont Universités, CNRS UMR 6158, LIMOS, Clermont-Ferrand, France
alex.esbelin@univ-bpclermont.fr, remy.malgouyres@udamail.fr

Abstract. In many geometry processing applications, the estimation of differential geometric quantities such as curvature or normal vector field is an essential step. In this paper, we investigate new estimators for the first and second order derivatives of a real continuous function f based on convolution of the values of noisy digitalizations of f. More precisely, we provide both proofs of multigrid convergence of the estimators (with a maximal error $O\left(h^{1-\frac{k}{2n}}\right)$ in the unnoisy case, where $k = 1$ for first order and $k = 2$ for second order derivatives and n is a parameter to be choosed *ad libitum*). Then, we use this derivative estimators to provide estimators for normal vectors and curvatures of a planar curve, and give some experimental evidence of the practical usefullness of all these estimators. Notice that these estimators have a better complexity than the ones of the same type previously introduced (cf. [4] and [8]).

Keywords: derivative estimation, curvature estimation, discrete derivation, convolution.

1 Introduction

In the framework of shape analysis, a common problem is to estimate derivatives of functions, or normals and curvatures of curves and surfaces, when only some (possibly noisy) sampling of the function or curve is available. This problem has been investigated through finite difference methods, scale-space methods, and discrete geometry, etc. ... For detailed informations about the state of the art, the reader is refered to [1], [5] and [7].

This paper focuses on estimating the derivatives on the boundary of digital planar shapes. Suppose that the digital data is distributed around the true sample of the Euclidean shape according to some noise. The curvature estimation is provided to be as close as possible to the curvature of the underlying Euclidean shape before digitization. More precisely, provided some formal models of the noise, the quality of the estimation should be improved as the digitization step gets finer and finer. This property is called the multigrid convergence(see [3], [2], [6] and [10]).

Our objective is to design estimators of successive derivatives for digital data which are provably multigrid convergent, accurate in practice, computable in an exact manner, robust to perturbations.

The first section provides definitions for first and second order discrete derivatives of digital curves. These discrete derivatives are proved to provide multigrid

E. Barcucci et al. (Eds.): DGCI 2014, LNCS 8668, pp. 384–395, 2014.

convergent estimators for the corresponding continuous derivatives, even in a noisy case.

The second section explains how to use these discrete differential notions to estimate continuous normal field and curvature.

The third section gives some experimental evidence to the multigrid convergence of all these estimators.

2   Discrete Derivatives for Discrete Functions

Throughout this paper, we call *discrete function* a function from $\mathbb{Z}$ to $\mathbb{Z}$ or $\mathbb{Z}^2$.. Let $f : \mathbb{R} \mapsto \mathbb{R}$ be a continous function. We say that the discrete function $\Gamma : \mathbb{Z} \mapsto \mathbb{Z}$ is a discretization of f on the interval $[a; b]$ for the discretization step $h > 0$ with error ϵ_h if, for any integer i such that $a \leq ih \leq b$, we have $h\Gamma(i) = f(ih) + \epsilon_h(i)$. We consider here the uniform noise case: $0 \leq \|\epsilon_h\|_\infty \leq Kh^\alpha$ with $0 < \alpha \leq 1$ and K a positive constant. Note that the rounded case and the floor case are particular cases of uniform noise with $\alpha = 1$.

2.1   First Order Derivatives

Definition 1. *Let m be a positive integer. The Taylor-optimal mask of size m for discrete first order derivation (TO_m mask for short) is the finite sequence of rational numbers $\mathbf{a_m}$ defined by $a_{m,0} = 0$ and for $-m \leq i < 0$ or $0 < i \leq m$ by*

$$a_{m,i} = \frac{(-1)^i}{i} \frac{\binom{m}{i}}{\binom{m+i}{i}} = \frac{(-1)^i}{i} \frac{\binom{2n}{n+i}}{\binom{2m}{m}}. \tag{1}$$

Definition 2. *Let $\Gamma : \mathbb{Z} \mapsto \mathbb{Z}$ be a discrete function. Let $p \in \mathbb{N}^*$ and $i_0 \in \mathbb{Z}$. The m-Taylor-optimal first order discrete derivative of Γ at point i_0 with step p is*

$$(\Delta_{m,p}\Gamma)(i_0) = \frac{1}{p} \sum_{i=-m}^{i=m} a_{m,i}\Gamma(i_0 - ip). \tag{2}$$

In order to show that the discrete derivative of a discretized function provides an estimate for the continuous derivative of the real function, we would like to evaluate the difference between $(\Delta_{m,p}\Gamma)(i_0)$ and $f'(i_0h)$.

Lemma 1. $\displaystyle\sum_{i=-m}^{i=m} ia_{m,i} = -1$ *and for $k = 0$ or $2 \leq k \leq 2m$, $\sum_{i=-m}^{i=m} i^k a_{m,i} = 0$.*

Proof. Let us first notice that the lemma is trivial for the even values of k because of the equality $a_{m,-i} = -a_{m,i}$.

Let us now consider the odd case. Let f be a polynomial and define the classical finite difference operator δ_+ by $\delta_+(f)(x) = f(x+1) - f(x)$. Iterating this operator leads to the well known equality:

$$\delta_+^{2m}(f)(x) = \sum_{i=0}^{2m} \binom{2m}{i}(-1)^{2m-i} f(x+i). \tag{3}$$

Notice that applying the operator δ_+ to a polynomial of degree d leads to a polynomial of degree $d - 1$. Applying equality 3 to the polynomials defined by $(x - n)^{k-1}$ for $1 \le k \le 2m$, we get

$$0 = \sum_{i=0}^{2m} \binom{2m}{i} (-1)^{2m-i} (i + x - m)^{k-1}. \tag{4}$$

Let now $x = 0$ in equality 4, then for all $1 \le k \le 2m$ we have

$$0 = \sum_{i=0}^{2m} \binom{2m}{i} (-1)^{2m-i} (i - m)^{k-1} = (-1)^m \sum_{i=-m}^{m} \binom{2m}{m+i} (-1)^i i^{k-1}.$$

Hence, for all odd k such that $1 \le k \le 2m$, we have

$$\sum_{i=-m}^{i=m} i^k a_{m,i} = \frac{1}{\binom{2m}{m}} \left(\sum_{i=-m, i \ne 0}^{i=m} i^{k-1} (-1)^i \binom{2m}{m+i} \right) = \begin{cases} 0 & \text{if } k > 1 \\ -1 & \text{if } k = 1 \end{cases}.$$

Theorem 1. *Let $k \in \mathbb{N}$ with $k \ge 2$ and let f be C^k on $\mathbb{R}$ and let $k_0 = Min\{k, 2m + 1\}$. Let $p = \lfloor h^{-1+\frac{\alpha}{k_0}} \rfloor$. Then*

$$|(\Delta_{m,p}\Gamma)(i_0) - f'(i_0 h)| = O\left(h^{\alpha - \frac{\alpha}{k_0}}\right). \tag{5}$$

Proof. The difference between the discrete derivative of Γ and the continuous derivative of f may be obviously seen as the sum of $EM(f, \Gamma, m, p, i_0)$ called method's error and $ED(f, \Gamma, m, p, i_0)$ called discretization's error respectively defined as

$$EM(f, \Gamma, m, p, i_0) = \left(\frac{1}{ph} \sum_{i=-m}^{i=m} a_{m,i} f((i_0 - ip)h) \right) - f'(i_0 h)$$

and

$$ED(f, \Gamma, m, p, i_0) = \frac{1}{ph} \sum_{i=-m}^{i=m} a_{m,i} \epsilon(i_0 - ip).$$

We intend to bound both errors. The choice of p will appear to equalize the speeds of growth of both bounds. From Taylor formula and lemma 1, we first majorize the method's error. Let us denote $\sum_{i=-m}^{i=m} a_{m,i} f((i_0 - ip)h))$ by S. We have $S = \sum_{i=-m}^{i=m} a_{m,i} \left(\sum_{j=0}^{j=k_0-1} \frac{f^{(j)}(i_0 h)}{j!} (-iph)^j + \frac{f^{(k_0)}(x_{i,j})}{k_0!} (-iph)^{k_0} \right)$ for some $x_{i,j}$ between $i_0 h$ and $(i_0 - ip)h$. Hence

$$S = \sum_{j=0}^{j=k_0-1} \frac{f^{(j)}(i_0 h)}{j!} (-ph)^j \left(\sum_{i=-m}^{i=m} a_{m,i} i^j \right) + \frac{(-ph)^{k_0}}{k_0!} \left(\sum_{i=-m}^{i=m} a_{m,i} i^{k_0} f^{(k_0)}(x_{i,j}) \right)$$

and from lemma 1, we have $S = f'(i_0 h) + \frac{(-ph)^k}{k!} \left(\sum_{i=-m}^{i=m} a_{m,i} i^{k_0} f^{(k_0)} (x_{i,j}) \right)$, which leads to the following majoration:

$$|EM(f, \Gamma, \mathbf{a}, p, i_0)| \leq \frac{\|f^{(k_0)}\|_\infty}{k_0!} \left(\sum_{i=-m}^{i=m} |i^{k_0} a_{m,i}| \right) h^{k_0 - 1} p^{k_0 - 1}$$

$$|EM(f, \Gamma, \mathbf{a}, p, i_0)| \leq \frac{\|f^{(k_0)}\|_\infty}{k_0!} \left(\sum_{i=-m}^{i=m} |i^{k_0} a_{m,i}| \right) h^{\alpha - \frac{\alpha}{k_0}}$$

Now we staightforwardly majorize the discretization's error: $|ED(f, \Gamma, m, p, i_0)|$ $\leq \frac{\|\epsilon\|_\infty}{ph} \sum_{i=-m}^{i=m} |a_i| \leq K \left(\sum_{i=-m}^{i=m} |a_i| \right) h^{\alpha - \frac{\alpha}{k_0}} \frac{1}{1 - h^{1 - \frac{\alpha}{k_0}}}$ and the proof is complete.

In practice, $4 \leq m \leq 10$. It is easy to check that, for such mask sizes, we have $2 < \sum_{i=-m}^{i=m} |a_{m,i}| < 2.93$ and for all $0 \leq k \leq 2m + 1$, we have $\frac{5}{2} < \frac{1}{k!} \sum_{i=-m}^{i=m} i^k |a_{m,i}| \leq 5$. Hence, for such mask sizes, we have for h great enough $|(\Delta_{m,p}\Gamma)(i_0) - f'(i_0 h)| \leq (5\|f^{(k_0)}\|_\infty + 3K) h^{\alpha - \frac{\alpha}{k_0}}$

2.2   Second Order Derivatives

Definition 3. *Let m be a positive integer. The Taylor-optimal mask of size m for discrete second order derivation (TO_m^2 mask for short) is the finite sequence of rational numbers $\mathbf{b_m}$ defined for $i \neq 0$ by*

$$b_{m,i} = \frac{(-1)^{i+1}}{i^2} \frac{\binom{m}{i}}{\binom{m+i}{m}} = \frac{(-1)^{i+1}}{i^2} \frac{\binom{2m}{m+i}}{\binom{2m}{m}} \tag{6}$$

$b_{m,0}$ is such that $\sum_{i=-m}^{i=m} b_{m,i} = 0$

Definition 4. *Let $\Gamma : \mathbb{Z} \mapsto \mathbb{Z}$ be a discrete function. Let $p \in \mathbb{N}^*$ and $i_0 \in \mathbb{Z}$. The m-Taylor-optimal second order discrete derivative of Γ at point i_0 with step p is*

$$(\Delta_{m,p}^2 \Gamma)(i_0) = \frac{2}{p^2 h} \sum_{i=-m}^{i=m} b_{m,i} \Gamma(i_0 - ip) \tag{7}$$

Lemma 2. $\sum_{i=-m}^{i=m} i^2 b_{m,i} = 1$ *and for $0 \leq k \leq 1$ or $3 \leq k \leq 2m$, we have* $\sum_{i=-m}^{i=m} i^k b_{m,i} = 0.$

Proof. This lemma comes easily from lemma 1.

Theorem 2. *Let* $k \in \mathbb{N}$ *with* $k \geq 3$ *and let* f *be* C^k *on* $\mathbb{R}$ *and let* $k_0 = Min\{k, 2m+1\}$. *Let* $p = \lfloor h^{-1+\frac{\alpha}{k_0}} \rfloor$. *Then*

$$\left| (\Delta^2_{m,p}\Gamma)(i_0) - f''(i_0 h) \right| = O\left(h^{\alpha - \frac{2\alpha}{k_0}} \right) \tag{8}$$

Proof. The proof is analoguous to the one for theorem 1.

3   Normal Vectors and Curvature Estimation

In order to provide estimators for the normal vectors and the curvatures of a parametrized curve $g = (g_1, g_2) : (a, b) \mapsto \mathbb{R} \times \mathbb{R}$, we shall use the classical definitions and properties: for each real t_0 such that $a < t_0 < b$, the normal vector is $Ng(t_0) = (g'_2(t_0), -g'_1(t_0))$ and the curvature may be computed using $Ng(t_0) = \frac{g'_1(t_0)g''_2(t_0) - g'_2(t_0)g''_1(t_0)}{(g'^2_1(t_0) + g'^2_2(t_0))^{3/2}}$.

3.1   Normal Vectors Estimation

Definition 5. *Let* $\Gamma = (\gamma_1, \gamma_2) : \mathbb{Z} \mapsto \mathbb{Z}^2$ *be a discrete function. Let* $p \in \mathbb{N}^*$ *and* $i_0 \in \mathbb{Z}$. *The m-Taylor-optimal discrete normal vector of* Γ *at point* i_0 *with step* p *is*

$$(N_{m,p}\Gamma)(i_0) = ((\Delta_{m,p}\gamma_2)(i_0), -(\Delta_{m,p}\gamma_1)(i_0)) \tag{9}$$

We assume now that a planar simple closed C^1 parameterized curve C (i.e., the parameterization is periodic and injective on a period) is given together with a family of parameterized discrete curves $(\Sigma_h)_{h \in H}$ with Σ_h contained in a tube with radius $H(h)$ around C. We estimate the continous normal vector at a point of C by a discrete normal vector at a not too far point of Σ_h. The following theorem gives a bound to the error of this estimation, and in particular shows that this error uniformly converges to 0 with h.

Theorem 3. *Let* $g = (g_1, g_2)$ *be a* C^k *parameterization of a simple closed curve* C *with* $k \geq 2$. *Let* $Ng = (g'_2, -g'_1)$ *be a normal vector field of* g. *Suppose that for all* i *we have* $\|g(ih) - h\Sigma_h(i)\|_\infty \leq Kh^\alpha$. *Let* $p = \lfloor h^{-1+\frac{\alpha}{k_0}} \rfloor$ *and* $k_0 = Min\{k, 2m+1\}$. *Then*

$$\|(N_{m,p}\Gamma)(i_0) - Ng\| = O\left(h^{\alpha - \frac{\alpha}{k_0}} \right) \tag{10}$$

Proof. The proof is straightforward from theorem 1.

3.2   Curvature Estimation

Definition 6. *Let* $\Gamma = (\gamma_1, \gamma_2) : \mathbb{Z} \mapsto \mathbb{Z}^2$ *be a discrete function. Let* $p \in \mathbb{N}^*$ *and* $i_0 \in \mathbb{Z}$. *The m-Taylor-optimal discrete curvature of* Γ *at point* i_0 *with step* p *is*

$$(C_{m,p}\Gamma)(i_0) = \frac{(\Delta_{m,p}\gamma_2)(i_0)\left(\Delta^{(2)}_{m,p}\gamma_1\right)(i_0) - (\Delta_{m,p}\gamma_1)(i_0)\left(\Delta^{(2)}_{m,p}\gamma_1\right)(i_0)}{\left(((\Delta_{m,p}\gamma_1)(i_0))^2 + ((\Delta_{m,p}\gamma_2)(i_0))^2\right)^{\frac{3}{2}}} \tag{11}$$

Under the same assumption than in the previous subsection, we estimate the continous curvature at a point of C by a discrete curvature at a not too far point of Σ_h.

4   Experimental Evaluation

In this section, we present an experimental evaluation of our various differential estimators. We need to compare the estimated differential quantities values with expected Euclidean ones on graphs of functions or on parametric curves on which such information is known. The considered shapes are simple continous shapes such as the sine function graph, discs or ellipses. These continuous objects have been digitized, with an eventually additional uniform noise. In the 2D shapes cases, we have got a 8-connexe parametrization of the eventually noisy boundary (without ouliers). Then we compare the values of the discrete differential quantity with the corresponding continuous one computed for close points. Considering the empirical multigrid convergence, we always compute the worst case error for a familly of points on the curves for various resolution steps. In the noisy cases, we consider five random curves and compute the average of the worst case errors for these five curves.

4.1   First Order Derivation

First Order Derivative of the Sine function. Figure 1 shows the estimated values of the first order discrete derivatives of noisy digitizations of the sine function graph on the interval $[2; 2.25]$. The discretization step is $h = \frac{1}{1000}$. We use the estimator $\Delta_{7,200}$. The noise is uniform on a set of values $\{-n, ..., +n\}$, with $n = 0, 1, 2, 5$.

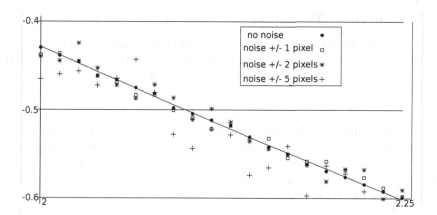

Fig. 1. Estimations of the first order derivatives of the sine function using $\Delta_{7,200}$ for digitizations with $h = \frac{1}{1000}$ as a discretization step and with various levels of noise

Empirical Multigrid Convergence. Here we compute estimations for the first order derivative of the sine function for various resolution step, using Δ_{7,h^n} for various values of n. For each resolution step, the computation is achieve for one hundred points with abcissae $(x_{h,k} = 2 + \frac{k}{h})_{0 \le k < 100}$. Then we evaluate the maximal errors for these points:

$$Max\left\{\left|(\Delta_{7,h^n}\Gamma)(x_{h,k}) - cos(x_{h,k})\right|; 0 \le k < 100\right\}$$

The graph of the function defined by $2h^{0.9}$ which is less than the best theoretical bounding function ($8h^{0.9}$, see theorem 1) is drawn on the figure 2.

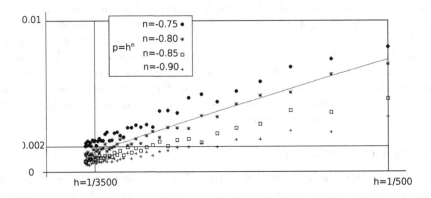

Fig. 2. Maximal error for approximations of the first order derivatives of the sine function using Δ_{7,h^n} at one hundred points for various n

4.2   Second Order Derivation

Second Order Derivative of the Sine Function. Figure 3 shows the computed values of the second order discrete derivatives of rounded digitalizations of the sine function using $\Delta^2_{7,h^{-0.9}}$ for the same values of h than the one considered in [9].

Empirical Multigrid Convergence. Figure 4 shows the estimations of the second order derivative of the sine function for various resolutions, using Δ_{7,h^n} for various n. For each resolution step, the computation is achieve for two hundred points with abcissae $(x_{h,k} = 2 + \frac{k}{h})_{0 \le k < 200}$. Then we evaluate the maximal absolute errors for these points:

$$Max\left\{\left|(\Delta^2_{7,h^n}\Gamma)(x_{h,k}) + sin(x_{h,k})\right|; 0 \le k < 200\right\}$$

4.3   Curvatures

Here we compare the proposed 2D curvature estimators with the continuous one.

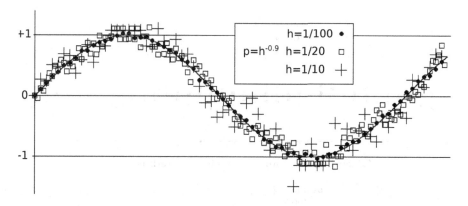

Fig. 3. Estimations of the second order derivative of the sine function using $\Delta_{7,h^{-0.9}}$ at different resolutions

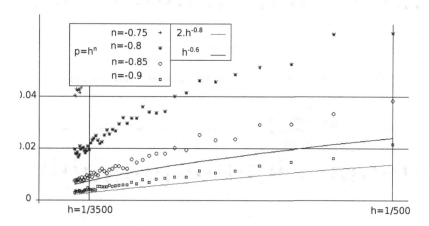

Fig. 4. Maximal error for approximations of the second order derivatives of the sine function using Δ_{7,h^n}^2 for various n at two hundred points

Curvature of a Noisy Circle. Figure 5 shows the computed values of the discrete curvature of noisy digitizations of a circle of radius $\frac{1}{2}$. These values have been computed for forty points around the discrete curve. The considered digitization step is $h = \frac{1}{1000}$. We have introduced various uniform noises. Each graph presents results for different values of the computation steps p. The parametrized discrete curves have been obtained by drawing a noisy disc and an ellipse, then extracting the boundary (hence eliminating the outliers).

Empirical Multigrid Convergence Here we compute the curvature of digitizations of a circle of radius 1 and of an ellipse of equation $4x^2 + y^2 = 1$ for various digitization steps, using the $C_{7,p}$ mask for various computations steps. The

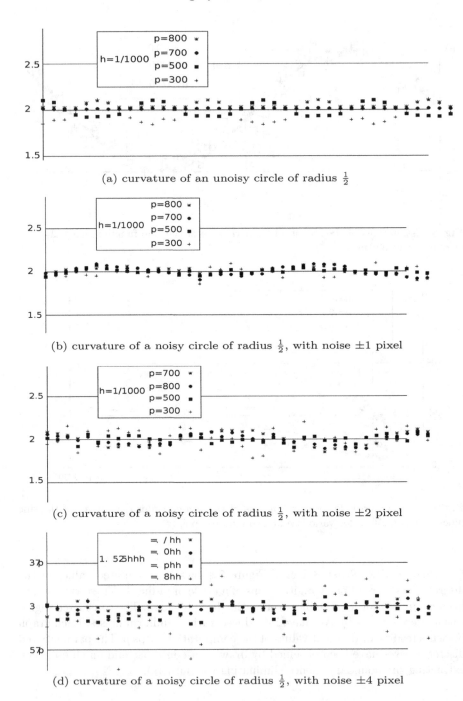

(a) curvature of an unoisy circle of radius $\frac{1}{2}$

(b) curvature of a noisy circle of radius $\frac{1}{2}$, with noise ± 1 pixel

(c) curvature of a noisy circle of radius $\frac{1}{2}$, with noise ± 2 pixel

(d) curvature of a noisy circle of radius $\frac{1}{2}$, with noise ± 4 pixel

Fig. 5. Estimations of the curvature of a circle of radius $\frac{1}{2}$ for digitization step $h = \frac{1}{1000}$ and various steps p and noises

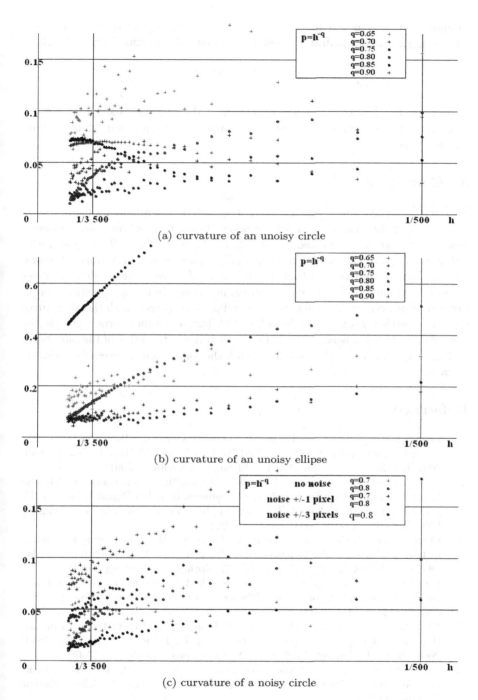

(a) curvature of an unoisy circle

(b) curvature of an unoisy ellipse

(c) curvature of a noisy circle

Fig. 6. Maximal relative error for curvature approximation of (a) the unnoisy circle of radius 1, (b) the unnoisy ellipse of equation $4x^2 + y^2 = 1$ (c) a noisy circle of radius 1, using the $C_{7,p}$ estimators for various computation steps p

computation is achieved for twenty points $(p_{h,k})_{0 \leq k < 20}$ around the discrete curves in the same quadrant. Then we evaluate the maximal error for these points:

$$Max \left\{ \left| \frac{(C_{7,p}\varGamma)(p_{h,k}) - c(\bar{p}_{h,k})}{c(\bar{p}_{h,k})} \right| ; 0 \leq k < 20 \right\}$$

(see Figure 6). Here $c(\bar{p}_{h,k})$ is the curvature of the continuous shape at a point $\bar{p}_{h,k}$ close to the point $p_{h,k}$ (namely the point of the continuous shape having the same abcissa).

5   Conclusion

We have presented a new way for estimating differential quantities using convolution. The main idea is to use sparse nodes. Using this technic allows a better complexity than the other known convolution-based methods. The use of rational numbers as coefficients of the convolution mask is not very heavy, because they all have a common constant denominator. Moreover, this method provides a theoretical multigrid convergence and simulations show a good estimation in practice. However, we have to compare carefully our method with the alternative ones in a further work. A bottleneck of the Taylor optimal kernels is that the step of discretization needs to be known to determine the value of the parameter p. This is generally not the case. We thank the anonymous referees for valuable remarks.

References

[1]   Coeurjolly, D., Miguet, S., Tougne, L.: Discrete Curvature Based on Osculating Circle Estimation. In: Arcelli, C., Cordella, L.P., Sanniti di Baja, G. (eds.) IWVF 2001. LNCS, vol. 2059, pp. 303–312. Springer, Heidelberg (2001)
[2]   Coeurjolly, D., Lachaud, J.-O., Roussillon, T.: Multigrid Convergence of Discrete Geometric Estimators. In: Brimkov, V., Barneva, R. (eds.) Digital Geometry Algorithms, Theoretical Foundations and Applications of Computational Imaging. LNCVB, vol. 2, pp. 395–424. Springer (2012)
[3]   Coeurjolly, D., Lachaud, J.-O., Levallois, J.: Integral Based Curvature Estimators in Digital Geometry. In: Gonzalez-Diaz, R., Jimenez, M.-J., Medrano, B. (eds.) DGCI 2013. LNCS, vol. 7749, pp. 215–227. Springer, Heidelberg (2013)
[4]   Esbelin, H.-A., Malgouyres, R., Cartade, C.: Convergence of Binomial-Based Derivative Estimation for 2 Noisy Discretized curves. Theoretical Computer Science 412, 4805–4813 (2011)
[5]   Kerautret, B., Lachaud, J.-O., Naegel, B.: Comparison of Discrete Curvature Estimators and Application to Corner Detection. In: Bebis, G., et al. (eds.) ISVC 2008, Part I. LNCS, vol. 5358, pp. 710–719. Springer, Heidelberg (2008)
[6]   Klette, R., Rosenfeld, A.: Digital Geometry: Geometric Methods for Digital Picture Analysis. Series in Computer Graphics and Geometric Modeling. Morgan Kaufmann (2004)
[7]   Lachaud, J.-O., Vialard, A., de Vieilleville, F.: Analysis and Comparative Evaluation of Discrete Tangent Estimators. In: Andrès, É., Damiand, G., Lienhardt, P. (eds.) DGCI 2005. LNCS, vol. 3429, pp. 240–251. Springer, Heidelberg (2005)

[8] Malgouyres, R., Brunet, F., Fourey, S.: Binomial Convolutions and Derivatives Estimation from Noisy Discretizations. In: Coeurjolly, D., Sivignon, I., Tougne, L., Dupont, F. (eds.) DGCI 2008. LNCS, vol. 4992, pp. 370–379. Springer, Heidelberg (2008)

[9] Provot, L., Gérard, Y.: Estimation of the Derivatives of a Digital Function with a Convergent Bounded Error. In: Debled-Rennesson, I., Domenjoud, E., Kerautret, B., Even, P. (eds.) DGCI 2011. LNCS, vol. 6607, pp. 284–295. Springer, Heidelberg (2011)

[10] Roussillon, T., Lachaud, J.-O.: Accurate Curvature Estimation along Digital Contours with Maximal Digital Circular Arcs. In: Aggarwal, J.K., Barneva, R.P., Brimkov, V.E., Koroutchev, K.N., Korutcheva, E.R. (eds.) IWCIA 2011. LNCS, vol. 6636, pp. 43–55. Springer, Heidelberg (2011)

On Finding Spherical Geodesic Paths and Circles in $\mathbb{Z}^3$

Ranita Biswas and Partha Bhowmick

Department of Computer Science and Engineering
Indian Institute of Technology, Kharagpur, India
{biswas.ranita,bhowmick}@gmail.com

Abstract. A *discrete spherical geodesic path* between two voxels s and t lying on a discrete sphere is a/the 1-connected shortest path from s to t, comprising voxels of the discrete sphere intersected by the real plane passing through s, t, and the center of the sphere. We show that the set of sphere voxels intersected by the aforesaid real plane always contains a *1-connected cycle* passing through s and t, and each voxel in this set lies within an isothetic distance of $\frac{3}{2}$ from the concerned plane. Hence, to compute the path, the algorithm starts from s, and iteratively computes each voxel p of the path from the predecessor of p. A novel number-theoretic property and the 48-symmetry of discrete sphere are used for searching the *1-connected* voxels comprising the path. The algorithm is *output-sensitive*, having its time and space complexities both linear in the length of the path. It can be extended for constructing 1-connected discrete 3D circles of arbitrary orientations, specified by a few appropriate input parameters. Experimental results and related analysis demonstrate its efficiency and versatility.

Keywords: Discrete sphere, geodesic path, geometry of numbers, discrete 3D circles.

1   Introduction

The shortest path between two points on a curved surface is called *geodesic*. There exist several works related to geodesics on a 3D triangulated surface, e.g., the fast marching technique [8]. This technique and *Polthier's straightest geodesics theory* [13] are used in [11] for finding *approximate geodesics* on triangulated surfaces. For exact geodesics, a cubic-time *line-of-sight algorithm* is proposed in [1].

The first algorithm to solve the discrete geodesic problem as the shortest path (SP) between a source and a destination point on an arbitrarily polyhedral surface is referred in the literature as MMP [12]. The discrete surface points are first preprocessed and stored in a suitable data structure in $O(n^2 \log n)$ time, and then the actual SP is reported by *continuous Dijkstra's algorithm* in $O(k + \log n)$ time, where $n = \#$edges on the surface and $k = \#$faces crossed by SP. Improving MMP to $O(n^2)$ time complexity is done in CH algorithm [4] using a set of

E. Barucci et al. (Eds.): DGCI 2014, LNCS 8668, pp. 396–409, 2014.
© Springer International Publishing Switzerland 2014

windows on the polyhedron edges for encoding the shortest paths. However, it is shown in [14] that MMP, in practice, runs faster than CH. Later, it has been shown in [16] that CH can be made to run faster than MMP, using priority queue and filtering out the useless windows. Recently, a parallel version of CH is proposed in [19]. Further developments with graph-theoretic and numerical methodologies may be seen in [17, 18].

Problems related to geodesic paths and their characterization in the digital space have gained significant attention in recent time. In [5], a new geodesic metric and the A^* *algorithm* are used to find the shortest path between a source and a destination voxel. In [3], *rubberband algorithm* is proposed for computation of minimum-length polygonal curves in cube-curves in 3D space. The idea can be extended to solve various Euclidean shortest path (ESP) problems inside of a simple cube arc, inside of a simple polygon, on the surface of a convex polytope, or inside of a simply-connected polyhedron [10].

In $\mathbb{R}^3$, a *spherical geodesic path* is defined between two points $p \in \mathbb{R}^3$ and $q \in \mathbb{R}^3$ lying on a real sphere $S_r^{\mathbb{R}}$ of radius r. The path always lies along the intersection circle of $S_r^{\mathbb{R}}$ and the 3D plane passing through p, q, and the center of $S_r^{\mathbb{R}}$. We make an analogous definition for *discrete spherical geodesic path* $\pi_r^{\mathbb{Z}}(s,t)$ from a point (voxel) $s \in \mathbb{Z}^3$ to another point $t \in \mathbb{Z}^3$ lying on the discrete sphere, $S_r^{\mathbb{Z}}$, of radius r. W.l.o.g., we fix the center of $S_r^{\mathbb{Z}}$ at $o(0,0,0)$, and consider r as a positive integer. Then, $\pi_r^{\mathbb{Z}}(s,t)$ is defined as a/the 1-connected shortest path from s to t, comprising only those voxels of $S_r^{\mathbb{Z}}$ which lie sufficiently close to the real plane $\Pi_r^{\mathbb{R}}(s,t)$ passing through s, t, and o.

We first show that there always exists a 1-connected cycle in the set $I_r^{\mathbb{Z}}(s,t)$ comprising the voxels of $S_r^{\mathbb{Z}}$ intersected by $\Pi_r^{\mathbb{R}}(s,t)$. The set $I_r^{\mathbb{Z}}(s,t)$ admits the characterization that all its voxels lie within an isothetic distance of $\frac{3}{2}$ from $\Pi_r^{\mathbb{R}}(s,t)$. Subsequently, $\pi_r^{\mathbb{Z}}(s,t)$ becomes a subset of $I_r^{\mathbb{Z}}(s,t)$, and is efficiently obtained by a prioritized Breadth-First-Search algorithm on the underlying graph corresponding to $I_r^{\mathbb{Z}}(s,t)$. For computation of $I_r^{\mathbb{Z}}(s,t)$, $S_r^{\mathbb{Z}}$ is defined as the irreducible 2-separable set of voxels (3D integer points) that are uniquely identified by certain number-theoretic properties. The algorithm computes the set $I_r^{\mathbb{Z}}(s,t)$ using these properties, without considering the entire set $S_r^{\mathbb{Z}}$. Figure 1 shows a result of our algorithm, where the search space of BFS, its 18 neighborhood on $S_r^{\mathbb{Z}}$, and the final geodesic path $\pi_r^{\mathbb{Z}}(s,t)$ are shown in different colors.

The rest of the paper is organized as follows. Section 2 explains certain elementary number-theoretic properties of a digital sphere, used for computing $I_r^{\mathbb{Z}}(s,t)$. Section 3 contains characterization of discrete spherical geodesic path and circle. The algorithm to compute the geodesic path from a point s to a point t lying on $S_r^{\mathbb{Z}}$ is presented in Section 4. Section 5 contains some test results, and Section 6 the concluding notes.

2 Digital Sphere

We first introduce definitions and properties of digital sphere related to this work. These are subsequently used to design the algorithms for finding geodesic

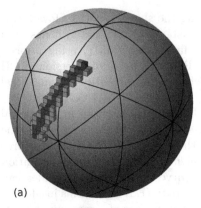 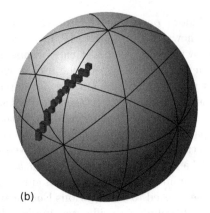

(a)   (b)

Fig. 1. A geodesic path reported by the proposed algorithm for $r = 17$. (a) Red: $s(-6, -1, 16)$ and $t(2, 14, 10)$; blue: $I_r^{\mathbb{Z}}(s, t)$; yellow: 18-neighborhood of the breadth-first search space. (b) The geodesic path $\pi_r^{\mathbb{Z}}(s, t)$ shown in red.

paths and 3D circles in $\mathbb{Z}^3$. The first point to observe is that, opposed to a real sphere, a digital sphere has only nine planes of symmetry. Three of these are the planes containing the *great circles* parallel to three coordinate planes; and for each of these three planes, there exist two more planes aligned at $+45^0$ and -45^0 to it. These nine planes of symmetry give rise to eight coordinate octants, called *c-octants*. Each c-octant contains 6 Möbius triangles [7], thus dividing the sphere into 48 *quadraginta octants* or *q-octants*.

2.1   Representation

The c-octants and the q-octants are uniquely represented by 3-tuples (see Appendix), which are carefully prepared for efficient implementation of our algorithm. Each c-octant $\mathbb{C}_i$ is represented by a 3-tuple of signs of coordinate axes, namely $C_i := \left(c_i^{(1)}, c_i^{(2)}, c_i^{(3)}\right)$. For example, $C_1 = (+, +, +)$, $C_2 = (-, +, +)$, and so forth. The 3-tuple for each q-octant, on the contrary, represents the three signed coordinate axes. In particular, in the 3-tuple $Q_i := \left(q_i^{(1)}, q_i^{(2)}, q_i^{(3)}\right)$ representing $\mathbb{Q}_i$, each element $q_i^{(\cdot)}$ has two variables, namely ω and σ. The variable ω contains a literal (name of the coordinate axis) from $\{\mathbf{x}, \mathbf{y}, \mathbf{z}\}$, and the variable σ contains the sign of the corresponding coordinate. With this representation, we have $Q_1 = (+\mathbf{x}, +\mathbf{y}, +\mathbf{z})$, $Q_2 = (+\mathbf{y}, +\mathbf{x}, +\mathbf{z})$, $Q_3 = (+\mathbf{y}, +\mathbf{z}, +\mathbf{x})$, $\ldots, Q_{24} = (-\mathbf{x}, +\mathbf{z}, -\mathbf{y})$, $\ldots, Q_{48} = (-\mathbf{x}, -\mathbf{z}, -\mathbf{y})$. That is, for Q_{24} as an instance, we have $\omega[q_{48}^{(1)}] = \mathbf{x}$, $\sigma[q_{48}^{(1)}] = $ '$-$', $\omega[q_{48}^{(2)}] = \mathbf{z}$, etc. Our representation ensures the following.

1. $\mathbb{C}_a = \{\mathbb{Q}_b : b = 6(a - 1) + c, c = 1, 2, \ldots, 6\}$.
2. Two q-octants $\mathbb{Q}_i$ and $\mathbb{Q}_j$ lie in the same c-octant if and only if $\lceil i/6 \rceil = \lceil j/6 \rceil$ (with $\mathbb{C}_{\lceil i/6 \rceil}$ as their common c-octant).
   Equivalently, $\mathbb{Q}_i$ and $\mathbb{Q}_j$ lie in the same c-octant if and only if $\sigma[q_i] = \sigma[q_j]$ $\forall (q_i, q_j) \in \{(q_i', q_j') : ((q_i', q_j') \in Q_i \times Q_j) \wedge (\omega[q_i'] = \omega[q_j'])\}$.

3. Let $w = 0$ be one of the three coordinate planes, with $w \in \{x, y, z\}$. Then two c-octants $\mathbb{C}_i$ and $\mathbb{C}_j$ lie in two different half-spaces defined by $w = 0$ if and only if the elements in C_i and C_j corresponding to w are different.

Example 1. $C_1(+, +, +)$ and $C_2(-, +, +)$ have their 1st element different, which implies they are in two different half-spaces defined by the coordinate plane $x = 0$; however, their 2nd and 3rd elements being both '+', either of them lies in the half-space $y \geqslant 0$ and in the half-space $z \geqslant 0$.

2.2 Metrics

We define *x-distance* and *y-distance* between two (real or integer) points, $p(i, j)$ and $p'(i', j')$, as $d_x(p, p') = |i - i'|$ and $d_y(p, p') = |j - j'|$, respectively. In $\mathbb{R}^3$ or in $\mathbb{Z}^3$, we have also *z-distance*, given by $d_z(p, p') = |k - k'|$, for $p(i, j, k)$ and $p'(i', j', k')$. Using these inter-point distances, we define the respective *x-*, *y-*, and *z*-distances between a point $p(i, j, k)$ and a surface S as follows. Let $d_x(p, S)$ be the *x*-distance between a point $p(i, j, k)$ and a surface S. If there exists a point $p'(x', y', z')$ in S such that $(y', z') = (j, k)$, then $d_x(p, S) = d_x(p, p')$; otherwise, $d_x(p, S) = \infty$. The other two distances, i.e., $d_y(p, S)$ and $d_z(p, S)$, are defined in a similar way; note that the metric $d_z(p, S)$ is not defined in 2D. These metrics are used to define the *isothetic distance* as follows.

Definition 1. *Between two points $p_1(i_1, j_1)$ and $p_2(i_2, j_2)$, the isothetic distance is taken as the Minkowski norm [9], $d_\infty(p_1, p_2) = \max\{d_x(p_1, p_2), d_y(p_1, p_2)\}$; between a point $p(i, j)$ and a curve C, it is $d_\perp(p, C) = \min\{d_x(p, C), d_y(p, C)\}$, where $d_x(p, C)$ and $d_y(p, C)$ are defined similar to $d_x(p, S)$ and $d_y(p, S)$ respectively; between a 3D point $p(i, j, k)$ and a surface S, it is $d_\perp(p, S) = \min\{d_x(p, S), d_y(p, S), d_z(p, S)\}$.*

2.3 Topology

A *voxel* is an integer point in 3D space, and equivalently, a 3-cell [9]. Two voxels are said to be *0-adjacent* if they share a vertex (0-cell), *1-adjacent* if they share an edge (1-cell), and *2-adjacent* if they share a face (2-cell). Thus, two distinct voxels, $p_1(i_1, j_1, k_1)$ and $p_2(i_2, j_2, k_2)$ are *1-adjacent* if and only if $|i_1 - i_2| + |j_1 - j_2| + |k_1 - k_2| \leqslant 2$ and $\max\{|i_1 - i_2|, |j_1 - j_2|, |k_1 - k_2|\} = 1$; *2-adjacent* if and only if $|i_1 - i_2| + |j_1 - j_2| + |k_1 - k_2| = 1$; and *0-adjacent* if and only if $|i_1 - i_2| = |j_1 - j_2| = |k_1 - k_2| = 1$. Clearly, 0-adjacent (1-adjacent) voxels are not considered as adjacent while considering 1-neighborhood (2-neighborhood) connectivity. Note that the 0-, 1-, and 2-neighborhood notations, as adopted in this paper and also in [15], correspond respectively to the classical 26-, 18-, and 6-neighborhood notations used in [6].

Based on above definitions, a digital sphere is said to be *2-separating* if it does not contain any *2-tunnel*, that is, its interior and exterior are not connected by a *2-connected path* [6]. A 2-separating digital sphere is *irreducible* if and only if it does not contain any *simple voxel*, that is, removal of any voxel violates its topological property of 2-separableness [6]. We use $S_r^{\mathbb{R}}$ to denote the real sphere of radius r and centered at o, use $S_r^{\mathbb{Z}_1}$ to denote the part of $S_r^{\mathbb{Z}}$ lying in $\mathbb{Q}_1$, and use $p \in S_r^{\mathbb{Z}}$ when a voxel p belongs to (voxel set) $S_r^{\mathbb{Z}}$. Our work is based on the following definition of digital sphere.

Definition 2. *A digital sphere $S_r^{\mathbb{Z}}$ is an irreducible 2-separating subset of the voxel set having isothetic distance less than $\frac{1}{2}$ from $S_r^{\mathbb{R}}$.*

Note that in [15], the only strict k-separating or irreducible digital sphere results from outer Gaussian digitization, but its voxels are not limited by a maximum isothetic distance of $\frac{1}{2}$ from $S_r^{\mathbb{R}}$. The closed centered 2-separating digitized sphere is another proposition in [15], which is not necessarily irreducible.

2.4   Characterization

The characterization of $S_r^{\mathbb{Z}}$ is required to decide in constant time whether a particular voxel (i, j, k) belongs to $S_r^{\mathbb{Z}}$. We start with the following lemmas.

Lemma 1. $d_{\perp}\left(p, S_r^{\mathbb{R}}\right) = \left|k - \sqrt{r^2 - (i^2 + j^2)}\right| \; \forall \, p(i, j, k) \in S_r^{\mathbb{Z}_1}.$

Proof. Let $p(i, j, k) \in S_r^{\mathbb{Z}_1}$, and $(x, j, k), (i, y, k)$, and (i, j, z) be the respective points on $S_r^{\mathbb{R}}$ taken along the lines parallel to x-, y-, and z-axes, and passing through p. Observe that the points (x, j, k) and (i, y, k) may be nonexistent, but the point (i, j, z) always exists. If all three exist, then

$$x^2 + j^2 + k^2 = i^2 + y^2 + k^2 = i^2 + j^2 + z^2 = r^2, \text{or}, \; k^2 - z^2 = j^2 - y^2 = i^2 - x^2$$
$$\text{or}, \; (k+z)(k-z) = (j+y)(j-y) = (i+x)(i-x). \tag{1}$$

In $\mathbb{Q}_1$, $i \leqslant j \leqslant k$ and $x \leqslant y \leqslant z$, or, $i + x \leqslant j + y \leqslant k + z$; so, from Eq. 1,

$$|k - z| \leqslant |j - y| \leqslant |i - x|. \tag{2}$$

If one or both (x, j, k) and (i, y, k) do not exist, then also $|k - z|$ remains the minimum. Hence, from Eq. 2, $d_{\perp}\left(p, S_r^{\mathbb{R}}\right) = |k - z| = \left|k - \sqrt{r^2 - (i^2 + j^2)}\right|.$ □

Lemma 2. $d_{\perp}(p, S_r^{\mathbb{R}}) < \frac{1}{2} \; \forall \, p \in S_r^{\mathbb{Z}}.$

Proof. If possible, let, w.l.o.g., $p(i, j, k) \in S_r^{\mathbb{Z}_1}$, such that $\left|k - \sqrt{r^2 - (i^2 + j^2)}\right| = \frac{1}{2}$, or, w.l.o.g., $k - \sqrt{r^2 - (i^2 + j^2)} = -\frac{1}{2}$, which implies $S_r^{\mathbb{R}}$ has $p'(i, j, k + \frac{1}{2})$ as the point of intersection in $\mathbb{Q}_1$ with the 3D straight line $(x = i, y = j)$. Since $(i, j, k + \frac{1}{2})$ lies on $S_r^{\mathbb{R}}$, we have $i^2 + j^2 + \left(k + \frac{1}{2}\right)^2 = r^2$, which is a contradiction, since r, i, j, k are all integers. □

Lemma 2 helps in characterizing a voxel $p \in S_r^{\mathbb{Z}}$, as stated next.

Theorem 1. $p(i, j, k) \in S_r^{\mathbb{Z}}$ *if and only if* p *is not simple and* $i^2 + j^2 + k^2 \in \left[r^2 - \max\{|i|, |j|, |k|\}, \; r^2 + \max\{|i|, |j|, |k|\} - 1\right].$

Proof. Let, w.l.o.g., $p \in \mathbb{Q}_1$. So, $\max\{|i|, |j|, |k|\} = k$. Hence, by Lemma 1 and Lemma 2, $p \in S_r^{\mathbb{Z}_1}$ if and only if p is not simple and

$$-\frac{1}{2} < k - \sqrt{r^2 - (i^2 + j^2)} < \frac{1}{2} \tag{3}$$

$$\Leftrightarrow k^2 - k + \frac{1}{4} < r^2 - (i^2 + j^2) < k^2 + k + \frac{1}{4}. \tag{4}$$

Since $k^2 - k$, $r^2 - (i^2 + j^2)$, $k^2 + k$ are integers, Eq. 4 is true if and only if

$$k^2 - k < r^2 - (i^2 + j^2) \leqslant k^2 + k$$
$$\Leftrightarrow r^2 - k \leqslant i^2 + j^2 + k^2 < r^2 + k, \tag{5}$$

and hence the proof for 1st q-octant. For other q-octants, the proof is similar. □

Now, to obtain the necessary and sufficient condition of deciding whether a voxel is simple, we need the following theorem.

Theorem 2. *A voxel $p(i, j, k)$ with $d_\perp(p, S_r^{\mathbb{R}}) < \frac{1}{2}$ is simple if and only if $i^2 + j^2 + k^2 = r^2 + \max\{|i|, |j|, |k|\} - 1$ and $\mathrm{mid}\{|i|, |j|, |k|\} = \max\{|i|, |j|, |k|\}$, where $\mathrm{mid}\{\cdot\}$ denotes the median element.*

Proof. As in the proof of Theorem 1, let, w.l.o.g., $p \in \mathbb{Q}_1$; so, $\mathrm{mid}\{|i|, |j|, |k|\} = j$ and $\max\{|i|, |j|, |k|\} = k$. Let also, $d_\perp(p, S_r^{\mathbb{R}_1}) < \frac{1}{2}$, which implies p satisfies Eq. 5 by Lemma 1 and Lemma 2.

Now, we prove that $p(i, j, k)$ lies on $S_r^{\mathbb{Z}_1}$ and cannot be a simple voxel if $j < k$. For this, first observe that $(i, j, k - 1)$ and $(i, j, k + 1)$ lie in $\mathbb{Q}_1$, as $j \leqslant k - 1$. Next, observe that for any $(i', j') \in \mathbb{Z}^2$, there can be at most one integer value of k' so that (i', j', k') satisfies Eq. 3. This implies that $(i, j, k - 1)$ lies in the interior and $(i, j, k + 1)$ in the exterior of $S_r^{\mathbb{Z}}$. Hence, discarding p would violate the 2-separableness of $S_r^{\mathbb{Z}}$.

Now, the conditions $i^2 + j^2 + k^2 = r^2 + \max\{|i|, |j|, |k|\} - 1$ and $\mathrm{mid}\{|i|, |j|, |k|\} = \max\{|i|, |j|, |k|\}$ imply $(i^2 + k^2 + k^2) = (r^2 + k - 1)$, which is true if and only if

$$(i^2 + (k - 1)^2 + k^2) = (r^2 - k)$$
$$\Leftrightarrow (i, k - 1, k) \in S_r^{\mathbb{Z}} \text{ by Theorem 1, and } (i, k, k - 1) \in S_r^{\mathbb{Z}}$$
$$\Leftrightarrow (i, k, k) \text{ is simple.}$$

For p lying in some other octant, the proof follows a similar way. □

Using Theorem 1 and Theorem 2, we get a mathematically refined definition of digital sphere, as stated in the following theorem.

Theorem 3. *The voxel set of the digital sphere $S_r^{\mathbb{Z}}$ is given by*

$$\left\{ \begin{array}{c} (i, j, k) \in \mathbb{Z}^3 : r^2 - \max\{|i|, |j|, |k|\} \leqslant i^2 + j^2 + k^2 < r^2 + \max\{|i|, |j|, |k|\} \\ \wedge \left(\begin{array}{c} (i^2 + j^2 + k^2 \neq r^2 + \max\{|i|, |j|, |k|\} - 1) \\ \vee (\mathrm{mid}\{|i|, |j|, |k|\} \neq \max\{|i|, |j|, |k|\}) \end{array} \right) \end{array} \right\}.$$

3   Discrete Spherical Geodesic Path and Circle

Theorem 3 is used to decide in constant time whether a voxel $p(i, j, k)$ belongs to $S_r^{\mathbb{Z}}$. For generating the discrete spherical geodesic path $\pi_r^{\mathbb{Z}}(s,t)$ from a voxel $s \in S_r^{\mathbb{Z}}$ to a voxel $t \in S_r^{\mathbb{Z}}$, we consider the real plane $\Pi_r^{\mathbb{R}}(s,t)$ that passes through s, t, and the center of $S_r^{\mathbb{Z}}$. Considering voxels as 3-cells, let $I_r^{\mathbb{Z}}(s,t)$ be the set of voxels of $S_r^{\mathbb{Z}}$ intersected by $\Pi_r^{\mathbb{R}}(s,t)$. We have the following lemma for $I_r^{\mathbb{Z}}(s,t)$.

Lemma 3. $d_{\perp}(p, \Pi_r^{\mathbb{R}}(s,t)) \leqslant \frac{3}{2} \ \forall \ p \in I_r^{\mathbb{Z}}(s,t)$.

Proof. Let $\delta_e := d_e(p, \Pi_r^{\mathbb{R}}(s,t))$ be the real (Euclidean) distance of the point p from $\Pi_r^{\mathbb{R}}(s,t)$. If $\Pi_r^{\mathbb{R}}(s,t)$ intersects the voxel p, then $\delta_e \leqslant \frac{\sqrt{3}}{2}$.

Now, let $\delta_x = d_x(p, \Pi_r^{\mathbb{R}}(s,t)), \delta_y = d_y(p, \Pi_r^{\mathbb{R}}(s,t)), \delta_z = d_z(p, \Pi_r^{\mathbb{R}}(s,t))$. Observe that $\delta_x = \frac{\delta_e}{\cos \theta_x}, \delta_y = \frac{\delta_e}{\cos \theta_y}, \delta_z = \frac{\delta_e}{\cos \theta_z}$, where, $\cos^2 \theta_x + \cos^2 \theta_y + \cos^2 \theta_z = 1$. Here, $\cos \theta_x$ is the angle between the x-axis-parallel line through p and the perpendicular on $\Pi_r^{\mathbb{R}}(s,t)$ dropped from p, etc. So, the supremum of $d_{\perp}(p, \Pi_r^{\mathbb{R}}(s,t)) := \min\{\delta_x, \delta_y, \delta_z\}$ corresponds to the infimum of the largest element in $C_\theta := \{\cos \theta_x, \cos \theta_y, \cos \theta_z\}$, and hence to the infimum of the largest element in $C_\theta^{(2)} := \{\cos^2 \theta_x, \cos^2 \theta_y, \cos^2 \theta_z\}$, subject to $\cos^2 \theta_x + \cos^2 \theta_y + \cos^2 \theta_z = 1$. Clearly, the largest element in $C_\theta^{(2)}$ is at least $\frac{1}{3}$, or, the largest element in C_θ is at least $\frac{1}{\sqrt{3}}$, whence $d_{\perp}(p, \Pi_r^{\mathbb{R}}(s,t)) \leqslant \delta_e / \frac{1}{\sqrt{3}} = \frac{3}{2}$.   □

Theorem 4. *For any two voxels $s \in S_r^{\mathbb{Z}}$ and $t \in S_r^{\mathbb{Z}}$, there always exist two 1-connected paths, $\pi_r^{\mathbb{Z}}(s,t)' \subset I_r^{\mathbb{Z}}(s,t)$ and $\pi_r^{\mathbb{Z}}(t,s)'' \subset I_r^{\mathbb{Z}}(s,t)$, such that $\pi_r^{\mathbb{Z}}(s,t)' \cup \pi_r^{\mathbb{Z}}(t,s)''$ forms a 1-connected simple cycle in $I_r^{\mathbb{Z}}(s,t)$.*

Proof. Given a continuous surface A, there is a unique *supercover* of A, defined as the set of all voxels intersecting A [6]. Hence, if $\Pi_r^{\mathbb{Z}}(s,t)$ denotes the *supercover* of $\Pi_r^{\mathbb{R}}(s,t)$, then all the voxels—conceived as 3-cells—that are intersected by $\Pi_r^{\mathbb{R}}(s,t)$, comprise the set $\Pi_r^{\mathbb{Z}}(s,t)$. As shown in [2], the supercover of a plane is 2-separable.

We define $\mathcal{S}_{r-}^{\mathbb{Z}}$ and $\mathcal{S}_{r+}^{\mathbb{Z}}$ as the respective interior and exterior of $S_r^{\mathbb{Z}}$. So, by Definition 2, the sets $\mathcal{S}_{r-}^{\mathbb{Z}}$ and $\mathcal{S}_{r+}^{\mathbb{Z}}$ are disconnected in 2-neighborhood. Also, let $\Pi_{r-}^{\mathbb{Z}}(s,t) = \Pi_r^{\mathbb{Z}}(s,t) \cap \mathcal{S}_{r-}^{\mathbb{Z}}$ and $\Pi_{r+}^{\mathbb{Z}}(s,t) = \Pi_r^{\mathbb{Z}}(s,t) \cap \mathcal{S}_{r+}^{\mathbb{Z}}$. Note that $\Pi_{r-}^{\mathbb{Z}}(s,t)$ is a non-empty set and always contains o for $r \geqslant 1$, since $\Pi_r^{\mathbb{R}}(s,t)$ passes through o. This yields

$$\Pi_r^{\mathbb{Z}}(s,t) = \Pi_{r-}^{\mathbb{Z}}(s,t) \cup I_r^{\mathbb{Z}}(s,t) \cup \Pi_{r+}^{\mathbb{Z}}(s,t) \tag{6}$$

where, $\Pi_{r-}^{\mathbb{Z}}(s,t), I_r^{\mathbb{Z}}(s,t)$, and $\Pi_{r+}^{\mathbb{Z}}(s,t)$ are pairwise disjoint.

Now, as $\mathcal{S}_{r-}^{\mathbb{Z}}$ and $\mathcal{S}_{r+}^{\mathbb{Z}}$ are not 2-connected, their respective subsets $\Pi_{r-}^{\mathbb{Z}}(s,t)$ and $\Pi_{r+}^{\mathbb{Z}}(s,t)$ are also not 2-connected. So, by Eq. 6, the set $I_r^{\mathbb{Z}}(s,t)$ forms a 2-separating set between $\Pi_{r-}^{\mathbb{Z}}(s,t)$ and $\Pi_{r+}^{\mathbb{Z}}(s,t)$, or, equivalently, $I_r^{\mathbb{Z}}(s,t)$ is a 1-connected set that also 2-separates $S_r^{\mathbb{Z}}$. Therefore, there always exists a 1-connected simple path $\pi_r^{\mathbb{Z}}(s,t)' \in I_r^{\mathbb{Z}}(s,t)$ from s to t, and another 1-connected simple path $\pi_r^{\mathbb{Z}}(t,s)'' \in I_r^{\mathbb{Z}}(s,t)$ from t to s, where $\pi_r^{\mathbb{Z}}(s,t)' \cap \pi_r^{\mathbb{Z}}(t,s)'' = \{s,t\}$. Hence, there always exists a 1-connected simple cycle $(\pi_r^{\mathbb{Z}}(s,t)' \cup \pi_r^{\mathbb{Z}}(t,s)'')$ in $I_r^{\mathbb{Z}}(s,t)$ containing any two voxels $s \in S_r^{\mathbb{Z}}$ and $t \in S_r^{\mathbb{Z}}$.   □

From Theorem 4, it is clear that for two given voxels $s \in S_r^{\mathbb{Z}}$ and $t \in S_r^{\mathbb{Z}}$, we get at least two 1-connected paths, $\pi_r^{\mathbb{Z}}(s,t)'$ and $\pi_r^{\mathbb{Z}}(s,t)''$, in $I_r^{\mathbb{Z}}(s,t)$, having no voxels in common, excepting s and t. The discrete 3D (integer) circle passing through two given voxels s and t is, therefore, given by $C_r^{\mathbb{Z}}(s,t) = \pi_r^{\mathbb{Z}}(s,t)' \cup \pi_r^{\mathbb{Z}}(t,s)''$. Note that specifying only $s \in S_r^{\mathbb{Z}}$ and $t \in S_r^{\mathbb{Z}}$ would suffice to get $\pi_r^{\mathbb{Z}}(s,t)$, and hence $C_r^{\mathbb{Z}}(s,t)$, since a unique value of r would satisfy Theorem 1 for each of s and t.

4   Algorithm DSGP

We define *inter-octant distance* $d_{i,j}^{(8)}$ corresponding to $\mathbb{C}_i$ and $\mathbb{C}_j$. With $s \in \mathbb{C}_i$ and $t \in \mathbb{C}_j$, it is given by the count of q-octants crossed by $\pi_r^{\mathbb{Z}}(s,t)$ before entering $\mathbb{C}_j$. Mathematically,

$$d_{i,j}^{(8)} = 1 + \sum_{u=1}^{3} 2^{u-1} \left(c_i^{(u)} \oplus c_j^{(u)} \right) \tag{7}$$

where, $c_i^{(u)} \oplus c_j^{(u)} = 1$ if $c_i^{(u)} \neq c_j^{(u)}$, and 0 otherwise. If $i = j$, then $d_{i,j}^{(8)} = 0$; otherwise, the value of $d_{i,j}^{(8)}$ lies in the interval $[1, 7]$. The maximum value $d_{i,j}^{(8)} = 7$ is obtained when $\mathbb{C}_i$ and $\mathbb{C}_j$ are *diametrically opposite*, i.e., $c_i^{(u)} \neq c_j^{(u)}$ for $u = 1, 2, 3$. The pair (s, t) becomes *antipodal* if their c-octants are diametrically opposite and s, o, t are collinear. Then $\Pi_r^{\mathbb{R}}(s, t)$ has no fixed orientation, and so a third point q on $S_r^{\mathbb{Z}}$ needs to be specified, which would lie in $\pi_r^{\mathbb{Z}}(s, t)$.

Similarly, we define the *intra-octant distance* $d_{i,j}^{(6)}$ between two q-octants, $\mathbb{Q}_i$ and $\mathbb{Q}_j$, when they lie in same c-octant. It provides the count of q-octants containing the geodesic path from any point $s \in \mathbb{Q}_i$ to any point $t \in \mathbb{Q}_j$. According to our representation, it is given by one plus the minimum number of swaps among the elements in $\mathbb{Q}_i$, so that, after swaps, the transformed 3-tuple is identical with $\mathbb{Q}_j$. Two elements are swapped in $\mathbb{Q}_i$ or in any of its intermediate configurations only if the elements are consecutive in $\mathbb{Q}_i$ or in that configuration (i.e., 3-tuple). Using $d_{i,j}^{(8)}$ and $d_{i,j}^{(6)}$, we compute the *q-octant distance* $d_{i,j}^{(48)}$ between s and t. It gives the count of q-octants containing the geodesic path from s to t, irrespective of their positions on the sphere. Its measure turns out to be

$$d_{i,j}^{(48)} = d_{i,j}^{(8)} + d_{i,j}^{(6)} - 1. \tag{8}$$

Combining the above, we simplify the rule of determining the sequence of q-octants containing $\pi_r^{\mathbb{Z}}(s,t)$ as follows. Let $\mathbb{Q}_i$ and $\mathbb{Q}_j$ be the q-octants containing s and t, respectively. Then the sequence of q-octants through which $\pi_r^{\mathbb{Z}}(s,t)$ passes, is given by a/the minimum-length sequence of transformations applied on $\mathbb{Q}_i$ to attain the configuration $\mathbb{Q}_j$. Following are the rules of transformation.

T_1. Change the sign of the first element $q_i^{(1)}$ in $\mathbb{Q}_i$ (or its intermediate configuration) only if $\sigma[q_i^{(1)}] \neq \sigma[q_j^{(1)}]$. This signifies transition from one half-space (or, c-octant) to another half-space.

T_2. Swap two elements in Q_i (or its intermediate configuration) only if they are consecutive. This signifies transition from one q-octant to its adjacent q-octant.

From the sequence of q-octants obtained by the required transformations, we determine the q-octant $\mathbb{Q}_{i'}$ immediately next to the q-octant $\mathbb{Q}_i$ of s. We use the 3-tuples corresponding to $\mathbb{Q}_i$ and $\mathbb{Q}_{i'}$ for computing the direction vector $\mathbf{d}_s := (d_s^{(1)}, d_s^{(2)}, d_s^{(3)}) \in \{+1, -1, \pm1\}^3$. It is required to find the candidate voxels that are 1-adjacent to s ($\mathbb{A}_1(s)$), belong to $I_r^{\mathbb{Z}}(s,t)$, and is directed towards the shorter between the two possible geodesics from s to t (Theorem 4). The elements $d_s^{(1)}, d_s^{(2)}, d_s^{(3)}$ correspond to the moves along x-, y-, z-axes, respectively. The notation $+1$ signifies that there can be a unit move or no move (from s) along the positive axis of the corresponding coordinate; similarly, -1 signifies a unit move or no move along the negative axis, and ±1 signifies no move or a unit move along positive or negative axis. In case of more than one minimum-length sequence of q-octants from $\mathbb{Q}_i$ to $\mathbb{Q}_j$, we consider the q-octant nearest to $\mathbb{Q}_i$ and common to these sequences, for computing $\mathbf{d}_s$. The rationale is that only one of these sequences would be intersected by $\Pi_r^{\mathbb{R}}(s,t)$, and hence the q-octant common to these sequences is used. The following examples clarify the idea.

Example 2. See Fig. 2. Given $s(10, -2, 6) \in \mathbb{Q}_{15}$ and $t(-3, 10, 6) \in \mathbb{Q}_{12}$, their respective 3-tuples are $Q_{15} := (-y, +z, +x)$ and $Q_{12} := (-x, +z, +y)$. The minimum-length sequence of transformations corresponding to $\boldsymbol{\pi}_r^{\mathbb{Z}}(s,t)$ is:

$(-y, +z, +x) \xrightarrow{T_1} (+y, +z, +x) \xrightarrow{T_2} (+y, +x, +z) \xrightarrow{T_2} (+x, +y, +z) \xrightarrow{T_2} (+x, +z, +y)$
$\xrightarrow{T_1} (-x, +z, +y)$, or, $Q_{15} \xrightarrow{T_1} Q_3 \xrightarrow{T_2} Q_2 \xrightarrow{T_2} Q_1 \xrightarrow{T_2} Q_6 \xrightarrow{T_1} Q_{12}$.

Notice that there is another minimum-length sequence: $(-y, +z, +x) \xrightarrow{T_1} (+y, +z, +x)$ $\xrightarrow{T_2} (+z, +y, +x) \xrightarrow{T_2} (+z, +x, +y) \xrightarrow{T_2} (+x, +z, +y) \xrightarrow{T_2} (-x, +z, +y)$, which implies $Q_{15} \xrightarrow{T_1} Q_3 \xrightarrow{T_2} Q_4 \xrightarrow{T_2} Q_5 \xrightarrow{T_2} Q_6 \xrightarrow{T_2} Q_{12}$.

Either of these implies that the q-octant next to $\mathbb{Q}_{15}$ through which $\boldsymbol{\pi}_r^{\mathbb{Z}}(s,t)$ passes, is $\mathbb{Q}_3$. Since $Q_{15} = (-y, +z, +x)$ and $Q_3 = (+y, +z, +x)$, the y-coordinate of each voxel $p \in \mathbb{A}_1(s) \cap I_r^{\mathbb{Z}}(s,t)$, cannot ever decrease. On the contrary, the x- and the z-coordinates of p have no such restriction. Hence, the direction vector $\mathbf{d}_s$ is chosen as $(\pm1, +1, \pm1)$.

Example 3. Let $s \in \mathbb{Q}_1$ and $t \in \mathbb{Q}_4$. So, $Q_1 = (+x, +y, +z)$ and $Q_4 = (+z, +y, +x)$. We have two minimum-length sequence of transformations: (i) $Q_1 \xrightarrow{T_2} Q_2 \xrightarrow{T_2} Q_3 \xrightarrow{T_2} Q_4$; (ii) $Q_1 \xrightarrow{T_2} Q_6 \xrightarrow{T_2} Q_5 \xrightarrow{T_2} Q_4$.

Contrary to Example 2, here the q-octants following $\mathbb{Q}_1$ in two cases are different: $\mathbb{Q}_2$ for (i) and $\mathbb{Q}_6$ for (ii). So, we look ahead until there is a matching q-octant, i.e., $\mathbb{Q}_4$ in this case. We compute $\mathbf{d}_s$ as the relative shifts in positions of the coordinate values in Q_4. In Q_1, the 1st element is $+x$, which is shifted to 3rd position in Q_4. So, the 1st element in $\mathbf{d}_s$ becomes $+1$, and by similar reasoning with the 2nd and the 3rd elements, we get $\mathbf{d}_s = (+1, \pm1, -1)$.

Analysis. See Algorithm 1 and its demonstration in Fig. 2. The adjacency list L of the underlying undirected graph $G(V, E)$ is prepared based on 1-adjacency of the voxels in $I_r^{\mathbb{Z}}(s,t)$. The vertices adjacent to each $u \in V$ are inserted in the

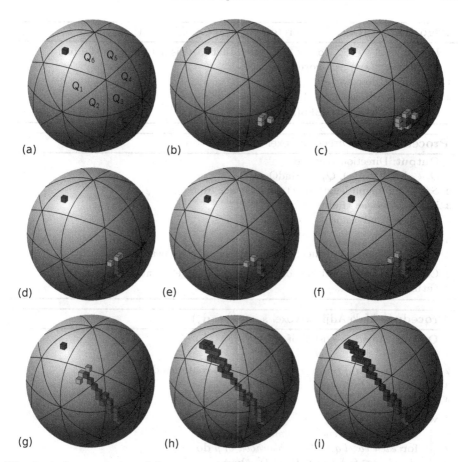

Fig. 2. A demonstration of the proposed algorithm for $r = 12$. (a) $s(10, -2, 6) \in \mathbb{Q}_{15}$, $t(-3, 10, 6) \in \mathbb{Q}_{12}$. (b) Yellow: $S_r^{\mathbb{Z}} \cap \mathbb{A}_1(s)$. (c) Blue: $S_r^{\mathbb{Z}} \cap \mathbb{A}_1(s) \cap I_r^{\mathbb{Z}}(s, t)$, Yellow: $\{p \in \mathbb{A}_1(q) : q$ is Blue$\}$. (d-h) Blue: Progress of Procedure MakeAdjList for $I_r^{\mathbb{Z}}(s, t)$. (i) Red: $\pi_r^{\mathbb{Z}}(s, t) \subset I_r^{\mathbb{Z}}(s, t)$.

adjacency chain $L[u]$ of u in non-increasing order of their isothetic distances from $\Pi_r^{\mathbb{R}}(s, t)$, (MakeAdjList, Line 9). This is needed to maintain locally minimum isothetic distance from $\Pi_r^{\mathbb{R}}(s, t)$ while running Prioritized-BFS (Algorithm 1, Line 3). In Line 8 of MakeAdjList, Theorem 1 is used to determine the voxels that are 1-adjacent with the current voxel and belong to $S_r^{\mathbb{Z}}$, in constant time. Thus, MakeAdjList and Prioritized-BFS consumes $O(n)$ time each, where n is the number of voxels comprising $\pi_r^{\mathbb{Z}}(s, t)$. The direction vector $\mathbf{d}_s$ is computed from the sequence(s) in no more than $O(n)$ time complexity. Hence, the total time complexity of Algorithm DSGP is linear in the length of $\pi_r^{\mathbb{Z}}(s, t)$.

5   Results

The proposed algorithm is implemented in C in Ubuntu 12.04 32-bit, Kernel Linux 3.2.0-31-generic-pae, GNOME 3.4.2, Intel® Core™ i5-2400 CPU 3.10GHz.

Algorithm 1: DSGP (Discrete Spherical Geodesic Path).

Input: voxel $s \in S_r^{\mathbb{Z}}$, voxel $t \in S_r^{\mathbb{Z}}$, such that (s, t) is not an antipodal pair
Output: $\pi_r^{\mathbb{Z}}(s, t)$ as a voxel sequence

1 $\mathbf{d}_s \leftarrow$ FindDirection(s, t)
2 $L \leftarrow$ MakeAdjList$(s, t, \mathbf{d}_s)$
3 $\pi_r^{\mathbb{Z}}(s, t) \leftarrow$ Prioritized-BFS(s, t, L)

Procedure FindDirection(voxel s, voxel t)

Output: Direction vector $\mathbf{d}_s$

1 $Q_i \leftarrow$ FindQoct(s), $Q_j \leftarrow$ FindQoct(t)
2 $\mathcal{S}_{i,j} \leftarrow$ minimum-length q-octant sequence from Q_i to Q_j
3 **if** $\mathcal{S}_{i,j}$ *is unique* **then**
4 $\quad\lfloor$ $Q_{i'} \leftarrow$ 2nd element (q-octant) in $\mathcal{S}_{i,j}$

5 **else**
6 $\quad\lfloor$ $Q_{i'} \leftarrow$ common element in the sequences $\{\mathcal{S}_{i,j}\}$ nearest to Q_i

7 Compute $\mathbf{d}_s$ from positions of the corresponding elements in Q_i and $Q_{i'}$
8 **return** $\mathbf{d}_s$

Procedure MakeAdjList(voxel s, voxel t, $\mathbf{d}_s$)

Output: Adjacency list L of $I_r^{\mathbb{Z}}(s, t)$

1 $visited[s] \leftarrow$ TRUE
2 $Q \leftarrow \{q : (q \in S_r^{\mathbb{Z}}) \wedge (q \in \Pi_r^{\mathbb{Z}}(s, t)) \wedge ((s, q) \text{ conforms } \mathbf{d}_s)\}$
3 **for** *each* $q \in Q$ **do**
4 $\quad\lfloor$ $visited[q] \leftarrow$ FALSE

5 **while** $Q \neq \emptyset$ **do**
6 $\quad$ voxel $p \leftarrow$ Dequeue(Q), $visited[p] \leftarrow$ TRUE
7 $\quad$ **for** *each voxel* q *in 1-neighborhood of* p **do**
8 $\quad\quad$ **if** $(q \in S_r^{\mathbb{Z}}) \wedge (q \in \Pi_r^{\mathbb{Z}}(s, t))$ **then**
9 $\quad\quad\quad\lfloor$ insert q in $L[p]$ in non-increasing order of $d_\perp(q, \Pi_r^{\mathbb{R}}(s, t))$
10 $\quad\quad$ **if** $visited[q] =$ FALSE **then**
11 $\quad\quad\quad\lfloor$ Enqueue(Q, q)

12 **return** L

As the algorithm is of linear time complexity and readily implementable with primitive operations in the integer space, it computes the spherical geodesic paths and 3D circles in $\mathbb{Z}^3$ quite fast and efficiently. To demonstrate this, a summary of some experimental results is given in Appendix. For radius r ranging from 10 to 1000, different source and destination points are chosen, and their geodesic paths are computed. For each path $\pi_r^{\mathbb{Z}}(s, t)$, its length $|\pi_r^{\mathbb{Z}}(s, t)|$, measured in terms of number of voxels comprising the path, is shown, along with the corresponding q-octant distance, $d_{i,j}^{(48)}$. The CPU time, measured in milliseconds, reflects the linear-time behavior of the algorithm, as explained in Section 4.

The figure in Appendix shows a set of discrete spherical geodesics and their corresponding circles produced by the algorithm. Note that a discrete geodesic circle can be obtained by taking the union of the path $\pi_r^{\mathbb{Z}}(s,t)$ with its complementary path, i.e., $\pi_r^{\mathbb{Z}}(t,s)$, taken in the same order of cyclic movement. Clearly, such a circle would always include s and t. However, the inclusion of t is not ensured if we ignore t during Prioritized-BFS and moves forward until the traversal returns to s, although the resultant geodesic circle would comprise voxels lying within an isothetic distance of $\frac{3}{2}$ from $\Pi_r^{\mathbb{R}}(s,t)$.

6 Conclusion

We have shown how number-theoretic characterization helps in developing efficient algorithms related to discrete geodesics on a spherical surface. The problems of finding *iso-contours* and of *geodesic distance query*, defined and attempted in 3D real space [17, 18], are also pertinent in 3D digital space. The technique introduced in this paper may be extended to solve such problems with efficiency and theoretical guarantee.

References

[1] Balasubramanian, M., Polimeni, J.R., Schwartz, E.L.: Exact geodesics and shortest paths on polyhedral surfaces. IEEE TPAMI 31, 1006–1016 (2009)
[2] Brimkov, V., Coeurjolly, D., Klette, R.: Digital planarity—A review. Discrete Appl. Math. 155, 468–495 (2007)
[3] Bülow, T., Klette, R.: Digital curves in 3D space and a linear-time length estimation algorithm. IEEE TPAMI 24, 962–970 (2002)
[4] Chen, J., Han, Y.: Shortest paths on a polyhedron. In: Proc. SoCG, pp. 360–369 (1990)
[5] Coeurjolly, D., Miguet, S., Tougne, L.: 2D and 3D visibility in discrete geometry: An application to discrete geodesic paths. PRL 25, 561–570 (2004)
[6] Cohen-Or, D., Kaufman, A.: Fundamentals of surface voxelization. GMIP 57, 453–461 (1995)
[7] Coxeter, H.S.M.: Regular Polytopes. Dover Pub. (1973)
[8] Kimmel, R., Sethian, J.A.: Computing geodesic paths on manifolds. Proc. Natl. Acad. Sci. USA, 8431–8435 (1998)
[9] Klette, R., Rosenfeld, A.: Digital Geometry: Geometric Methods for Digital Picture Analysis. Morgan Kaufmann, San Francisco (2004)
[10] Li, F., Klette, R.: Analysis of the rubberband algorithm. Image Vision Comput. 25, 1588–1598 (2007)
[11] Martínez, D., Velho, L., Carvalho, P.C.: Computing geodesics on triangular meshes. Computers & Graphics 29, 667–675 (2005)
[12] Mitchell, J.S.B., Mount, D.M., Papadimitriou, C.H.: The discrete geodesic problem. SIAM J. Comput. 16, 647–668 (1987)
[13] Polthier, K., Schmies, M.: Straightest geodesics on polyhedral surfaces. In: ACM SIGGRAPH 2006 Courses, pp. 30–38 (2006)
[14] Surazhsky, V., Surazhsky, T., Kirsanov, D., Gortler, S.J., Hoppe, H.: Fast exact and approximate geodesics on meshes. ACM TOG 24, 553–560 (2005)
[15] Toutant, J.L., Andres, E., Roussillon, T.: Digital circles, spheres and hyperspheres: From morphological models to analytical characterizations and topological properties. Discrete Appl. Math. 161, 2662–2677 (2013)
[16] Xin, S.Q., Wang, G.J.: Improving Chen and Han's algorithm on the discrete geodesic problem. ACM TOG 28, Art. 104 (2009)

[17] Xin, S.Q., Ying, X., He, Y.: Constant-time all-pairs geodesic distance query on triangle meshes. In: Proc. I3D 2012, pp. 31–38 (2012)

[18] Ying, X., Wang, X., He, Y.: Saddle vertex graph (SVG): A novel solution to the discrete geodesic problem. ACM TOG 32, Art. 170 (2013)

[19] Ying, X., Xin, S.Q., He, Y.: Parallel Chen-Han (PCH) algorithm for discrete geodesics. ACM TOG 33, Art. 9 (2014)

Appendix

Table. C-octants and Q-octants

C-oct	Q-octants	Notation	Q-oct	Notation	Q-oct	Notation	Q-oct	Notation
C_1	$Q_1, \ldots, Q_6$	$+++$	Q_1	$(+x, +y, +z)$	Q_2	$(+y, +x, +z)$	Q_3	$(+y, +z, +x)$
C_2	$Q_7, \ldots, Q_{12}$	$-++$	Q_7	$(-x, +y, +z)$	Q_8	$(+y, -x, +z)$	Q_9	$(+y, +z, -x)$
C_3	$Q_{13}, \ldots, Q_{18}$	$+-+$	Q_{13}	$(+x, -y, +z)$	Q_{14}	$(-y, +x, +z)$	Q_{15}	$(-y, +z, +x)$
C_4	$Q_{19}, \ldots, Q_{24}$	$--+$	Q_{19}	$(-x, -y, +z)$	Q_{20}	$(-y, -x, +z)$	Q_{21}	$(-y, +z, -x)$
C_5	$Q_{25}, \ldots, Q_{30}$	$++-$	Q_{25}	$(+x, +y, -z)$	Q_{26}	$(+y, +x, -z)$	Q_{27}	$(+y, -z, +x)$
C_6	$Q_{31}, \ldots, Q_{36}$	$-+-$	Q_{31}	$(-x, +y, -z)$	Q_{32}	$(+y, -x, -z)$	Q_{33}	$(+y, -z, -x)$
C_7	$Q_{37}, \ldots, Q_{42}$	$+--$	Q_{37}	$(+x, -y, -z)$	Q_{38}	$(-y, +x, -z)$	Q_{39}	$(-y, -z, +x)$
C_8	$Q_{43}, \ldots, Q_{48}$	$---$	Q_{43}	$(-x, -y, -z)$	Q_{44}	$(-y, -x, -z)$	Q_{45}	$(-y, -z, -x)$

Q-oct	Notation	Q-oct	Notation	Q-oct	Notation
Q_4	$(+z, +y, +x)$	Q_5	$(+z, +x, +y)$	Q_6	$(+x, +z, +y)$
Q_{10}	$(+z, +y, -x)$	Q_{11}	$(+z, -x, +y)$	Q_{12}	$(-x, +z, +y)$
Q_{16}	$(+z, -y, +x)$	Q_{17}	$(+z, +x, -y)$	Q_{18}	$(+x, +z, -y)$
Q_{22}	$(+z, -y, -x)$	Q_{23}	$(+z, -x, -y)$	Q_{24}	$(-x, +z, -y)$
Q_{28}	$(-z, +y, +x)$	Q_{29}	$(-z, +x, +y)$	Q_{30}	$(+x, -z, +y)$
Q_{34}	$(-z, +y, -x)$	Q_{35}	$(-z, -x, +y)$	Q_{36}	$(-x, -z, +y)$
Q_{40}	$(-z, -y, +x)$	Q_{41}	$(-z, +x, -y)$	Q_{42}	$(+x, -z, -y)$
Q_{46}	$(-z, -y, -x)$	Q_{47}	$(-z, -x, -y)$	Q_{48}	$(-x, -z, -y)$

Table. Summary of results

r	s and its q-octant		t and its q-octant		$|\pi_r^{\mathbb{Z}}(s,t)|$	$d_{i,j}^{(48)}$	Time (μs)
10	$(0,3,10)$	$\mathbb{Q}_1$	$(4,6,7)$	$\mathbb{Q}_1$	7	0	53
10	$(4,-4,8)$	$\mathbb{Q}_{13}$	$(2,-9,4)$	$\mathbb{Q}_{18}$	8	1	61
10	$(7,1,7)$	$\mathbb{Q}_2$	$(-3,7,-6)$	$\mathbb{Q}_{36}$	21	6	81
20	$(1,5,19)$	$\mathbb{Q}_1$	$(11,12,12)$	$\mathbb{Q}_1$	16	0	111
20	$(-4,10,17)$	$\mathbb{Q}_7$	$(-20,3,2)$	$\mathbb{Q}_{10}$	24	3	145
20	$(-7,3,-18)$	$\mathbb{Q}_{32}$	$(4,-10,17)$	$\mathbb{Q}_{13}$	52	8	330
50	$(0,-12,49)$	$\mathbb{Q}_{13}$	$(8,-18,46)$	$\mathbb{Q}_{13}$	11	0	84
50	$(30,1,40)$	$\mathbb{Q}_2$	$(46,18,8)$	$\mathbb{Q}_4$	40	2	261
50	$(35,-35,-4)$	$\mathbb{Q}_{41}$	$(-12,-13,47)$	$\mathbb{Q}_{19}$	81	4	445
100	$(24,61,-76)$	$\mathbb{Q}_{25}$	$(57,58,-58)$	$\mathbb{Q}_{25}$	34	0	167
100	$(-39,-48,-79)$	$\mathbb{Q}_{43}$	$(-88,-17,-45)$	$\mathbb{Q}_{45}$	66	2	315
100	$(-11,78,61)$	$\mathbb{Q}_{12}$	$(98,-17,7)$	$\mathbb{Q}_{16}$	170	6	1000
200	$(116,115,115)$	$\mathbb{Q}_3$	$(176,62,73)$	$\mathbb{Q}_3$	93	0	392
200	$(33,33,194)$	$\mathbb{Q}_1$	$(199,14,11)$	$\mathbb{Q}_4$	242	3	1292
200	$(46,161,110)$	$\mathbb{Q}_6$	$(-87,-2,180)$	$\mathbb{Q}_{20}$	230	4	1677
500	$(-13,406,291)$	$\mathbb{Q}_{12}$	$(-250,340,268)$	$\mathbb{Q}_{12}$	239	0	1178
500	$(50,-494,58)$	$\mathbb{Q}_{18}$	$(171,-226,412)$	$\mathbb{Q}_{13}$	439	1	2925
500	$(-31,433,248)$	$\mathbb{Q}_{12}$	$(117,-171,-455)$	$\mathbb{Q}_{37}$	1142	8	33347
1000	$(25,-929,368)$	$\mathbb{Q}_{18}$	$(539,-637,551)$	$\mathbb{Q}_{18}$	628	0	5159
1000	$(384,917,-104)$	$\mathbb{Q}_{29}$	$(110,504,-857)$	$\mathbb{Q}_{25}$	892	2	7771
1000	$(932,300,-204)$	$\mathbb{Q}_{28}$	$(-637,705,311)$	$\mathbb{Q}_{11}$	1889	5	61852

Figure. Discrete spherical geodesics and their corresponding circles for $r = 30$. The sequence of red voxels is $\pi_r^{\mathbb{Z}}(s,t)$ with $s(8,25,14) \in \mathbb{Q}_6$ and $t(29,3,6) \in \mathbb{Q}_3$, which, when combined with $\pi_r^{\mathbb{Z}}(t,s)$, shown in yellow, yields the discrete 3D geodesic circle passing through s, t, and centered at o. Shown in blue are 16 longitude circles produced by extending the geodesics from source points taken from the discrete great circle on zx-plane to destination point $t(0,30,0)$ for each.

Discrete Curve Evolution on Arbitrary Triangulated 3D Mesh

Sergii Poltaretskyi[1,2,3], Jean Chaoui[1,2,3], and Chafiaa Hamitouche-Djabou[1,3]

[1] Télécom-Bretagne, 655 Avenue du Technopole, 29200 Plouzané, France
sergii.poltaretskyi@telecom-bretagne.eu
[2] IMASCAP, 65 Place Copernic, 29280 Plouzané, France
[3] Laboratoire de Traitement de l'Information Médicale, INSERM UMR 1101,
29609 Brest, France

Abstract. Discrete Curve Evolution (DCE) algorithm is used to eliminate objects' contour distortions while at the same time preserve the perceptual appearance at a level sufficient for object recognition. This method
is widely used for shape similarity measure, skeleton pruning and salient
point detection of binary objects on a regular 2D grid. Our paper aims at
describing a new DCE algorithm for an arbitrary triangulated 3D mesh.
The difficulty lies in the calculation of a vertex cost function for an object contour, as on a 3D surface the notion of Euclidean distance cannot
be used. It is also very difficult to compute a geodesic angle between lines
connecting vertices. We introduce a new cost function for border vertex
which is only based on geodesic distances. We apply the proposed algorithm on vertex sets to compute an approximation of original contours,
extract salient points and prune skeletons. The experimental results justify the robustness of our method with respect to noise distortions.

Keywords: discrete curve evolution, landmark points, vertices sets, triangulated mesh, skeleton pruning.

1 Introduction

Image Processing is one of the most rapidly evolving areas of information technology today, with growing applications in all areas of business, defense, health,
space, etc. It also forms a core area of research within the computer science and
engineering disciplines with more than 30 percent of the scientific publication
volume in the world. It forms the basis for all kinds of future visual automation.

Very often image processing deals with object recognition, analysis and matching. A wide range of techniques that are described in the current literature operate with objects represented on regular grids, such as 2D pixel or 3D voxel
images. In contrast, very few work is dedicated to object analysis on a non-
regular grids; for example a 3D mesh represented with triangles, where each
vertex could have an arbitrary number of connections.

The purpose of this study is to propose an efficient method to determine
characteristic landmarks of binary sets of vertices lying on an arbitrary triangulated surface (blue points on Fig.1(a)). A very interesting method is described

E. Barcucci et al. (Eds.): DGCI 2014, LNCS 8668, pp. 410–421, 2014.

by Latecki et al. [5,6,7] where authors proposed iteratively evolve a boundary of a 2D object by deleting one point at each iteration. A point that is deleted at a current step is the one with the lowest cost function. The cost function is designed so that it reflects a contribution of a boundary vertex to the global object shape and it is computed for all boundary vertices. As a result we obtain a subset of vertices that best represents the shape of a given contour and can be called landmarks. Authors named their algorithm as Discrete Curve Evolution (DCE) as at every step it evolves a discrete curve to represent only the most significant parts.

The cost function for DCE is very simple and involves computing only Euclidean distances and angles. The proposed method is fast, easy to implement and demonstrates impressive results. All these facts inspire us to apply this technique to a shape defined on a triangulated surface. Unfortunately, on an arbitrary triangulated mesh the notion of Euclidean distances and angles do not exist which makes impossible to compute the cost function, and hence use the DCE. Tothe best of our knowledge, there is no publication in the literature concerning the DCE for vertices sets on polygonal mesh. In this paper we introduce a new cost function that can be easily computed on a triangulated surface and yields better results on regular 2D grids compared with the classical DCE.

One very important application of the DCE is skeleton pruning. It is well described in the literature for a regular 2D grid [3], but there is no publication regarding sets of vertices on triangulated surfaces. This paper describes how the DCE can be used on polygonal meshes for skeleton pruning. First we propose to address a problem of skeleton extraction for sets of vertices as it is not a trivial task. Only few methods in the literature have been dedicated to this problem on triangulated surfaces. Rossl et al. [2] have presented method in which some mathematical morphology operators have been developed and applied to a set of vertices on triangulated meshes. The proposed method is very efficient and simple to implement, however it has several drawbacks leading to disconnected skeleton. A recent work described in [1] improves the previously proposed method of skeleton extraction. Authors claim that in [2], vertex classification is not sufficient as there are still unmarked vertices that are not considered in the skeletonization. To overcome the problem, a new class of vertices is proposed and new rules for topological thinning process are introduced. The proposed method has been tested on relatively homogeneous and on irregular meshes. Obtained skeletons reflect correctly the topology and geometry of input vertices' sets that proofs robustness of the approach.

An important factor of skeleton computing algorithm is its sensitivity to object's boundary deformation, as a minor noise or variation of boundary often generates redundant branches. To demonstrate it, we apply a method described in [1] to a vertices set. We show the obtained results on Fig.1(b) where lot of unnecessary branches were produced, that in turns could hamper further processing. Second major contribution of this paper is an application of our DCE algorithm for skeleton pruning on arbitrary triangulated surfaces.

Skeleton pruning is a well-known problem in 2D binary image processing. These are many different techniques proposed in the current literature; we will not list all of them as it is not the scope of current paper, an interested reader can find a good overview in [3]. We are looking for a method that yields excellent results and can be implemented for a set of vertices on a triangulated surface. We found two algorithms [3,10] that helped us to develop a solution for our problem. The first method is called Skeleton Pruning by contour partition [3]. The main idea consists in partitioning an object contour into segments and remove skeleton vertices whose generating points all lie on the same contour segment. According to Blum's definition of the skeleton [4], every skeleton point is linked to boundary points that are tangential to its maximal circle, so called generating points. The most important question is how to partition a boundary contour into segments as it plays a key role on resulting skeleton. Authors proposed to use the DCE and demonstrate excellent results. The second method [10] also uses the DCE algorithm, but for a difference purpose. Here, authors detect landmark points that actually represent skeleton ending points. Then these points are propagated inside an object with a condition that always maintains equal distance to objects' boundaries.

In this work we propose a solution for skeleton pruning that is also based on the DCE; but in our case we use a specific DCE algorithm that can also work on a triangulated surface.

Therefore, for the sake of understanding in Section 2 we briefly describe the DCE algorithm for contours on a regular 2D grid. Section 3 describes a new DCE algorithm that can be used on arbitrary triangulated surfaces. In Section 4 we show an application of the newly created DCE method to skeleton pruning on a polygonal mesh and we demonstrate obtained results.

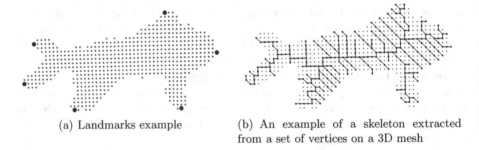

(a) Landmarks example

(b) An example of a skeleton extracted from a set of vertices on a 3D mesh

Fig. 1. Examples of landmarks and skeleton for a set of vertices on a triangulated mesh

2   Discrete Curve Evolution

Discrete Curve Evolution (DCE) algorithm introduced in [5,6,7], was developed to neglect the distortions of objects' contours in digital images while at the same time preserve the perceptual appearance at a sufficient level for object recognition. An obvious solution is to eliminate these distortions by contour approximation (or curve evolution). The basic idea of the method is very simple. At every evolution step, a contour point with the smallest relevance measure is removed, and two of its neighbor points become connected and form a contour's line segment. The relevance measure K is given by:

$$K(s1, s2) = \frac{\beta(s1, s2)\ l(s1)\ l(s2)}{l(s1) + l(s2)} \qquad (1)$$

where β is the turn angle at the common vertex of two contour's segments $s1$ and $s2$ and $l(.)$ is the length function, normalized with respect to the total length of a polygonal curve. The main property of this measure is that higher the value of $K(s1, s2)$, the larger the contribution of the curve of arc $s1 \cup s2$ to the shape [5,6].

To demonstrate the DCE and to be able to compare it with an algorithm proposed in this paper we created a 2D shape on a triangulated grid shown on Fig.2(a). To be able to use the relevance measure defined as (1), our triangulated grid is located on a plane so that Euclidean distances and angles can be easily computed. Figure 2 shows 3 stages of boundary evolution where an input image clearly has noisy borders. At first, this method allows us to smooth object's borders by removing noisy vertices (compare Fig.2(a) with Fig.2(b) and Fig.2(c)).

If we continue to evolve the curve, we will linearize digital arcs that are relevant to the curve shape, which will result in a successive simplification of the curve shape. Since in every evolution step, the number of digital line segments in the curve decomposition decreases by one, the evolution converges to a convex polygon, which defines the highest level in the shape hierarchy.

Figure2(d) shows an iteration stage where we left with 6 vertices that represent boundary contour. We can notice that the algorithm removed a top vertex that is an important point for the shape and normally should be presented for further image processing tasks such as object recognition.

3 DCE on an Arbitrary Triangulated Surface

To apply DCE algorithm to a contour formed by vertices on a triangulated surface we propose to use the same idea as for contours on a regular 2D grid. The main difference between these two cases is that on a 3D surface the notions of Euclidean distances as well as Euclidean angles do not exist. We need to find a new relevance measure that can be easily computed on a 3D surface.

If we simply try to translate a relevance measure from a regular 2D grid onto a triangulated domain we need to find equivalent measurements. A Euclidean distance could be easily replaced with a geodesic distance and can be efficiently computed with the help of Fast Marching Method (FMM) [8]. Another parameter that needs to be translated is the Euclidean angle. Compare to distances, a computation of geodesic angles is not a trivial problem. To overcome this issue

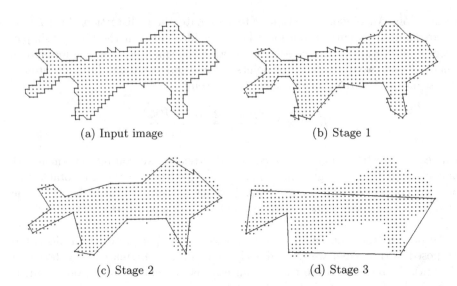

(a) Input image   (b) Stage 1

(c) Stage 2   (d) Stage 3

Fig. 2. Three stages of Discrete Curve Evolution with the relevance measure proposed in [6]

we propose a new measure that uses only geodesic distances to compute a cost function for each vertex.

To obtain our formula we analyze four cases shown on Fig.3. Figures 3(a), 3(b), 3(c) have spikes formed by two segments with exactly the same length but different distances between non-connected vertices that form these spikes. We observe that a spike on Fig.3(a) is not relevant compared to overall boundary while spikes on Fig.3(b) and Fig.3(c) are important. Figures 3(c) and 3(d) contain spikes with the same distances between non-connected vertices but different segments' lengths that form these spikes; where we can observe that a pick on Fig.3(d) more important than on a Fig.3(c). We conclude that a vertex cost is directly proportional to a sum of length of two segments that are connected to the studying vertex and is inversely proportional to a distance between non-connected vertices of these segments. We define a formula for a cost function as the next:

$$K = \frac{a + b}{\sqrt{c}} \tag{2}$$

where a, b, and c are shown on Fig.4(a).

It is important to note that to find a distance c between non-connected vertices of a spike we search for the shortest path between these points only inside an object that is defined by the studied contour, Fig.4(a). When a vertex is found in a concave region a distance c will be equal to the sum of segments that form a cavity, Fig.4(b):

$$c = a + b \tag{3}$$

A very important point is why we use a square root of c in (2) instead of just c. As we can see, the last condition implies that we are no longer able to distinguish between "concave" vertices. Let us imagine two different concave regions:

a) $a = 10$ cm, $b = 20$ cm and $c = 30$ cm
b) $a = 1$ cm, $b = 0.5$ cm and $c = 1.5$ cm

A cost function without the square root will give a cost value 1 for both cases, while the case "a" is more important as it represents bigger segment. A cost function with the square root will give a cost value 4.48 for the case "a" and 1.22 for the case "b". Here we can easily decide to remove a vertex from the case "b".

We now have a relative measure that can be used on an arbitrary triangulated surface as it is very easy to compute geodesic distances between points.

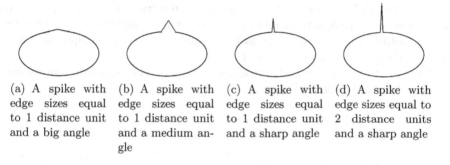

(a) A spike with edge sizes equal to 1 distance unit and a big angle

(b) A spike with edge sizes equal to 1 distance unit and a medium angle

(c) A spike with edge sizes equal to 1 distance unit and a sharp angle

(d) A spike with edge sizes equal to 2 distance units and a sharp angle

Fig. 3. The influence of a vertex connecting two segments on the shape of the curve depends on segment lengths that form a spike and the distance between the non-connected vertices of these segments

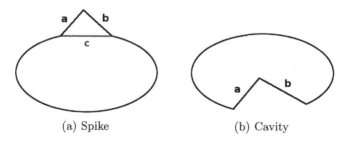

(a) Spike

(b) Cavity

Fig. 4. Two types of contour topology

3.1   Application of DCE on a Planar Triangulated Surface

To validate our cost function we first intend to apply it to a contour shown on Fig.2(a). This contour is defined around an object that is drawn on a triangulated surface where all triangles, that define this surface, lie on the same plane. This fact allows us to apply a relevance measure defined in [5] and a measure developed in this paper to the same boundary. Figure5 shows result obtained with the current method. For a good visual comparison we demonstrate several contour evolution stages that contain exactly the same number of boundary vertices.

With the first stage both contours contain 30 vertices. Comparing our results Fig.5(b) with previous algorithm Fig.2(b) we can clearly state that our method generates a contour that is a lot smoother and approximates the input shape in a better way.

With the second stage we generate contours with 20 vertices, Fig.5(c) and Fig.2(c). Both methods create a good approximation of the input shape but we still can say that our approach generates a slightly smoother boundary.

With the third stage we produce contours with 6 vertices, Fig.5(d) and Fig.2(d). Our approach converged to a convex polygon that approximates in a good way the input contour. Previous approach lost an important part of the input boundary and did not achieve a convex shape, Fig.2(d).

We can state that method introduced in this paper outperform previously proposed algorithm on a 2D grid.

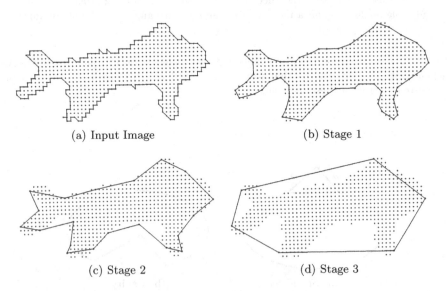

(a) Input Image   (b) Stage 1

(c) Stage 2   (d) Stage 3

Fig. 5. Three stages of the proposed Discrete Curve Evolution algorithm

3.2   Application of DCE on an Arbitrary Triangulated Surface

After comparing our results with the previous method, we apply our algorithm on objects defined on arbitrary triangulated surfaces. Figure 6 shows 4 iterations of a contour evolution. As we can see all points that are removed from the contour are less significant compared to points that approximate a boundary at a current iteration.

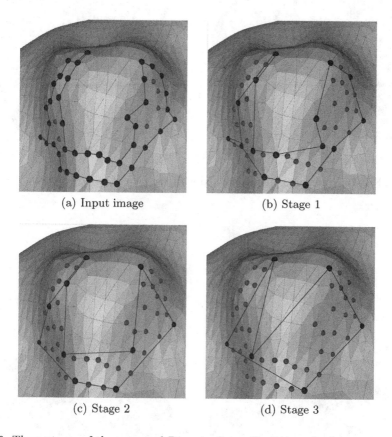

(a) Input image   (b) Stage 1

(c) Stage 2   (d) Stage 3

Fig. 6. Three stages of the proposed Discrete Curve Evolution algorithm on an arbitrary triangulated surface

Figures 7, 8 show four binary sets of vertices that lie on different meshes. With green points we highlight landmarks detected with the help of our algorithm. Obtained results demonstrate high robustness and efficiency of the proposed method.

4   Application to Skeleton Pruning

The skeleton is important for object representation and recognition in various areas such as image retrieval, computer graphics, character recognition, image

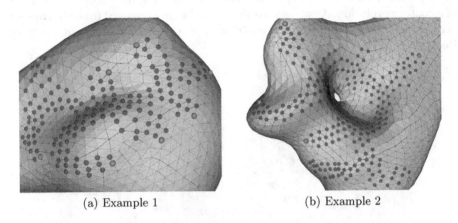

(a) Example 1   (b) Example 2

Fig. 7. Landmarks points detected with the help of proposed method

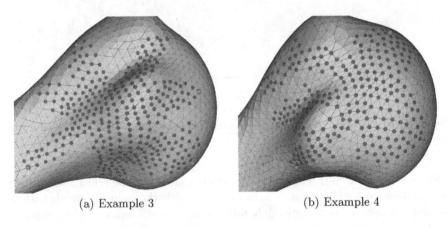

(a) Example 3   (b) Example 4

Fig. 8. Landmarks points detected with the help of proposed method

processing and biomedical image analysis [4]. Skeleton-based representations are the abstraction of objects that contains shape features and topological structures. Great work have been done by lot of researchers to recognize the generic shape by matching skeleton structures, where the most significant problem is the skeleton's sensitivity to an object's boundary deformation. Little noise or variation of the boundary often generates redundant skeleton branches that may seriously disturb the topology of the skeleton's graph [3]. Lot of efforts have been made to develop methods for skeleton pruning; a good overview could be found in [3]. All these methods focused on skeleton extraction and pruning for a well defined topological grid. To the best of our knowledge, no method exists for skeleton pruning on a polygonal mesh. This section describes a new algorithm to prune a skeleton on an unstructured triangulated grid.

We are inspired by two different publications [3] and [10]. In [3] authors propose to use the DCE algorithm for skeleton pruning. The main idea is to remove all skeleton points whose generating points lie on the same contour segment. Another method is proposed in [10] where authors suggest to find salient points of an object contour with the help of DCE. These points represent the stable endpoints of the skeleton. Authors use these points to propagate the skeleton inside the object by selecting each time a point from the object that has equal distances to the contour parts.

We suggest to use the DCE algorithm to extract salient points of an object's boundary. We then connect these points to the closest points on the computed skeleton. The last step is the actual pruning where we remove all skeleton branches that are not connected to salient points.

Here we present few results of the proposed pruning on triangulated surface. First, we apply a method described in [1] to extract a skeleton from a set of vertices on polygonal mesh and we demonstrate results on Fig.9(a) where we observe many unnecessary branches caused by noisy boundary. Our next step is to find shape landmarks (or salient points) with the help of the proposed DCE algorithm and to connect these points to the closest skeleton point. Then, we remove all branches that do not start with the found landmarks. Figure 9(a) shows the obtained results, where we highlight detected landmarks in green. We can state that the obtained skeleton is 100 percent clean and represents complete topology and geometry of the input shape.

Now, we demonstrate few more results for different shapes on an arbitrary triangulated mesh. Figure 9(a) shows a worm shape with the calculated skeleton and landmarks. Extracted skeleton contains redundant branches that need to be removed. After applying our algorithm we obtain results shown on Fig.9(b).

Figure 9 contains a starfish form which is more complicated than a worm shape. We can see that our method correctly identified shape landmarks and the pruned skeleton reflects the topology of an input shape.

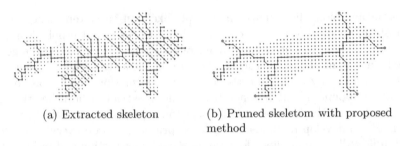

    (a) Extracted skeleton   (b) Pruned skeletom with proposed
                                              method

Fig. 9. Skeleton pruning technique for a set of vertices on a triangulated surface proposed in this paper

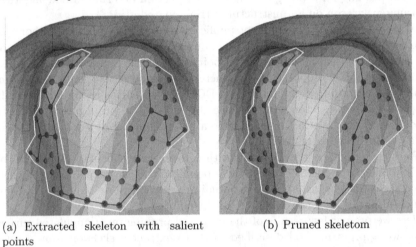

(a) Extracted skeleton with salient   (b) Pruned skeletom
points

Fig. 10. Skeleton pruning technique for a shape of a worm on a 3D surface

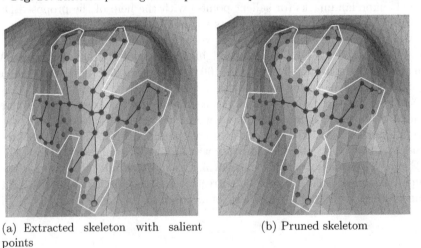

(a) Extracted skeleton with salient   (b) Pruned skeletom
points

Fig. 11. Skeleton pruning technique for a shape of a starfish on a 3D surface

5   Conclusion

In this paper we developed a new algorithm to detect characteristic landmarks of any shape defined as a set of vertices on an arbitrary triangulated surface or a set of pixels on a regular grid. The proposed method iteratively removes boundary vertices with smallest contribution to the global object shape so that at the end we keep only points that maximally represent a shape contour. We apply the proposed algorithm for several objects represented on triangulated surfaces. The obtained results justify the good reliability of our approach.

We also showed a useful application of the proposed algorithm for skeleton pruning. The demonstrated results clearly show efficiency of this approach. Extracted skeletons represent correctly objects' geometry and topology.

References

1. Kudelski, D., Viseur, S., Mari, J.-L.: Skeleton extraction of vertex sets lying on arbitrary triangulated 3D meshes. In: Gonzalez-Diaz, R., Jimenez, M.-J., Medrano, B. (eds.) DGCI 2013. LNCS, vol. 7749, pp. 203–214. Springer, Heidelberg (2013)
2. Rossl, C., Kobbelt, L., Seidel, H.P.: Extraction of feature lines on triangulated surfaces using morphological operators. In: AAAI Spring Symposium on Smart Graphics, vol. 00-04, pp. 71-75 (March 2000)
3. Bai, X., Latecki, L.J., Liu, W.-Y.: Skeleton Pruning by Contour Partitioning with Discrete Curve Evolution. IEEE Transactions on Pattern Analysis and Machine Intelligence 29(3), 449–462 (2007)
4. Blum, H.: Biological Shape and Visual Science (Part I). J. Theoretical Biology 38, 205–287 (1973)
5. Latecki, L.J., Lakämper, R.: Convexity Rule for Shape Decomposition Based on Discrete Contour Evolution. Computer Vision and Image Understanding (CVIU) 73, 441–454 (1999)
6. Latecki, L.J., Lakämper, R.: Polygon evolution by vertex deletion. In: Nielsen, M., Johansen, P., Fogh Olsen, O., Weickert, J. (eds.) Scale-Space 1999. LNCS, vol. 1682, pp. 398–409. Springer, Heidelberg (1999)
7. Latecki, L.J., Lakamper, R.: Shape similarity measure based on correspondence of visual parts. IEEE Trans. Pattern Analysis and Machine Intelligence (PAMI) 22(10), 1185–1190 (2000)
8. Kimmel, R., Sethian, J.A.: Computing Geodesic Paths on Manifolds. Proceedings of the National Academy of Sciences of the United States of America 95(15), 8431–8435 (1998)
9. Sethian, J.A.: Level Set Methods: Evolving Interfaces in Geometry, Fluid Mechanics, Computer Vision and Materials Sciences. Cambridge University Press (1996)
10. Yang, X., Bai, X., Yang, X., Zeng, L.: An Efficient Quick Algorithm for Computing Stable Skeletons. In: Congress on Image and Signal Processing (2009)

Author Index